DEVELOPMENTS IN SEDIMENTOLOGY 16

COMPACTION OF ARGILLACEOUS SEDIMENTS

FURTHER TITLES IN THIS SERIES

1. *L.M.J.U. VAN STRAATEN, Editor*
DELTAIC AND SHALLOW MARINE DEPOSITS

2. *G.C. AMSTUTZ, Editor*
SEDIMENTOLOGY AND ORE GENESIS

3. *A.H. BOUMA* and *A. BROUWER, Editors*
TURBIDITES

4. *F.G. TICKELL*
THE TECHNIQUES OF SEDIMENTARY MINERALOGY

5. *J.C. INGLE Jr.*
THE MOVEMENT OF BEACH SAND

6. *L. VAN DER PLAS Jr.*
THE IDENTIFICATION OF DETRITAL FELDSPARS

7. *S. DZULYNSKI* and *E.K. WALTON*
SEDIMENTARY FEATURES OF FLYSCH AND GREYWACKES

8. *G. LARSEN* and *G.V. CHILINGAR, Editors*
DIAGENESIS IN SEDIMENTS

9. *G.V. CHILINGAR, H.J. BISSELL* and *R.W. FAIRBRIDGE, Editors*
CARBONATE ROCKS

10. *P.McL.D. DUFF, A. HALLAM* and *E.K. WALTON*
CYCLIC SEDIMENTATION

11. *C.C. REEVES Jr.*
INTRODUCTION TO PALEOLIMNOLOGY

12. *R.G.C. BATHURST*
CARBONATE SEDIMENTS AND THEIR DIAGENESIS

13. *A.A. MANTEN*
SILURIAN REEFS OF GOTLAND

14. *K.W. GLENNIE*
DESERT SEDIMENTARY ENVIRONMENTS

15. *C.E. WEAVER* and *L.D. POLLARD*
THE CHEMISTRY OF CLAY MINERALS

17. *M.D. PICARD* and *L.R. HIGH*
SEDIMENTARY STRUCTURES OF EPHEMERAL STREAMS

DEVELOPMENTS IN SEDIMENTOLOGY 16

COMPACTION OF ARGILLACEOUS SEDIMENTS

BY

HERMAN H. RIEKE, III

School of Mines, West Virginia University, Morgantown, W. Va. (U.S.A.)

AND

GEORGE V. CHILINGARIAN

Petroleum Engineering Department, University of Southern California, Los Angeles, Calif. (U.S.A.)

ELSEVIER SCIENTIFIC PUBLISHING COMPANY
Amsterdam London New York 1974

ELSEVIER SCIENTIFIC PUBLISHING COMPANY
335 JAN VAN GALENSTRAAT
P.O. BOX 1270, AMSTERDAM, THE NETHERLANDS

AMERICAN ELSEVIER PUBLISHING COMPANY, INC.
52 VANDERBILT AVENUE
NEW YORK, NEW YORK 10017

Library of Congress Card Number: 74-190682

ISBN: 0-444-41054-6

With 217 illustrations and 49 tables.

Printed in The Netherlands

This book is dedicated to

ACADEMICIAN N.B. VASSOEVICH

for his important contributions
to the field of compaction of sediments

to our inspirer

PROFESSOR K.O. EMERY

and to our parents

MARTHA D. OTTE
HERMAN H. RIEKE, Jr.
and
KLAVDIA GORCHAKOVA
VAROS KH. CHILINGARIAN

PREFACE

From time to time there appears a book that contains in a readily-digestible manner a great wealth of information which relates to a particular subject. *Compaction of Argillaceous Sediments* by H.H. Rieke, III and G.V. Chilingarian is just such a text. This is an advanced text, which is designed especially for teachers, researchers, industrial scientists and students who are engaged in careful and detailed studies of fine-grained sediments, as well as in a study of the processes of compaction that convert clays, silts, and related fine-textured materials into sedimentary rocks. A vast body of objective data, as well as attending interpretations, which relate to sedimentation and subsequent compaction of the sediments has been marshalled into this fine book. The authors have succeeded in synthesizing data from hundreds of diverse sources into one well-organized book. They have also added much objective data accrued from their own extensive research work. To those geologists and engineers, whose interests lie in the realm of investigating the effects, and results, of compaction in argillaceous and related sediments, this book will provide many answers. Furthermore, it should prove to be a means of lending encouragement, and of pointing out suggested routes, for further study and research. This book is a welcome addition to geologic literature, because it not only summarizes the significant and definitive work of hundreds of investigators, but also presents the viewpoints of the authors, who represent two of the profession's young and dynamic students of fine-grained sediments and of the processes of compaction that change such materials into sedimentary rocks. The book is well organized with data being presented in eight well-written chapters.

In their introductory chapter, Professors Rieke and Chilingarian comfortably introduce the reader to this interesting field of study and discuss the composition of argillaceous sediments. Some fundamental fluid mechanics concepts are also presented in order that one can understand and appreciate the fluid-statics problems. From there, the discussion moves smoothly and quickly through the topic of the mechanics of compaction, fortified with laboratory and field data on compaction. The authors then review the diagenesis–metamorphism transition phase and properties of argillaceous sediments as related to compaction. This is followed by introduction to interstitial fluids, subsidence, and abnormal fluid pressures. After the well-organized introductory chapter, Rieke and Chilingarian carry their readers into a provocative discussion of the interrelationships that exist among density, porosity, remaining moisture content, pressure, and depth of both recent and ancient sediments. Results of the compaction of recent sediments, muds and clays, as determined experimentally both in the field and in the laboratory, are presented and superbly documented with objective data in the form of graphs and tables. This is an integral part, a strong rubric, of not only this but all other chapters of *Compaction of Argillaceous Sediments.* The authors have succeeded in laying before the interested reader the objective data that document their main thesis.

Chapter 3, reviewed by Dr. Leons Kovisars, provides the reader with an insight into the theory of consolidation and of the state of stress in compacting shales, and presents an analysis of pressure–depth–density relationships. Another signal accomplishment is a discussion of differential compaction and an estimation of the maximum effective pressure that has ever existed in an argillaceous sediment. Effect of compaction on various properties of fine-grained sediments is thoroughly treated in Chapter 4.

The discussion of chemical alterations and of the behavior of interstitial fluids during compaction, is amply documented in Chapter 5. A new wealth of information is presented that relates to composition and classification of interstitial solutions, processes that affect the chemical composition of interstitial waters, and fluid release mechanisms. Effects of compaction on the chemistry of solutions that are squeezed out of clays during compaction are meticulously treated, from both a theoretical and an experimental approach. What do we know concerning diagenetic changes of pore waters? A thorough and up-to-date discussion of answers to this complex question is given in Chapter 5. The latter chapter also delves into squeezing of oil and bitumens out of muds and shales. The authors point out, however, that a considerable amount of research work still remains to be done in this field.

Subsidence of a depocenter in which fine-grained sediments dominate, and the resultant compaction of these fine-textured materials, is a subject normally overlooked (or perhaps avoided) in most texts of this nature. Rieke and Chilingarian, however, have devoted a full chapter to this subject; adequately treating such tropics as the origin of sedimentary basins and geosynclines, the hydrogeologic cycle involved, subsidence caused by the fluid withdrawal, and a mathematical analysis of subsidence. Indeed, this chapter alone fully justifies the publication of this text in the assessment of the writer. As the reader closes this chapter, he embarks on a pleasant and informative journey into an investigation of abnormal geopressures. The authors discuss the origin of abnormal geopressures due to compaction, faulting, phase changes in minerals during compaction, and many other causes. They also present the anatomy of a high-pressure zone. Great interest in the mechanics of compaction has been created as a result of occurrence of "over-pressured" zones, which pose great problems during drilling operations in the Gulf Coast of the United States and many other areas around the world.

In the final chapter, Rieke and Chilingarian award the reader with a superb discussion on equipment and techniques used in compaction studies. At this point, readers will realize that their libraries are enhanced with an unusual and outstanding text. It is not a one-reading experience, but rather is a source of much-needed data that relate to compaction in fine-textured sediments such as clay and silt. The geologic profession is engaged in a never-ending study of compaction and lithification of sediments that give rise to sedimentary rock; and because shales and related sedimentary rocks comprise such a vast body of rocks on the surface of the earth, they demand the rigors of disciplined investigation. This need becomes even greater, if man is to extract more oil and gas, and also coal, as sources of energy. Various oil shales are great storehouses of kerogen (Green River Formation of Utah, Colorado, and Wyoming, for example), and one day they will, of necessity, have to be exploited. What were the effects of compaction on such sediments, and many others as well, and what can man reasonably deduce by studying them? Please read *Compaction of Argillaceous Sediments* by Rieke and Chilingarian with pleasure and edification. It may not provide all the answers to questions that are posed, but it goes a long way in the right direction.

H.J. BISSELL
Brigham Young University
Provo, Utah

CONTENTS

Chapter 1

INTRODUCTION

Compaction of sediments under the influence of their own weight has long been a known geologic phenomenon. As pointed out by Hedberg (1926), in the seventeenth century Steno recognized that variations in the attitude of sedimentary strata might be due to compaction. In his excellent paper entitled "The application of quantitative methods to the study of the structure and history of rocks", Sorby (1908) presented original data on the porosity of natural sediments, discerned an inverse relationship between porosity and age, and recognized that the compaction of sediments is primarily a change in porosity. The structural effects of compaction were mentioned by Shaw and Munn (1911), whereas the idea that the compaction of sediments may have played a part in the origin of oilfield structures has been championed by McCoy (1934), Shaw (1918), and Mehl (1919). In 1926, Hedberg attempted a quantitative evaluation of the compaction of sediments. His work aroused great interest and inspired further research on the subject. Nevin (1931) presented an excellent discussion of geologic information on the compaction of sediments available up to 1931. Early geologic explanations on the settling of sediments and on the gravitational and other compaction theories have been presented by Blackwelder (1920), Monnett (1922), Lahee (1923), Teas (1923), Terzaghi (1925), Hedberg (1926, 1936), and Athy (1930a). Gravitational compaction is defined as the expulsion of pore fluids and the pore volume decreases in a sedimentary column as a result of normal and shear-compressional stresses due to the overburden load. Differential compaction refers to the gravitational compaction of sediments over and around a positive buried geomorphological feature such as a hill or a reef. The compaction of argillaceous sediments is treated in the present book, whereas the compaction of coarse-grained sediments is to be presented in a forthcoming volume of the series *Developments in Sedimentology*.

According to Adrian F. Richards, Director of Marine Geotechnical Laboratory of Lehigh University (personal communication, 1971), the term *consolidation* appears to be gaining widespread acceptance in geologic circles, because the term *compaction* is never used in synonymy by engineers for consolidation. The basic laws covering consolidation are those of soil mechanics. The writers, however, chose the term *compaction* in preference to *consolidation*, because the latter term is understood by many geologists to include such diagenetic processes as cementation.

Associated with the compaction, burial, and diagenesis of argillaceous sediments is the genesis of the clay minerals. Clay minerals which comprise argillaceous sediments owe their origin to three principal processes: (1) detrital inheritance from preexisting rocks and sediments; (2) transformation of the clays from one form to another in the sedimentary environment owing to their instability at low temperatures and various ranges of pH and Eh; (3) neoformation of clay minerals in situ. The latter two processes are very complex and at the present time it is very difficult to evaluate the precise role of compaction in creating an environment favorable for clay-mineral changes. The term *transformation*, in this book, applies to those changes that modify a clay mineral from one type to another, e.g., montmorillonite to illite; whereas *neoformation* applies to formation of clay minerals from non-clay minerals and the synthesis of clay minerals in situ.

COMPOSITION OF ARGILLACEOUS SEDIMENTS

The most important allogenic components of argillaceous sediments include: (1) various clay minerals including gibbsite, (2) quartz, (3) feldspars, (4) carbonates, (5) amorphous silica and alumina, (6) pyroclastic material, and (7) organic matter. As pointed out by Müller (1967), biogenic materials which form in the basin itself include carbonates, amorphous silica, and organic matter.

The essential part of the clay fraction ($<2\mu$) in the Atlantic Ocean is composed of illite, kaolinite, chlorite, montmorillonite, random mixed-layer mineral, and gibbsite. As a rule, illite is the dominant clay mineral, and in the northern Atlantic it comprises more than 90% of the clay-mineral fraction (Biscayne, 1964). In the South Atlantic, montmorillonite amounts to as much as 40% of the total clay-mineral fraction, whereas in the North Atlantic area only up to 20% of clay fraction is montmorillonite. In the southwestern area of the Indian Ocean, montmorillonite constitutes over 80% of the clay fraction, possibly owing to diagenetic transformation of volcanic material (Biscayne, 1964).

There is a large increase in kaolinite content (up to more than 50%) and of gibbsite (more than 10%) in the sediments of the Atlantic Ocean adjacent to the tropical rivers in South America and Africa. In the sediments of the Indian Ocean around Madagascar, the gibbsite content is very high: more than 30%. The chlorite content is commonly less than 20% of the clay mineral fraction; however, larger amounts are found in Antarctic regions as well as off Newfoundland. According to Biscayne (1964), there are no mixed-layer minerals east of the Mid-Atlantic Ridge. Frequency of these minerals decreases from north to south, west of the Mid-Atlantic Ridge.

According to Griffin and Goldberg (1963), illite, montmorillonite, chlorite, kaolinite, and, to a much lesser extent, halloysite are the main clay minerals in the Pacific Ocean. Illite was found to be abundant in all samples from the North Pacific area. Kaolinite is confined to nearshore sediments. Montmorillonite is also generally more abundant in nearshore sediments. Chlorite content in nearshore sediments increases with increasing latitude.

The most abundant clay mineral in the South Pacific area is montmorillonite (mainly nontronite), which is associated with phillipsite. Illite, kaolinite and chlorite are less abundant.

Most of the samples of recent marine and fresh-water argillaceous sediments contain 2–5% of organic matter. This organic matter consists mainly of chemically undefined "organic residue" (60–80%) plus less than 10% of amino acids (Degens, 1967).

Holmes and Hern (1942) determined the illite, montmorillonite and kaolinite contents in recent alluvium deposited by the Mississippi River and its tributaries: 60–75%, 10–15%, and 10–20%, respectively, in the eastern tributaries: 40–60%, 10–45%, and 5–15%, respectively, for the Mississipi River; and 40–60%, 20–45%, and 10–20%, respectively, in the western tributaries. Numerous analyses of the shales and mudstones also show that illite is usually the most abundant clay mineral; the commonly present kaolinite and montmorillonite clays compose less than 50% of the clay-mineral fraction.

Inasmuch as the thickness and shape of particles of different types of clay influence their behavior during compaction, the physical properties and characteristics of various clay minerals are presented in Table I.

The initial water content of most argillaceous muds is approx. 50–80% which corresponds to a porosity of 70–90% (Müller, 1967, p.135). In sands, on the other hand, porosity is only 30–50%, which corresponds to 20–30% water content. According to Serruya (1969), the upper 200 cm of the lemanic sediments of Lake Geneva contain an average of about 150% water on a dry-weight basis. The water content of the very first layer of sediments, when measured on an undisturbed sample, reaches 250%.

TABLE I
Physical properties and characteristics of various clay minerals
(After Warner, 1964, p.14)

Property or characteristic	*Clay mineral*		
	kaolinite	*illite*	*montmorillonite*
Thickness of unit cell (Å)	7	10	9.6
Thickness of clay plate (Å)	500–20,000	> 30	10–80
Surface area (m^2/g)	≈ 15	≈ 90	800 (theoretical)
Surface diameter (μ) *	0.3–4	0.1–0.3 or larger	0.01–0.1
Density (g/cm^3)	2.60–2.68	2.64–2.69	2.2–2.7
Cation exchange capacity (meq/100g)	3.15	10–40	80–150
Shape	hexagonal	hexagonal(?) plate or lath	plate-shape(?)

* 1 μ (micron) = 10,000 Å (Ångstrom units).

SOME FUNDAMENTAL FLUID MECHANICS CONCEPTS

Fundamental equation of fluid statics

The fundamental equation of fluid statics states that pressure increases with depth, the increment per unit length being equal to the weight per unit volume (Binder, 1962, p.13):

$$dp = -\rho g dz \tag{1-1}$$

where dp is increment in pressure; dz is increment in depth (z is a vertical distance measured positively in the direction of decreasing pressure); ρ is density (mass per unit volume); and g is gravitational acceleration. The minus sign indicates that pressure decreases with increasing z. The above relationship can be clearly understood on examining Fig.1, which shows vertical forces on the infinitesimal element in the body of a static

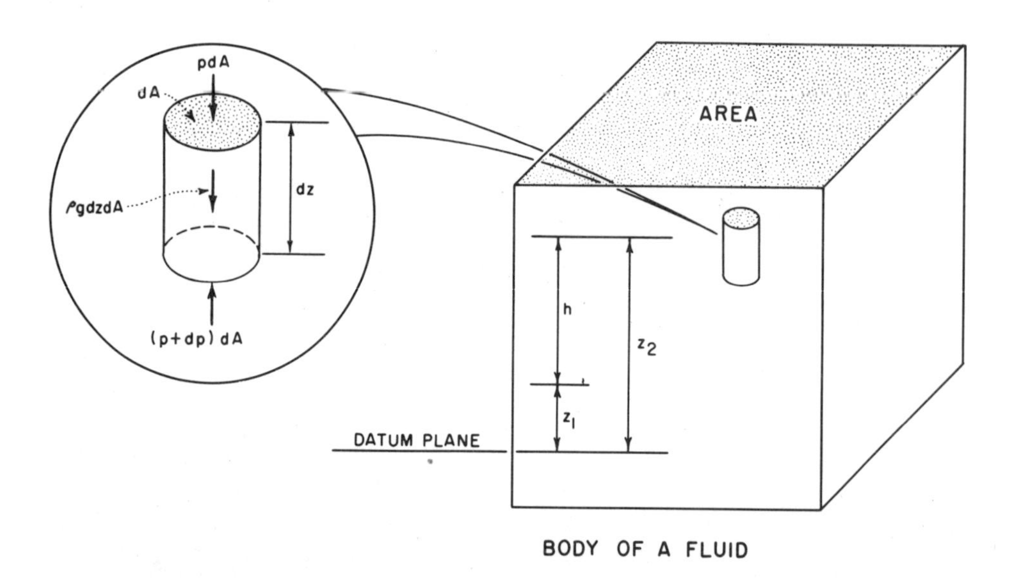

Fig.1. Schematic diagram of vertical forces on an infinitesimal element in body of any fluid. (Modified after Binder, 1962, fig.2-2, p.13.)

fluid. In this figure, dA represents an infinitesimal cross-sectional area, p is the pressure on the top surface of the element and $(p + dp)$ is the pressure on the bottom surface. Inasmuch as the pressure increase is due to the fluid weight, the weight of the element (ρg dz dA) is balanced by the force due to pressure difference (dp dA):

$$dp\, dA = -\rho g\, dz\, dA \tag{1-2}$$

or:

$$dp = -\rho g \, dz$$

In integral form, the above equation can be expressed as follows (see Fig.1):

$$\int_1^2 \frac{dp}{\rho g} = -\int_1^2 dz = -(z_2 - z_1) \quad (1\text{-}3)$$

If ρ is assumed to be constant, eq.1–3 becomes:

$$p_2 - p_1 = -\rho g(z_2 - z_1) \quad (1\text{-}4)$$

or:

$$\Delta p = \gamma h \quad (1\text{-}5)$$

where h is the difference in depth between two points, which is commonly referred to as the "pressure head"; and $\gamma(= \rho g)$ is the specific weight. On expressing γ in lb./cu. ft. and h in ft., pressure difference Δp is found in lb./sq. ft.

Buoyancy

When a body is completely or partly immersed in a static fluid, there is an upward vertical buoyant force on this body equal in magnitude to the weight of displaced fluid. This force is a resultant of all forces acting on the body by the fluid. The pressure is greater on the parts of the fluid more deeply immersed. The pressures at different points on the immersed body are independent of the body material. For example, if the same fluid is substituted for the immersed body, this fluid will remain at rest. This means that the buoyant, upward force on the substituted fluid is equal to its weight.

If the immersed body is in static equilibrium, the buoyant force and the weight of the body are equal in magnitude and opposite in direction, passing through the center of gravity of the body. For a comprehensive treatment of fluid statics, the reader is referred to an excellent book on fluid mechanics by Binder (1962).

The buoyancy principle is illustrated in Fig.2. The weight of a solid body, having a volume of 1 cu. ft. and a specific weight of 162.24 lb./cu. ft. (sp.gr. = 2.6), is registered on the scale as 162.24 lb. (Fig.2A). If the container is filled with water, the buoyant force acting upwards will be equal to the weight of displaced fluid (= 62.4 lb.) and consequently the weight registered on the scale will be equal to 99.84 lb. (= 162.24 – 62.4 = 99.84 lb.), as shown in Fig.2B.

If a porous medium, having a porosity of 20%, is saturated with water and the specific gravity of the solid grains is equal to 2.6, the weight registered on the scale is equal to the

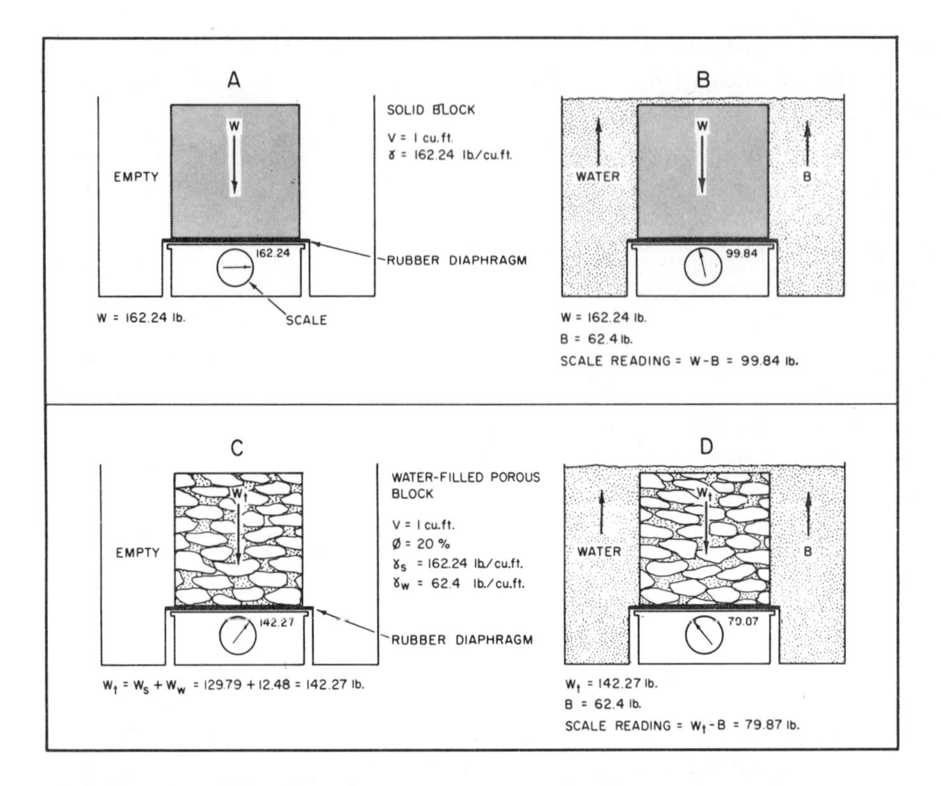

Fig.2. Illustration of the effect of buoyancy on the weight of solid and water-filled porous materials. W = weight in lb.; B = buoyant force in lb.; V = volume in cu. ft.; γ = specific weight in lb./cu. ft.; ϕ = porosity. Subscripts: t = total, s = solids, and w = water. Inside container, solid blocks are placed on top of the rubber diaphragm, which, in turn, rests over the scale.

sum of the weights of the solid grains and of the fluid, i.e., 2.6 × 62.4 (1–0.20) + 62.4(0.20) = 142.27 lb. (Fig.2C). The weight registered with water inside the container is equal to 142.27–62.4 = 79.87 lb., i.e., the weight of the saturated porous medium (142.27 lb.) acting downwards minus the buoyant force (62.4 lb.), which acts upwards and is equal to the weight of the displaced fluid.

The latter experiment also demonstrates that Archimedes principle can be applied to the porous rocks underground. The force of uplift owing to pore pressure is not proportional to surface porosity, and the fluid pressure p_p is effective over the entire area of any surface passed through the system, irrespective of the fraction of the area lying within the solid-filled part of the space (see stimulating discussions by Laubscher, 1960, and Hubbert and Rubey, 1960).

MECHANICS OF COMPACTION

According to Rowell (1965), there are two kinds of porosity in shales: (1) intercrystalline porosity which comprises the voids between the clay particles and depends directly upon the degree of orientation and packing of the clay particles; and (2) intracrystalline porosity which is composed of the bound water volume on the external and internal surfaces of the clay minerals. The intracrystalline porosity depends upon the surface area of the clay mineral and the thickness of the water layer coating it. This thickness depends upon the presence of exchangeable cations on the surface which tend to hydrate and form an electric double layer.

Argillaceous muds with an initial porosity of 80%, for example, undergo a marked physical change during the compaction. There is a continuous decrease in porosity with increasing depth of burial. This occurs very rapidly to a depth of about 500 m and more slowly below that depth. An excellent discussion on diagenesis of argillaceous sediments has been presented by Müller (1967). The term *diagenesis* is commonly applied to all changes which take place in a freshly deposited sediment until it reaches the stage of metamorphism. The writers, however, define the term diagenesis as all those processes which change a fresh sediment into a stable rock of a substantial hardness, under conditions of pressure and temperature not widely removed from those existing on the earth's surface in various depocenters. All changes occurring subsequent to lithification, but prior to metamorphism, are termed *epigenetic* or *catagenetic*.

According to Müller (1967), only grain size, clay-mineral composition, and temperature play an important role in the porosity reduction below a depth of about 500 m (deep-burial stage). At this depth, the total volume of the sediment decreases by about 50% and clay mud becomes a mudstone (or shale if fissile) with a porosity reduction of about 30%. With a further decrease of porosity, the mudstone (or shale) becomes an argillite with a porosity of only about 4–5%. Slate is the product of metamorphism.

Millot (1949), on the other hand, defined *argillite* as a common rock composed for the most part of one or several clay minerals and without any notable bedding, whereas the term *shale* refers to layered argillite. Some so-called *argillites*, however, exhibit distinct layering on examination of rock thin sections under the microscope; this presents a nomenclatural problem. Many geologists prefer the term *mudstone* to that of *argillite*.

As sediments accumulate on the sea floor, the underlying layers are compacted. If interstitial water can escape, it mingles with the waters in overlying sediments and finally comes in contact with sea water. Thus, the fluid pressure at any point is hydrostatic. Porosity decreases as compaction occurs. On the other hand, if the escape of interstitial water is inhibited, the matrix cannot be packed more closely and the porosity and fluid pressure will be abnormally high.

Void-volume reductions in an argillaceous formation as a result of the overburden stress depend on the following factors: (1) lithology, (2) initial thickness, (3) maximum effective stress, (4) rate of loading, (5) decomposition of organic material, and (6) recrystallization and transformation of clay minerals. In addition, Meade (1968) listed the following factors which influence the water content of clayey sediments under applied loads: (1) particle size, (2) type of clay minerals, (3) adsorbed cations, (4) interstitial

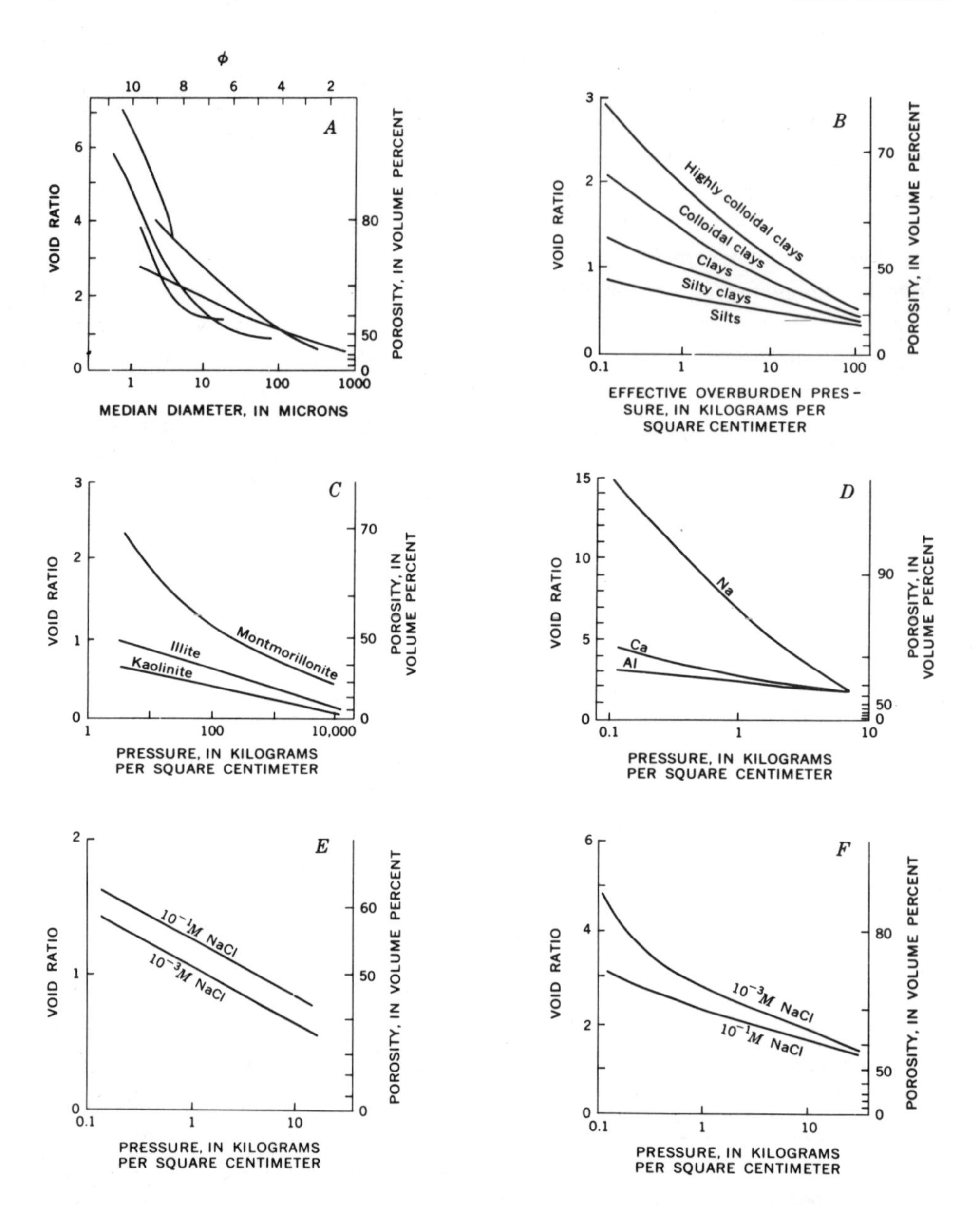

Fig.3. Influence of different factors on the relationship between void ratio and pressure in clayey materials. A. Relationship between void ratio and median particle diameter at overburden pressures less than 1 kg/cm^2 (after Meade, 1964, p.B6). B. Generalized influence of particle size (modified from Skempton, 1953, p.55). C. Influence of clay-mineral species (modified from Chilingar and Knight, 1960, p.104). D. Influence of cations adsorbed by montmorillonite (modified from Samuels, 1950). E. Influence of NaCl concentrations in unfractionated illite, about 60% of which was coarser than 2 μ in size (modified from Mitchell, 1960, fig.M3). F. Influence of NaCl concentration in illite finer than 0.2 μ (modified from Bolt, 1956, p.92). (After Meade, 1968, fig.1, p. D4.)

electrolyte solutions, (5) acidity, and (6) temperature. Because of the intimacy of the water–clay relationship, the reduction in pore volume of clays with increasing overburden pressure can best be analyzed in terms of water removal. The effects of particle size, clay minerals, adsorbed cations, and interstitial electrolyte solutions are presented in Fig.3. Compaction of sediments is accomplished, chiefly, by elimination of pore space subsequent to expulsion of the interstitial fluids. Decomposition of organic matter and recrystallization and transformation of the clays may also play a major role.

The volume of organic matter in argillaceous sediments is reduced through two basic processes: (1) decomposition of organic matter, as a result of which water and volatile gases such as methane and carbon dioxide are given off; and (2) physical compaction of organic matter. Observations indicate that organic-rich mud is more porous than ordinary mud and that under high pressure carbonaceous shales are more compacted and have less porosity than ordinary shales (Weller, 1959, p.301). The early transformation of vegetable matter into peat and then to lignite is accompanied by a tremendous reduction in bed thickness. Falini (1965, p.1335) estimated that reduction in bed thickness when peat is transformed to lignite would be about tenfold (final thickness = one-tenth of the original thickness). As inorganic impurities in the vegetable matter increase, the percentage loss of volume owing to volatilization will decrease; the less volatile residue remains behind in all cases (Hedberg, 1926, p.1038). The weight of the overlying beds drives off water and gas occupying the pore space. Partly decomposed organics and some of the decomposition products appear to form a complex series of hydrogels, the presence of which introduces several complex problems not duplicated in other lithologies (Weller, 1959, p.301).

Time is an important factor in coalification, as pointed out by Weller (1959, p.304). There are few Paleozoic coals that have not attained the bituminous rank, whereas most Tertiary coals have passed through the sub-bituminous stage. Criteria indicative of the rank of coal permit a better evaluation of different stages of coal diagenesis than is possible in the case of diagenesis of other sediments. Weller believes that the anthracite rank of coal cannot be the result of burial alone, inasmuch as these coal beds are always associated with intrusive igneous rocks and deformational activity. Teichmüller and Teichmüller (1967, p.400) pointed out that the rank of coal increases with depth if the geothermal temperature is greater than 50°C. High geothermal temperatures accelerate geochemical coalification. A strong heating of short duration can have the same effect as low heating over longer periods of time. Thus, the rank of a coal is indicative of temperature only if the duration of the heating of the coal is known (Teichmüller and Teichmüller, 1967). It should be noted that bituminous material follows a different diagenetic path from that which leads to hydrocarbon accumulations, even though coal is commonly associated with thick shale sequences and petroliferous sandstones.

On examining thick muds (clay-in-water suspensions) the writers observed the following possible arrangements of clay platelets (see Fig.4): (a) deflocculated-dispersed (or completely dispersed, not aggregated); (b) deflocculated with edge-to-surface contacts

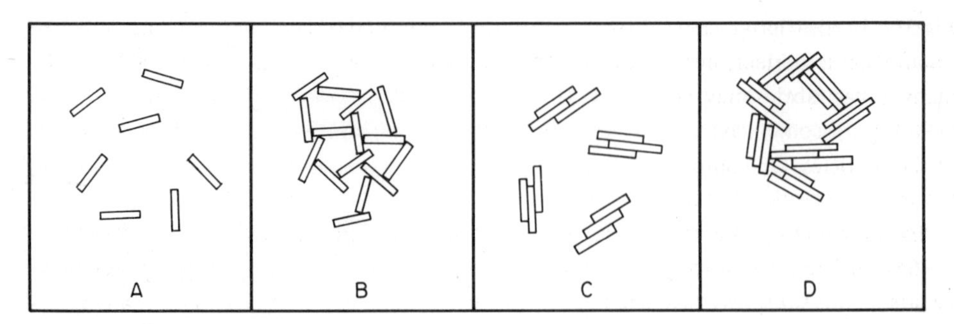

Fig.4. Possible arrangements of clay platelets in thick clay-in-water suspensions. A. Deflocculated-dispersed. B. Deflocculated with edge-to-surface contacts. C. Flocculated-dispersed. D. Flocculated with edge-to-surface contacts.

(edge is positively charged, whereas surface is negative); (c) flocculated-dispersed (or aggregated but not flocculated); and (d) flocculated (or aggregated) with edge-to-surface contacts.

As a result of pore-volume reduction during compaction, clay particles move closer together. The most widely accepted arrangement is a preferred orientation because of the characteristically platy shape of clay-mineral particles (Fig. 5A). Another possible packing arrangement, called turbostratic by Aylmore and Quirk (1960, p.1046), is presented in

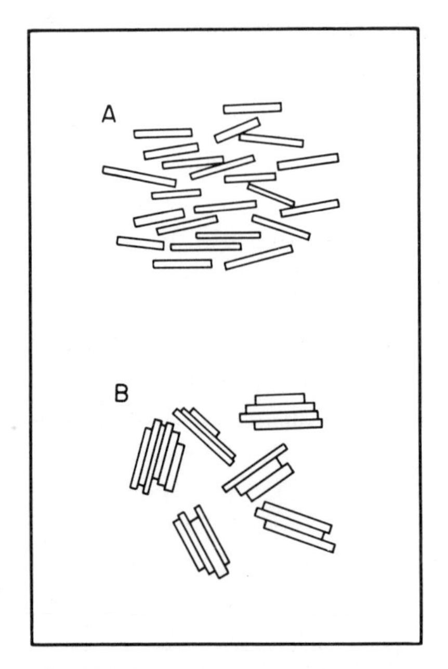

Fig.5. Idealized arrangement of clay-mineral particles that may be formed during compaction. A. Preferred orientation. B. Turbostratic orientation. (After Meade, 1968, fig.3, p.D6.)

Fig.5B. In the latter case, within the packets or domains of clay particles the planar orientation is perfect, but there is a random orientation between these packets. Probably, many other fabrics may be formed as a result of compaction. As pointed out by Meade (1968, p.6), some clays may be compacted without developing any regular arrangement of the particles. On compaction, the system presented in Fig.4A probably would give rise to the most impermeable rock, whereas that presented in Fig.4D will be least impermeable.

Experiments by O'Brien (1963, pp.20–40) on illite and kaolinite pastes and by Martin (1965) on kaolinite slurries indicate that most of the reorientation of clay particles that occurs at pressures less than about 100 kg/cm^2 takes place during the very early stages of compaction, provided there is a sufficient initial water content. According to Martin (1965), a pressure of 1 kg/cm^2 produces a very marked preferred orientation, which does not change upon further consolidation up to 32 kg/cm^2. Low pressures on the order of a few kilograms per square centimeter can produce marked preferred orientation in kaolinite and illite but do not seem to produce much orientation in montmorillonite. In the laboratory, pressures around or greater than 100 kg/cm^2, however, can produce preferred orientation in any platy clay mineral, including montmorillonite (Meade, 1968, p.D7). Larger clay particles seem to be more susceptible to preferred orientation than the smaller ones (Chilingar et al., 1963; Meade, 1968, p.D7).

In 1959, Hubbert and Rubey clearly showed in their classical paper that the overburden pressure, S or σ, is supported jointly by the effective grain-to-grain stress, δ or σ', of the clay aggregates and by the pore-fluid pressure, p_p:

$$S = \delta + p_p \tag{1-6}$$

or:

$$\sigma = \sigma' + p_p \tag{1-7}$$

If the overburden stress, S, is in lb./sq. ft.*, then it is equal to:

$$S = \gamma_b \times D \tag{1-8}$$

where γ_b is the bulk specific weight of water-saturated overlying sediment in lb./cu. ft. and D is depth in ft. Inasmuch as effective stress, δ, increases continuously with decreasing porosity, ϕ, it is a function of either porosity or the remaining moisture content, M:

$$\delta = f(\phi) \tag{1-9}$$

or:

$$\delta = f(M) \tag{1-10}$$

* In this book, pressure is mostly expressed either in terms of pounds per square inch (lb./sq. inch) or kilograms per square centimeter (kg/cm^2); some conversion factors are presented in Appendix A.

The specific weight, γ, is the weight per unit volume and can be expressed in terms of lb./cu. ft. The density ρ, which is mass per unit volume and is equal to γ/g, where g is gravitational acceleration (= 32.174 ft. sec^{-1} sec^{-1}), can be expressed in terms of slugs/cu. ft. For example, pure water, which has a specific weight of 62.4 lb./cu. ft., has a density of 1.94 (= 62.4/32.2) slugs/cu. ft. In other words, the mass ρ is attracted by the earth with a force of magnitude ρg.

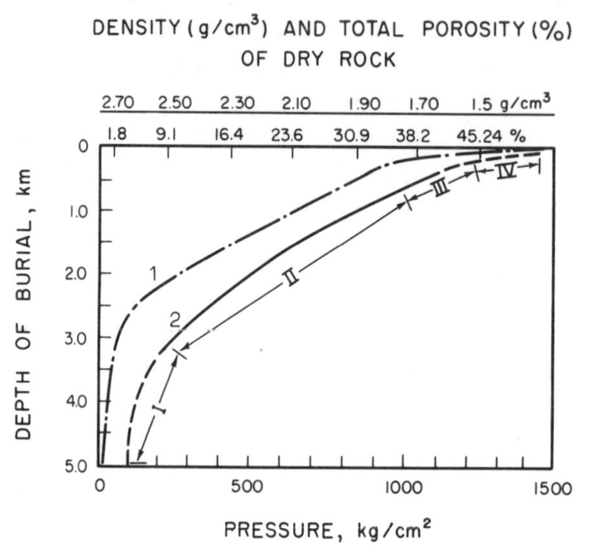

Fig.6. Variation in porosity and density of argillaceous sediments with increasing depth of burial. Compaction stages: *I* = with very great difficulty; *II* = with great difficulty; *III* = with difficulty, and *IV* = with ease. *1* = After Weller (1961); *2* = after Vassoevich (1960). (In: Golovin and Legoshin, 1970, fig.3, p.136.)

Several different compaction models are presented in Chapter 3. The one developed by Vassoevich (1960; also in: Golovin and Legoshin, 1970) seems to be quite satisfactory (see Fig.6). As shown in Fig.6, with increasing depth of burial the rate of compaction decreases. For example, at a depth of 1 km the density increases by 0.05 g/cm^3 per 100 m in burial depth, whereas at a depth of 2 km the density increase is only 0.025 g/cm^3. Vassoevich recognized four compaction stages during which compaction occurs: (I) with very great difficulty; (II) with great difficulty; (III) with difficulty; and (IV) with ease.

LABORATORY AND FIELD DATA ON COMPACTION

Laboratory observations related to the compaction of argillaceous sediments can be divided into two main types:

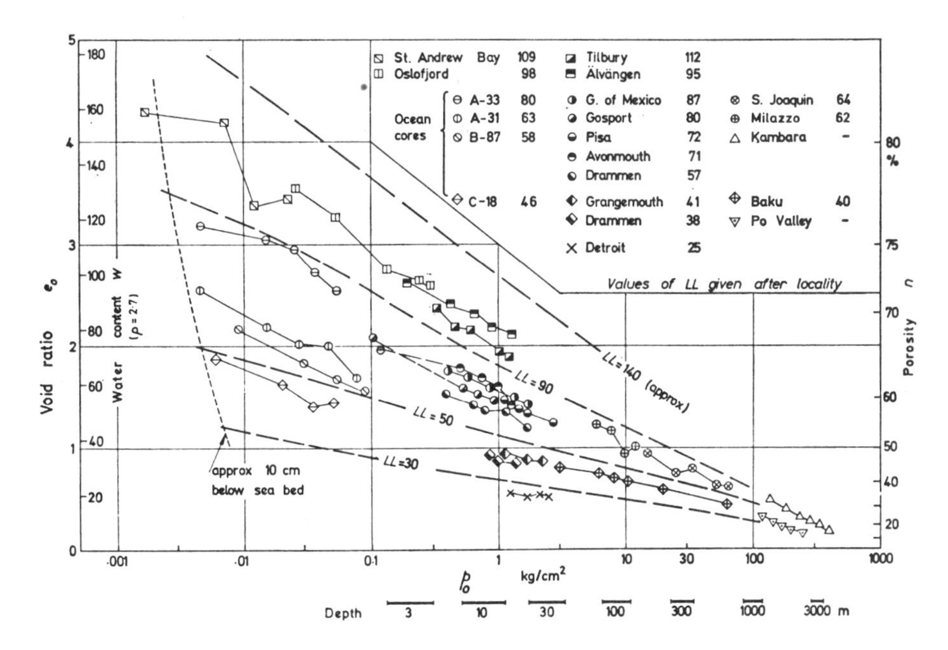

Fig.7. Sedimentation–compression curves for normally consolidated argillaceous sediments. *LL* = Liquid limit. (After Skempton, 1970, p. 403. Courtesy of the *Journal of the Geological Society of London.*)

(1) The first consists of laboratory results of compression tests on loose soil-like materials at low and high pressures. These results are sometimes not well suited for geologic applications. Applied pressures in low-pressure tests ranged up to approximately 1,000 p.s.i., which would be commensurable with the total overburden stress exerted by about 1,000 ft. of sediment. Skempton (1970) presented sedimentation–compaction curves for a wide lithological range of argillaceous deposits, relating void ratio to effective overburden pressures (effective pressure = total pressure – pore pressure) at low pressures (Fig.7). These low-pressure experiments have value in that they establish the broad pattern of response of clay and soil matrix to compaction loads. High-pressure experiments (> 1,000 p.s.i. and up to 500,000 p.s.i.) have been used to obtain data on the basic chemical, physical, and material properties of argillaceous sediments (e.g., Kryukov and Komarova 1954; Chilingar and Knight, 1960; Chilingar et al., 1963; Chilingar and Rieke, 1970).

(2) The second type of data is provided by measurements on large sandstone and shale core samples from oil and gas wells. The volume reductions and changes in compressibility, permeability, porosity, and density have been investigated in relation to simulated burial depth (e.g., Fatt, 1953; Knutson and Bohor, 1963; and Rieke et al., 1969). Athy (1930a) first discussed the concept that compressibility of sediments has been an important factor in developing differential compaction structures in the midcontinent area of the United States.

Both laboratory compaction tests and core analyses, and comparison between them, are important in better understanding compaction processes. Although the porosity–pressure relations are more basic than porosity–depth relations, they are not as useful in their geologic applications. Unfortunately, not all of the factors are known or have been scaled properly in the laboratory experiments to insure a proper understanding of the results. The mechanisms by which volume reductions take place in indurated and non-indurated sediments and the factors which are effective in the vertical deformation of rocks are explored both quantitatively and qualitatively in this book. The various properties of rocks (elasticity, strength, porosity, permeability, density, etc.) are incorporated into physical–mathematical approaches which model both the compaction and subsidence processes in argillaceous sediments and rocks. A considerable amount of field data on relationships among density, porosity, pressure, and depth of burial is presented in Chapter 2.

The need for laboratory data on gravitational compaction at high pressures was stressed by Hubbert and Rubey (1959) and Rubey and Hubbert (1959) in their classic papers. As pointed out by Rubey and Hubbert (1959, p.174), in geology the problem involves the compaction of clays which are deposited as sediments in geosynclinal basins; with progressive sedimentation the imposed loads could range from zero p.s.i. up to the weight of several miles of superposed sediments (overburden stresses of 14,000–36,000 p.s.i.). Rubey and Hubbert (1959, p.174) further stated that "the magnitude of the stresses involved, and the degree of compaction produced, is thus in a range far beyond the limits for which laboratory data are available." Laboratory apparatuses have been developed by several investigators (e.g., by Chilingar and Knight, 1960; Chilingar et al., 1963; and Sawabini et al., 1971) to study the various physical properties of saturated clays under high uniaxial and triaxial stresses (see Fig.8). At this point, one should ask a very important question: What type of compaction equipment (hydrostatic, triaxial, or uniaxial) could best reproduce the conditions existing in nature? In the opinion of the writers, during the initial stages of sedimentation, the pressure is probably hydrostatic. As sedimentation progresses, the pressure becomes triaxial. Finally, as the overburden load becomes large enough, the pressures are probably uniaxial.

At present, experimental studies of clays and clay-mineral assemblages which have varying stability ranges, together with analytical field investigations, are providing a more quantitative understanding of the various diagenetic processes and the migration of interstitial fluids. These modern high-pressure techniques make it possible to subject fluid-saturated clayey material to static pressures as high as 500,000 p.s.i. ($\approx$ 34.5 kb) at room temperature and up to 50,000 p.s.i. ($\approx$ 3.45 kb) at temperatures up to 400°F. New and advanced equipment, however, is continuously being developed at various research laboratories.

Different types of equipment and techniques that are employed in the study of sediment compaction are presented in Chapter 8. In addition, as an aid to earth scientists involved in designing compaction equipment, the related apparatuses used in the investi-

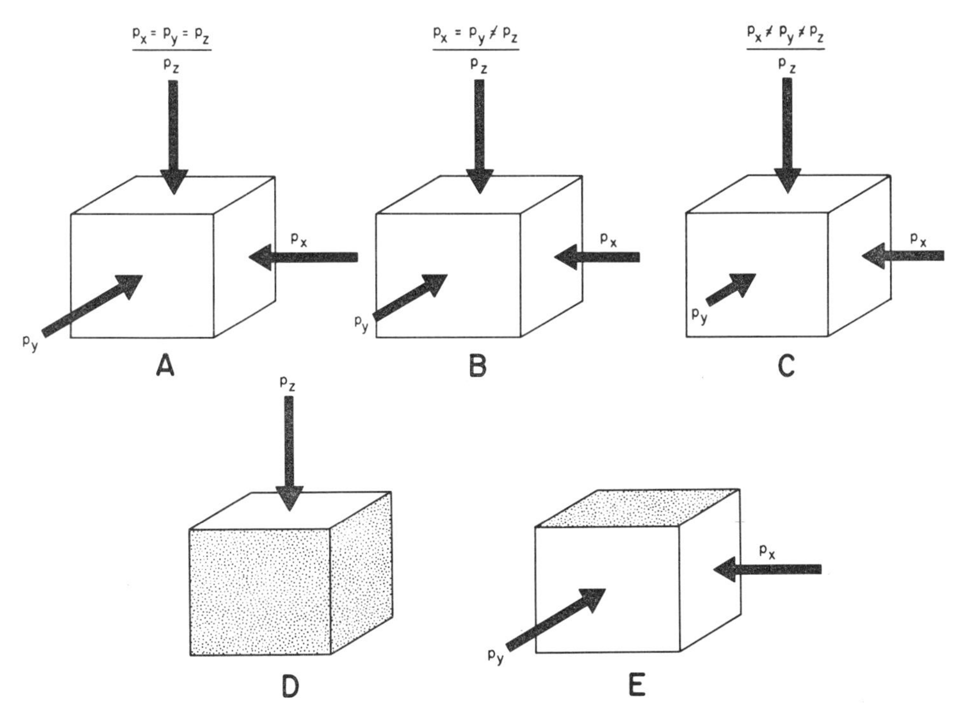

Fig.8. Different loading arrangements on sediments in the laboratory experiments. A. Hydrostatic loading (three principal stresses are equal: $p_x = p_y = p_z$). B. Triaxial loading (two out of three principal stresses are equal: $p_x = p_y \neq p_z$). C. Polyaxial loading (all three principal stresses have different magnitude: $p_x \neq p_y \neq p_z$). D. Uniaxial loading (force is perpendicular to one face of the cube, while the four faces perpendicular to this face are stationary). E. Biaxial loading (two principal stresses in the same plane are equal, whereas the two faces of the cube parallel to these stresses are stationary). (Classification proposed by C.T. Sawabini, personal communication, 1971.)

gation of associated topics, such as stress–strain and pressure–volume–temperature relationships and geochemical changes in various minerals and rocks with increasing pressure and temperature, are also discussed.

Inasmuch as successful laboratory research on the compaction of sediments largely depends on proper design of apparatus, the importance of Chapter 8 cannot be overemphasized. This chapter may also stimulate thinking on possible directions to follow in future designs of high-pressure and high-temperature autoclave devices for sedimentological studies.

DIAGENESIS–METAMORPHISM TRANSITION PHASE

The use of high pressures and temperatures in the laboratory permits studying the diagenesis–metamorphism transition phase and investigating the effects of a pressure greater than the overburden pressure on rocks. In high-pressure experiments, strict dependence of the properties of a deforming, saturated clay on the applied stress can be established only when the parameters being determined are referred to a particular point in the sample. From the physical–mathematical viewpoint, all the experimental stresses and strains are determined for some imaginary volumetric body (cylindrical or cubic) isolated in the sediment sample. In reality, however, all measurements represent average values for the test sample. The assumption that the pressure under which a rock recrystallizes is equal to that generated by the overburden load alone cannot be strictly correct, and it is possible that the departure from normal hydrostatic conditions as explained herein is significant. Metamorphosed rocks show abundant evidence of deformation by plastic flow. The total applied pressure during metamorphism is, therefore, equal to the stress generated by the overburden load plus an increment of stress sustained by the strength of rocks flowing under a tectonic compressional stress. This latter stress increment has been called the "tectonic overpressure" by Clark (1961), and its role in forming metamorphosed rocks such as the jadeite–aragonite-bearing rocks of the Franciscan Formation (California) from massive graywackes and shales was investigated by Brace et al. (1970). Based on laboratory experiments, however, these authors concluded that the possibility of tectonic overpressure to increase the mean pressure to the minimum values necessary for the development of mineral assemblages characteristic of Franciscan jadeite–aragonite facies seems remote.

Jamieson (1963, p.1067) pointed out that the possible effects of pressures as high as ten or more times greater than those corresponding to "depth of burial" must be considered at least locally and ephemerally in geologic structures. One such an occurrence would be the folding of a thin incompetent layer of rock confined between two highly competent layers in a sinusoidal fashion. Another case would be sudden faulting or folding with the sudden application of effective overburden to unconsolidated thin sediment layers.

On the geologic time scale, high stresses in sedimentary basins may be relieved by creep and pore-fluid leakage, and in older depositional basins there may be no present evidence that high stresses have existed. There seem to be two posssible indicators of prior existence of high pressure:

(1) If clays and other minerals transform to high-pressure forms and the back reactions are sufficiently slow or nonexistent, a high-pressure mineral and associated interstitial fluid assembly may be preserved.

(2) The possible preservation of certain unstable mineral forms instead of high-pressure forms. The latter case exists if a polymorphic transformation from $A \rightarrow B$ proceeds as certain pressure and temperature conditions are maintained. When the pressure is reduced, B becomes unstable with respect to A and also to C; the latter is another possible low-pressure mineral end product, but one which is less stable at low pressures than A. Thus, the mineral form B might transform to C (Jamieson, 1963, p.1096). If the reaction $C \rightarrow A$ is nonexistent or slow, then C will be found in the sediments, even though it did not

form directly from A. This is paradoxical unless B is also present (or its prior existence is recognized) or C originated outside the basin of deposition.

The role of tectonic stress in shale compaction is discussed in Chapter 3.

PROPERTIES OF ARGILLACEOUS SEDIMENTS AS RELATED TO COMPACTION

The effects of compaction on various properties of argillaceous sediments, such as compressibility and permeability, are presented in Chapter 4. It appears that a considerable amount of research still remains to be done in this area.

Recent research on the elastic properties of rocks and minerals has been mainly concerned with (1) the study of the effects of fabric, porosity and fluid content on the elastic properties of rocks, (2) the measurement of the effects of overburden pressure and temperature on the elastic properties of minerals and rocks, and (3) the development of new experimental techniques and equipment for simultaneous measurement of various physical properties of rocks and minerals at high pressures and temperatures. It was determined that, in general, the strength of argillaceous sediments is determined by the texture and increases with increasing degree of orientation of clay particles.

Studies on the elastic properties of natural rocks must necessarily take into consideration the problem of anisotropy produced by (a) random orientation of grains, (b) nonuniform porosity, (c) the presence of fractures, and (d) the fluid content within the pores or cracks. Liebermann and Schreiber (1971) presented an excellent review of the current state of study on the elastic properties of minerals.

INTERSTITIAL FLUIDS

Large amounts of pore fluids are squeezed out of continental and marine sediments during compaction. There is a lack of knowledge about the origin of these expelled fluids, the direction of their movement, and their chemical character. In the past, little attention has been given in the literature to the problem of determining the chemical composition of interstitial fluids in argillaceous rocks. Two important questions must be answered: (1) Will the chemical composition of these fluids change during the long process of compaction? (2) If so, what chemical changes take place? The answers to these questions will result in a better understanding of the diagenetic changes in marine muds as they are transformed into shales. Noble (1963) observed that lateral secretion of interstitial fluids expelled from sediments during compaction could be a possible explanation of the genesis of certain vanadium–uranium deposits in sandstones and lead–zinc deposits in carbonates common to the Colorado Plateau and Mississippi Valley, U.S.A., respectively.

Vine and Tourtelot (1970) stated that a better understanding of the enrichment of minor elements in thick black shales may eventually lead to the discovery of economically valuable concentrations of minor elements. They postulated that the availability of metals in pore fluids, which have come into contact with organic matter throughout the history of the fluids, is probably the most significant factor in determining the suite of enriched metals present. In some cases, the ore enrichment process may be an accumulative one, in which the organic matter in the shales plays a never-ending role from deposition, through compaction and diagenesis, to tectonic uplift. For many ore deposits, however, the most significant period of enrichment may be during the circulation and expulsion of the fluids during gravitational compaction.

The widespread occurrence of marine sediments in most of the petroliferous sedimentary basins suggests that the pore waters are ultimately derived from the sea. There is a striking similarity between the amino-acid spectra in fossil brines and in present-day sea water. Similarities in solutes also seem to bear this out. Yet, systematic chemical differences exist between the ancient and modern connate waters. Changes in content of various electrolytes may be caused by a variety of processes and mechanisms such as dissolution and precipitation, pH–Eh variations, microbial activities, base exchange, chemisorption, and dolomitization.

Similarities between isotope characteristics of oil-field brines and present-day sea water lead to the conclusion that in many cases the concentration of inorganic salts has not been accomplished by syngenetic evaporation. Instead, it probably occurred as a result of compaction and ion filtration by charged-net clay membranes. Changes in the chemistry of migrating fluids on gravitational compaction may also shed some light on the origin and migration of oil (Degens and Chilingar, 1967). Chilingar (1961), Chilingarian and Rieke (1968), and Weaver (1967) suggested that the controlling mechanism in oil migration is the presence or lack of hydratable clay minerals which would release large amounts of water during compaction (Chilingar and Knight, 1960). Oil is expelled by this available water. Many oil shales containing non-hydratable clays remain as such because of lack of water to push the oil out.

On migrating through muds as a result of compaction, subterranean waters may increase considerably in salinity (Degens and Chilingar, 1967; Degens et al., 1962). The solubility of hydrocarbons is enhanced by the presence of higher quantities of electrolytes; however, brines saturated with certain inorganic salts will eventually release their hydrocarbons. In the case of an emulsion transfer, salts will act as de-emulsifying agents. In both cases, an organic or oil phase will instantaneously develop as soon as the solubility product of NaCl is exceeded (Degens et al., 1962).

Based on extensive research on desalinization of water by reverse osmosis, one can state that an ion-filtration mechanism can feasibly operate in a geologic horizon (Van Everdingen, 1968). If water is transported by molecular diffusion through the membrane, a desalting effect results because of the high-concentration and high-diffusion coefficients of water and low-concentration and low-diffusion coefficients of salts. The energy necessary to force waters through the clay–shale membrane can be supplied by the compaction of sediments. In addition, movement can occur as a result of thermal or electrical potential gradients. At the present time, the most widely accepted theory is the semipermeable clay–shale-membrane or salt-sieve theory (Russell, 1933; De Sitter, 1947; Siever et al., 1965), which has not been either proved or disproved (Collins, 1970).

Experimental data on the chemistry of solutions expelled due to compaction

There are three major methods of extracting interstitial fluids from sediments involving (1) centrifuging, (2) squeezing out, and (3) washing out. The first two methods are based on the same principle of compressing the sediment particles together and thereby forcing the fluid out of the pore space. V. Marchig (personal communication, 1971) stated that the main difference between the first two methods is the force that generates the compressive stress, and it is possible to use them jointly. In using these two methods one must keep in mind the fact that only a portion of the interstitial fluid has been expelled from the pores and that the chemical composition of the expelled fluid probably differs from that of the portion remaining in the sediment. Marchig centrifuged his samples for 15 min at 3,400 r.p.m. and used a drying temperature of 105°C.

The third method of fluid extraction differs from the other two in principle. Basically, this technique consists of two steps after dividing the sample into two parts: (1) determination of the volume of interstitial fluid present in the sediment sample by drying and (2) extraction of all the soluble salts from the second wet sample by washing them out with distilled water or solvent. After analyzing the washed-out solution, the composition of the interstitial water is calculated, correcting for dilution effects.

V. Marchig (personal communication, 1971) compared the centrifuging method with the washing-out method, using a series of recent coarse-grained to silty marine sands from the North Sea and the Bay of Biscay and fine-grained calcareous argillaceous sediments from a deep-sea drilling site in the Atlantic Ocean. The analytical results indicate that the washed-out interstitial waters will show a greater salinity than the water expelled by squeezing out or centrifuging. This difference in salinity was attributed by Marchig to selective filtration of the ions by the clay minerals present in the sediment samples during compaction or centrifuging. In addition, one could expect that certain cations will exchange between the interstitial fluids and the clays during centrifuging and squeezing-out procedures. In washing-out techniques, on the other hand, part of the cations adsorbed on the clays may go into solution. V. Marchig (personal communication, 1971) attributed the enrichment of Ca^{2+} cations in washed-out solutions partially to the dissolution of carbonates owing to the greater volume of wash water used, or to the lowering of pH values as a result of oxidation of sulfide to sulfate. It is likely that any sulfide present in the sediment samples was oxidized to sulfate by atmospheric oxygen, thereby causing the pH of the interstitial solution to decrease. This results in the dissolution of the carbonates, freeing the Ca^{2+} cations. Magnesium was slightly enriched in Marchig's centrifuged interstitial waters, whereas it was strongly impoverished in washed-out waters. Apparently, during the washing-out process, Mg^{2+} cations replace univalent cations having a lower tendency for adsorption.

The reproducibility of the results is fair when using the centrifuging or squeezing-out method. On the other hand, the results obtained by using the washing-out technique are difficult to reproduce. Marchig found errors of up to 16% for total salinity between different runs, with errors of up to 30% for individual ions. In argillaceous sediments, the amount of clays present greatly affects the results, because cations are being deadsorbed with each subsequent wash. At the present time, it is not possible to determine what portions of the cations are deadsorbed and what portions are contributed by the redissolved salts.

Von Engelhardt and Gaida (1963) compressed pure montmorillonite and kaolinite clay muds saturated with solutions having different concentrations of NaCl and $CaCl_2$ at different pressures ranging from 30 to 3,200 atm. Their results show that for a given clay the equilibrium porosity which is reached at a distinct overburden pressure does not depend on electrolyte concentration. For pressures between 30 and 800 atm, the concentration of electrolyte in pore fluids diminishes in montmorillonite clay with increasing compaction. This was explained by Von Engelhardt and Gaida (1963) as due to the electrochemical properties of base-exchanging clays. If the pore fluids contain an electrolyte, the liquid immediately surrounding the clay particle will contain less electrolyte than fluid farther away from the double layer. Base-exchanging clays suspended in an electrolyte solution adsorb a certain amount of pure water which is bound in double layers around each particle (Von Engelhardt and Gaida, 1963, p.929). During compression, the electrolyte-rich solution is removed, and the fluid of the double layers poor in electrolyte content, is left behind. At higher overburden pressures (from 800 to 3,200 atm), an increase of salt concentration within the remaining pore water may be caused by inclusion of small droplets of fluids in the highly compressed clay, acting as a barrier to the movement of ions. The passage of anions through the double layer is retarded by the fixed negative surface charges on the clay particles. Ion blocking increases with increasing ion exchange capacity and compression of the clay. Apparently, ion blocking is greater for dilute solutions than for concentrated ones.

Experimental results obtained by Rieke et al. (1964) and Chilingar et al. (1969) show that the salinity of squeezed-out solutions from argillaceous sediments progressively changes with increasing overburden pressure. Both marine muds and clays plus sea-water mixtures were used. As pointed out by Rieke et al. (1964), most of the salts present in the waters, which are trapped during sedimentation, are squeezed-out during the initial stages of compaction. The salinity of squeezed-out solutions progressively decreases with increasing overburden pressure. Thus, these authors concluded that the mineralization of interstitial solutions in shales should be lower than that in the associated sandstones. In rapid compaction experiments (raising pressure to a desired value immediately), the salinity of squeezed-out solutions increased with increasing overburden pressure (Chilingar et al., 1969). Inasmuch as solutions were squeezed out rapidly, however, the rate of squeezing may be an important factor that has to be considered. On squeezing rapidly, the portion of fluid near the water vent is squeezed out at lower pressures, whereas at higher pressures the fluids deeper inside the clay body also have a chance to contribute, but only their more saline portion . As discussed in Chapter 5, other investigators disagree with these findings and claim that the chemistry of extruded interstitial solutions from clays does not change with increasing overburden pressure. Some researchers suggest (e.g., Sawabini and Chilingarian, 1972) that at low pressures the salinity of expelled interstitial solutions first increases (goes through a maximum) before starting to decrease with increasing overburden pressure. Only future research and development of accurate leaching techniques will show whether during the latter stages of compaction various salts are left behind in compacting argillaceous sediments as solids or not.

Long et al. (1970) attempted to show that there is a relationship between the composition of the interstitial water of a shale and that of the associated sandstones. They pointed out that the electric double layer is composed of a layer of ions adsorbed on the clay plate (Stern layer) and a diffuse double layer of ions (Gouy layer). If the salinity of the interstitial fluids is greater than 0.4 M/l, the diffuse layer is destroyed, leaving only the Stern layer (Norrish, 1954). The thickness of this layer is approximately 5 Å. This value corresponds to the thickness of four layers of water with a unit thickness of about 2.5 Å, if present between the opposite surfaces of two crystals or of two plates of clay, in the case of expandable clay minerals (Long et al., 1970, p.192). Van Olphen (1954) showed that the energy of the electric double layer, as a function of its thickness, goes through a minimum at about 5 Å.

A gradual expulsion of the four layers of interplanar water, at constant temperature, takes place at different pressures. According to Van Olphen (1963), at 25°C the first, second, third, and fourth layers of water are removed at pressures of 5,400, 2,500, 1,300, and 200 atm, respectively. At 50°C, the first and second layers are removed at lower pressures of 4,450 and 660 atm, respectively. According to Von Engelhardt and Gaida's (1963) experimental data, the first and second water layers are removed at 3,000 and 800 kg/cm^2. These two pressures correspond to porosities of 33% and 20%. Dickinson (1951) assigned a pressure value of 550 kg/cm^2 for the removal of the second water layer. Thus, Long et al. (1970, p.193) concluded that van Olphen's theoretical values agree satisfactorily with Von Engelhardt and Gaida's (1963) experimental data.

According to Von Engelhardt and Gaida (1963, p.929, fig.14), as the porosity is lowered from 30 to 20% the concentration of the pore solution begins to rise. These results were interpreted by Long et al. (1970) as due to the presence of chloride ions in the last two water layers, which are expelled with difficulty when the second water layer is removed. Consequently, the salinity of the interstitial fluid could increase instead of diminishing.

Possibly, the anions can be accomodated in the Stern layer. Grahame (1947) suggested that the anions, having smaller ionic radii than cations, are located in a plane even closer to the clay surface (inner Helmholtz plane) than the cations (second outer plane—outer Helmholtz plane). Grahame stated that in the Stern layer only the anions are really chemisorbed with loss of a part of their hydration shells, whereas the cations remain hydrated and are only attracted by electrostatic forces (Kruyt, 1952).

According to Long et al. (1970, p.199), the composition of the interstitial water in a shale is determined by the contributions of both the pore water (intercrystalline porosity) and the adsorbed water (intracrystalline porosity). These authors summarized the flow system through shales as follows:

(1) The water that migrates from the aquifer into the shale voids maintains its composition.

(2) The exchange position sites on the clay surface are preferentially saturated with certain cations depending on the composition of water present in the intercrystalline pores (Kruyt, 1952).

(3) If there is no longer any intercommunication between the intercrystalline pores or they disappear completely owing to compaction, the water that migrates into the shales will flow only through the double layers and thus will modify their composition.

(4) Brine filtration through a shale membrane results in modification of both the concentration and composition of the salts. The bivalent cations, which are more strongly adsorbed on the clay surfaces, are less mobile; whereas the monovalent cations are more mobile. Consequently, the solutions coming out of shales are enriched in the latter.

According to the model proposed by Warner (1964), and which is based on Donnan theory, any change in the electrolyte content of expelled interstitial water will depend only on surface potential and particle spacing. Consequently, the salinity of extruded pore water should decrease with increasing compaction pressure in the case of montmorillonite and illite clays, which have high and moderate base exchange capacities, respectively. On the other hand, in the case of kaolinite clay, which has a low exchange capacity, according to Von Engelhardt and Gaida (1963) and Warner (1964) there are no changes in the concentration of waters expelled during the compaction process.

Methods of studying the chemistry of interstitial solutions in the field are still difficult and expensive. Foster and Whalen (1966, p.165) presented a technique for estimating formation pressures from electric logs. This technique is especially applicable in the offshore Louisiana area in detecting abnormally high reservoir pressures. These authors, however, assumed that the water resistivity values in shales are equal to, or of the same magnitude as, that calculated for nearby sands. Consequently, the solution of this unsolved problem (i.e., what are the differences between the chemistry of interstitial water in shales and that in associated sandstones) is of great practical importance.

SUBSIDENCE

Land subsidence, which is discussed in Chapter 6, can be defined as the sinking of the earth's surface owing to adjustments in subterranean material. In a geologic sense, virtually the entire earth's surface has undergone subsidence and uplift at various times during the history of this planet. Subsidence has been attributed to ground-water extraction, oil and gas production, and salt, sulphur, and coal mining. In some instances, it is a combination of some of the above causes (Allen et al., 1971). Sinking of the land's surface has also been well documented during earthquakes. There is a clear distinction between compaction of shales and associated sediments caused by natural forces and man-induced aquifer-system compaction.

During recent times accurate and detailed elevation changes in the land surface have been recorded throughout the world which are concurrent with population expansion and urbanization. According to Allen et al. (1971), man-induced subsidence can be classified into the following categories:

(1) Ground-water subsidence which is caused by the removal of water from an aquifer.

(2) Hydrocompaction (hydroconsolidation) which is caused by the addition of water to water-deficient surficial deposits.

(3) Oil-field subsidence which is caused by the extraction of oil, gas, and associated brines.

(4) Near-surface compaction which is caused by surface loading or vibration of sediments.

(5) Mining subsidence which is caused by caving and slumping into mine excavations such as tunnels or caverns (for example, as a result of Frasch process).

Many examples of subsidence have been documented – see *Proceedings of the International Symposium on Land Subsidence*, 1969, sponsored by I.A.S.H. (International Association of Scientific Hydrology) and UNESCO. Subsidence owing to oil and gas production from unconsolidated to poorly lithified and poorly sorted sands has mainly occurred in the Goose Creek area near Galveston, Texas; Wilmington Field, Long Beach, California; Inglewood Field, Los Angeles, California; Niigata Gas Field, Japan; Po River delta gas-producing area, Italy; and in the oil fields along the Bolivar Coast of Lake Maracaibo, Venezuela. Water removal from confined aquifers composed of relatively unconsolidated lacustrine and alluvial sediments of Late Cenozoic age has produced ground subsidence in (1) Tokyo, Nagoya, Osaka, and Nagasaki, Japan; (2) Mexico City, Mexico; (3) Venice, Italy; (4) London, England; (5) Far West Rand area in the Republic of South Africa; (6) Savannah, Georgia; the Houston–Galveston area of Texas; Eloy-Picacho, Arizona; Denver, Colorado; and the San Joaquin and Santa Clara valleys in California, U.S.A.; and (7) the Taipei Basin, Taiwan. Studies of man-induced subsidence cases may help to understand better the geologic subsidence and compaction processes.

ABNORMAL FLUID PRESSURES

Shales adjacent to permeable sands containing fluids at high pressures are much more porous than is normal for shales at comparable depths. This can be attributed to both stratigraphic and structural conditions. Expulsion of water from shale pores during compaction is sometimes restricted by overlying lithified (e.g., as a result of $CaCO_3$ deposition) or impermeable strata and by faults. Such conditions give rise to "overpressured" (or "abnormally high-pressure") reservoirs, whereas the associated shales are termed "undercompacted" or "overpressured". Abnormally high fluid pressures are discussed in Chapter 7. There is water movement from the overpressured shales into the sands as hydrocarbons are depleted owing to production. According to Hottman and Johnson (1965), fluid pressures within the pore space of shales can be determined by using data obtained from both acoustic and resistivity logs.

During the past decade, problems related to the development of deep-drilling techniques and hydrocarbon production from reservoir rocks, mainly associated with thick sequences of argillaceous sediments in Tertiary basins, have become increasingly urgent in the petroleum industry. Successful drilling to depths greater than 20,000 ft. and subsequent production depend to a great extent on knowledge of the basic mechanical properties and deformation characteristics of the encountered formations, especially shales, and the chemistry of their associated fluids. Changes in the physical and chemical properties of rocks and interstitial fluids with depth and the ability to understand and predict these changes are fundamental to successful petroleum exploration and development.

The relative salinities of interstitial waters in well-compacted and undercompacted shales and their associated sandstones, as envisioned by the writers, are presented in Fig.9, if one assumes that fluids in the center of the capillaries (1) are more saline than fluids adjacent to the capillary walls and (2) are expelled first. In both cases (undercompacted and well-compacted shales), the salinities of interstitial fluids in shales should be fresher than those in associated sandstones if changes in all other variables are identical. Interstitial water in undercompacted shales, however, should be more saline than that in well-compacted shales, because in the former case a smaller portion of the more saline fluid present in the center of capillaries is squeezed into adjacent sandstones. Future field and laboratory investigations will show whether these relationships hold true or not.

In the worldwide search for oil and gas, fluids at pressures much higher than normal hydrostatic pressures have been encountered in many countries (see Fig.10). Abnormally high fluid pressures, which are often misnamed "geopressures", are defined by some geologists and petroleum engineers as any fluid pressure which exceeds the hydrostatic pressure of a column of water containing 80,000 p.p.m. total solids (Fertl and Timko, 1970). Many deviations from this generalization, which holds true for the Gulf Coast depositional basin, readily occur in other areas. High hydrostatic gradients (change in pore pressure per unit depth), with respect to the normal brine gradient of 0.465 p.s.i./ft. or 0.106 bar/m, exist throughout the entire pre-Tertiary sedimentary sequence in many basins such as the Williston Basin of North America. Formation waters with approximately 356,000 p.p.m. total solids occur in the latter basin, giving rise to an overall hydrostatic pressure gradient of 0.512 p.s.i./ft. (Finch, 1968). In order for fluid pressures to be classified as abnormal in the Williston Basin, they would have to yield a gradient larger

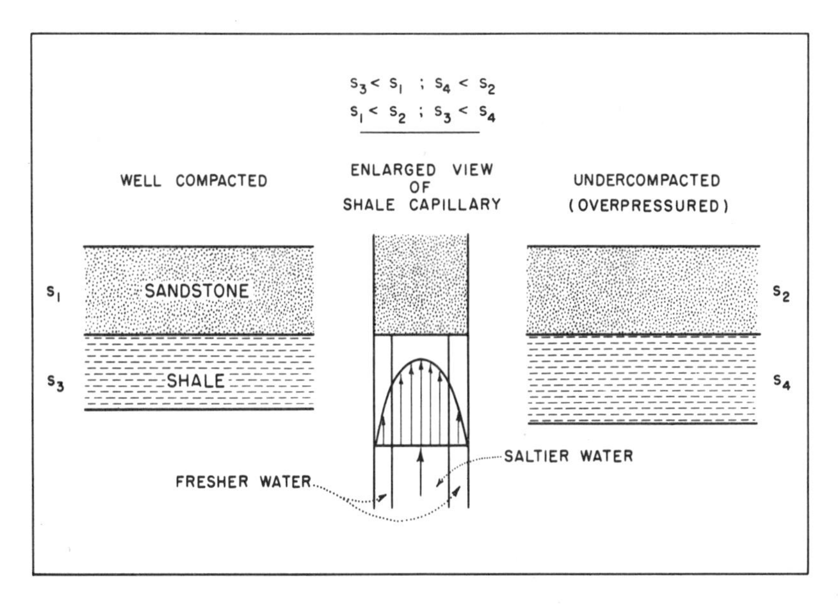

Fig.9. Relative salinities of interstitial fluids in well-compacted and undercompacted (overpressured) shales, as envisioned by the writers. Arrows show direction of flow and velocity profile in an enlarged view of a capillary in a vertical direction. S = Salinity.

than 0.512 p.s.i./ft. A more comprehensive definition of abnormal pressure, therefore, is a fluid pressure that materially exceeds the weight of a column of interstitial fluid ($p_p = \gamma_f h$, where p_p is pressure in lb./sq. ft., γ_f is specific weight of interstitial fluid in lb./cu. ft., and h is height of column in ft.).

It is not sufficient merely to recognize the presence of abnormal pressures. The magnitude of these fluid pressures must be known in order that drilling fluids can be selected to (a) prevent blow-out, (b) optimize penetration rate, (c) recognize hydrocarbon shows from productive formations, (d) lessen damaging fluid entry into potential pay sands, and (e) permit more reliable interpretation of formation log data (Griffin and Bazer, 1968).

Areas in the United States where abnormal fluid pressures have been reported in the literature include the Gulf Coast Basin (post-Cretaceous sediments), Anadarko Basin in Oklahoma (Pennsylvanian Morrowan sediments), Williston Basin in North Dakota (Devonian Sanish zone in the Antelope Oil Field), and in the Ventura area of California. Outside the United States, abnormal fluid pressures are known to exist in the Arctic Islands; Africa (Algeria, Morocco, Mozambique and Nigeria); Europe (Austria, the Carpathian region of eastern Europe, Rumania, France, Germany, The Netherlands, Ukraine, and in the Urals and Caucasian region of U.S.S.R.); Far East (Burma, China, India, Indonesia, Japan, Malaysia, and New Guinea); Middle East (Iran, Iraq, and Pakistan); and South America (Argentina, Colombia, Trinidad, and Venezuela).

In the Gulf Coast area of Louisiana, normal hydrostatic pressures occur to depths ranging from 8,000 to 12,000 ft. (Wallace, 1965, p.124). There is normally a short

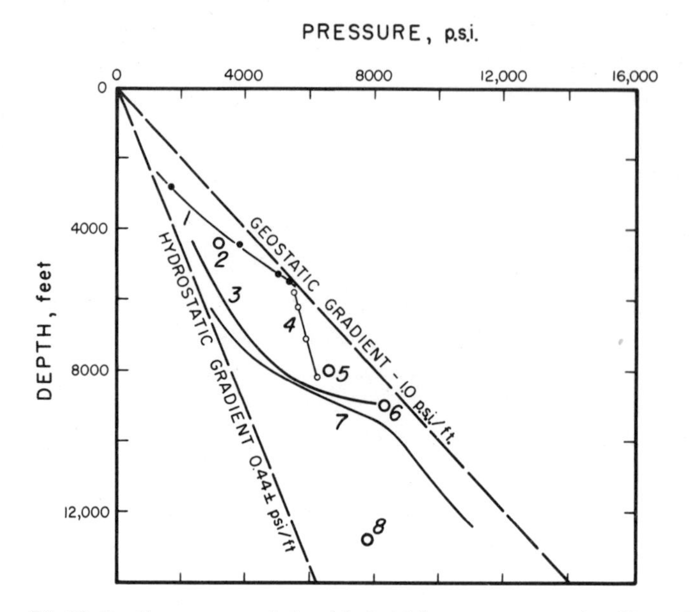

Fig.10. Depth–pressure relationship in high-pressure areas. *1* = Khaur Field, Pakistan; *2* = Lost Hills, California; *3* = Ventura Avenue, California (pressures measured near crest); *4* = Chia-Surkh, Iraq (after Cooke, 1955); *5* = Iran; *6* = D-7 zone of Ventura Avenue Field, California; *7* = Gulf Coast fields, Texas–Louisiana; *8* = Church Buttes, Wyoming. Pressure gradient of 1 p.s.i./ft. corresponds to overburden specific gravity of 2.31. Hydrostatic gradient of water having specific weight of 62.4 lb./cu. ft. is equal to 0.433 p.s.i./ft. ($p = \gamma h = (62.4 \times 1)/144 = 0.433$ p.s.i.). (After Watts, 1948, fig.2, p.144. Courtesy of the S.P.E. of A.I.M.E.)

pressure-transition zone at this interval followed by a drastic increase in the wellbore pressure. High abnormal pressures in these oil or gas wells usually result in difficulties during drilling operations. In many of the oil fields, therefore, detection of abnormal formation pressures has become imperative. Geologists and petroleum engineers need to discover a way to detect and predict high fluid pressures in advance of the bit in such areas so that protective casing strings can be set in the wells during drilling. In many cases wells either have been abandoned before reaching the desired formation or have blown out, often burning the drilling rigs and injuring or killing rig personnel. The numerous methods employed in the industry to detect and evaluate high-pressure zones are thoroughly discussed by Fertl and Timko (1970).

Thick Tertiary strata in the California basins have entirely different geologic characteristics from those along the Gulf Coast. In Californian basins, very thick and highly porous petroleum-saturated sandstones are interbedded with shaley sandstones having low permeability. In California, the problem of high pressures is not as paramount as along the Gulf Coast, but the problem of subsidence in connection with fluid withdrawal is of great concern to oil companies and local governmental agencies. In the opinion of the writers, all these problems are mainly related to compaction processes and can be resolved with increasing knowledge on this subject.

REFERENCES

Allen, D.R., Chilingar, G.V., Mayuga, M.N. and Sawabini, C.T., 1971. Studio e previsione della subsidenza. In: A. Mondadori (Editor), *Enciclopedia della Scienzá e della Tecnica.* Annuario della EST, Milano, Italy, pp.281–292.

Athy, L.F., 1930a. Density, porosity, and compaction of sedimentary rocks. *Bull. Am. Assoc. Pet. Geologists,* 14(1): 1–24.

Athy, L.F., 1930b. Compaction and oil migration. *Bull. Am. Assoc. Pet. Geologists,* 14(1): 25–35.

Aylmore, L.A.G. and Quirk, J.P., 1960. Domain of turbostratic structure of clays. *Nature,* 187: 1046–1048.

Binder, R.C., 1962. *Fluid Mechanics.* Prentice-Hall, Englewood Cliffs, N.J., 4th ed., 453 pp.

Biscayne, P.E., 1964. Mineralogy and sedimentation of the deepsea. Sediment fine fraction in the Atlantic Ocean and adjacent seas and oceans. *Yale Univ., Dep. Geol., Geochem. Tech. Rep.,* 8: 86 pp.

Blackwelder, E., 1920. The origin of the central Kansas oil domes. *Bull. Am. Assoc. Pet. Geologists,* 4(10): 39–94.

Bolt, G.H., 1956. Physico-chemical analysis of the compressibility of pure clays. *Geotechnique (Lond.),* 6(1): 86–93.

Brace, W.F., Ernst, W.G. and Kallberg, R.W., 1970. An experimental study of tectonic overpressure in Franciscan rocks. *Geol. Soc. Am. Bull.,* 81(5): 1325–1338.

Chilingar, G.V., 1961. Notes on compaction. *J. Alberta Soc. Pet. Geologists,* 9: 158–161.

Chilingar, G.V. and Knight, L., 1960. Relationship between pressure and moisture content of kaolinite, illite, and montmorillonite clays. *Bull. Am. Assoc. Pet. Geologists,* 44: 101–106.

Chilingar, G.V., Rieke III, H.H. and Robertson Jr., J.O., 1963. Relationship between high overburden pressures and moisture content of halloysite and dickite clays. *Geol. Soc. Am. Bull.,* 74(8): 1041–1048.

Chilingar, G.V., Rieke III, H.H., Sawabini, S.T. and Ershaghi, I., 1969. Chemistry of interstitial solutions in shales versus that in associated sandstones. *Soc. Pet. Eng. Am. Inst. Min. Metall. Engrs., 44th Ann. Fall Meet., Denver, Colo., S.P.E. Paper* no.2527: 8 pp.

Chilingarian, G.V. and Rieke III, H.H., 1968. Data on consolidation of fine-grained sediments. *J. Sediment. Petrol.,* 38(3): 811–816.

Clark, S.P., 1961. A redetermination of equilibrium relations between kyanite and sellimanite. *Am. J. Sci.,* 259(9): 641–650.

Collins, A.G., 1970. Geochemistry of some petroleum-associated waters from Louisiana. *U.S. Dep. Inter., Bur. Mines Rep. Invest.,* 7326: 31 pp.

Cooke, P.W., 1955. Some aspects of high weight muds used in drilling abnormally high pressure formation. *Proc. World Pet. Congr., 4th, Rome,* Sec. II: 43–57.

Degens, E.T., 1967. Diagenesis of organic matter. In: G. Larsen and G.V. Chilingar (Editors), *Diagenesis in Sediments.* Elsevier, Amsterdam, pp.343–390.

Degens, E.T. and Chilingar, G.V., 1967. Diagenesis of subsurface waters. In: G. Larsen and G.V. Chilingar (Editors), *Diagenesis in Sediments.* Elsevier, Amsterdam, pp.477–502.

Degens, E.T., Chilingar, G.V. and Pierce, W.D., 1962. Sobre el origen del petroleo dentro de concreciones de carbonato de calcio de edad Miocenica, formadas en agaus dulces. *Bol. Asoc. Mexicana Geol. Pet.,* 14(11, 12): 275–292.

De Sitter, L.U., 1947. Diagenesis of oil-field brines. *Bull. Am. Assoc. Pet. Geologists,* 31(11): 2030–2040.

Dickinson, G., 1951. Geological aspects of abnormal reservoir pressures in the Gulf Coast region of Louisiana, U.S.A. *Proc. World Pet. Congr.,* 3rd, The Hague, Sect. 1: 1–17.

Falini, F., 1965. On the formation of coal deposits of lacustrine origin. *Geol. Soc. Am. Bull.,* 76(12): 1317–1346.

Fatt, I., 1953. The effect of overburden pressure on relative permeability. *Trans. Am. Inst. Min. Metall. Engrs.,*198: 325–326.

Fertl, W.H. and Timko, D.J., 1970. Occurrence and significance of abnormal-pressure formations. *Oil Gas J.,* 68(1): 97–108.

Finch, W.C., 1968. Abnormal pressure in the Antelope Field, North Dakota. *Soc. Pet. Eng., 43rd Ann. Meet., Houston, Texas, S.P.E.* Paper no. 2227: 4 pp.

Foster, J.B. and Whalen, H.E., 1966. Estimation of formation pressures from electrical surveys – offshore Louisiana. *J. Pet. Tech.,* 17(2): 165–171.

Golovin, E.A. and Legoshin, V.P., 1970. Concerning epigenetic (superimposed) processes in sedimentary rocks. In: A.V. Sidorenko et al. (Editors), *Status and Problems of Soviet Geology.* Nauka, Moscow, 1: 130–147.

Grahame, D.C., 1947. The electrical double layer and the theory of electrocapillarity. *Chem. Revs.,* 41(3): 441–501.

Griffin, D.G. and Bazer, D.A., 1968. A comparison of methods for calculating pore pressure and fracture gradients from shale density measurements using the computer. *Soc. Pet. Eng., 43rd Ann. Fall Meet., Houston, Texas, S.P.E.* Paper no.2166: 12 pp.

Griffin, J.J. and Goldberg, E.D., 1963. Clay-mineral distribution in the Pacific Ocean. In: M.N. Hill (General Editor), *The Sea, Ideas and Observations in Progress in the Study of the Seas. 3. The Earth Beneath the Sea.* Interscience, New York, N.Y., pp.728–741.

Hedberg, H.D., 1926. The effect of gravitational compaction on the structure of sedimentary rocks. *Bull. Am. Assoc. Pet. Geologists,* 10(10): 1035–1072.

Hedberg, H.D., 1936. Gravitational compaction of clays and shales. *Am. J. Sci.,* 5th Ser., 31(184): 241–287.

Holmes, R.S. and Hern, W.E., 1942. Chemical and physical properties of some important alluvial soils of the Mississippi drainage basin. *U.S. Dep. Agric. Tech. Bull.,* 833: 82 pp.

Hottman, C.E. and Johnson, R.K., 1965. Estimation of formation pressures from log-derived shale properties. *J. Pet. Tech.,* 16(6): 717–722.

Hubbert, M.K. and Rubey, W.W., 1959. Role of fluid pressure in mechanics of overthrust faulting. I. Mechanics of fluid-filled porous solids and its application to overthrust faulting. *Bull. Geol. Soc. Am.,* 70(2): 115–166.

Hubbert, M.K. and Rubey, W.W., 1960. Role of fluid pressure in mechanics of overthrust faulting, a reply to discussion by H.P. Laubscher. *Bull. Geol. Soc. Am.,* 71: 617–628.

Jamieson, J.C., 1963. Possible occurrence of exceedingly high pressures in geological processes. *Geol. Soc. Am. Bull.,* 74(8): 1067–1070.

Knutson, C.F. and Bohor, B.F., 1963. Reservoir rock behavior under moderate confining pressure. In: C. Fairhurst (Editor), *Rock Mechanics.* Pergamon, New York, N.Y., pp.627–658.

Kruyt, H.R., 1952. *Irreversible Systems. Colloid Science,* I. Elsevier, Amsterdam, 368 pp.

Kryukov, P.A. and Komarova, N.A., 1954. Concerning squeezing out of water from clays at very high pressures. *Dokl. Akad. Nauk S.S.S.R.,* 99(4): 617–619.

Lahee, F.H., 1923. *Field Geology.* McGraw-Hill, New York, N.Y., 651 pp.

Laubscher, H.P., 1960. Role of fluid pressure in mechanics of overthrust faulting: discussion. *Bull. Geol. Soc. Am.,* 71: 611–616.

Liebermann, R.C. and Schreiber, E., 1971. Elastic properties of minerals. *Trans. Am. Geophys. Union, I.U.G.G. Rep.,* Pt. 2, 52(5): 142–147.

Long, G., Neglia, S. and Rubino, E., 1970. Pore fluid in shales and its geochemical significance. In: G.D. Hobson and G.C. Speers (Editors), *Advances in Organic Geochemistry.* Pergamon, New York, N.Y., pp.191–217.

Martin, R.T., 1965. Quantitative measurements of wet clay fabric (abstract). *North Am. Clay Miner. Conf., 14th, Berkeley,* p.31.
McCoy, A.W., 1934. An interpretation of local structural development in midcontinent areas associated with deposits of petroleum. In: E.L. Wrather and F.H. Lahee (Editors), *Problems of Petroleum Geology.* Am. Assoc. Pet. Geologists, Tulsa, Okla., pp.581–627.
Meade, R.H., 1964. Removal of water and rearrangement of particles during the compaction of clayey sediments – review. *U.S. Geol. Surv., Prof. Pap.,* 497–B: 1–23.
Meade, R.H., 1968. Compaction of sediments underlying areas of land subsidence in central California. *U.S. Geol. Surv., Prof. Pap.,* 497D: 1–39.
Mehl, M.J., 1919. The influence of the differential compression of sediments on the attitude of bedded rocks. Paper presented at meeting of *Am. Assoc. Adv. Sci.,* St. Louis, Dec., 1919. Abst. published in *Science,* New Ser., 51(May, 1920): 520.
Millot, J., 1949. Relations entre la constitution et la genèse des roches sédimentaires argileuses. *Thèse Sci. Nancy et Géol. Appl. Prospec. Min.,* 2(2–4): 1–352.
Mitchell, J.K., 1960. The application of colloidal theory to the compressibility of clays. In: R.H.G. Parry (Editor), *Interparticle Forces in Clay–Water–Electrolyte Systems.* Commonwealth Sci. Ind. Res. Organization, Melbourne, pp.2-92–2-97.
Monnett, V.E., 1922. Possible origin of some of the structures of the midcontinent oil field. *Econ. Geol.,* 17: 194–200.
Müller, G., 1967. Diagenesis in argillaceous sediments. In: G. Larsen and G.V. Chilingar (Editors), *Diagenesis in Sediments.* Elsevier, Amsterdam, pp.127–177.
Nevin, C.M., 1931. *Principles of Structural Geology.* Wiley, New York, N.Y., 303 pp.
Noble, E.A., 1963. Formation of ore deposits by water of compaction. *Econ. Geol.,* 58: 1145–1156.
Norrish, K., 1954. The swelling of montmorillonite. *Faraday Soc. (Lond.) Discuss.,* 18: 120–134.
O'Brien, N.R., 1963. *A study of Fissility in Argillaceous Rocks.* Ph.D. Thesis, Univ. Illinois, Urbana, Ill., 80 pp.
Rieke III, H.H., 1970. *Compaction of Argillaceous Sediments* (20–500,000 p.s.i.). Thesis., Univ. Southern California, Los Angeles, Calif., 682 pp.
Rieke III, H.H., Chilingar, G.V. and Robertson Jr., J.O., 1964. High-pressure (up to 500,000 p.s.i.) compaction studies on various clays. *Int. Geol. Congr., 22nd, New Delhi,* Part 15: 22–38.
Rieke III, H.H., Ghose, S.K., Fahhad, S.A. and Chilingar, G.V., 1969. Some data on compressibility of various clays. *Proc. Int. Clay Conf., Tokyo,* 1: 817–828.
Rowell, D.L., 1965. Influence of positive charge on the inter- and intra-crystalline swelling of oriented aggregates of Na montmorillonite in NaCl solutions. *Soil Sci.,* 100(5): 340–347.
Rubey, W.W. and Hubbert, M.K., 1959. Role of fluid pressure in mechanics of overthrust faulting. II. Overthrust belt in geosynclinal area of western Wyoming in light of fluid-pressure hypothesis. *Bull. Geol. Soc. Am.,* 70(2): 167–206.
Russell, W.L., 1933. Subsurface concentration of chloride brines. *Bull. Am. Assoc. Pet. Geologists,* 17: 1213–1228.
Samuels, S.G., 1950. The effect of base exchange on the engineering properties of soils. *Build. Res. Stn.* (Great Britain), Note C 176: 16 pp.
Sawabini, C.T. and Chilingarian G.V., 1972. Changes in chemistry of interstitial solutions extruded out of montmorillonite clay saturated in sea water with increasing overburden pressure. In preparation.
Sawabini, C.T., Chilingar, G.V. and Allen, D.R., 1971. Design and operation of a triaxial, high-temperature, high-pressure compaction apparatus. *J. Sediment. Petrol.,* 41(3): 871–881.
Serruya, C., 1969. Problems of sedimentation in the Lake of Geneva. *Verh. Int. Ver. Limnol.,* Stuttgart, 17: 209–218.
Shaw, E.W., 1918. Anomalous dips. *Econ. Geol.,* 13: 598–610.
Shaw, E.W. and Munn, M.J., 1911. Coal, oil, and gas of the Foxburg Quadrangle, Pa. *U.S. Geol. Surv. Bull.,* 454: 40 pp.
Siever, R., Beck, K.C. and Berner, R.A., 1965. Composition of interstitial waters of modern sediments. *J. Geol.,* 73(1): 39–73.
Skempton, A.W., 1953. Soil mechanics in relation to geology. *Proc. Yorkshire Geol. Soc.,* 29: 33–62.

Skempton, A.W., 1970. The consolidation of clays by gravitational compaction. *Q. J. Geol. Soc. Lond.*, 125(3): 373–411.
Sorby, H.C., 1908. On the application of quantitative methods to the study of the structure and history of rocks. *Q. J. Geol. Soc. Lond.*, 64: 171–232.
Teas, L.P., 1923. Differential compacting the cause of certain Clairborne dips. *Bull. Am. Assoc. Pet. Geologists*, 9: 370–373.
Teichmüller, M. and Teichmüller, R., 1967. Diagenesis of coal (coalification). In: G. Larsen and G.V. Chilingar (Editors), *Diagenesis in Sediments.* Elsevier, Amsterdam, pp.391–415.
Terzaghi, K., 1925. Principles of soil mechanics. II. Compressive strength of clays. *Eng. News Rec.*, 95: 796–800.
Van Everdingen, R.O., 1968. Mobility of main ion species in reverse osmosis and the modification of subsurface brines. *Can. J. Earth Sci.*, 5(8): 1253–1260.
Van Olphen, H., 1954. Interlayer forces in bentonite. *Proc. Natl. Conf. Clays Clay Miner., 2nd, 1953 – Natl. Acad. Sci., Natl. Res. Counc. Publ.*, 327: 418–438.
Van Olphen, H., 1963a. *An Introduction to Clay Colloid Chemistry.* Wiley, New York, N.Y., 301 pp.
Van Olphen, H., 1963b. Compaction of clay sediments in the range of molecular particle distances. *Clays Clay Miner. – Proc. Natl. Conf. Clays Clay Miner., 11th*, pp.178–187.
Vassoevich, N.B., 1960. Experiment in constructing typical curve of gravitational compaction of clayey sediments. *Nov. Neft. Tekh.* (News of Petroleum Technology), *Geol. Ser.*, 1960(4): 11–15.
Vine, J.D. and Tourtelot, E.B., 1970. Geochemistry of black shale deposits – A summary report. *Econ. Geol.*, 65: 253–272.
Von Engelhardt, W. and Gaida, K.H., 1963. Concentration changes of pore solutions during the compaction of clay sediments. *J. Sediment. Petrol.*, 33(4): 919–930.
Wallace, W.E., 1965. Application of electric log measured pressure to drilling problems and a new simplified chart for wellsite pressure computation. *The Log Analyst*, 6: 4–10.
Warner, D.L., 1964. *An Analysis of the Influence of Physical–Chemical Factors Upon the Consolidation of Fine-Grained Clastic Sediments.* Ph. D. Diss., Univ. California, Berkeley, Calif., 136 pp..
Watts, E.V., 1948. Some aspects of high pressures in the D–7 zone of the Ventura Avenue Field. *Trans. Am. Inst. Min. Metall. Engrs., Pet. Div.*, 174: 191–200.
Weaver, C.E., 1967. The significance of clay minerals in sediments. In: B. Nagy and U. Colombo (Editors), *Fundamental Aspects of Petroleum Geochemistry.* Elsevier, Amsterdam, pp.37–75.
Weller, J.M., 1959. Compaction of sediments. *Bull. Am. Assoc. Pet. Geologists*, 43(2): 273–310.

Chapter 2

INTERRELATIONSHIPS AMONG DENSITY, POROSITY, REMAINING MOISTURE CONTENT, PRESSURE AND DEPTH

INTRODUCTION

This chapter is concerned primarily with the interrelationships among density, porosity, depth, and overburden pressure. Experimental data on the relationship between the remaining interstitial fluid content and applied pressure are also presented. According to Strakhov (1954, p.595), the rapid escape of water from semiconsolidated clays occurs at depths down to 250–300 m. With continued subsidence, sediments become consolidated rocks and there is only a very slow escape of water and further compaction.

Müller (1967) pointed out that once a certain overburden load has been reached and sediment is compacted, the process is irreversible; that is, even after a later uplift and erosion of the upper layers, with consequent release of pressure, the porosity attained at the maximum burial depth does not change subsequently. Thus, the maximum depth of burial experienced by the argillaceous sediment can be estimated from the porosity of unweathered rocks.

It has been shown by Chilingar and Knight (1960a) that at any particular high overburden pressure the fluid content of a laboratory compacted clay is related to the type of clay mineral present. In recent years a great deal of new data has been accumulated on the compaction of muds, especially owing to the encounter of the petroleum industry with abnormally high fluid pressures in deeply-buried shale sequences. Some of these data are presented here.

DENSITY–DEPTH–PRESSURE RELATIONSHIPS

During the gravitational compaction of marine sediments, there is generally a rapid increase in bulk density within the first few hundred feet of burial. Bulk density is defined here as the density of the sediment in its natural state. It is commonly expressed as the mass of the undisturbed sample together with interstitial fluids divided by its external volume. This external volume includes not only the volume of the solid portion of the sample but also its permeable and impermeable pore space. At depth, the bulk densities of rocks tend to approach the weighted average of the grain densities because of decreasing pore volume and the expulsion of the interstitial fluids.

The change in the bulk density of porous rocks under compaction is dependent upon the changes in pore volume. Dobrynin (1962, p.362) studied bulk-density changes in water-saturated sandstones up to 20,000 p.s.i. He concluded from the experimental data that changes in the density of rock matrix in sands and the contained water at room temperature could be ignored as factors affecting bulk-density values at these overburden loads.

The wet bulk density of fluid-saturated porous rocks is given by the equation:

$$\rho_{bw} = \rho_g - (\rho_g - \rho_f)\phi \tag{2-1}$$

where ρ_{bw} is wet bulk density, ρ_g is matrix (grain-mineral) density, ρ_f is fluid density, and ϕ is fractional porosity. The dry bulk density ρ_{bd}, can be calculated from the following basic expression:

$$\rho_{bd} = \rho_g(1-\phi) \tag{2-2}$$

Dobrynin (1962) expressed the relative changes in the bulk density of porous rocks as:

$$\frac{\Delta\rho_b}{\rho_b} = \left[\frac{\rho_g - \rho_f}{\dfrac{\rho_g}{\phi} - (\rho_g - \rho_f)}\right]\frac{\Delta\phi}{\phi} \tag{2-3}$$

On considering that the pore compressibility of consolidated sandstones generally varies between $0.5 \cdot 10^{-5}$ p.s.i.$^{-1}$ and $2.0 \cdot 10^{-5}$ p.s.i.$^{-1}$, and on assuming that $\rho_g = 2.65$ g/cm^3 and $\rho_f = 1.0$ g/cm^3, Dobrynin (1962) calculated the relative changes of density under pressure using mathematical equations (Fig.11). According to this figure, the average decrease in porosity for net overburden pressures of 0 up to 20,000 p.s.i. is 0–10%, whereas the increase in density is in the range of 0–2%. Graphs similar to Fig.11 should be prepared for argillaceous sediments.

Dallmus (1958, p.912) pointed out that an increase in bulk density as a result of compaction in sandstones, limestones, chemically precipitated rocks, and other competent rocks is very moderate. This is not the case in fine-grained clastics (Fig.12), where bulk density increases rapidly on compaction. It is a function of overburden and tectonic stresses, temperature, time, loading rate, and, in part, of grain-size distribution, secondary cementing material, trapped salts in the pores, and mineralogy of the non-clay fraction. Morgan (1969) observed from data on fresh-water clayey sediments in Lake Erie that there was no simple, clear correlation between the median diameter of the grains and bulk density. Bulk density of argillaceous sediments and rocks can vary extensively with depth from one region to another and even within the same stratigraphic unit in a depositional basin. Dana (1967) investigated the lateral and vertical variations of bulk density within a Miocene sandstone and shale sequence in the San Bernardino Mountains in California. This author did not find any noticeable systematic variation in bulk densities. The study

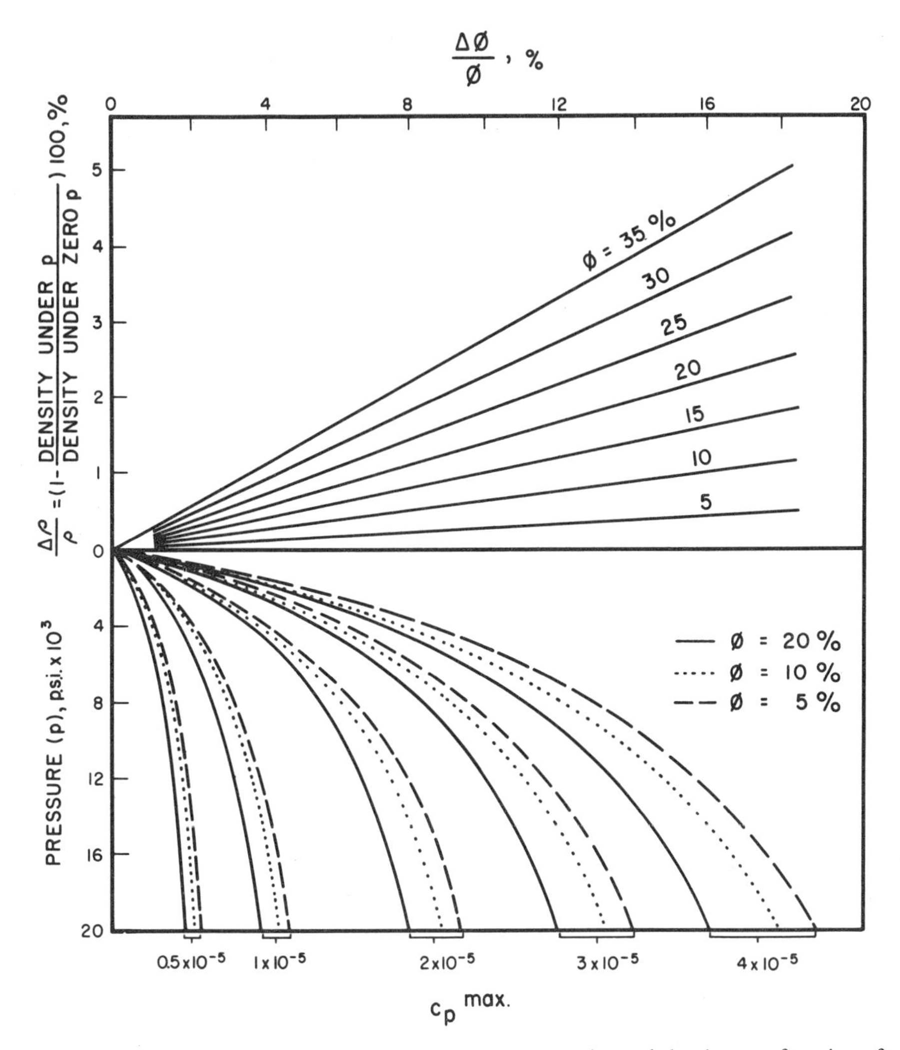

Fig.11. Dobrynin's graph for determining changes in porosity and density as a function of net overburden pressure. The pore compressibility, $c_p = -(1/V_p)[dV_p/dp]$, where V_p = pore volume and p = net overburden pressure. (After Dobrynin, 1962, fig.3, p.363.)

did show, however, that there can be a considerable variation in bulk density within a short distance in rocks. For these reasons, no universal bulk shale density curve can be constructed to characterize a specific type of argillaceous sediment or rock. Marked deviations in bulk shale density measurements with depth are used by geologists to help predict abnormal fluid pressure zones in Gulf Coast Tertiary sediments and in other areas, and to detect major unconformities.

Cebell and Chilingarian (1972) determined the specific weight of halloysite, hectorite and illite clay compacts directly (volume–weight measurements) at various pressures and

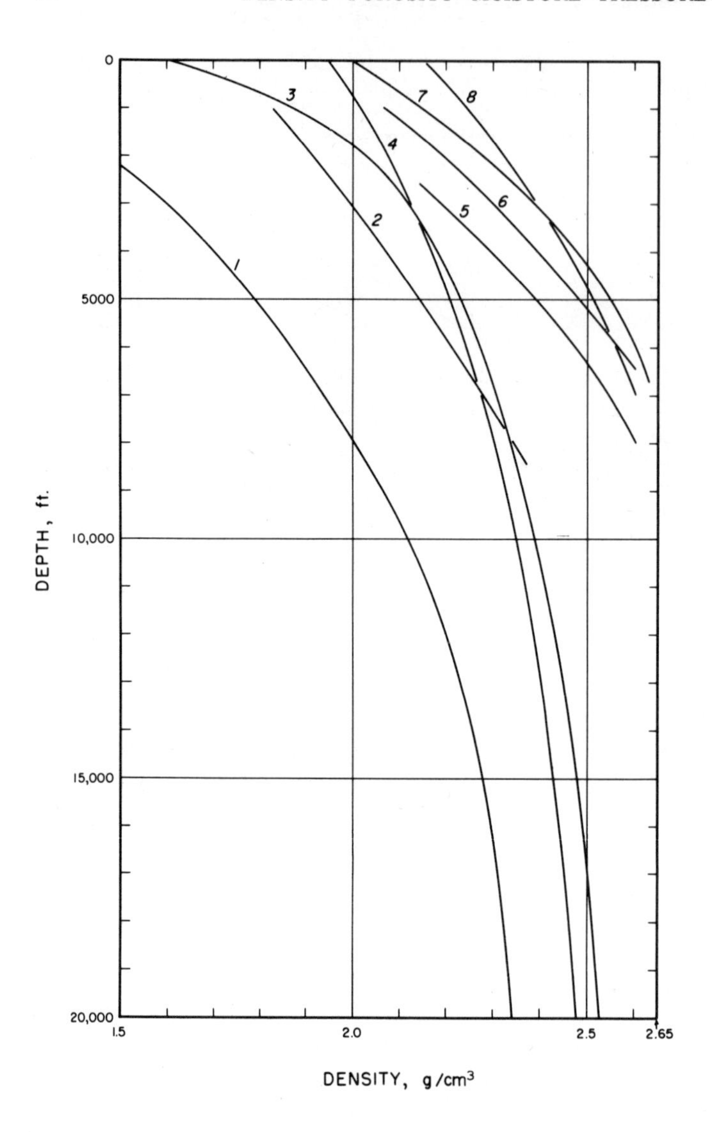

Fig.12. Variation of shale bulk densities with depth in sedimentary basins. *1* = methane-saturated clastic sedimentary rock (probable minimum density) (after McCulloh, 1967, p.A19); *2* = mudstone–Po Valley Basin, Italy (after Storer, 1959); *3* = average Gulf Coast shale densities; values derived from geophysical data (after Dickinson, 1953, p.427); *4* = average Gulf Coast shale densities derived from density logs and formation samples (after Eaton, 1969); *5* = Motatan-1 – Maracaibo Basin, Venezuela (after Dallmus, 1958, p.916); *6* = Gorgeteg No.1 – Hungary; calculated wet density values (after Skeels, 1943); *7* = Pennsylvanian and Permian dry shale density values, Oklahoma and Texas; Athy's adjusted curve (after Dallmus, 1958, p.913); *8* = Las Ollas-1 – eastern Venezuela (after Dallmus, 1958, p.918).

TABLE II

Densities of halloysite, hectorite and illite clays at different pressures and drying temperatures (After Cebell and Chilingarian, 1972)

Pressure (p.s.i.)	*Dehydrating temperature (°C)*	*Duration of dehydrating (hours)*	*Density (g/cm³)*		
			halloysite	*hectorite*	*illite*
2,600	130	24		2.59 (two runs)[1]	
3,400	130	24	2.51		
9,800	130	24			2.59
34,000	130	24	2.34	2.40	2.475 *
9,800	30	72			2.40 **
34,000	30	72			2.37 **
2,000	105	24			2.68
14,300	105	24			2.53
34,000	105	24			2.38

[1] One sample contained 62.3% $CaCO_3$ and the other, 58% $CaCO_3$. Densities in the case of hectorite clay were not corrected for $CaCO_3$ content.

* Average of two runs: 2.47 and 2.48 g/cm^3.

** The drying temperature of 30°C is probably so low that some of the pore water remains in the clay compact.

using different drying temperatures (Table II). The specific weights of solid grain-matrix decreased or the specific volumes (1/specific weight) increased with increasing overburden pressure. They explained this anomaly as possibly due to the lower specific weight of the remaining water or increasing amounts of adsorbed or interlayer water ("driving-in" of water effect). It is also possible that at higher pressure the lattice water in clays may be in a much more orderly arrangement than interstitial water and, consequently, is more compact. In that case, there would be a tendency for water to penetrate the clay lattice as the pressure increases. The results of Cebell and Chilingarian (1972) should be kept in mind in studying density–depth–pressure relationships.

Density variations of recent sediments and shales with depth

Preiss (1968) measured the bulk density of water-saturated sediments of the sea floor in situ by gamma-ray transmission equipment. With the transmission technique, the vertical thickness of the sediment layer "seen" is equal to the thickness of the scintillation (detection) crystals, which is usually between 2 and 5 cm. Fig.13 shows a typical profile observed. The r.m.s. difference between 57 comparisons of core and in-situ density was 1.2%.

In 1930, Athy presented the results of some 2,200 bulk-density (dry) determinations and 200 porosity determinations of Permian and Pennsylvanian shale samples from northeastern Oklahoma and Texas (Fig.14). As shown in this figure, the bulk densities are not constant for a given depth of burial. The solid curve represents the average bulk density–

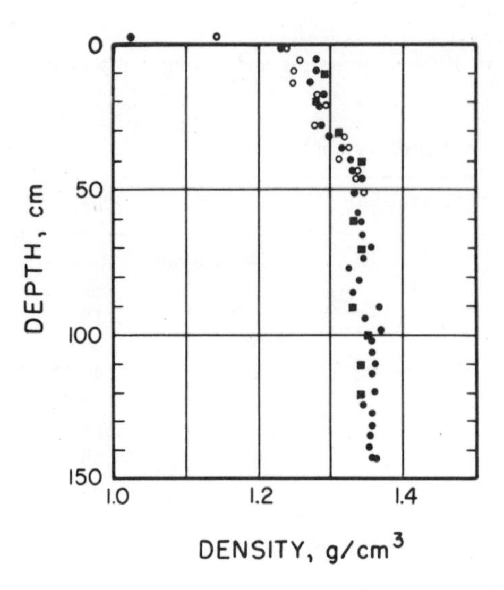

Fig.13. Relationship between bulk density and depth in deep-sea sediment. Solid circles = in-situ readings; open circles = repeated in-situ readings; solid squares = measurements made on cores by laboratory gamma-ray apparatus. No allowance was made for instrument tower settlement or core shortening. (After Preiss, 1968, fig.3, p.640.)

depth relationship for the samples measured in the laboratory by Athy. The dotted portion is a hypothetical extension of the curve to 1.4 g/cm^3, which was taken by him to be the average bulk density of the surface clay. Athy's bulk density data, if applied to Tertiary sediments, would be questioned on the basis of geologic age and tectonic stress history. Dallmus (1958, p.913) adjusted Athy's curve so that it would give a more reasonable surface bulk density value ($\approx$2 g/cm^3) (Fig.12, curve *7*).

Hedberg (1936) analyzed formation samples of Tertiary shales taken from a large geosynclinal basin in Venezuela. The sequence was undisturbed, nearly horizontal and removed from areas of major tectonic disturbance. Absence of major unconformities in the section made it possible to assume that existing overburden stresses are essentially the maximum overburden loads experienced by the sediments. Hedberg's data have been questioned because of his technique in measuring the bulk densities.

Dickinson (1953, p.426) reported that shallow clay deposits have a bulk density of about 1.8 g/cm^3. The available data on the in-situ density measurements and estimates employed by geophysicists for the Tertiary Gulf Coast sediments form the basis of Dickinson's curve (Fig.12, curve *3*). These values agree closely with those of Ham (1966) and Foster and Whalen (1966) in the lower-pressure region, whereas at higher overburden pressures (depth of 9,000 ft.), Dickinson's bulk density values are lower.

The dry-shale bulk densities reported by Kerr and Barrington (1961) are in close accord with those of Dickinson's in the depth interval of 6,000–10,000 ft. Ham's (1966) data is based on shale density measurements from density logs. Eaton (1969) compiled

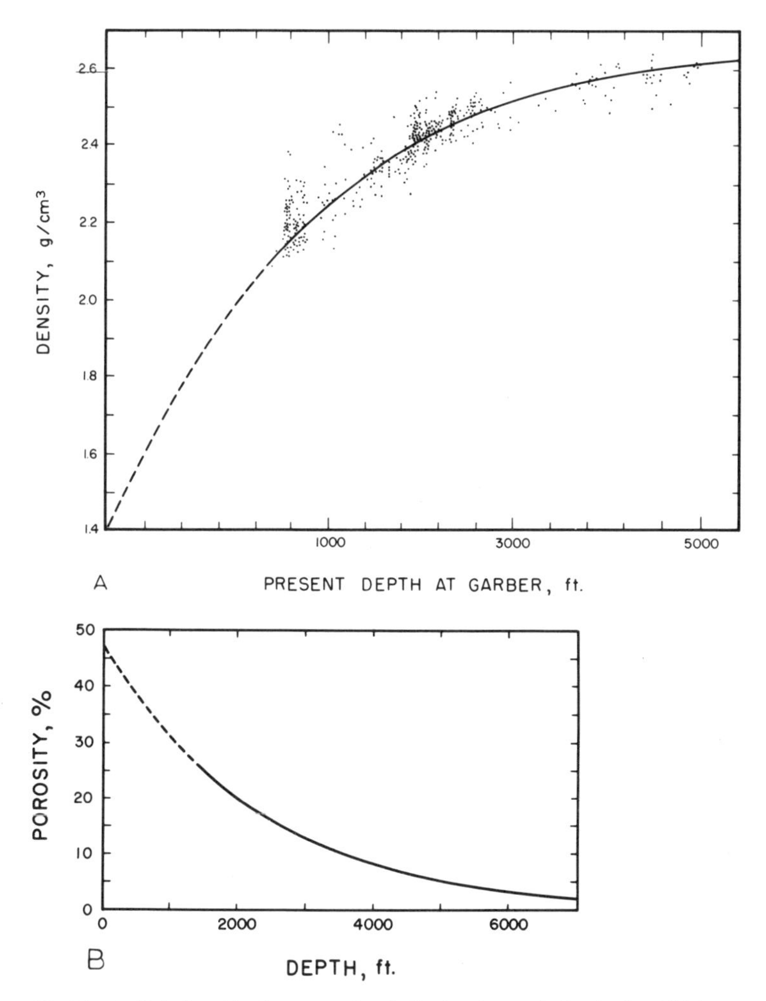

Fig.14. A. Relationship between dry bulk density and depth for Oklahoma shales. B. Relationship between porosity and depth for Oklahoma shales. (After Athy, 1930, fig.2 and 3, pp.12 and 13. Courtesy of American Association of Petroleum Geologists.)

numerous density values from density logs and shale cuttings which differed from Dickinson's data (Fig.12, curve *4*). A good presentation of Skeel's unpublished data (Fig.12, curve *4*) is given by Johnson (1950). The densities of rocks saturated with petroleum fluids, ranging in composition from methane to 30° API* crude, do not fall in the area of the densities of water-saturated rocks and shales. The densities of gas- and oil-saturated rocks range from the probable minimum density of clastic sedimentary rocks (Fig.12, curve *1*)

$$* {}^{\circ}\text{API} = \frac{141.5}{\text{sp. gr. at } 60^{\circ}\text{F}} - 131.5$$

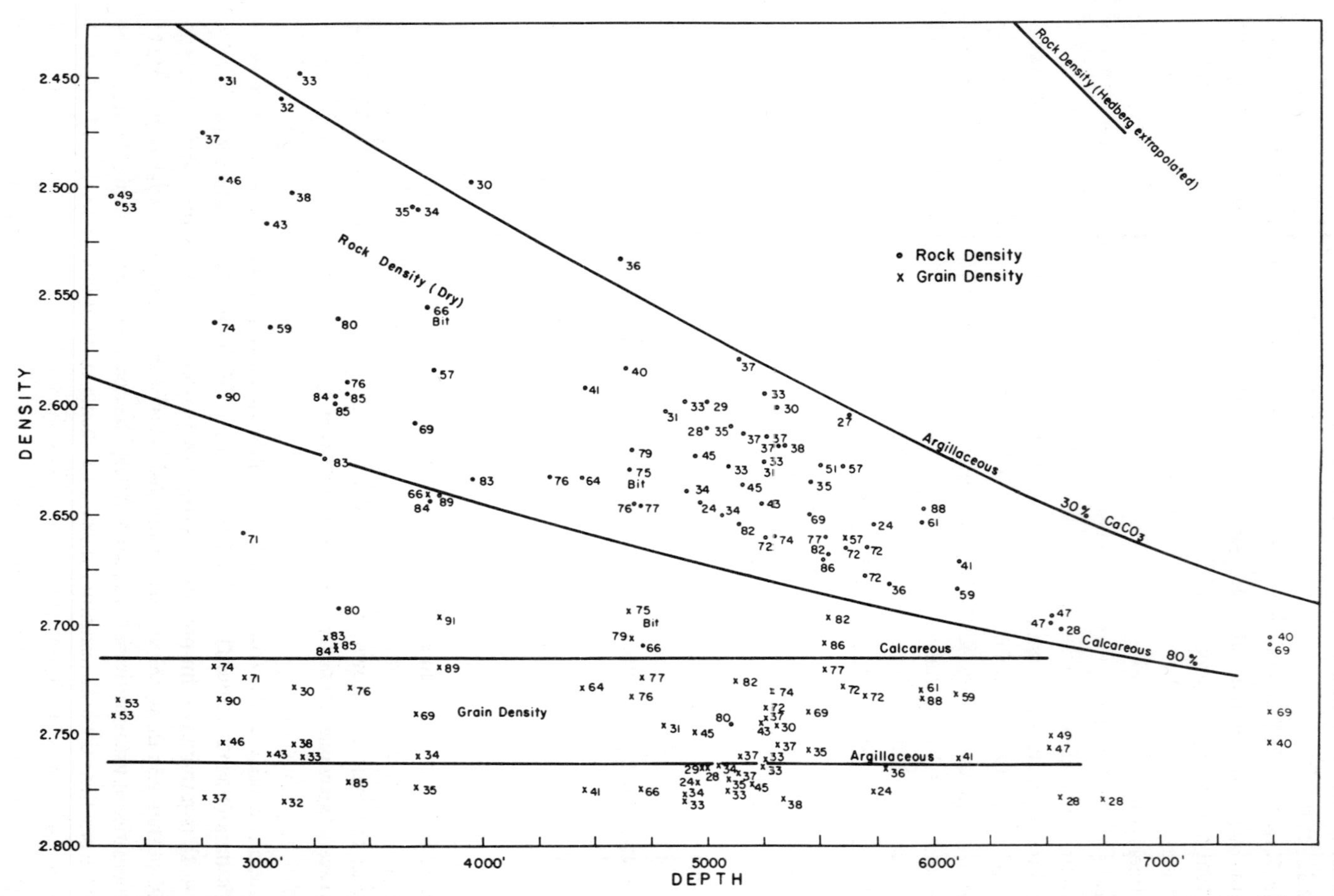

Fig.15. Relationship between density and depth of Ireton rocks. Carbonate content of each sample is noted in percentage by weight. Rock densities – upper part of graph; grain densities – lower part of graph. (After McCrossan, 1961, fig.12, p.459. Courtesy of American Association of Petroleum Geologists.)

as determined by McCulloh (1967, p.A19) to the average Gulf Coast shale density values (Fig.12, curve *3*).

McCrossan (1961, p.458), who studied the Upper Devonian inter-reef calcareous-illitic Ireton shales of central Alberta, Canada, found that shale samples high in carbonate content have a slightly lower average density (2.72 g/cm^3) than those samples having a low carbonate content (2.77 g/cm^3) (Fig.15). The differences may be related to (1) presence of finely divided pyrite associated with the finer-grained, more argillaceous rocks; (2) density of clay minerals being greater than that of calcite; or (3) greater adsorption of pycnometer fluid on the grains of the argillaceous rocks. McCrossan (1961) suggested that the illite grain densities, which are in the mica range (2.7–2.8 g/cm^3), are greater than that of calcite (2.71 g/cm^3). Boswell (1953) attributes these high values to the presence of iron compounds in the illite clay, such as mica and chlorite; however, chlorite was not detected in the Ireton shales using X-ray examination. The bulk densities of argillaceous strata are lower than those of calcareous beds because of the greater porosity of the former (McCrossan, 1961, p.458). The classical shale density–depth relation, as shown in Fig.12, does not always hold true in sedimentary basins where shale bulk-density anomalies are brought about by secondary mineralization. An excellent example of such a condition is where finely dispersed authigenic siderite in shales masks the wet bulk density readings in offshore wells in Indonesia (personal observations). Pyritization could also cause significant increase in shale densities (Johnson, 1950, p.333).

Porosity–density relationship

Porosity is an inverse function of pressure in homogeneous muds and shales (Weller, 1959, p.285). This is a fundamental relationship which is considered in all compaction studies. Overburden load cannot be measured directly in wells, but it can be calculated as shown in Chapter 3. Porosity–depth relationships are similar to those of bulk density–depth. Shale porosity values can be calculated from shale density values measured from whole cores, side-wall samples, and formation cuttings if the fluid density and grain density of the shale are known. Normally an average grain density of 2.65 g/cm^3 is assumed; however, this value does not hold true in many cases, especially where the clay-size particles of heavy minerals are present. Athy's (1930) data show that shale porosities can be as variable as bulk-density values at a given depth; Fig.14B shows an average curve relating porosity and depth of burial.

As shown in Fig.16, the porosity of clays rapidly decreases with depth: 35% at a depth of 400–500 m; less than 20% at a depth of 2,000 m; and less than 10% at a depth of 3,000 m. The density gradient varies from 0.05 to 0.02 g/cm^3 per 100 m of depth.

Comparison of bulk density–depth and porosity–depth relationships is difficult should one consider all the possible variables. First, there are basic differences in the methods used to obtain shale bulk-density values. Density measurements on formation samples can be made while the samples are fresh and water-saturated (Hedberg, 1936), or

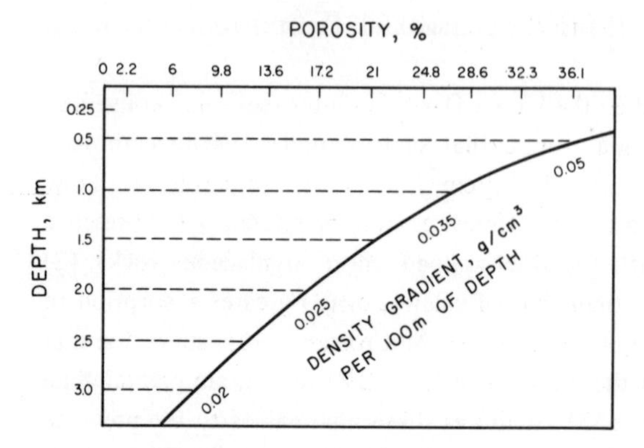

Fig.16. Typical compaction curve of clays, showing variation of porosity and density gradient with depth. (After Vassoevich, 1958, in: Tkhostov et al., 1970, fig.1, p.34.)

after drying them (Athy, 1930; Kerr and Barrington, 1961). Igelman and Hamilton (1963) and Bush and Jenkins (1970) pointed out that the drying temperature to which the mud or shale sample is heated is critical. Temperatures greater than 180°F in an unhumidified oven will practically remove all of the clay-mineral adsorbed water and can cause oxidation or other chemical changes in the component minerals. McCulloh (1965) demonstrated that whole cores of compact shales of Paleozoic age can be used with confidence to determine dry bulk density at depths of a few thousand feet. His conclusion was based on a comparison of precise laboratory core measurements with density values calculated from field gravimeter measurements.

Secondly, bulk-density measurements can be obtained in situ by using geophysical methods (Nafe and Drake, 1957; McCulloh, 1965) and well-logging tools (Ham, 1966). The accuracy of both methods, however, generally suffers from not knowing accurately the values for the initial wet bulk density, grain density, and interstitial fluid density. Differences in densities of the sediments are caused by factors such as geologic age, compaction history, and mineral and organic composition.

The in-situ fractional porosity, ϕ, can be expressed in terms of parameters partly derived from formation samples and logs as follows:

$$\phi = \frac{\rho_{bw} - \rho_{bd}}{\rho_f} \tag{2-4}$$

where ρ_{bw} is the wet bulk density of sediment in its in-situ condition, ρ_{bd} is the dry bulk density, and ρ_f is the average density of the interstitial fluids filling the shale pores under natural conditions. Other porosity expressions are:

$$\phi = \frac{\rho_g - \rho_{bw}}{\rho_g - \rho_f} \tag{2-5}$$

and

$$\phi = 1 - \frac{\rho_{bd}}{\rho_g} \tag{2-6}$$

where ρ_g is the average grain density.

POROSITY–DEPTH RELATIONSHIPS

Many authors have been concerned with variations of porosity in the shales as a function of depth. Papers by Athy (1930), Hedberg (1936), Hammer (1950), Dallmus (1958), Weller (1959), Von Engelhardt (1960), McCrossan (1961), Woollard (1962), and McCulloh (1965) are the most noteworthy. There has been a considerable amount of laboratory investigation of porosity–pressure relations, mostly on unconsolidated sediments (Kryukov and Komarova, 1954; Laughton, 1957; Chilingar and Knight, 1960a; Von Engelhardt and Gaida, 1963; Robertson, 1967; and Chilingarian and Rieke, 1968). Change in porosity is mainly a function of maximum overburden stress and of time, although it is also affected by lithology, depositional environment, overpressured fluid zones, diagenesis, and tectonic stress. The interrelation between these factors is quite complex and, consequently, pronounced variations in porosity–depth curves occur from place to place in porous sedimentary rocks.

Porosity variations with depth

Porosity decreases with increasing depth and, as shown in Fig.17, there is a marked decrease in porosity at shallow depths. The greatest porosity loss in sediments occurs in the first few hundreds of feet of burial. Mechanical compaction of muds may occur in geologically short periods of time if fluid expulsion occurs as porosity decreases (Magara, 1968).

In Fig.17 a comparison can be made of the porosity–depth relations from several different areas. The figure illustrates: (*1*) Proshlyakov's (1960) published porosity data from the Ciscaucasus, U.S.S.R.; (*2*) Meade's (1966) data as adapted by Brown (1969); (*3*) Athy's (1930) curve based on bulk densities of 200 samples of Pennsylvanian and Permian shales in northern Oklahoma; (*4*) Hosoi's (1963) curve based on density measurements of Tertiary mudstones in Akita and Yamagata Prefectures, Japan; (*5*) Hedberg's (1936) curve based on bulk densities of samples of Tertiary shales in Venezuela; (*6*) Dickinson's (1953) average bulk density curve for Tertiary Gulf Coast shales; (*7*) Magara's (1968) curve from the Shiunji gas field, Nagaoka Plain, Japan; (*8*) Weller's (1959) combined curve; (*9*) a curve derived from Ham's (1966) data, employing eq.2–5; and (*10*) Foster and Whalen's (1966) data. Difference in the position of the curves in Fig.17 is due to the factors such as (1) the amount of induration in the sediment, (2) geologic age, (3) compositional and textural variations, and (4) tectonic stress history.

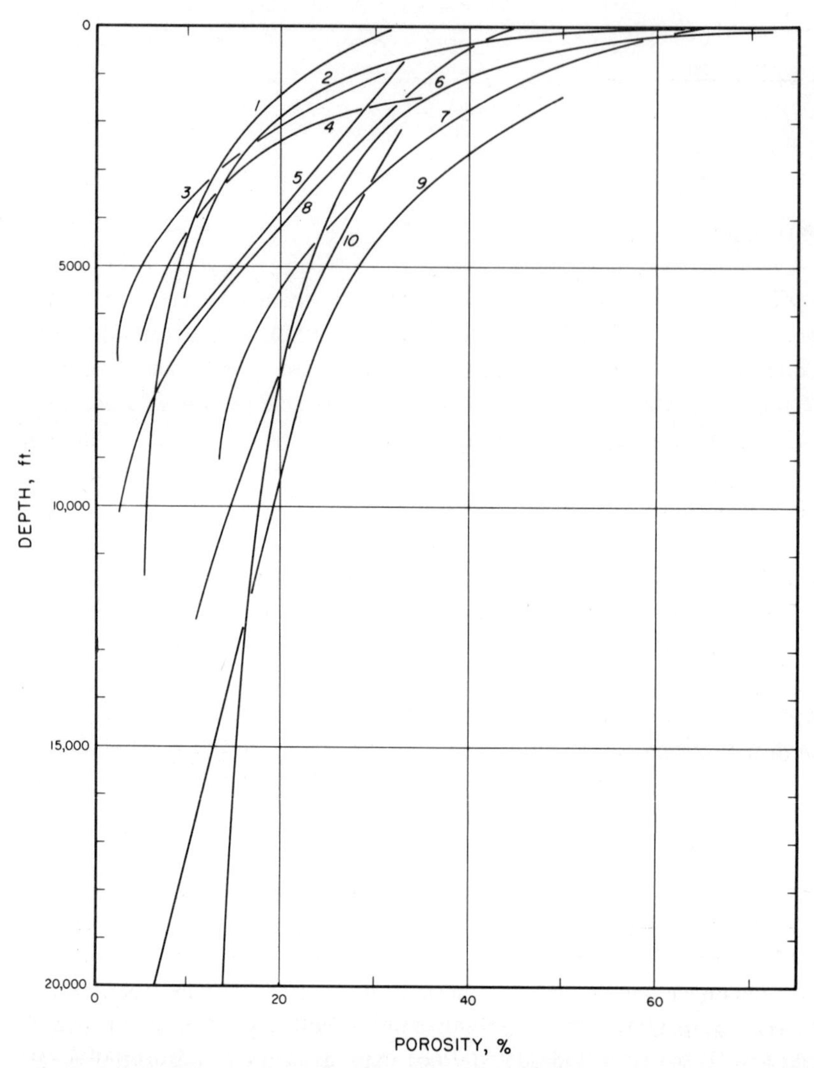

Fig.17. Relationship between porosity and depth of burial for shales and argillaceous sediments. *1* = Proshlyakhov (1960); *2* = Meade (1966); *3* = Athy (1930); *4* = Hosoi (1963); *5* = Hedberg (1936); *6* = Dickinson (1953); *7* = Magara (1968); *8* = Weller (1959); *9* = Ham (1966); *10* = Foster and Whalen (1966).

Weller (1959) constructed the porosity versus depth curve (Fig.17, curve *8*; and Fig.18), which consisted of a combination of results from three earlier studies by Terzaghi (1925b), Athy (1930) and Hedberg (1936). Terzaghi's data was matched with Hedberg's data by adjusting the latter along the horizontal depth scale to the point where a smooth junction could be made. This shift of 500 ft. to the right suggested to Weller that this amount of overburden may have been removed by erosion. Athy's curve re-

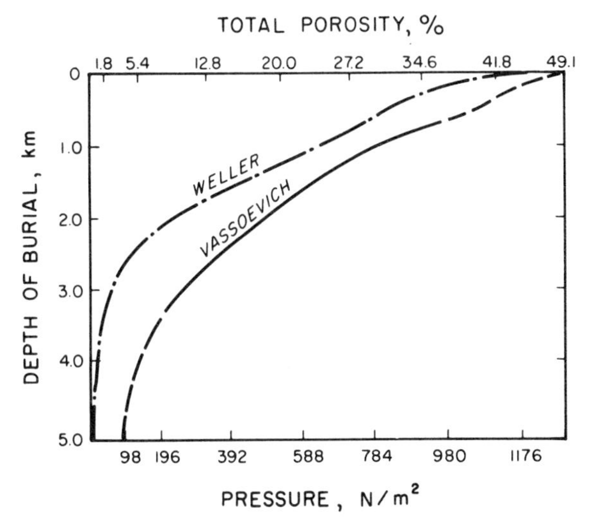

Fig.18. Relationship between porosity, depth of burial and overburden pressure. N = unit of force (Newton) = 102 g-force = 10^5 dynes. (After Weller and Vassoevich, in: Kartsev et al., 1969.)

quired a 3,000-ft. shift to the right in order to obtain a smooth fit. Athy's (1930) data, however, can be questioned on the basis of the geologic age and structural deformation. It seems that Weller's density curve is only of limited value, because it was derived by combining the different data sets which may not be compatible with each other. Dickinson's data (Fig.17, curve *6*) agree reasonably well with those of Ham (Fig.17, curve *9*) and Foster and Whalen (Fig.17, curve *10*) in the low-pressure range, but appear to give higher porosity values at depths greater than 8,000 ft.

Relationship between total porosity and depth of burial, based on data presented by Vassoevich (1960) and Weller (1959), is presented in Fig.18.

McCulloh (1967) plotted total-porosity values found in a large variety of sedimentary rocks in order to obtain a curve of maximum porosity versus depth (Fig.19). As shown by Fig.19, the maximum ranges of shale porosity differences vary from about 16% at 10,000 ft. to about 25.5% at 1,000 ft. The ranges appear to be the greatest among the youngest rocks and narrower for the older rocks. McCulloh (1967, p.A8) pointed out that his limits will no doubt be revised as more determinations of total porosity of sedimentary rocks accumulate.

Lapinskaya and Proshlyakov (1970, p.116) showed relationships between porosity and depth of burial for (a) sandy-silty rocks and (b) clayey rocks in the Pre-Caucasus area and Pre-Caspian Depression (Fig.20). In sandy-silty rocks, containing 15–20% cement, the total (pycnometer) porosity is decreasing 6–9% for every 1,000 m of depth and at a depth of 3,500–4,000 m constitutes 5–10%. Initially, clayey rocks are compacted rapidly and at a depth of 2,000 m their porosity decreases to 20–26%. The subsequent decrease in porosity at greater depth occurs at a lower rate. At a depth of 4,000 m, the

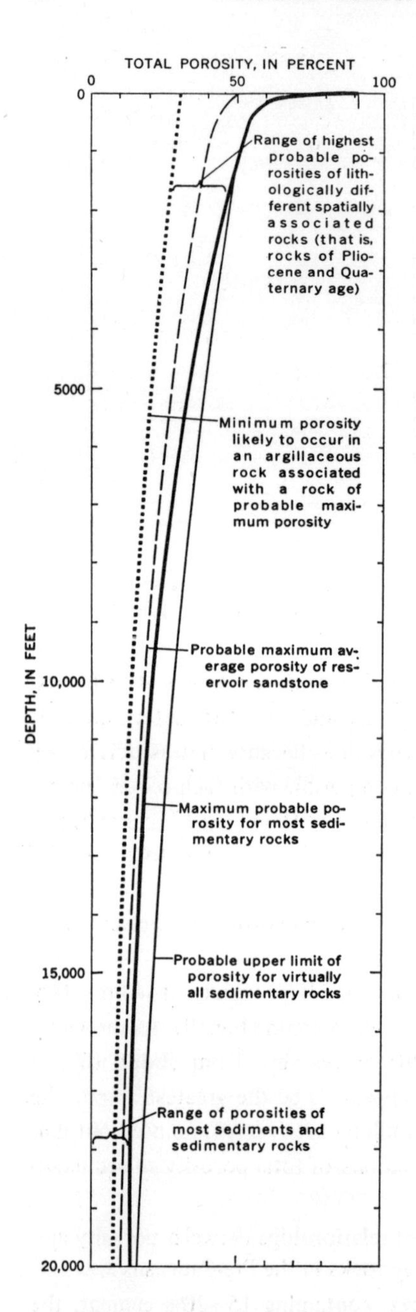

Fig.19. Relationship between total porosity and depth of burial for sedimentary rocks. Based on more than 4,000 laboratory measurements of cores from Los Angeles and Ventura basins in California, and other scattered localities in the United States, and the Po Basin of Italy. (After McCulloh, 1967, fig.4, p.A9.)

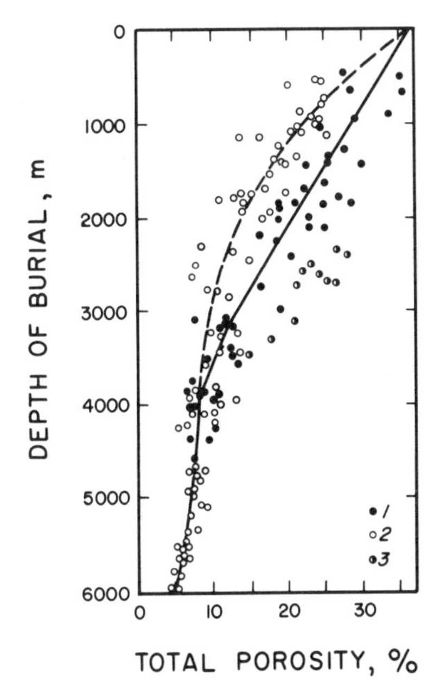

Fig.20. Dependence of porosity of sandy-silty and clayey rocks on depth of burial in Mesozoic and Upper Paleozoic deposits of Pre-Caucasus and Pre-Caspian Depression. *1* = sandy-silty rocks with 60–80% content of coarse-grained material (silt + sand); *2* = clays and argillites; *3* = epigenetically altered sandy-silty rocks. (After Lapinskaya and Proshlyakov, 1970, fig.1, p.116.)

total porosities of various terrigenous rocks become very close to each other and constitute 5–10%. In the Aralsorskiy borehole (U.S.S.R.) the porosity is equal to 3–3.5% at a depth of 6,000 m.

Relationship between porosity and present depth of burial for Liassic shales in northwestern Germany is presented in Fig.21. The porosities of shales obtained from deep boreholes in zones *I–V* are more or less what is to be expected at the present depth of burial, whereas those in zones *A, C, Ho* and *He* are too high in relation to the present depth. The latter can be explained by the fact that during the uplift of salt plugs, the Liassic sediments were raised from a level shown by the arrows in Fig.22 (Füchtbauer, in: Von Engelhardt, 1960).

Teodorovich and Chernov (1968) presented two valuable diagrams (Figs.22 and 23) on the relationship between porosity, depth of burial, and sonic velocity. They derived the following approximate formulas relating porosity, ϕ; permeability, k; and depth, D:

For siltstones:

$$\phi = 29.50 - 0.003\,D \qquad (0\text{–}4{,}500\text{ m}) \qquad (2\text{-}7)$$

or:

$$\phi = 31.53\exp(-0.000152\,D) \qquad (\text{Fig.22, curve 2}) \qquad (2\text{-}8)$$

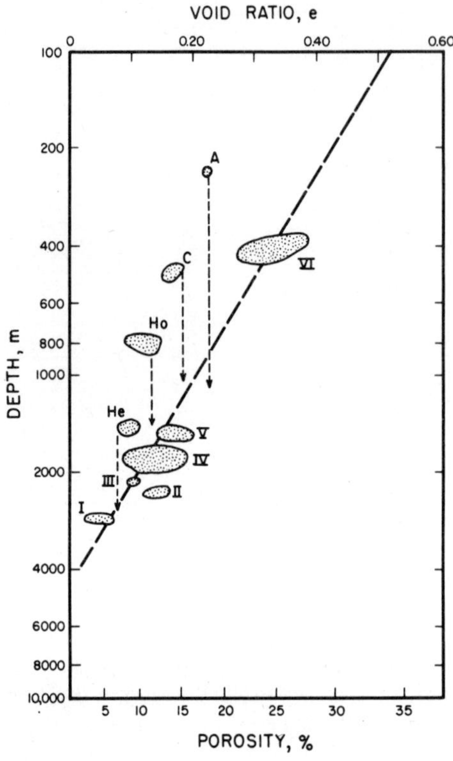

Fig.21. Porosity and void ratio of Liassic shales as related to present depth. The arrows indicate the maximum depth of burial. (After H. Füchtbauer, cf. Von Engelhardt, 1960.)

and:

$$\log k = 2.961 \exp(-0.000147\,D) \tag{2-9}$$

or:

$$\log k = 2.87 - 0.003\,D \qquad (500\text{–}4,500 \text{ m}) \tag{2-10}$$

For sandstones:

$$\phi = 28.21 \exp(-0.000122\,D) \tag{2-11}$$

or:

$$\phi = 28.74 - 0.003\,D \qquad \text{(Fig.22, curve 3)} \tag{2-12}$$

and:

$$\log k = 2.803 \exp(-0.000074\,D) \tag{2-13}$$

or:

$$\log k = 2.88 - 0.0002\,D \tag{2-14}$$

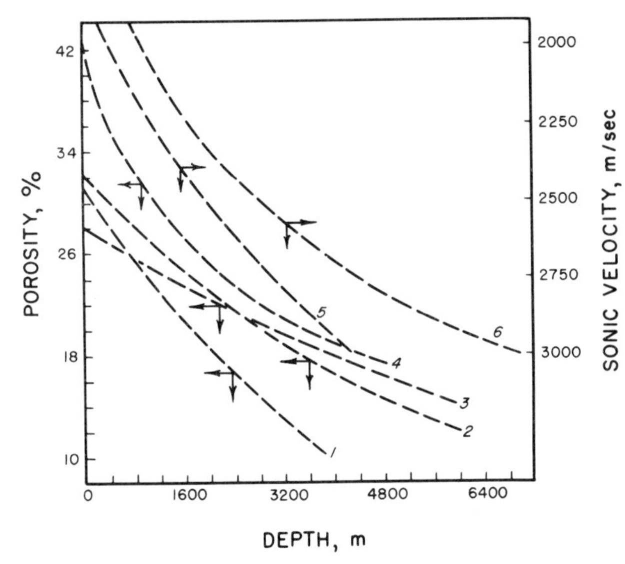

Fig.22. Relationship between porosity, depth of burial, and sonic speed. (After Teodorovich and Chernov, 1968, fig.3, p.86.) *1* = clays (Apsheron Peninsula, Azerbayjan S.S.R., U.S.S.R.); *2* = siltstones (Apsheron Peninsula); *3* = sandstones (Apsheron Peninsula); *4* = Quaternary and Upper Pliocene clays (Near-Kurinskaya Depression); *5* = Apsheron Peninsula (after R.M. Gadzhiev, 1965); *6* = Lower Kurinskaya Depression (after R.M. Gadzhiev, 1965).

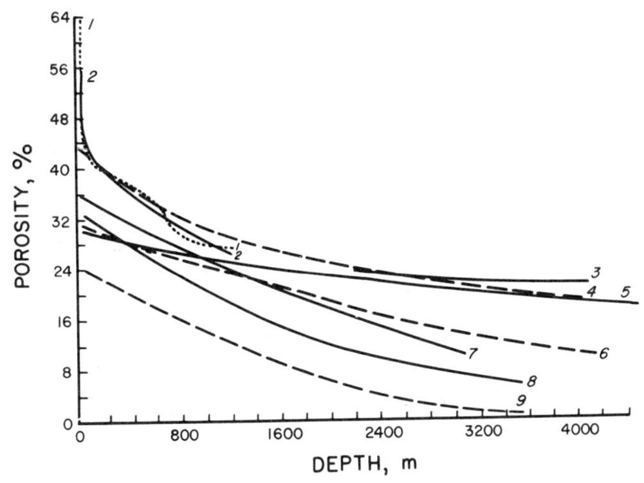

Fig.23. Relationship between porosity of various terrigenous deposits and depth of burial. (After Teodorovich and Chernov, 1968, fig. 6, p.89.) *1* = clays and *2* = siltstones from Recent to top of Middle Pliocene, Alyaty-Sea (after Koperina and Dvoretskaya, 1965); *3* = fine-grained and medium-grained sands with carbonate-clayey cement up to 15%, Apsheron Peninsula; *4* = Quaternary and Upper Pliocene clays, Near-Kurinskaya Depression (after Prozorovich and Sultanov, 1961); *5* = fine- and medium-grained sands with carbonate-clayey cementing material up to 30%, Apsheron Peninsula; *6* = clays, Apsheron Peninsula (after Prozorovich and Sultanov, 1961); *7* = fine-grained quartz sandstones, Devonian, southeastern slope of Voronezh Arch (after Karpov, 1964); *8* = quartz siltstones, Devonian, southeastern slope of Voronezh Arch (after Karpov, 1964); *9* = shales, Devonian, southeastern slope of Voronezh Arch (after Karpov, 1964).

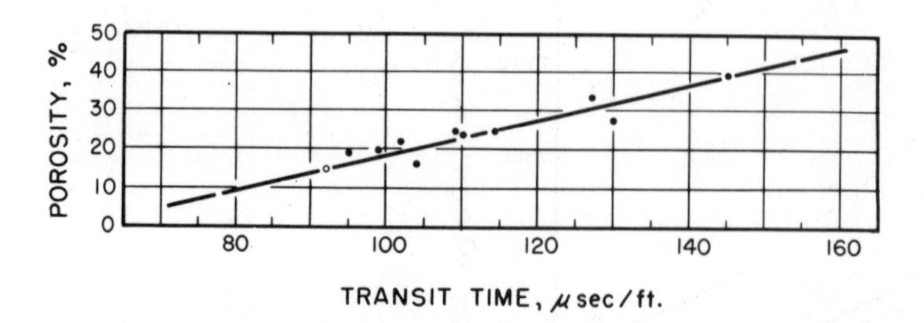

Fig.24. Relationship between mudstone porosity, ϕ, and transit time, $\Delta t_{\log}$. (After Magara, 1968, fig.6, p.2474. Courtesy of American Association of Petroleum Geologists.)

Relationship between porosity and sonic transit time for Miocene mudstone, Nagaoka Plain, Japan, is presented in Fig.24. Stetyukha (1964) derived the following formula relating the porosity of shales to depth of burial in the Pre-Caucasus area:

$$\phi_D = \phi_i \, e^{-CD} \tag{2-15}$$

where ϕ_D = porosity at depth D; ϕ_i = initial porosity, which is equal to 37.89%; D = depth; and C = coefficient approximately equal to 10^{-4}. Linetskiy (1965) believes that at a depth of 1,500–2,000 m only oriented (adsorbed?) water is present. According to him, at these depths the porosity is of the order of 4–11%. On the other hand, according to Kartsev et al. (1969), porosity at 1,500–2,000 m depth is around 13–23%. It is important to note here that different drying temperatures were used by different investigators in determining the remaining moisture content; this could account for some differences. In addition, the type of clay would affect the results considerably.

Prozorovich (1962) showed that on using the formula $y = C - A\,e^{-Bx}$ (where y is the specific weight in g/cm^3, x is depth in m, C = 2.8 g/cm^3, A = 1.23 g/cm^3, and B = 0.43), the actual specific weight deviates from the calculated one from +0.05 to −0.06 g/cm^3 to a depth of about 3,000 m.

Von Engelhardt (1960) developed equations relating void ratio, e, and depth of burial, D, in m for Tertiary deposits of Venezuela (data after Hedberg, 1936), Tertiary sediments of the Po Basin (data after Storer, 1959), and the Liassic shales of northwestern Germany as follows:

$$e = e_1 - b \log D \quad \text{(general equation)} \tag{2-16}$$
$$e = 1.844 - 0.527 \log D \text{ (Venezuela)} \tag{2-17}$$
$$e = 1.700 - 0.481 \log D \text{ (Po Basin)} \tag{2-18}$$
$$e = 1.160 - 0.317 \log D \text{ (Liassic of Germany)} \tag{2-19}$$

where e_1 is the void ratio at a depth of 1 m and b = compressibility of clay. Porosities of these deposits at a depth of 1 m are 65, 63 and 54%, respectively.

TABLE III
Compaction equations for sediments of different geologic ages
(After Mukhin, 1968, p.46)

Age	*Region*	*Equation* [1]	*Initial porosity coefficient,* E_0
Lower Cambrian (500 million years)	northwestern area of Russian Platform	$E = 0.67 - 0.181 \log p$	1.39
Upper Jurassic (140 million years)	central area of Russian Platform	$E = 1.07 - 0.26 \log p$	2.11
Maykop	Pre-Caucasus	$E = 1.46 - 0.444 \log p$	3.13
Apsheron	Pre-Caucasus	$E = 1.26 - 0.45 \log p$	3.06
Tertiary (50–60 million years)	Venezuela	$E = 1.20 - 0.40 \log p$	2.80
Quaternary (1 million years)	southern part of Caspian Sea	$E = 1.10 - 0.384 \log p$	2.66

[1] E = porosity coefficient; it is the Russian equivalent of void ratio, e. p = pressure (kg/cm^2).

Mukhin (1968) presented compaction equations for sediments of different geologic ages (Table III). Relationship between the coefficient of porosity, E (E = Russian equivalent of void ratio, e), and pressure is presented for Quaternary deposits of the southern part of the Caspian Sea upon burial under overlying sediments to a depth of 800 m.

The following formula enables one to calculate the volume of squeezed-out solutions

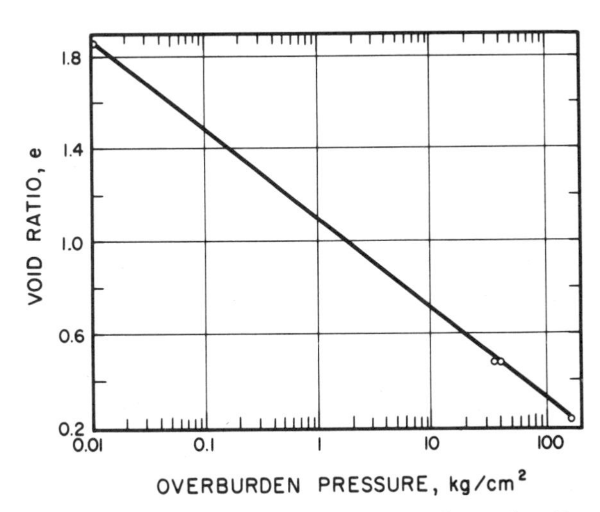

Fig.25. Relationship between coefficient of porosity, E, and pressure for Quaternary deposits of southern part of Caspian Sea. $E = e$ (void ratio). (After Mukhin, 1968, p.53.)

TABLE IV
Volumes of squeezed-out solutions from various depths of Quaternary deposits of the Caspian Sea (After Mukhin, 1968, p.54)

Depth of burial (m)	*Volume of squeezed-out pore solution*		*Depth of burial (m)*	*Volume of squeezed-out pore solution*	
	l/m³ of sediment	*l/m³/m*		*l/m³ of sediment*	*l/m³/m*
0.003–0.10	280	–	500–800	64	0.21
0.10–10	381	38.2	800–1,000	24.5	0.12
10–60	204	4.1	1,000–1,500	61	0.12
60–100	75	1.9	1,500–2,000	45.5	0.09
100–250	103	0.7	2,000–2,500	37.5	0.075
250–500	90	0.36	2,500–3,500	19.4	0.0385

during a certain compaction stage (Mukhin, 1968, p.51):

$$V_i = \frac{h_1}{1 + E_1}(E_{i-1} - E_i) \tag{2-20}$$

where V_i is the volume of interstitial solution squeezed out during a certain compaction stage; E_1 is the present-day average porosity coefficient of sediment; h_1 is the present-day thickness of deposit; and E_{i-1} and E_i are average coefficients of porosity at the beginning and end, respectively, of the compaction stage under investigation (Fig.25). As an example, the amounts of squeezed-out interstitial solutions from Quaternary deposits of the Caspian Sea from different depths are presented in Table IV.

Effect of carbonate content on porosity

McCrossan (1961, p.461) showed the effect of impurities on degree of compaction of Ireton shales, in which impurities primarily consist of carbonates which are relatively simple to determine.

Fig.26 shows a series of porosity–depth curves for samples having different carbonate content. In this graph, three variables (porosity, depth of burial, and carbonate content) are considered simultaneously, giving a series of curves which were drawn in by eye. These curves represent a surface of best fit through the points. As pointed out by Hedberg (1936), even for purer argillaceous sediments, pressure–porosity relations are better represented by a band than by a line. This is also true for calcareous shales. Each group on the graph (Fig.26) is represented by a band which slightly overlaps the bands on either side. The contours are drawn through the middle of the bands (McCrossan, 1961).

As shown in Fig.26, the more calcareous rocks have a lower initial porosity and the decrease in porosity of these rocks with depth is lower than in the case of less calcareous

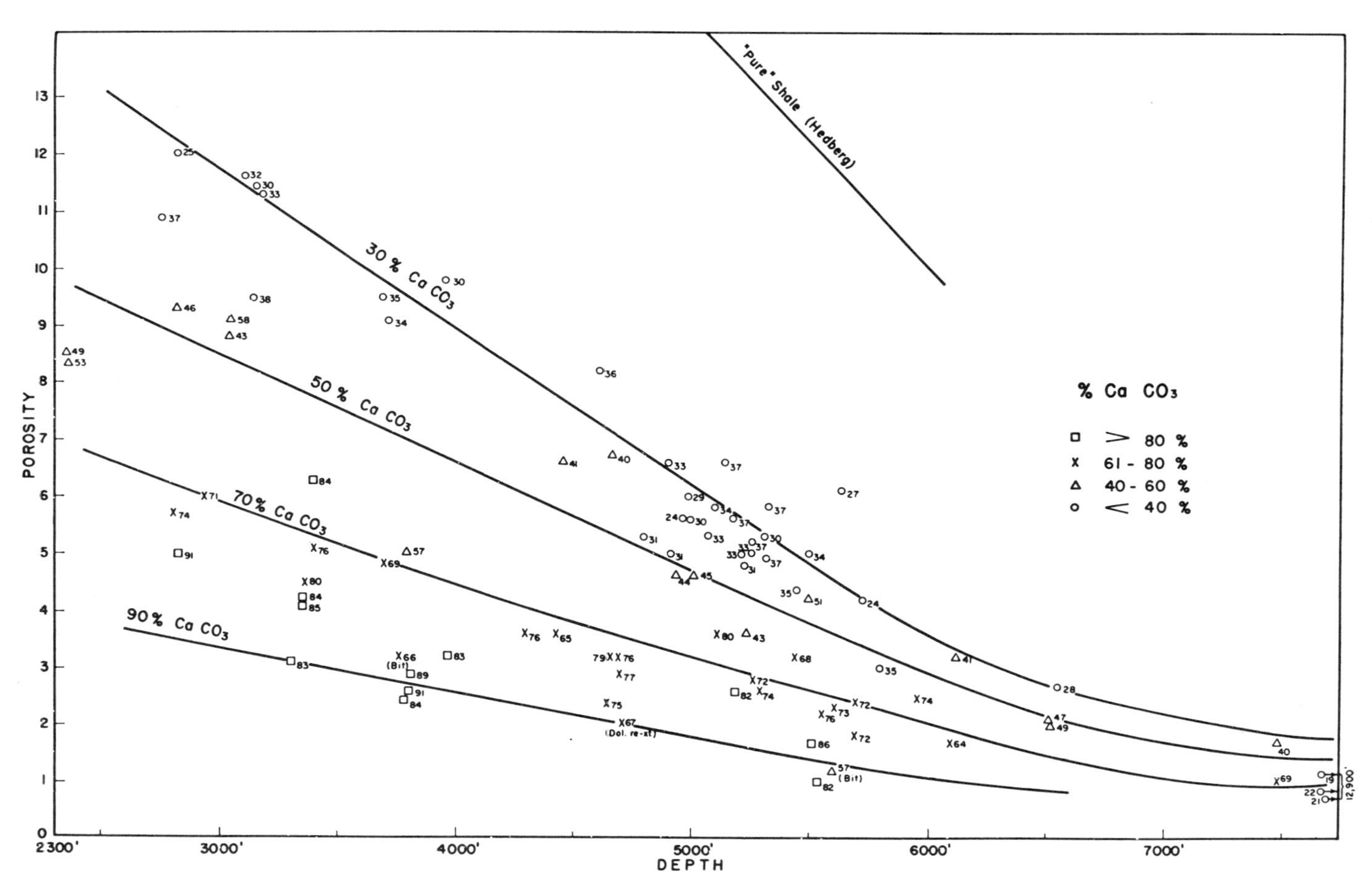

Fig.26. Relationship between porosity and depth of burial of Ireton rocks. Carbonate content shown in percentage by weight on the axis perpendicular to page. (After McCrossan, 1961, fig.13, p.461. Courtesy of American Association of Petroleum Geologists.)

rocks. Hedberg's (1936) curve for pure shale is also shown for comparison in the upper part of the figure. These relationships are probably due to the fact that the carbonate particles reduce the amount of compactable material. Even the 30% curve is also appreciably flatter than Hedberg's, even though the carbonate grains are in the coarse-clay size range. As pointed out by McCrossan (1961, p.460), comparison with Skempton's (1945) work is not advisable because the depth ranges of these two studies overlap only very slightly and extrapolation of either set of curves is dangerous.

EXPERIMENTAL COMPACTION OF RECENT SEDIMENTS, MUDS AND CLAYS

Nature and degree of hydration of clays, gums and silicic acid

Einstein's equation for the viscosity of a dilute dispersion of spherical particles is:

$$\mu_s = \mu_o(1 + 2.5R) \tag{2-21}$$

where μ_s is the viscosity of the dispersed system in centipoises, μ_o is the viscosity of the dispersion medium in centipoises, and R is the ratio of the volume of the dispersed phase to the volume of the sol.* Although this equation does not apply to viscous clay-water mixtures, it is useful when considering low clay contents (2–5% of clay by volume). Extensions to Einstein's equation that are presented in Kruyt (1949) take into consideration higher concentrations of both spherical and non-spherical particles (see also Robertson et al., 1965).

Kunitz (as quoted in: Abramson, 1934, p.129) showed that the following expression holds very well for R up to 0.50 for sugar solutions and sulfur suspensions:

$$\mu_s = \mu_o\left[\frac{1 + 0.5R}{(1 - R)^2}\right] \tag{2-22}$$

or:

$$\mu_s = \mu_o(1 + 4.5R + 12R^2 + \ldots)$$

Several other extensions of the Einstein equation to higher concentrations are presented in Kruyt (1952, pp.350–354) both for spherical and non-spherical particles. It is not known which one of these equations is the most accurate. Vand (as quoted in: Kruyt, 1952, p.351), for example, gave a seemingly reasonable equation for spherical particles on considering both hydrodynamic and mechanical interaction between particles:

---

* A colloidal suspension of a solid in a liquid.

$$\frac{\mu_s - \mu_o}{\mu_o} = 2.5R + 7.349R^2 \tag{2-23}$$

The knowledge on the dilute and concentrated suspensions of elongated particles is even scarcer, and the pertinent equations are complex (Kruyt, 1952; p.346, 353).

Inasmuch as charges on particles may influence the viscosity by making the apparent volume of the particles greater as a result of introducing an additional resistance owing to electrokinetic effects, Von Smoluchowski (as quoted in: Kruyt, 1949, p.205) gave an extension of Einstein's formula:

$$\frac{\mu_s - \mu_o}{\mu_o} = KR\left[1 + \frac{1}{\lambda\mu_o r^2}\left(\frac{D\zeta}{2\pi}\right)^2\right] \tag{2-24}$$

where λ = specific conductivity, r = particle radius, D = dielectric constant, K = constant, and ζ = electrokinetic potential (difference in potential between the layers of charges affixed to the particles and the layer of balancing charges in the liquid bathing the particles). Robertson et al. (1965) estimated that the increase in viscosity owing to charges on clay particles is probably of the order of 10%.

On determining the viscosity of clay dispersions of known concentration, one can calculate the approximate ratio of the volume of dispersed phase to the volume of the sol on using eq.2–21. Assuming that the viscosity of the dispersion medium, water, is 1 cp, Einstein's equation may be written as:

$$R = (\mu_s - 1)/2.5 \tag{2-25}$$

Thus, any discrepancy between the actual and calculated R enables one to estimate the increase in volume of the dispersed phase, namely, amount of clay swelling. For example, according to the results obtained during this study a sol containing 1.2% by weight of gum tragacanth (natural organic colloid) had a viscosity of 37 cp, whereas according to the Einstein formula the viscosity should be only 1.03 cp. This discrepancy indicated that the tragacanth gum has taken up large amounts of water, thus increasing the volume of the dispersed phase. The viscosity of the sol usually increases with time, indicating further swelling and expansion of the particles. In accordance with eq.2–25 in the foregoing case of 37 cp viscosity of the sol, it is necessary for the total volume of the particles of the dispersed phase to be about 1,200 times that of the dry tragacanth. It is not known with certainty whether the particles of tragacanth take up the water like a sponge or whether they become surrounded by thick layers of water molecules.

Hydration of clays

Similarly, various mud samples, containing a certain weight percent of clay were prepared. The laboratory procedure was described in detail by Chilingar et al. (1963) and Robertson et al. (1965). The swelling index, SI_v of clays, which is indicative of the degree of hydration, can be mathematically expressed as:

$$SI_v = \frac{R_c \text{ calculated from eq.2-25}}{R_a \text{ actual (dry basis)}} = \frac{\text{volume of hydrated clay}}{\text{volume of dry clay}} \qquad (2\text{-}26)$$

A mud sample containing 27% by weight of kaolinite clay had a viscosity of 3 cp, showing that the volume of the dispersed phase (hydrated clay) is about 7.3 times that of the dry clay. For a mud sample containing 2.7% by weight of bentonite, the viscosity was 2.9 cp. This indicated that the volume of hydrated bentonite is about 28.2 times that of the volume of dry clay. Table V shows the ratio of the volume of hydrated clay to dry clay to vary from 1.78 in the case of dickite to 3.96 in the case of halloysite. In the foregoing calculations a specific gravity of 2.5 g/cm^3 was assumed for all clays. Obviously, the calculated values are not absolute and would vary from clay to clay.

TABLE V
Measured and calculated (from Einstein's equation) viscosities of clay-water mixtures [1]
(After Chilingar et al., 1963)

Volume of water (cm^3)	*Weight of clay (g)*	*Measured viscosity (cp)*	*Calculated viscosity (cp) using R_{dry}*	*Vol. % of clay*		*Type of clay*	*Swelling index SI_v*
				actual (dry)	*calculated (hydrated)*		
95.0	5.4676	1.1	1.0563	2.25	4.0	dickite	1.78
90.0	9.1775	1.2	1.098	3.92	8.0	dickite	2.04
95.0	4.8973	1.2	1.0505	2.02	8.0	halloysite	3.96
90.0	10.0349	1.21	1.11	4.26	8.4	halloysite	1.97

[1] In these calculations, a specific weight of 2.5 g/cm^3 was assumed for clays.

In the opinion of Ralph E. Grim (personal communication, 1960) the water adsorbed immediately on the surfaces of the clay minerals is composed of oriented water molecules. The distance that such "oriented" water extends from the clay mineral surfaces depends on the nature of clay mineral, the nature of the adsorbed cations, and, perhaps, on the presence of organic matter. The water present in pores at considerable distances from the surface is not oriented, and is termed "liquid" water. In some cases there is a gradual transition from the non-oriented water on proceeding toward the clay mineral surface; whereas in the other cases there is a sharp transition in the character of the water. For example E. Grim (personal communication, 1960) believes that there is a gradual transition in the case of sodium montmorillonite and a more abrupt transition for calcium montmorillonite.

Van Olphen (1963a, b) has shown that the last few mono-molecular layers of oriented

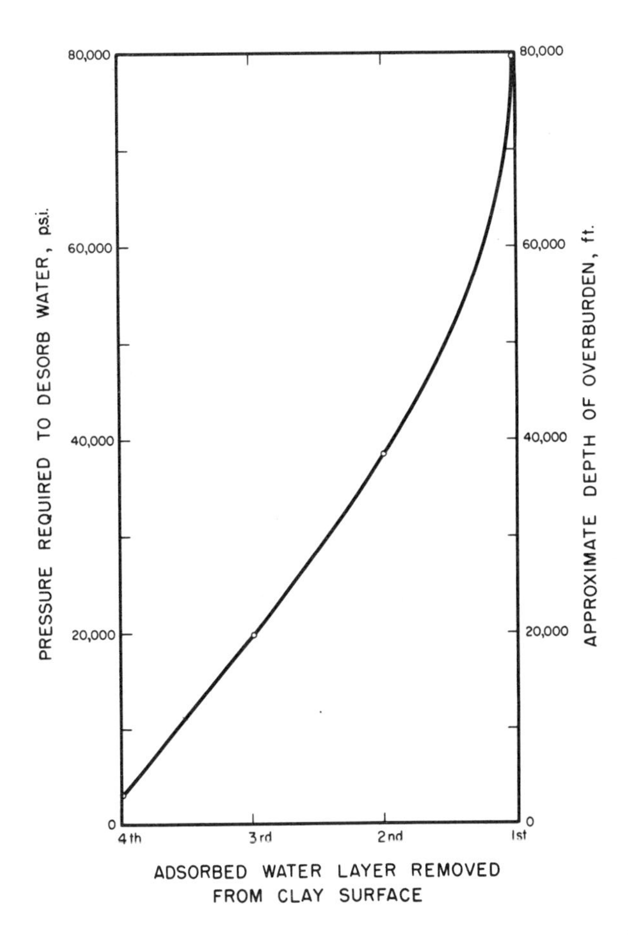

Fig.27. Pressure (or approximate equivalent of depth of burial) necessary to remove each of the four layers of water from interlayer positions on montmorillonite clay. (After Powers, 1967, fig.7, p.1243. Courtesy of American Association of Petroleum Geologists.)

water adsorbed on clay surfaces are not removed upon deep burial. According to Van Olphen (1963b), even on heating sodium Llano vermiculite clay to 50°C, it takes a pressure of about 65,000 p.s.i. to remove the last water layer. The resistance of clay to compaction is equal to the swelling pressure, which, in turn, is equal to the net repulsive pressure between the unit layers of clays. Fig.27, which is based on work-of-removal values reported by Van Olphen (1963b), was prepared by Powers (1967) to show the approximate pressures required for the removal of successive monomolecular layers of water from a calcium montmorillonite. The work-of-removal for water layers located between unit layers of montmorillonite was determined from desorption isotherms. Approximate depths of burial corresponding to the desorption pressures also are given in Fig.27 by Powers (1967).

The heat of adsorption of one water layer by Na-bentonite is 90 ergs/cm^2 (Van Olphen, 1954, p.426, 428), whereas the potential difference between one and two layers of water was calculated by Van Olphen to be 25 ergs/cm^2. The diameter of a water molecule is around 2.6 Å. The estimated forces necessary to remove each of the first two adsorbed layers of water from Na-bentonite are: (1) 49,400 p.s.i – first water layer, and (2) 13,700 p.s.i. – second water layer (Warner, 1964). According to Warner (1964) a void ratio of 0.3 still existed at 60,000 p.s.i.; and one layer of water between adjacent plates would lead to a void ratio of 0.26.

According to Powers (1967), there appear to be two ways in which the last four monomolecular layers of water may be desorbed from montmorillonite without using high pressures: (1) large electrostatic force and (2) heating sample to temperatures greater than 100°C. The repulsion force of clay unit layers may be exceeded by the electrostatic attraction force associated with the exchange of potassium for other ions, with resulting collapse of expanded clays to 10 Å and consequent desorption of water.

Desorption of the last four monomolecular layers of water, which requires high pressures, could occur at lower pressures as a result of potassium fixation during diagenesis of montmorillonite to illite. This begins at about 6,000 ft. and continues at an increasing rate to about 9,000–12,000 ft. (Powers, 1967, p.1243).

Hydration of silicic acid

According to Carman (1940, in: Iler, 1955, p.39), the surfaces of the colloidal silica particles are so large that the hydration is measurable; and the limit of subdivision is a single silicon-oxygen tetrahedron corresponding to a molecule of H_4SiO_4. He also assumed that the hydration of silica is confined to the surface and is due to the completion of the surface tetrahedron. Radczewski and Richter (1941, in: Iler, 1955, p.40) have observed in the electron microscope that spherical colloidal particles of silica are spongy in character and consist of smaller silica particles (1/100th of the diameter of the aggregate particles).

Kruyt and Postma (1925, in: Iler, 1955, p.98) found that their least viscous sol, containing 0.78% by weight of SiO_2 (0.35% by volume), had a relative viscosity of approx. 1.2. According to Einstein's equation for the viscosity of dilute dispersions, this corresponds to a volume fraction of the dispersed phase of about 8%. Consequently, the hydrated particles would contain only about 5% by volume of SiO_2 and 95% by volume of water occluded in the spongy network structure of the colloidal particles.

According to Dzis'ko, Vishnevskaya and Chesalova (1950, in: Iler, 1955, p.236), the physically adsorbed water is removed by drying to a constant weight at 115°C, whereas the water remaining on silica gel at 115°C is present as a layer of hydroxyl groups. The amount of this "bound water" is proportional to the surface area of the gel; and when the silica is heated to 500–600°C this layer is partly removed without the sintering of silica.

Relationship between residual moisture content and pressure

The escape of water from consolidating clays is rapid at depths down to 250–300 m. With further subsidence, sediments become consolidated rocks and there is only slow escape of water and further compaction. V.D. Lomtadze (1953, in: Eremenko, 1960, pp.536–537) studied compaction of various clays and pure quartz sand up to pressures of 5,000 kg/cm^2 (71,115 p.s.i.). Some results obtained by Lomtadze (1954) are shown in Table VI. On increasing pressures from 3,000 to 5,000 kg/cm^2, the water content in Cambrian clay and kaolin decreased from 6–7 to 3–4%. At a pressure of 3,000 kg/cm^2, bentonite still contained more than 20% of water. At high pressures water can also be squeezed out from a clean quartz sand; this is associated with its mechanical breakdown.

TABLE VI
Relationship between moisture content (%) of some natural clays and pressure
(After Lomtadze, 1954)

Pressure (kg/cm^2)	*Finland Bay mud*	*Ribbon clay, Leningrad, U.S.S.R.*	*Ioldian clay, Vyg River, U.S.S.R.*	*Ioldian clay, Kuman River, U.S.S.R.*	*Cambrian clay, Leningrad, U.S.S.R.*
0	71.2	31.4	45.8	33.0	11.7
60	30.8	17.1	28.6	21.1	–
150	23.5	14.0	22.8	17.2	–
500	17.9	11.5	19.5	12.8	–
1000	12.2	8.5	14.3	9.7	8.7
2000	–	–	–	–	6.8
3000	–	–	–	–	5.2

Relationship between moisture content and pressure for clays as presented by Sergeev (1970), is shown in Fig.28; whereas the experimental data obtained by Lomtadze (1953, in: Klubova, 1965) is presented in Fig.29.

Chilingar and Knight (1960a) and Chilingarian and Rieke (1968) studied the relationship between the moisture content (in %) and the applied pressure, ranging from 40 to 400,000 p.s.i., for various standard American Petroleum Institute clays, soils, and marine muds. For comparison purposes, two natural organic colloids (gum ghatti and gum tragacanth) and silicic acid were studied by Chilingar and Knight (1960a, b).

Relationship between the moisture content (percent, dry weight basis) and pressure on semilogarithmic graph paper for various clays, clayey sediments, organic colloids and silicic acid, in the pressure range shown in Figs.30 and 31, possibly suggest that compaction is more or less a simple process where the curve is a straight line. The compaction of clays under laboratory conditions possibly occurs in three stages: (1) squeezing out of the interstitial fluid as the grains of clay minerals come in contact, (2) development of closer packing due to the rearrangement of grains, and (3) deformation of grains and further reduction of porosity. The resistance to compression appears to depend upon the strength

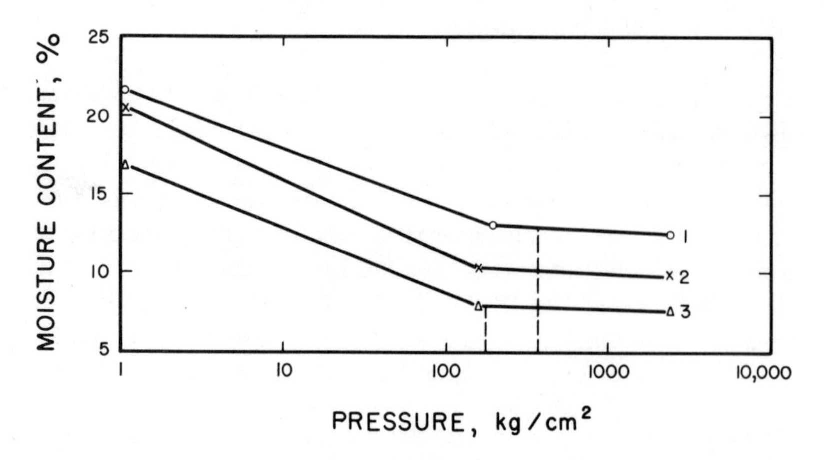

Fig.28. Relationship between volumetric moisture content (volume of water per unit of sediment bulk volume) and overburden pressure. *1* = clay having polymineral composition; *2* = kaolinite clay; *3* = average polymineral loam (containing sand and 25–50% clay). (After Sergeev, 1970, fig.1, p.191.)

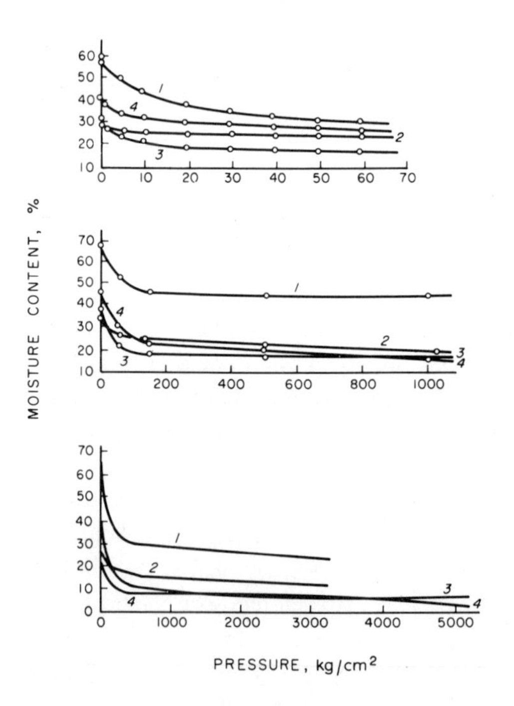

Fig.29. Relationship between moisture content of clays and overburden pressure. *1* = bentonite; *2* = marshalite; *3* = Cambrian clay; *4* = kaolinite. (After V.D. Lomtadze, 1953, in: Klubova, 1965, p.56.)

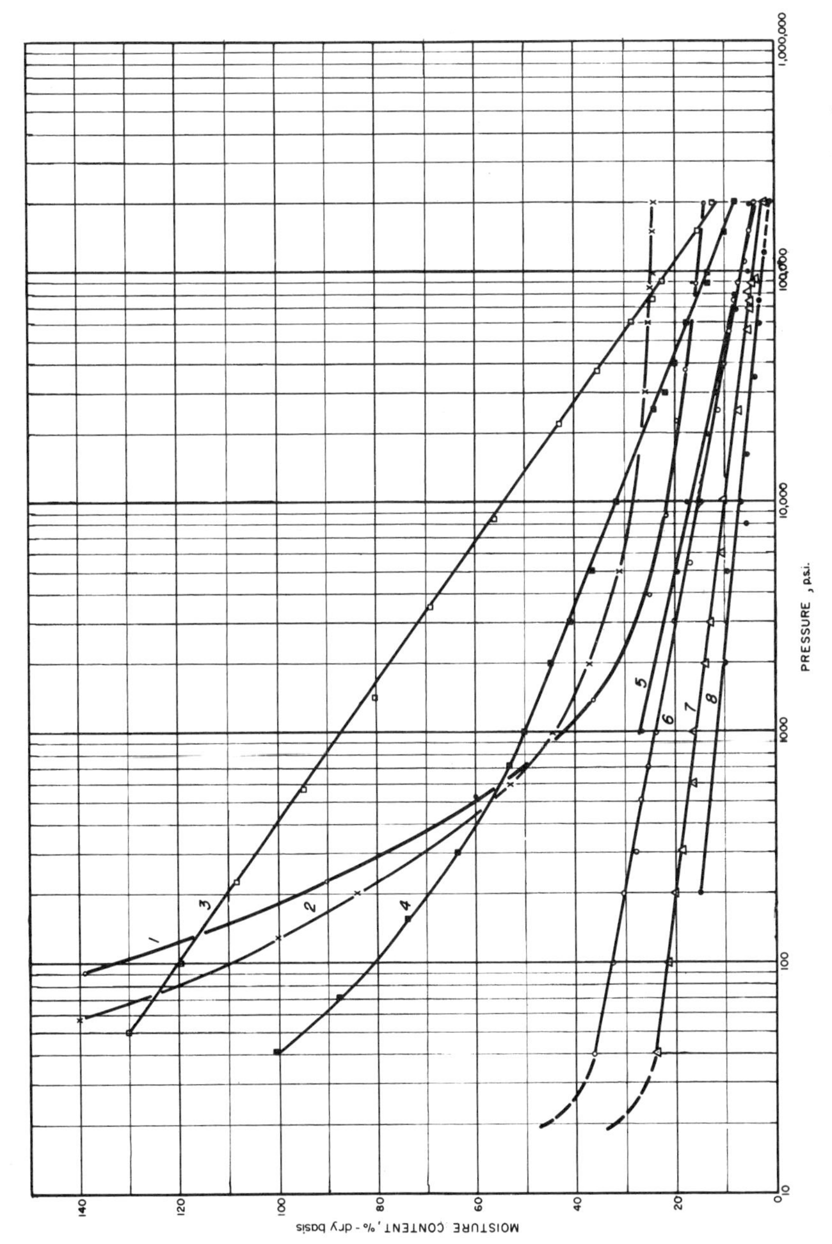

Fig.30. Relationship between moisture content M (% of dry weight) and pressure p (in p.s.i.) for various clays, gum ghatti, and gum tragacanth. *1* = gum ghatti (natural organic colloid); *2* = gum tragacanth (natural organic colloid); *3* = silicic acid – $M = 186 - 33 \log p$; *4* = montmorillonite no.25, Upton, Wyoming, John C. Lane Tract (bentonite) – $M = 104 - 18.06 \log p$ (straight-line portion of the curve); *5* = montmorillonite no.25 – $M = 58 - 10.2 \log p$ (hydrated in sea water); *6* = illite no.35, Fithian, Illinois – $M = 50 - 8.7 \log p$; *7* = kaolinite no.4, Macon, Georgia, Oneal Pit – $M = 33.9 - 5.96 \log p$; and *8* = dickite no.15, San Juanito, Chihuahua, Mexico – $M = 26.7 - 5.04 \log p$. (After Chilingarian and Rieke, 1968, fig.1, p.812.)

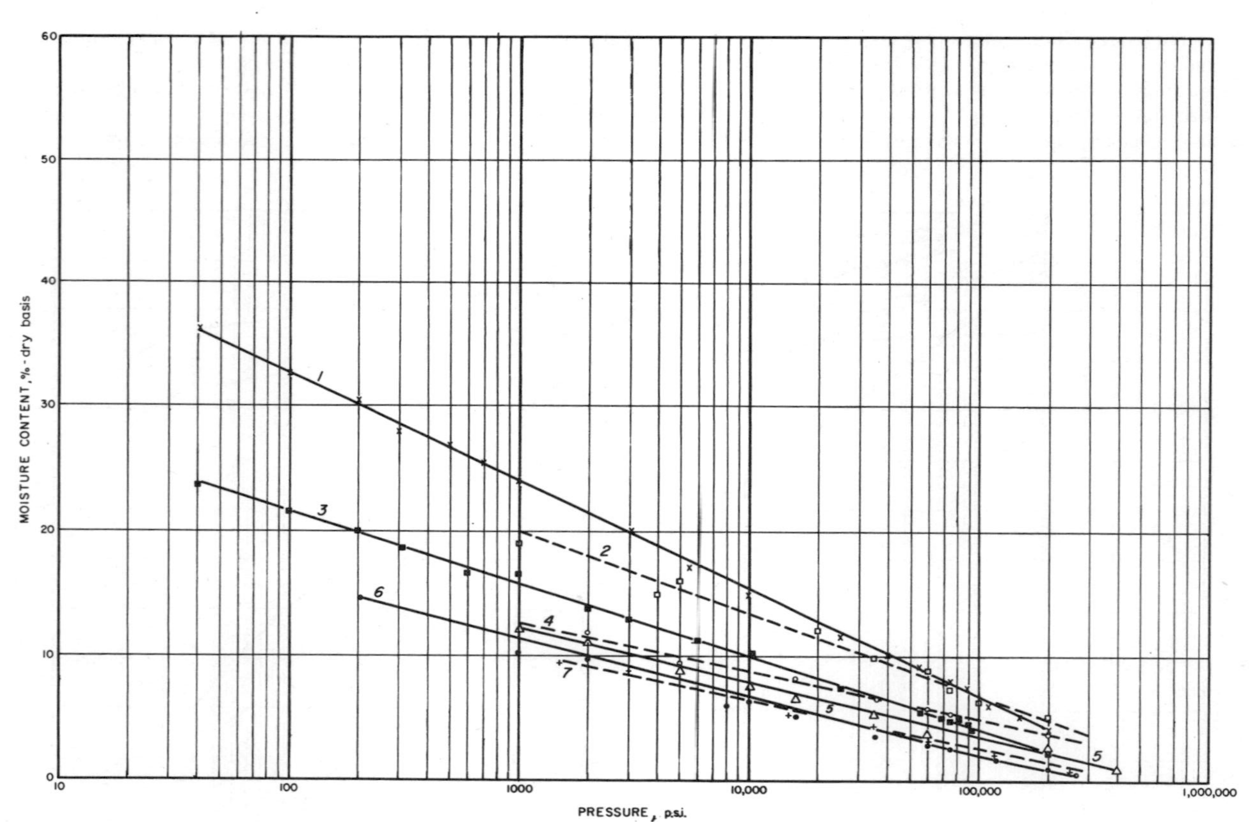

Fig.31. Relationship between moisture content M (% of dry weight) and pressure p (in p.s.i.) for various clays. *1* = illite no.35, Fithian, Illinois – $M = 50 - 8.7 \log p$; *2* = halloysite no.12, North Gardner Mine, Bedford (Huron), Indiana – $M = 40.1 - 6.66 \log p$; *3* = kaolinite no.4, Macon, Georgia, Oneal Pit – $M = 33.9 - 5.96 \log p$; *4* = P-95 dry lake clay, Buckhorn Lake, California – $M = 27 - 4.4 \log p$; *5* = hectorite no.34 from Hector, California, containing 50–58% by weight of $CaCO_3$ – $M = 26.5 - 4.55 \log p$; *6* = dickite no.15, San Juanito, Chihuahua, Mexico – $M = 26.7 - 5.04 \log p$; *7* = soil sample from weathered limestone terrain near Louisville, Kentucky – $M = 18.4 - 2.98 \log p$. (After Chilingarian and Rieke, 1968, fig.2, p.813.)

of the compacting mass, which increases with higher packing density and more coalesced structures.

For montmorillonite clay hydrated in distilled water, a break in the curve occurs at 1,000 p.s.i. According to Chilingar and Knight (1960a), possibly only oriented water is being removed from montmorillonite clay above 1,000 p.s.i. pressure. If initially the non-oriented (liquid) water and then the oriented water are being removed, for various clays having straight-line moisture-versus-pressure * relationships in the pressure range indicated, then there is indeed a very gradual transition from the oriented to the non-oriented water. The curves for illite and for kaolinite swing upward, from a straight line, at some pressures less than 40 p.s.i. (Fig.30, *6* and *7*, respectively); this was also noted by Meade (1964, p.B10). Possibly the pressures at which the pressure-versus-moisture content curves for various clays swing upward from a straight line indicate when the expulsion of oriented water begins. Prior to this, possibly only the free (liquid) water is removed.

The moisture-versus-pressure curve for montmorillonite clay hydrated in distilled water is a straight line between 1,000 and 200,000 p.s.i. As stated above, possibly the relatively loosely held water is lost below the pressure of 1,000 p.s.i., whereas at higher pressures (1,000–200,000 p.s.i.) the oriented water is being removed. In the case of montmorillonite hydrated in sea water, the moisture–pressure curve is a straight line lying between montmorillonite and illite fresh-water curves (Fig.30, *6*) in a 1,000–100,000 p.s.i. pressure range.

According to Chilingarian and Rieke (1968), the moisture-versus-pressure curve of hectorite clay [a three-layer type (silica–alumina–silica units) expanding lattice clay], containing 50–58% calcium carbonate, lies between those of kaolinite and dickite, instead of being close to that of montmorillonite (Fig.31, *5*). Thus, it appears that the calcium-carbonate content decreases the steepness of, and lowers, the moisture–pressure curve. This conclusion is in close accord with that reached by McCrossan (1960), who showed that the porosity of rocks decreases with increasing $CaCO_3$ content; as discussed previously, the slopes of the porosity–depth curves decreased with increasing $CaCO_3$ concentration (Fig.26).

Data on relationships between overburden pressure and remaining moisture content for halloysite and dickite are presented in Tables VII and VIII, whereas formulas expressing the relation between pressure and moisture content are given in Figs.30 and 31. For dickite clay there is a straight-line relationship between the moisture content and the pressure on semilogarithmic graph paper between 200 and 120,000 p.s.i. (Chilingar et al., 1963). At 200,000 p.s.i. the moisture content of dickite clay is equal to 0.61% and at 386,000 p.s.i. to 0.21%. Therefore, it appears that the moisture-versus-pressure curve for dickite is possibly not a straight line beyond 120,000 p.s.i. and approaches the pressure-

* In this section moisture-versus-pressure curve refers to plots of remaining moisture content (% dry basis) versus logarithm of pressure in p.s.i.

TABLE VII
Data on relationship between overburden pressure and remaining moisture content for halloysite clay (After Chilingar et al., 1963)

	Duration of compaction (days)	*Pressure (p.s.i.)*	*Moisture content (% of dry weight)*
Standard soil mechanics consolidometer	5	100	38.94*
	2	200	37.52*
	7	400	30.19*
High-pressure apparatus	2.5	1,000	20.90*
	2.5	2,000	20.00
	2.5	4,000	17.90
	2.5	7,000	17.10
	2.5	10,000	16.50
	2.5	15,000	15.80
	2.5	25,000	15.60
	2.5	35,000	14.00
	2.5	56,000	13.90
	2.5	76,000	13.50
	6	4,100	15.30*
	6	4,800	16.50
	6	120,000	7.45
	7	1,000	19.24**
	7	15,000	13.25**
	7	35,000	9.89**
	7	59,800	8.60**
	7	76,000	7.42**
	7	100,000	6.77**
	7	200,000	5.20**
	7	260,000	2.06**

* Result is questionable owing to factors such as evaporation.
** Equilibrium points.

axis asymptotically. Possibly this is the pressure at which movement of the last layers of bound water starts to occur. The moisture–pressure curve for dickite is below that of kaolinite (Fig.30). The fact that dickite crystals are usually larger than those of kaolinite clay may account for this phenomenon. The curve for halloysite clay lies above the kaolinite curve (Fig.31); this may possibly indicate that kaolinite (plates) particles move more freely with respect to each other (which results in closer packing) than do halloysite (tubes) particles.

The relationship between the moisture content in percent of dry weight and the logarithm of pressure in p.s.i. for P-95 clay (from Buckhorn Dry Lake) and a clay soil from weathered limestone terrain (from Louisville, Kentucky) is presented in Fig.31 (curves *4* and *7*). According to Rieke et al. (1964), a marine mud (AHF core No.6943) from a marine basin off southern California, which was dried and then redispersed in

TABLE VIII
Data on relationship between overburden pressure and remaining moisture content for dickite clay. (Repeatability of results ranges from 2–4%)
(After Chilingar et al., 1963)

	Duration of compaction (days)	*Pressure (p.s.i.)*	*Moisture content (% of dry weight)*
Standard soil mechanics consolidometer	5	200	20.50
	6	200	15.35
High-pressure apparatus	2	15,000	5.87
	5	1,000	8.10*
	5	386,000	0.21
	7	1,000	8.60*
	7	2,000	9.70**
	7	5,000	9.22**
	7	8,000	5.70*
	7	10,000	6.10*
	7	15,000	5.01**
	7	35,000	2.83**
	7	60,000	2.45**
	7	76,000	2.18**
	7	100,000	1.40**
	7	120,000	1.00**
	7	200,000	1.12
	10	200,000	0.61**
	12	15,000	5.13**

* Result is questionable due to factors such as evaporation.
** Equilibrium points.

distilled water, has a remaining moisture content of 6.2% at an applied overburden pressure of 15,000 p.s.i. (Table IX). A lake clay (derived from the Chattanooga Shale) which was hydrated in distilled water, had remaining moisture contents of 8.97% and 8.3% at 900 and 4,200 p.s.i., respectively (Table IX).

The steeper slopes of more-plastic clays (higher swelling) suggest that it is harder to squeeze a given volume of water out of the less-plastic clays; however, the percentage of water remaining in the more plastic clays is still greater. As shown in Fig.30, organic colloids release associated water more readily at low pressures than at high pressures and hold water at high pressures more strongly than the clays. Surface areas per unit weight (specific surface) of clay minerals influence the amount of water retained by the clay mineral under pressure (Meade, 1968, p.D3). The larger the specific surface area, the larger the amount of water retained by the clay under pressure. Thus, montmorillonite hydrated in distilled water, as indicated by Fig.30, has a larger specific surface area than does illite hydrated in distilled water; this is a well-established fact.

TABLE IX
Relationship between remaining moisture content and overburden pressure in p.s.i. for various clays (After Rieke et al., 1964)

Type of clay	*Duration of compaction (days)*	*Overburden pressure (p.s.i.)*	*Remaining moisture content (% of dry weight)*
P-95 (Buckhorn Dry Lake clay, California)	4	15,000	11.25*
	7	5,000	8.7
	7	35,000	7.02
	7	60,000	6.05
	8	1,000	10.27
	8	2,000	13.4
	8	5,000	9.8
	9	15,000	9.96
	9	76,000	3.2
	18	200,000	5.2
Hectorite No.34	2	15,000	6.59
	7	5,000	13.2
	7	15,000	6.71
	7	15,000	6.82
	7	60,000	3.2
	8	200,000	6.27
	9	150,000	7.34
	10	1,000	13.1
	10	1,000	14.2
	10	15,000	6.86
	10	35,000	5.9
	11	10,000	7.6
	11	200,000	3.0
	14	400,000	0.8
	16	2,000	12.05
Halloysite No.12	6	400,000	9.6*
	7	400,000	7.4*
	7	500,000	9.7*
Marine mud (California) (AHF 6943)	5	15,000	6.7*
	7	15,000	6.2
Lake clay (Louisville, Kentucky)	10	900	8.97
	7	4,200	8.30

* Non-equilibrium points.

The remaining moisture contents of halloysite hydrated in distilled water at 400,000 and 500,000 p.s.i. are abnormally high at these high pressures (Table IX; compare with Fig.31). This is possibly due to the fact that samples were squeezed too rapidly for the water to escape readily at low pressures and it was trapped as ice (?) at pressures above 100,000 p.s.i. Ice formation at these high pressures has been suggested by Bridgman (1958) and by I. Getting (personal communication, 1968).

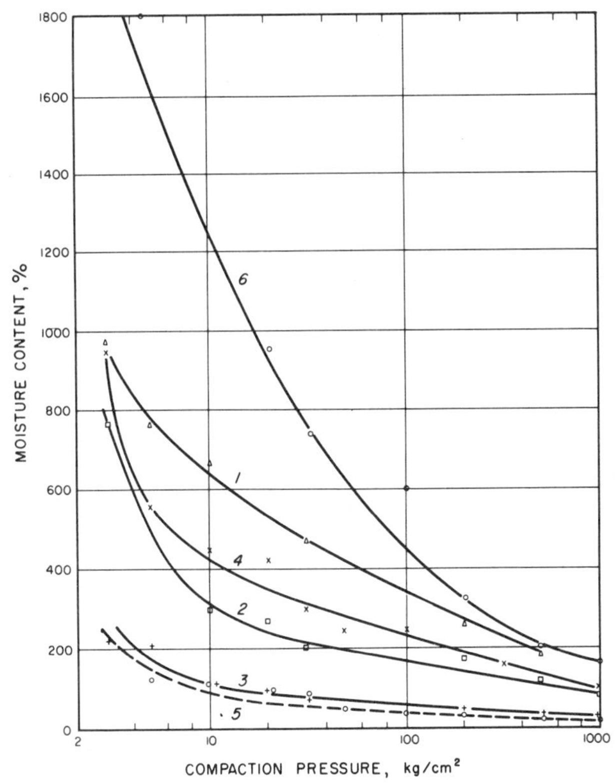

Fig.32. Relationship between moisture content of sapropels and squeezing pressure. *1* = upper layer; *2* = middle layer; *3* = lower layer; *4* = upper layer (frozen); *5* = lower layer (frozen); *6* = sapropel from Sudobl' Lake, U.S.S.R. 1–5 = sapropel from Sergeevskoe Lake, Minsk area, U.S.S.R. (After Lopotko et al., 1968, fig.1, p.91.)

The "hyperbolic-like" shape of the natural organic-colloid curves (Fig.30, *1* and *2*) is probably related to their greater plasticity than that of clays.

Lopotko et al. (1968) studied the relationship between the compaction pressure and remaining water content of sapropels. As shown in Fig.32, the processes of mechanical expulsion of free water from macropores occurs at pressures of 10–30 kg/cm^2. At higher pressures the more-or-less straight-line portions of curves asymptotically approach the abscissa. In this region, physically "adsorbed" (?) water is removed from micropores. The amount of water in sapropels varies between 1.5 and 32 g per g of dry material. Before compaction samples were centrifuged for 15 min ($p = 3$ kg/cm^2). A very small amount (only 0.04–0.6 μg/l) of vitamin B-12, which is present in the dry material in the amounts of 8 to 110 μg/kg, is squeezed out with the water.

On squeezing the silica gel, the particles are pressed together to the point where the Si–O–H groups of adjacent particles come in direct contact. As a result, there is polymerization and formation of Si–O–Si linkages, which weld the particles at the points of contact. The effect of squeezing is first to push the silica particles around so that they become more and more closely packed. At very high pressures, the spheres may be flattened at the points of contact (Iler, personal communication, 1960; also see Iler, 1955).

The results of this investigation also raise the question as to the nature of the water being expelled. Is it simply an ordinary water held within the pores of the gel or is it an oriented water? The first molecular layer of water next to the Si–O–H groups on the surface of the silica particles is held with relatively strong force (hydrogen bonding). Is the second layer of water essentially free water that is not different from the water still further from the surface, or is it oriented water? This question still remains unanswered.

Relationship between pressure and porosity of carbonate sediments

In recent years increased attention has been paid to field studies and experimental compaction studies of carbonate sediments (Terzaghi, 1940; Laughton, 1957; McCrossan, 1961; Fruth et al., 1966; Robertson, 1967; Ebhardt, 1968; and Coogan, 1970). Early compaction experiments by Terzaghi (1940) and Laughton (1957) were carried out under a limited pressure range in piston-sleeve type apparatuses. Laughton (1957) studied the velocity of propagation of compressional and transverse elastic waves in Globigerina oozes at different compacting pressures. E.C. Robertson (personal communication, 1966) conducted extensive experiments on aragonitic sediments from the Bahama Banks (Fig.33). He found that the consolidation curve of aragonite lies between the montmorillonite and

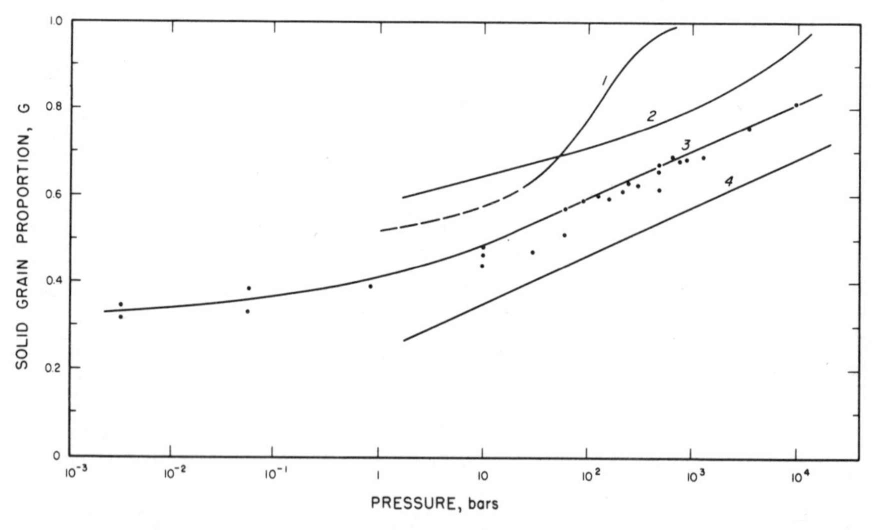

Fig.33. Relationship between solid grain proportion and pressure. *1* = shale (Athy, 1930); *2* = kaolinite (Chilingar and Knight, 1960a); *3* = fine-grained aragonitic sediment; *4* = montmorillonite (Chilingar and Knight, 1960a). *G* is defined on p.75. (After Robertson, 1967, fig.4, p.126.)

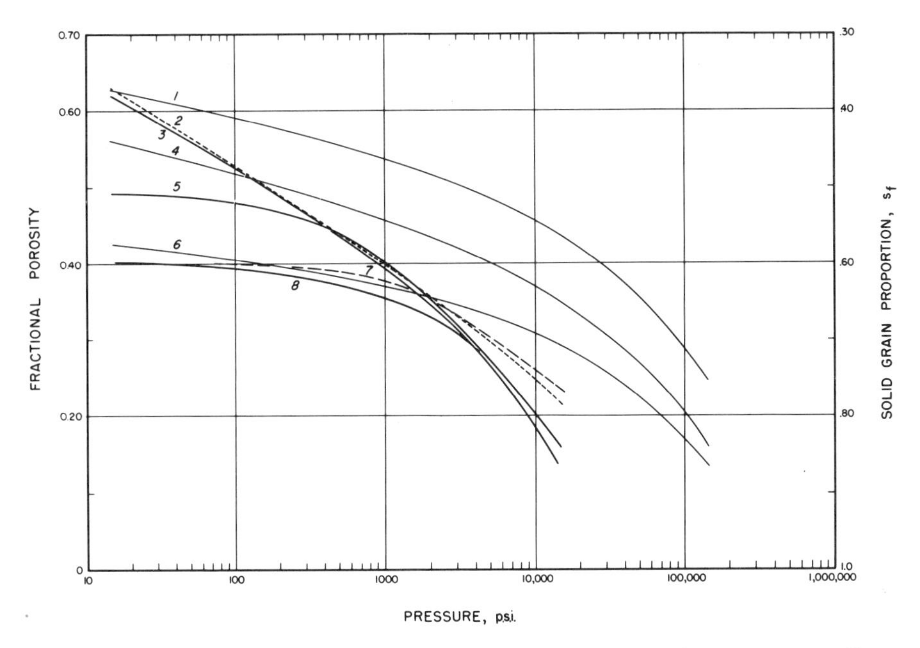

Fig.34. Relationship between fractional porosity and pressure for various carbonate sediments. *1* = aragonite sediment from the Bahamas with an initial water content of 80% (Robertson, 1967); *2* = clay-size aragonite grains (mud facies of modern Bahamian carbonate sediments) (Fruth et al., 1966); *3* = skeletal debris composed of foraminifer and mollusc shells (Fruth et al., 1966); *4* = aragonite sediment with an initial water content of 60% (Robertson, 1967); *5* = grapestone facies (Fruth et al., 1966); *6* = dry aragonite powder (Robertson, 1967); *7* = oolite facies composed of polished subspherical ooids and well-rounded grains (Fruth et al., 1966); and *8* = oolitic facies composed of well-rounded oolitically coated grains (Fruth et al., 1966).

kaolinite curves obtained by Chilingar and Knight (1960a) and is nearly parallel to them. This could be interpreted as meaning that the compaction mechanism is similar despite the difference in grain shapes (plates for the clays and needles for aragonite). Fig.34 shows that an initial difference in the porosity of Robertson's aragonite samples (Fig.34, curves *1*, *4* and *6*) is maintained through a large part of the pressure range. The inference is that the initial water content does not appreciably affect the change in compaction with increased pressure (porosity/pressure slope). Upon exposing a limestone core sample with 10% initial moisture content to a pressure of 100,000 p.s.i., 59.2% of the original moisture was squeezed out (Chilingarian and Rieke, 1968).

Ebhardt (1968) compacted recent carbonate sediments of different grain size and with different carbonate content up to pressures of 650 atm. The sediment samples tested by Ebhardt and their properties are presented in Table X, whereas experimental results are shown in Fig.35 and 36. The compaction curves for samples *P*, *Ad*, *I* and *Hz*, which are characterized by an increasing clay content, are similar to those of pure clays.

Elevated temperatures (90°C) cause more intense compaction and the void ratio is about 10% lower than for runs at room temperature (Fig.37).

TABLE X

Type, location and composition of samples analyzed by Ebhardt (1968, p.59)

Sample	*Provenance*	*Type of sediment*	*Content, %*				*Median diameter, Md (μ)*	*Sorting coefficient, $S_o = (Q_3/Q_1)^{1/2}$*
			carbonate	*sand*	*silt*	*clay*		
Ad IV	Adria	sand	86	96	3	1	450	1.6
Rm	Red Sea	sandy silt	75	20	80	–	20	1.3
P	Persian Gulf	clayey silt	65	17	56	27	6	3.7
F II	Florida Bay	clayey silt	86	8	61	31	7	3.8
Ad I	Adria	clayey silt	41	4	78	18	11	3.3
Hz	Persian Gulf	silty clay	36	–	48	62	2	–
$CaCO_3$	–	silt	100	–	100	–	22	1.5
Bd	Lake Constance	sandy silt	91	24	70	6	17	3.5

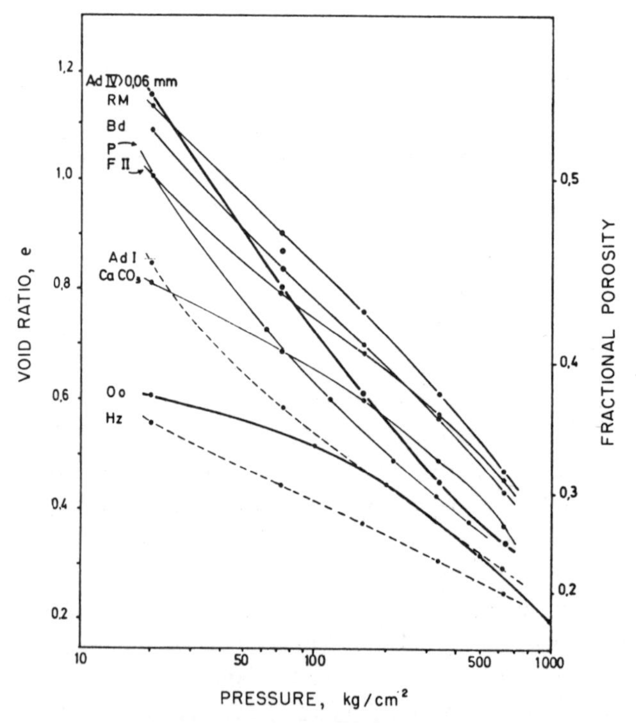

Fig.35. Relationship between void ratio and confining pressure for sediments having different carbonate content. (After Ebhardt, 1968, p.60.) Thick solid lines = sand-size sediments; dashed lines = clayey sediments; O_o = oolitic sediment, after Fruth et al. (1966).

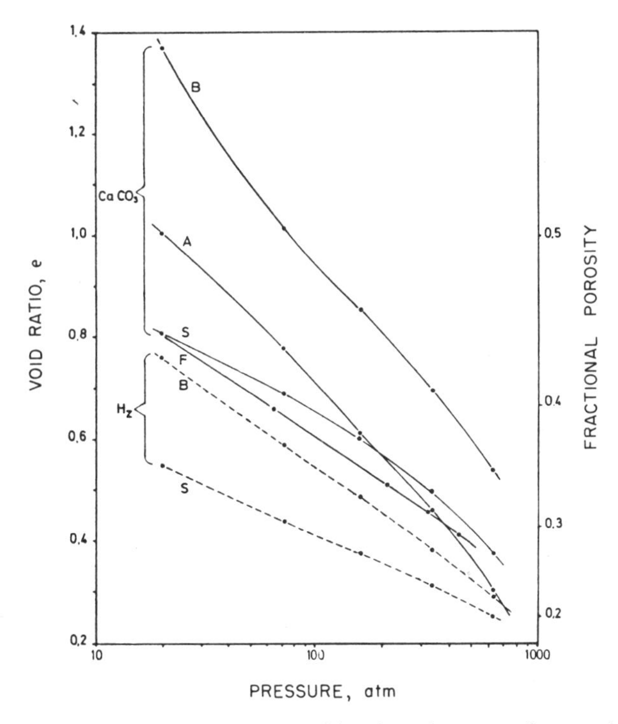

Fig.36. Relationship between void ratio and pressure for sample H_z and $CaCO_3$ samples, saturated in benzene (*B*), air (*A*), sea water (*S*), and fresh water (*F*). (After Ebhardt, 1968, fig.4, p.61.)

Fruth et al. (1966) designed a triaxial cell for systematic evaluation of pressure, temperature and strain-rate effects on calcareous sediments. The amount of compaction caused by increasing confining pressure on aragonite sediments from the Bahamas is shown in Fig.34. All compaction curves in this figure are consistent with the expected behavior of the sediment types: (1) negligible initial compaction of the well-sorted, hard oolite grains, (2) high compaction of the fine-grained mud, and (3) moderate compaction of the grapestone facies (a mixture of friable and hard grains and aggregates). Greater compaction of the skeletal type of carbonate sediment reflects initial voids and a more open structure created by the presence of shells, skeletal fragments, and large grains. With an increase in pressure, the fragments supporting the open structure yield and the void volume is reduced (Fruth et al., 1966, p.750). It seems that all five carbonate-sediment types presented in Fig.34 have nearly the same porosity values (≈35%) at a pressure of around 2,000 p.s.i. According to Fruth et al. (1966), the somewhat poorer sorting and initial compaction (in the field) of the oolitic facies may account for its lower porosity values than the oolite. The finer material fills the voids between the larger grains. The final porosity of the skeletal facies is lower than that of the aragonite mud, indicating that the skeletal facies compact better. Finer matrix grains tend to cushion the effects of compaction in oolitic facies. In the case of well-sorted oolite, with small amounts of

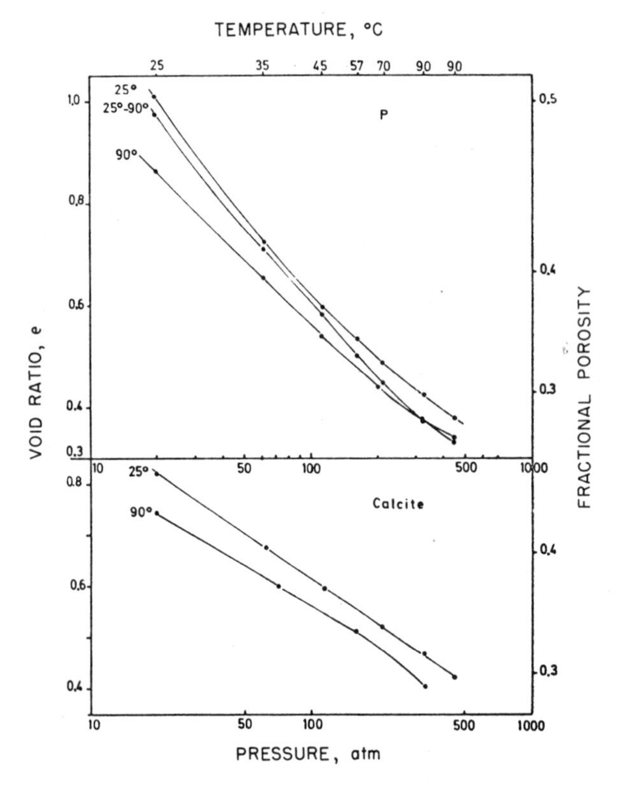

Fig.37. Relationship between void ratio and pressure at room temperature and at 90°C for sample *P* and $CaCO_3$ sample. In the case of sample represented by middle curve in the upper graph, the temperature was raised stepwise (25–90°C; see upper scale). (After Ebhardt, 1968, fig. 6, p. 62.)

matrix and hard grains predominating, there is a high degree of grain fracturing. There is a negligible initial compaction in the presence of small amounts of matrix and the grains are in contact with each other during compaction. Fractures tend to radiate from points of initial contact; however, in some grains a single fracture extends completely across the grain (Fruth et al., 1966, p.751).

It is claimed by many geologists that field observations on carbonate rocks often indicate that very little compaction has occurred. Often, the absence of the crushing of delicate fossils is brought forward as supporting evidence. Brown (1969, p.492) compacted pelecypod valves and whole shells in a carbonate-mud matrix up to a pressure of 15,000 p.s.i. Only rarely did some shells break, indicating that the presence of unbroken fossils surrounded by carbonate mud is not evidence by itself for the lack of compaction.

Moisture content versus time relationship

The moisture content-versus-pressure curves of clays with high swelling indices have

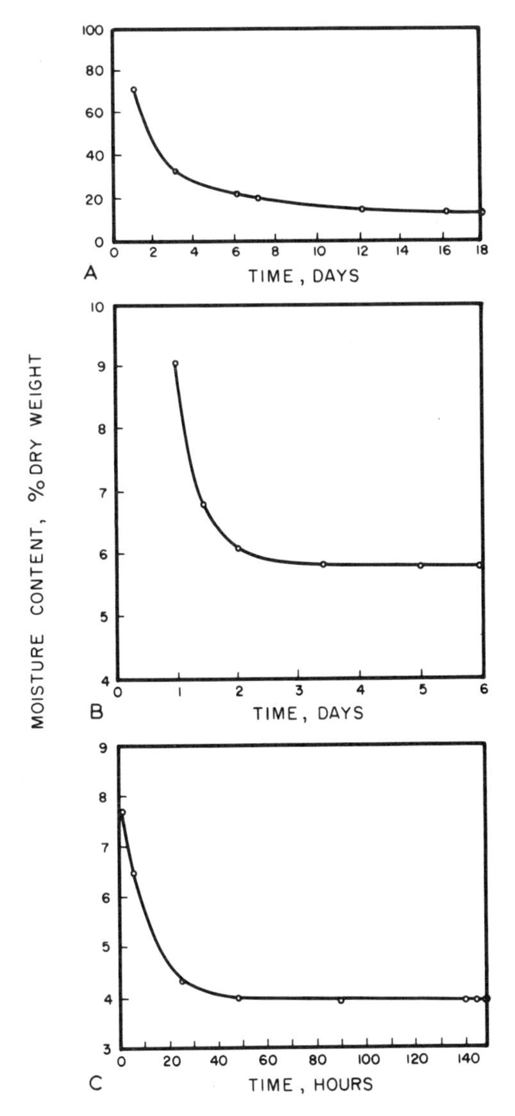

Fig.38. Moisture content versus time. A. Variation in moisture content (% dry weight) of montmorillonite clay (API no.25) with time (in days) at 88,500 p.s.i. B. Variation in moisture content (% dry weight) of illite clay (API no.35) with time (in days) at 91,000 p.s.i. C. Variation in moisture content (% dry weight) with time (in hours) for kaolinite clay (API no.4) at 88,500 p.s.i. (After Chilingar and Knight, 1960a, figs.4–6. Courtesy of American Association of Petroleum Geologists.)

steeper slopes than those of the clays which swell to a lesser degree; however, a longer time is needed to establish equilibrium for the former clays. Fig.38 and 39 show the relationship between the moisture content and the time of squeezing for various clays and silicic acid. Most of the water is squeezed out during a relatively short period of time

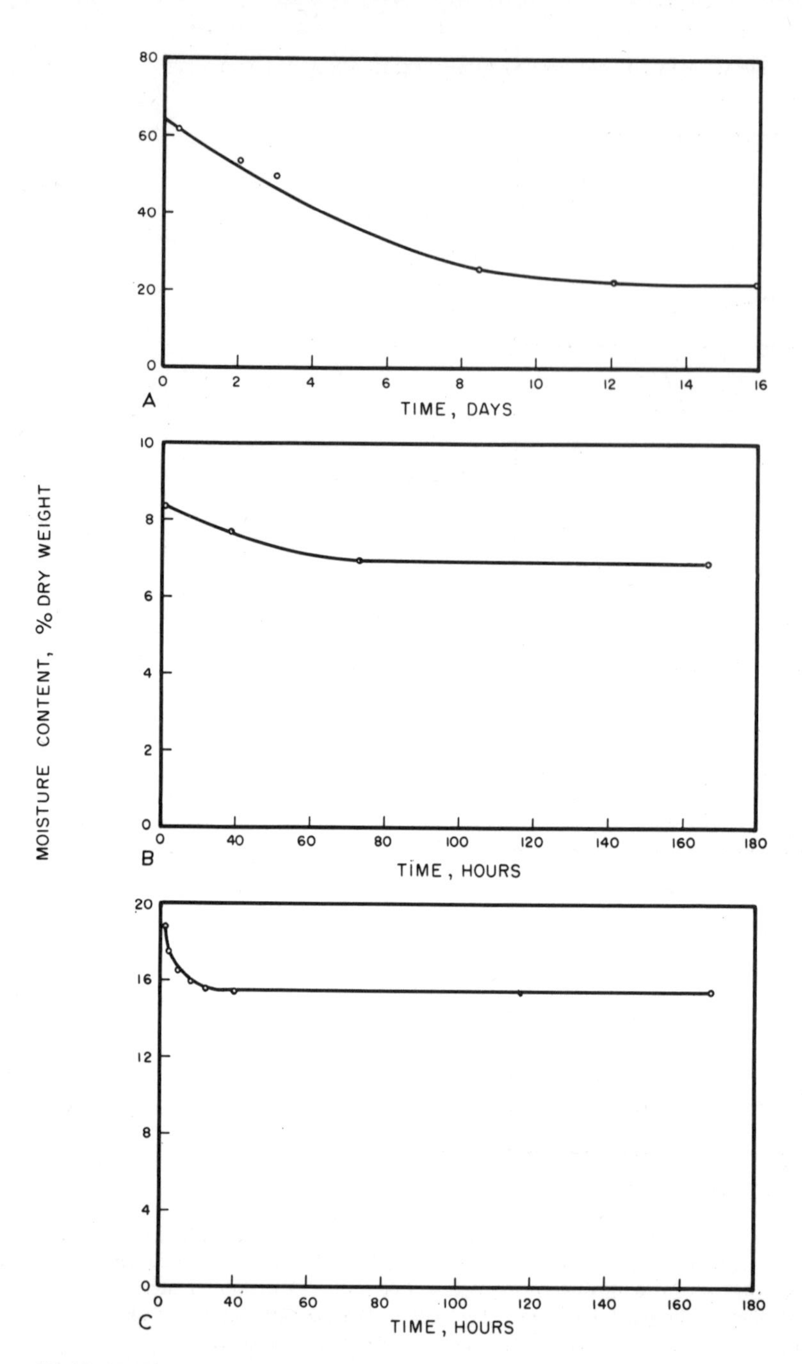

Fig.39. Moisture content versus time. A. Variation in moisture content (% dry weight) of silicic acid with time (in days) at 88,500 p.s.i. (After Chilingar and Knight, 1960b). B. Variation in moisture content (% dry weight) of montmorillonite (API no.25) hydrated in sea water with time (in hours) at 15,000 p.s.i. (After Rieke, 1970, p.173). C. Variation in moisture content (% dry weight) of soil from weathered limestone terrain (Louisville, Kentucky) with time (in hours) at 10,000 p.s.i. (After Rieke, 1970, p.173.)

(1–4 days), and equilibrium is reached after 7–60 days. Montmorillonite clay hydrated in distilled water requires more time to reach equilibrium than montmorillonite clay hydrated in sea water. Possibly, flocculated montmorillonite clays close-pack better and faster than do deflocculated ones. Illite, in turn, requires a longer interval than kaolinite for attainment of equilibrium at a given pressure.

Shishkina (1968) reported that on compacting diatomaceous marine mud having 70% moisture content at 405 kg/cm^2 (5,760 p.s.i.), 85% of all interstitial water is squeezed out after 4 hours. After 10 hours 90.0% of water was squeezed out at this pressure. Subsequent squeezing did not achieve any additional removal of fluids. On squeezing marine mud having initial moisture content of 49.5%, after 4 hours 87% of all interstitial fluids was squeezed out at 1,355 kg/cm^2. On increasing the pressure to 3,100 kg/cm^2, 93.5% of water was squeezed out after 7 hours and 96.4% after 10 hours. Subsequent squeezing for 6 hours did not produce any additional water.

At pressures of 400 kg/cm^2, depending on type of sediment and initial moisture content, up to 85% of all water is squeezed out after 4–6 hours. At 202 kg/cm^2 pressure, 80–85% of all water is squeezed out after 10 hours; remaining moisture content was around 16–23% (Shishkina, 1968).

E.C. Robertson (personal communication, 1967 and 1968) in analyzing the rate of

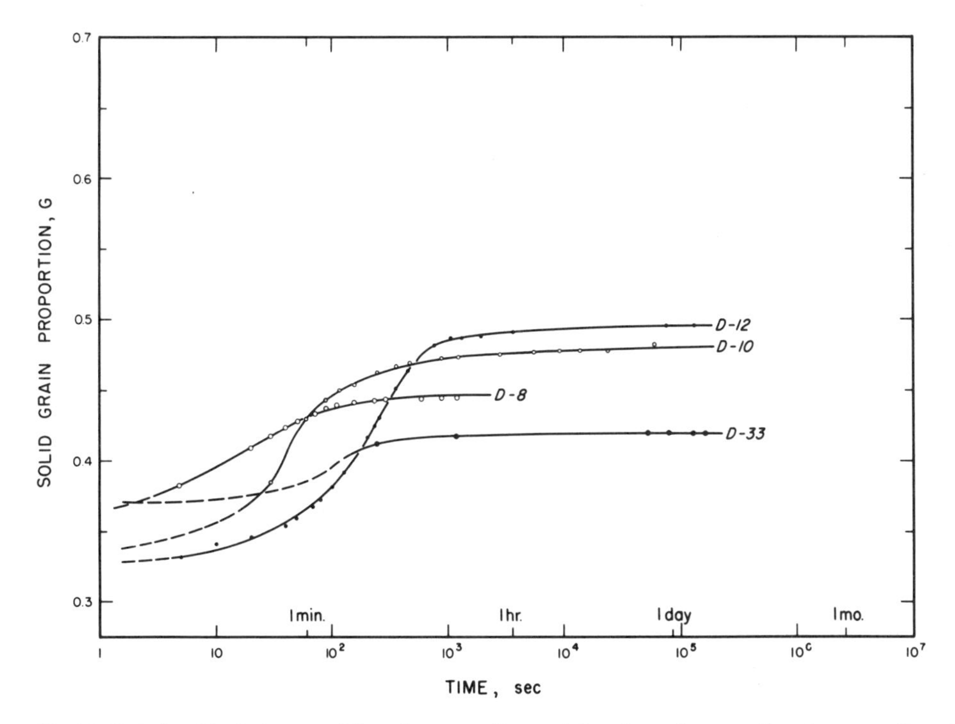

Fig.40. Relationship between solid grain proportion and duration of compaction for aragonite sediments. (After Robertson, 1967, fig.2, p.125.)

compaction of aragonite mud found that the practical minimum time required to approach maximum compaction was 3 hours (Fig.40), as shown by the flattening of the curves.

Calculation of compaction parameters

The transformation of moisture-content data into void ratio is based on the assumption that the interstitial fluid in the clay pores fully saturate all the pore space at any given pressure. The void ratio is expressed as:

$$e = V_v/V_s \qquad (2\text{-}27)$$

where V_v is the volume of the voids and V_s is the volume of the solids. These volumes can be expressed in terms of weights and specific weights:

$$V_v = W_w/\gamma_w \quad \text{and} \quad V_s = W_s/\gamma_s \qquad (2\text{-}28)$$

where W_w is the weight of the pore fluid, W_s is the weight of the solids, γ_w is the specific weight of water, and γ_s is the specific weight of the solids. Thus eq.2–27 becomes:

$$e = (W_w/\gamma_w)/(W_s/\gamma_s) \qquad (2\text{-}29)$$

Inasmuch as γ_s/γ_w ratio is equal to the specific gravity of clay, eq.2–29 can be written as:

$$e = (W_w/W_s)\,(\text{sp. gr.}) \qquad (2\text{-}30)$$

or:

$$e = (M \times \text{sp. gr.})/100 \qquad (2\text{-}31)$$

where M is the remaining moisture content in % (dry basis). Consequently, the specific weight of various clays should be accurately determined, in order to obtain accurate void ratio, e, values. The term density is commonly used to designate specific weight, which is not recommended by the writers.

Porosity (fractional or percent), ϕ, can be expressed in terms of void ratio as follows:

$$\phi = \frac{V_v}{V_b} = \frac{V_v}{V_v + V_s} = \frac{e}{1+e} \qquad (2\text{-}32)$$

where V_v is the void volume, V_s is the volume of solids, and V_b is the total (bulk) volume.

Robertson (1967) proposed a new compaction parameter called the solid-grain proportion, G, which can be expressed in terms of e and ϕ as follows:

$$G = \frac{V_s}{V_b} = \frac{\rho_{bd}}{\rho_s} = \frac{1}{1+e} = 1 - \phi \qquad (2\text{-}33)$$

Solid-grain proportion, which is a ratio of dry bulk density, ρ_{bd}, to grain bulk density, ρ_s, is a linear measure of the approach of the sediment's dry bulk density, ρ_{bd}, to its solid grain density at any stage of its compaction. The void ratio is an index of volume change in the sediment, whereas solid-grain proportion is an index of change in mass per unit volume (Robertson, 1967, p.119). Robertson (1967) prepared a graph illustrating the relation between e, ϕ, and G (Fig.41).

If it is assumed that the clay platelets are perfectly planar, have the same thickness, and are oriented parallel to one another without overlapping, then void ratio can be

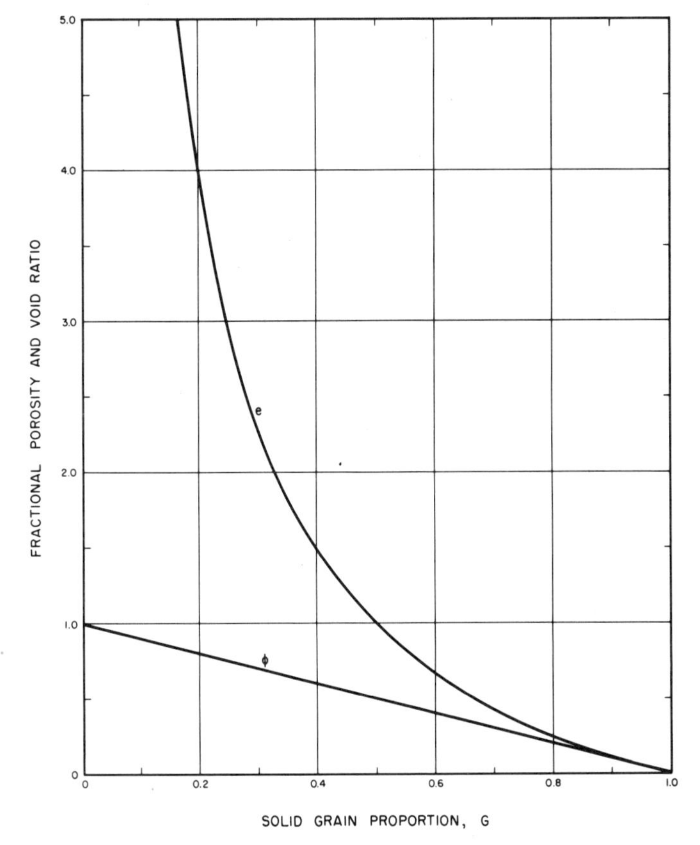

Fig.41. Relationship between void ratio, fractional porosity and solid-grain proportion. (Adapted after Robertson, 1967, fig.1, p.124.)

directly related to the interparticle spacing:

$$e = \rho_s \times s_s \times d \tag{2-34}$$

where s_s is specific surface of the clay and d is the half of the distance between the plates. Plots of void ratio versus pressure can be considered to be equivalent to plotting interparticle spacing versus pressure. Warner (1964, p.51) believes that this is significant because whatever mechanism controls porosity, the distance between the clay plates is a fundamental parameter. Eq.2–34 has some deficiencies, however, because clay plates are not of uniform thickness and are not generally oriented parallel to each other. In addition, only montmorillonite particles or the finest fractions of illite or kaolinite are likely to be close to being planar. In spite of these problems, eq.2–34 can be used to compute hypothetical interparticle spacing in clay systems (Warner, 1964).

Baldwin (1971) proposed calculation of a "decompaction number" which, when multiplied by the present thickness of a compacted interval, yields the original sediment thickness. The decompaction number, N_d, is the ratio of earlier thickness, h_e, to present thickness, h_p. Baldwin (1971) assumed that for a given bed, its thickness times the grain proportion remains constant during the compaction. Thus:

$$N_d = \frac{h_e}{h_p} = \frac{G_p}{G_e} = \frac{1-\phi_p}{1-\phi_e} \tag{2-35}$$

where G_e = earlier grain proportion, ϕ_e = earlier porosity, G_p = present grain proportion, and ϕ_p = present porosity. On assuming a solid-grain density of 2.66 g/cm^3, Baldwin (1971) prepared a diagram relating dry and wet bulk density, grain proportion, depth of burial, porosity, and void ratio (Fig.42). According to Baldwin (1971), the initial grain

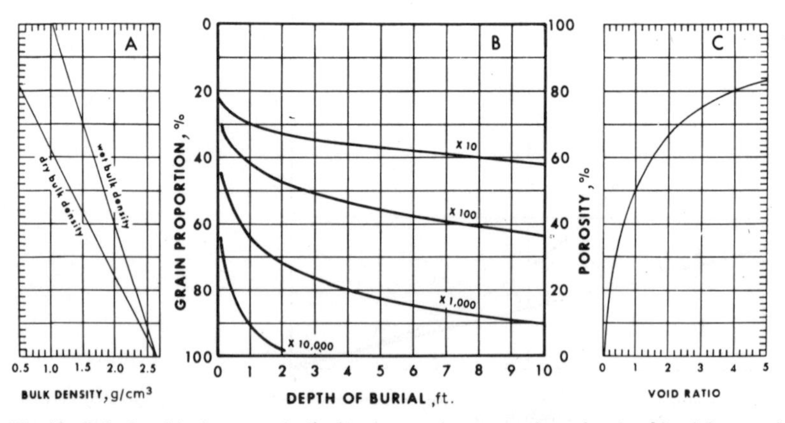

Fig.42. Relationship between bulk density, grain proportion, depth of burial, porosity and void ratio. For figure B, depth of burial values should be multiplied by the number shown for that curve. (After Baldwin, 1971, fig.1, p.294.)

proportion, G_e, averages 0.22 (ϕ = 78%). Data on solid-grain density and porosity obtained by Hamilton (1969a, b, c) for the deep-sea sediments of the North Pacific were cited by Baldwin (1971).

When the compacted muds are to be restored to the condition of zero depth of burial, G_e can be assumed to be equal to 22%. In the case of decompacting to a condition of intermediate depth of burial, the earlier grain proportion, G_e, can be inferred from Fig.42B.

The maximum amount of overburden can be estimated from Fig.42B, if the present grain proportion G_p, has been measured. Discrepancy between G_p and present overburden is a guide to the amount of erosion at a surface of unconformity (Johnson, 1950; Baldwin, 1971, p.295).

Effect of various electrolytes in interstitial water on compaction

Von Engelhardt and Gaida (1963, p.924) hydrated (1) kaolinite clay in NaCl-solution (0.6, 2 and 2.4 moles/l), and (2) montmorillonite clay in NaCl-solution (0.16 *N*, 0.58 *N*, 1.1 *N*, 2.3 *N*, 4.6 *N*) and in $CaCl_2$-solution having normality of 0.62 and 1.2. Within the pressure range of their experiments (441–47,027 p.s.i.), the salt content in the interstitial solution had no effect on the end values of the void ratio reached at distinct pressures. Bolt (1956) also reported that the void ratio of compressed montmorillonite clays is not influenced by the interstitial electrolytes at pressures above 10 atm (147 p.s.i.).

Von Engelhardt and Gaida (1963) noted that the compaction rate varied with salinity (Fig.43). They observed that compaction of clays proceeds more rapidly if the interstitial fluid is saline. The results presented in Figs.38A and 39B confirm this observation. As shown in Fig.36, for montmorillonite clay with greater concentration of salts in the

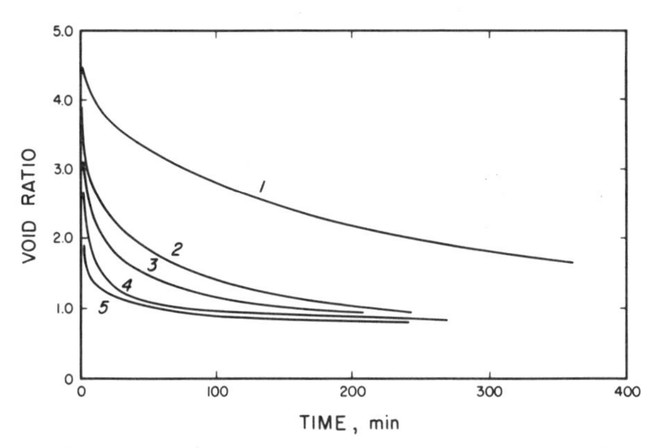

Fig.43. Relationship between void ratio and duration of compaction for montmorillonite clays at 126 atm pressure. *1* = pure water; *2* = 0.578 mole/l NaCl solution; *3* = 1.14 mole/l NaCl solution; *4* = 2.28 mole/l NaCl solution; and *5* = 4.57 mole/l NaCl solution. (After Von Engelhardt and Gaida, 1963, fig.1, p.925.)

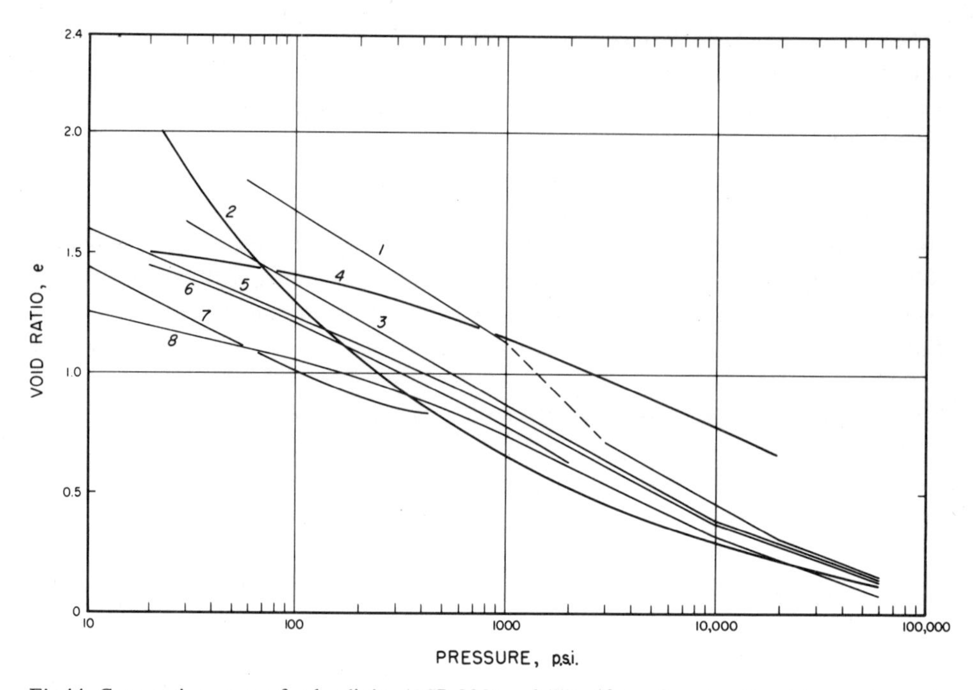

Fig.44. Compaction curves for kaolinite (ASP 200) and illite (from Grundy, Illinois) in various solutions. *1* = kaolinite in CCl_4; *2* = illite in various solutions (0.001 *M* NaCl, 0.55 *M* NaCl, 0.001 *M* $CaCl_2$); *3* = kaolinite in 0.001 *M* NaCl (pH = 6); *4* = quartz; *5* = kaolinite in 0.55 *M* NaCl (pH = 2.6 and 8.5); *6* = kaolinite in 0.001 *M* $CaCl_2$ (pH = 8.5); *7* = illite in CCl_4; and *8* = kaolinite in 0.001 *M* NaCl (pH = 8.5). (Data from Warner, 1964.)

interstitial fluids, a smaller amount of water is squeezed out for the same increment in pressure (also see Meade, 1964, p.B10).

Warner (1964, p.51) hydrated a homoionic Na-kaolinite (ASP* 200) in various solutions and then compacted the clay slurries at pressures up to 60,000 p.s.i. As shown by Fig.44, the void ratios of the slurries increased in the following order: (1) in 0.001 *M* NaCl solution (pH = 8.5); (2) in 0.001 *M* $CaCl_2$ solution (pH = 8.5); (3) in 0.55 *M* NaCl solution; (4) in 0.001 *M* NaCl solution (pH = 6); and (5) using CCl_4. The void ratio–pressure curves maintain their relative positions even to the highest test pressures. Differences in the void ratios at a given pressure can be attributed to a different packing arrangement of the kaolinite plates, as controlled by the chemistry of the hydrating fluid (Warner, 1964, p.53). Warner concluded that the compaction of kaolinite is not greatly influenced by double-layer repulsive forces; and forces resisting compaction are due to mechanical interference between the plates. In the deflocculated sample (0.001 *M* NaCl; pH = 8.5), rearranging and/or crushing of the particles constitute an important resistance to consolidation. In the flocculated (face to face) samples, on the other hand, electrical or chemical

* Minerals and Chemicals Phillip Corp., Menlo Park, N.J.

bonds between the randomly arranged particles could give rise to additional forces. The sharp drop exhibited by the CCl_4-saturated clay, at pressures between 1,000 and 3,000 p.s.i. was interpreted by Warner (1964, p.54) as due to a partial collapse of the edge-to-face structure.

Warner's (1964) experiments with a homoionic Na-illite from Grundy, Illinois, shown in Fig.44, gave results that were similar to the results obtained by Bolt (1956). There is very little difference between the illite hydrated in 0.55 *M* NaCl (pH = 6 and 8.5), 0.001 *M* NaCl (pH = 6 and 8.5), and 0.001 *M* $CaCl_2$ (pH = 8.5) solutions. The compaction curve for illite in CCl_4 lies below the other curves and merges with them at a pressure of around 500 p.s.i. Apparently, the illite particle arrangement as controlled by pH and electrolyte concentration did not greatly influence pressure–void-ratio relations. Warner (1964, p.58) stated that double-layer forces in conjunction with particle rearrangement are the important resistive forces below 1,000 p.s.i. At pressures >1,000 p.s.i. irregular surfaces of the larger particles or their Stern layers probably come in contact and provide the forces of resistance to compaction; volume reduction occurs as a result of particle crushing or the removal of tightly adsorbed water.

The void ratio–pressure data for a Wyoming Na-bentonite compacted in 0.001 *M* NaCl (pH = 6 and 8.5), 0.55 *M* NaCl (pH = 6 and 8.5), 0.001 *M* $CaCl_2$ (pH = 8.5) and 0.001 *M* NaCl (pH = 8.5 at 80°C) are presented in Fig.45. The pH of the bentonite slurries apparently did not influence the compaction curves. Warner (1964, p.65) showed that these

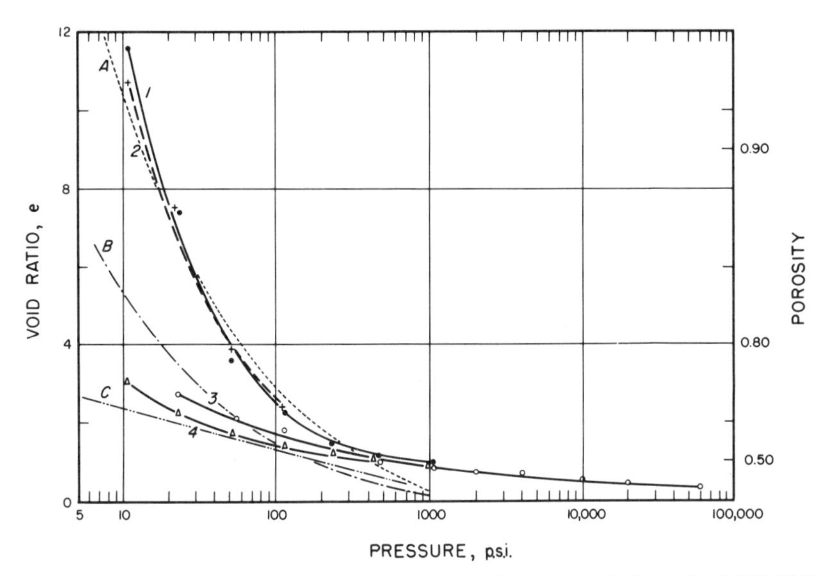

Fig.45. Compaction curves for Wyoming bentonite in various solutions. *1* = 0.001 *M* NaCl (pH = 6 and 8.5 at 25°C); *2* = 0.001 *M* NaCl (pH = 8.5 at 80°C); *3* = 0.55 *M* NaCl (pH = 6 and 8.5); *4* = 0.001 *M* $CaCl_2$ (pH = 8.5); *A* = theoretical curve for *1* and *2*; *B* = theoretical curve for *4*; and *C* = theoretical curve for *3*. (After Warner, 1964, fig.9.)

experimental curves agree reasonably well with curves predicted by Gouy-Chapman double-layer theory until the bentonite plates become separated by only two layers of adsorbed water on each particle (see Fig.45). The overburden pressure required to remove all water was roughly estimated by Warner to be 49,400 p.s.i.; however, one water layer still remained at a pressure of 60,000 p.s.i. Warner (1964, p.65) stated that this indicated that the void ratio–pressure relation for Wyoming Na-bentonite is controlled almost exclusively by physical–chemical factors with only a minor contribution from the mechanical interference.

An experimental compaction curve for quartz powder in 0.001 *M* NaCl solution is given in Fig.44 (Warner, 1964, p.66). The decrease in void ratio per increment of pressure is very small for quartz powder. Volume reduction of quartz is probably accomplished by crushing and rearrangement of the spherical grains. Bridging of the particles persists even at 20,000 p.s.i. (Warner, 1964, p.63), as indicated by a void ratio of about 0.7 and a

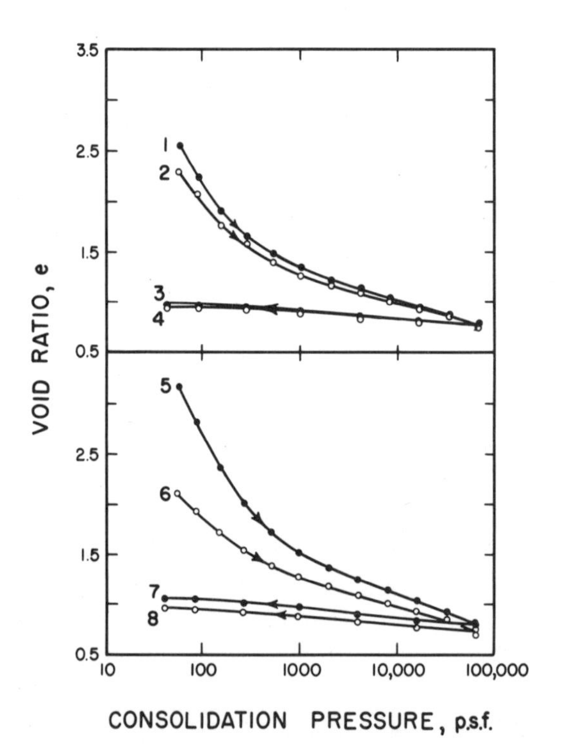

Fig.46. One-dimensional consolidation curves for kaolinite clay with wide variation in the electrolyte concentration of the pore water. *1* (compression) = 0.0001 *N* calcium; *2* (compression) = 1.0 *N* calcium; *3* (swelling) = 0.0001 *N* calcium; *4* (swelling) = 1.0 *N* calcium; *5* (compression) = 0.0001 *N* sodium; *6* (compression) = 1.0 *N* sodium; *7* (swelling) = 0.0001 *N* sodium; and *8* (swelling) = 1.0 *N* sodium. (After Olson and Mesri, 1970, fig.3, p.1868.)

porosity of about 40% at that pressure. At pressures in excess of 1,000 p.s.i. the consolidation curve for quartz powder lies above those of the clays.

On hydrating a sample of illite clay in NaCl-solution (5 g NaCl/100 cm^3 water) instead of distilled water, Chilingar and Knight (1960a) found that the moisture content at 10,000 p.s.i. was 16.6% as compared with 14.7% when hydrated with distilled water. Apparently the flocculated illite clay does not close-pack as much as deflocculated clay, which is contrary to the findings for montmorillonite clay. A sample of kaolinite clay hydrated in NaOH solution (0.007 *M* NaOH/g clay) had a moisture content of 17.1% at 200 p.s.i. as compared with 20.2% on using distilled water (Chilingar and Knight, 1960a). Tchillingarian (1952) had previously proved that OH^- ions adsorb on the surfaces of kaolinite clay particles. Possibly, the presence of OH^- ions may result in easier movement of clay particles with respect to each other, which causes closer packing. It should also be remembered that smaller amounts of NaOH deflocculate the clays, whereas greater concentrations result in flocculation.

Norrish and Quirk (1954) reported that in electrolyte solutions containing calcium, magnesium or aluminium, clays will not swell to interparticle distances greater than 9 Å,

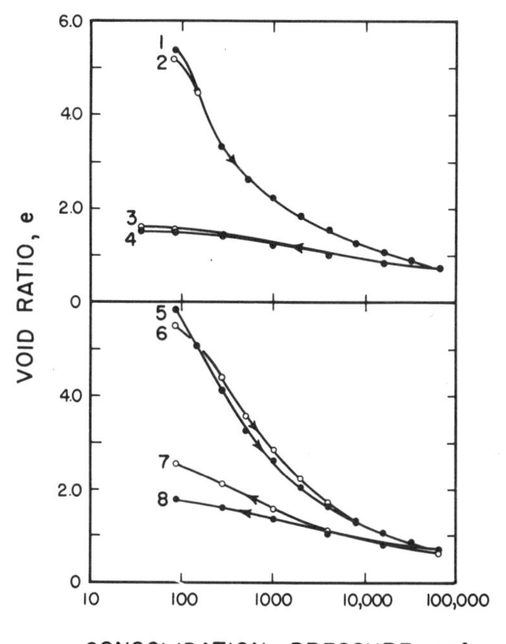

Fig.47. One-dimensional consolidation curves for illite clay with wide variation in the electrolyte concentration of the pore water. *1* (compression) = 1.0 *N* calcium; *2* (compression) = 0.001 *N* calcium; *3* (swelling) = 0.001 *N* calcium; *4* (swelling) = 1.0 *N* calcium; *5* (compression) = 1.0 *N* sodium; *6* (compression) = 0.001 *N* sodium; *7* (swelling) = 0.001 *N* sodium; and *8* (swelling) = 1.0 *N* sodium. (After Olson and Mesri, 1970, fig.5, p.1870.)

because of complete base exchange. This condition exists regardless of the concentration of the interstitial electrolyte solution.

Compaction studies of illite (API No.35) by Olson and Mitronovas (1962) at pressures below 430 p.s.i. showed that $MgCl_2$ and $CaCl_2$ create similar compression effects in the clay. Their experiments showed that the greatest resistance to compression occurred at a $MgCl_2$ concentration of 10^{-2} N and that there is a decrease in void ratio with decreasing concentration. The effects of the various electrolyte concentrations seem to disappear at around 430 p.s.i. This suggests a dominating effect of mechanical control of swelling.

Olson and Mesri (1970) conducted one-dimensional consolidation tests with wide variations in the electrolyte concentration in the pore water for kaolinite (Fig.46), illite (Fig.47), and smectite (Fig.48).

Variations in the electrolyte concentration had a negligible effect on the slopes of the swelling curves for kaolinite, indicating the applicability of Terzaghi's mechanical model (Fig.46). Variations in the positions of the virgin compression curves, particularly for sodium kaolinite, showed that double layers probably exerted an influence on the

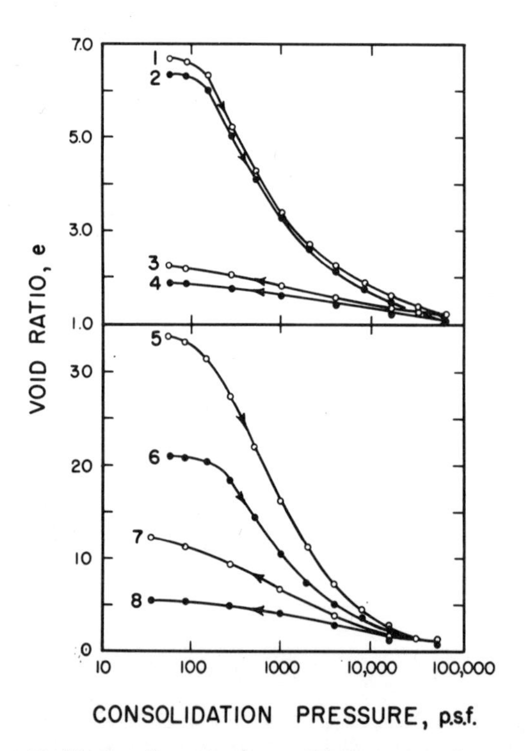

Fig.48. One-dimensional consolidation curves for smectite with wide variation in the electrolyte concentration of the pore water. *1* (compression) = 0.001 N calcium; *2* (compression) = 1.0 N calcium; *3* (swelling) = 0.001 N calcium; *4* (swelling) = 1.0 N calcium; *5* (compression) = 0.0005 N sodium; *6* (compression) = 0.1 N sodium; *7* (swelling) = 0.0005 N sodium; and *8* (swelling) = 0.1 N sodium. (After Olson and Mesri, 1970, fig.7, p.1872.)

properties of the original suspensions and affected the original geometric arrangement of the particles (Olson and Mesri, 1970, p.1866).

In the case of calcium-illite clay (Fig.47), the consolidation curves appear to be essentially independent of electrolyte concentration, suggesting a dominating mechanical control of swelling. In the case of sodium-illite, the concentration of electrolyte exerts only a minor effect on the position of the compression curve; whereas the position of the swelling curve is distinctly influenced, especially in the low values of effective stress range (Olson and Mesri, 1970, p.1868).

Also in the case of smectite* (Fig.48), the double layer effects seem to be much more significant when the ions are monovalent.

REFERENCES

Abramson, H.A., 1934. *Electrokinetic Phenomena and their Application to Biology and Medicine.* Chemical Catalog Co., New York, N.Y., 331 pp.

Athy, L.F., 1930. Density, porosity and compaction of sedimentary rocks. *Bull. Am. Assoc. Pet. Geologists,* 14(1): 1–24.

Avchyan, G.M. and Ozerskaya, M.L., 1968. Regularity in consolidation of sedimentary rocks with depth. *Izv. Akad. Nauk S.S.S.R., Ser. Geol.,* (2): 137–141.

Baldwin, B., 1971. Ways of deciphering compacted sediments. *J. Sediment. Petrol.,* 41(1): 293–301.

Bolt, G.H., 1956. Physico-chemical analysis of the compressibility of pure clays. *Geotechnique,* 6(2): 86–93.

Boswell, P.G.H., 1953. The natural moisture content, voids and compaction of unconsolidated clayey rocks. *Liverp. Manch. Geol. J.,* 1(3): 223–239.

Bridgman, P.W., 1958. *The Physics of High Pressure.* G. Bell and Sons, Ltd., 2nd ed., 445 pp.

Brown, P.R., 1969. Compaction of fine-grained terrigenous and carbonate sediments – a review. *Bull. Can. Pet. Geol.,* 17: 486–495.

Bush, D.C. and Jenkins, R.E., 1970. Proper hydration of clays for rock property determinations. *J. Pet. Technol.,* 22(6): 800–804.

Cebell, W.A. and Chilingarian, G.V., 1972. Some data on compressibility and density anomalies in halloysite, hectorite and illite clays. *Bull. Am. Assoc. Pet. Geologists,* 56(4): 796–802.

Chilingar, G.V. and Knight, L., 1960a. Relationship between pressure and moisture content of kaolinite, illite and montmorillonite clays. *Bull. Am. Assoc. Pet. Geologists,* 44(1): 101–106.

Chilingar, G.V. and Knight, L., 1960b. Relationship between overburden pressure and moisture content of silicic acid and gum Ghatti. *Proc. Int. Geol. Congr., 21st Sess., Norden,* Part 18: 384–388.

Chilingar, G.V., Rieke III, H.H. and Robertson Jr., J.O., 1963. Degree of hydration of clays. *Sedimentology,* 2(4): 341–342.

Chilingarian, G.V. and Rieke III, H.H., 1968. Data on consolidation of fine-grained sediments. *J. Sediment. Petrol.,* 38(3): 811–816.

Coogan, A.H., 1970. Measurements of compaction in oolitic grainstone. *J. Sediment. Petrol.,* 40(3): 921–929.

Dallmus, K.F., 1958. Mechanics of basin evolution and its relation to the habitat of oil in the basin. In: L.G. Weeks (Editor), *Habitat of Oil. Am. Assoc. Pet. Geologists, Mem.,* 36: 2071–2124.

Dana, S.W., 1967. Investigation of lateral and vertical variation of density within sedimentary rock. *Compass,* 44(4): 172–178.

---

* Montmorillonite.

Dickinson, G., 1953. Reservoir pressures in Gulf Coast Louisiana. *Bull. Am. Assoc. Pet. Geologists,* 37(2): 410–432.

Dobrynin, V.M., 1962. Effect of overburden pressure on some properties of sandstones. *Trans. Am. Inst. Min. Metall. Eng.,* 225: 360–366.

Eaton, B.A., 1969. Fracture gradient prediction and its application in oil-field operations. *J. Pet. Technol.,* 21(10): 1353–1360.

Ebhardt, G., 1968. Experimental compaction of carbonate sediments. In: G. Müller and G.M. Friedman (Editors), *Recent Developments in Carbonate Sedimentology in Central Europe.* Springer, Heidelberg, pp.58–65.

Eremenko, N.A. (Editor), 1960. *Petroleum Geology* (Ref. Book). Gostoptekhizdat, Moscow, 1: 592 pp.

Eremenko, N.A. and Neruchev, S.G., 1968. Primary migration during process of burial and lithogenesis of sediments. *Geol. Nefti i Gaza,* 1968(9): 5–8.

Foster, J.B. and Whalen, H.E., 1966. Estimation of formation pressures from electrical surveys – offshore Louisiana. *J. Pet. Technol.,* 18(2): 165–171.

Fruth Jr., L.S., Orme, G.R. and Donath, F.A., 1966. Experimental compaction effects in carbonate sediments. *J. Sediment. Petrol.,* 36(3): 747–754.

Füchtbauer, H. and Reineck, H.E., 1963. Porosität und Verdichtung rezenter, mariner Sedimente. *Sedimentology,* 2(4): 294–306.

Ham, H.H., 1966. New charts help estimate formation pressures. *Oil Gas J.,* 64(51): 58–63.

Hamilton, E.L., 1969a. *Sound Velocity, Elasticity, and Related Properties of Marine Sediments, North Pacific, I: Sediment Properties, Environmental Control, and Empirical Relationships.* Naval Undersea Res. and Dev. Center Tech. Publ. 143, San Diego, Calif., 58 pp.

Hamilton, E.L., 1969b. *Sound Velocity, Elasticity, and Related Properties of Marine Sediments, North Pacific, II: Elasticity and Elastic Constants.* Naval Undersea Res. and Dev. Center Tech. Publ. 144, San Diego, Calif., 64 pp.

Hamilton, E.L., 1969c. *Sound Velocity, Elasticity, and Related Properties of Marine Sediments, North Pacific, III: Prediction of in situ Properties.* Naval Undersea Res. and Dev. Center Tech. Publ. 145, San Diego, Calif., 79 pp.

Hammer, S., 1950. Density determinations by underground gravity measurements. *Geophysics,* 15(4): 637–652.

Hedberg, H.D., 1936. Gravitational compaction of clays and shales. *Am. J. Sci.,* 5th Ser., 231(184): 241–287.

Hosoi, H., 1963. First migration of petroleum in Akita and Yamagata Prefectures. *Jap. Assoc. Mineralogists, Petrologists, Econ. Geologists J.,* 49(2): 43–55; 49(3): 101–114.

Igelman, K.R. and Hamilton, E.L., 1963. Bulk densities of mineral grains from Mohole samples (Guadalupe Site). *J. Sediment. Petrol.,* 33(2): 474–477.

Iler, R.K., 1955. *The Colloid Chemistry of Silica and Silicates.* Cornell University Press, Ithaca, N.Y., 324 pp.

Johnson, F.W., 1950. Shale density analysis. In: L.W. LeRoy (Editor), *Subsurface Geologic Methods,* Colorado School of Mines, Golden, Colo., 2nd ed., 1166 pp.

Karpov, P.A., 1964. Some regularities in changes in porosity of terrigenous formations depending on depth of burial (on example of Devonian deposits of Volgagrad region). *Litol. y Paleznye Iskop.,* 1964(5): 118–121.

Kartsev, A.A., Vagin, S.B. and Baskov, E.A., 1969. *Paleohydrogeology.* Nedra, Moscow, 150 pp.

Kerr, P.F. and Barrington, J., 1961. Clays of deep shale zone, Caillou Island, Louisiana. *Bull. Am. Assoc. Pet. Geologists, 45(10): 1697–1712.*

Klubova, T.T., 1965. *Role of Clayey Minerals in Transformation of Organic Matter and Formation of Pore Spaces of Reservoirs.* Izd. Akad. Nauk S.S.S.R., Moscow, 107 pp.

Komarova, N.A. and Knyazeva, N.V., 1968. Extrusion of soil solutions by centrifuge method. In: G.V. Bogomolov et al. (Editors), *Pore Solutions and Methods of Their Study.* Nauka i Tekhnika, Minsk, pp.205–214.

Koperina, V.V. and Dvoretskaya, O.A., 1965. Density and porosity of clayey rocks. In: *Postsedimentation Changes of Quaternary and Pliocene Clayey Deposits of Bakinskiy Archipelago. Tr. Geol. Inst., Akad. Nauk S.S.S.R.* Nauka, Moscow, 1965(115): 115–123.

Kruyt, H.R. (Editor), 1949. *Reversible Systems. Colloid Science,* II. Elsevier, Amsterdam, 733 pp.

Kruyt, H.R. (Editor), 1952. *Irreversible Systems. Colloid Science,* I. Elsevier, Amsterdam, 389 pp.

Kryukov, P.A. and Komarova, N.A., 1954. Concerning squeezing out of water from clays at very high pressures. *Dokl. Akad. Nauk S.S.S.R.*, 99(4): 617–619.
Lapinskaya, T.A. and Proshlyakov, B.K., 1970. Problem of reservoir rocks during exploration for oil and gas at great depths. In: A.V. Sidorenko et al. (Editors), *Status and Problems of Soviet Lithology*, III. Nauka, Moscow, pp.115–121.
Laughton, A.S., 1957. Sound propagation in compacted ocean sediments. *Geophysics,* 22: 233–260.
Linetskiy, V.F., 1965. *Migration of Petroleum and Formation of Its Deposits.* Naukova Dumka, Kiev, 200 pp.
Lomtadze, V.D., 1953. Changes in moisture content of clays on compaction at high loading. *Zap. Leningrad. Gor. Inst.*, 29(2): 103–123.
Lomtadze, V.D., 1954. About role of compaction processes of clayey deposits in the formation of underground waters. *Dokl. Akad. Nauk S.S.S.R.*, 98(3): 451–454.
Lomtadze, V.D., 1955. Stages in the formation of properties of clayey rocks during their lithification. *Dokl. Akad. Nauk S.S.S.R.*, 102(4): 819–822.
Lopotko, M.Z., Mironov, A.M. and Puntus, F.A., 1968. Extruded pore waters of sapropels and their characteristics. In: Bogomolov et al. (Editors), *Pore Solutions and Methods of Their Study.* Nauka i Tekhnika, Minsk, pp.87–94.
Magara, K., 1968. Compaction and migration of fluids in Miocene mudstone, Nagaoka Plain, Japan. *Bull. Am. Assoc. Pet. Geologists,* 52(12): 2466–2501.
Meade, R.H., 1964. Removal of water and rearrangement of particles during the compaction of clayey sediments –review. *U.S. Geol. Surv., Prof. Pap.*, 497B: 23 pp.
Meade, R.H., 1966. Factors influencing the early stages of compaction of clays and sands – review. *J. Sediment. Petrol.*, 36: 1085–1101.
Meade, R.H., 1968. Compaction of sediments underlying areas of land subsidence in central California. *U.S. Geol. Surv., Prof. Pap.* 497D: 39 pp.
McCrossan, R.G., 1961. Resistivity mapping and petrophysical study of upper Devonian inter-reef calcareous shales of central Alberta, Canada. *Bull. Am. Assoc. Pet. Geologists,* 45(4): 441–470.
McCulloh, T.H., 1965. A confirmation by gravity measurements of an underground density profile based on core densities. *Geophysics,* 30(6): 1108–1132.
McCulloh, T.H., 1967. Mass properties of sedimentary rocks and gravimetric effects of petroleum and natural gas reservoirs. *U.S. Geol. Surv., Prof. Pap.*, 528A: 50 pp.
Morgan, N.A., 1969. Physical properties of marine sediments as related to seismic velocities. *Geophysics,* 34(4): 529–545.
Mukhin, U.V., 1968. Evaluation of the amount of pore solutions squeezed out from clayey sediments in natural conditions. In: G.V. Bogomolov et al. (Editors), *Pore Solutions and Methods of Their Study.* Nauka i Tekhnika, Minsk, pp.45–54.
Müller, G., 1967. Diagenesis in argillaceous sediments, In: G. Larsen and G.V. Chilingar (Editors), *Diagenesis in Sediments.* Elsevier, Amsterdam, pp.127–177.
Nafe, J.E. and Drake, C.L., 1957. Variation with depth in shallow and deep water marine sediments of porosity, density and the velocities of compressional and shear waves. *Geophysics,* 22(3): 523–552.
Norrish, K. and Quirk, J.P., 1954. Crystalline swelling of montmorillonite. *Nature,* 173: 255–257.
Olson, R.E. and Mitronovas, F., 1962. Shear strength and consolidation characteristics of calcium and magnesium illite. *Clays Clay Miner. Proc. Natl. Conf. Clays Clay Miner. 9th, 1960,* pp.185–209.
Olson, R.E. and Mesri, G., 1970. Mechanisms controlling compressibility of clays. *J. Soil Mech. Found. Div.*, 96(SM6): 1863–1878.
Powers, M.C., 1967. Fluid-release mechanisms in compacting marine mud rocks and their importance in oil exploration. *Bull. Am. Assoc. Pet. Geologists,* 51(7): 1240–1254.
Preiss, K., 1968. In situ measurement of marine sediment density by gamma radiation. *Deep-Sea Res.,* 15: 637–641.
Proshlyakov, B.K., 1960. Reservoir properties of rocks as a function of their depth and lithology. *Geol. Nefti i Gaza,* 1960(12): 24–29.
Prozorovich, E.A., 1962. About regularities in compaction of clays with depth (illustrated by Sarmatskiy Formation of Pre-Caucasus and Azerbayjan). *Nov. Neft. i Gaz. Tekh. (Geol.),* 1962(4): 62–64.
Prozorovich, E.A. and Sultanov, A.D., 1961. Density of clayey formations of some areas of Azerbayjan. *Dokl. Akad. Nauk Azerb. S.S.S.R.*, 17(4): 293–298.

Rieke III, H.H., 1970. *Compaction of Argillaceous Sediments* (20–500,000 p.s.i.). Thesis Univ. Southern California, Los Angeles, Calif., 682 pp.

Rieke III, H.H., Chilingar, G.V. and Robertson Jr., J.O., 1964. *Int. Geol. Congr., 22nd Sess., New Delhi*, Part 15: 22–38.

Robertson, E.C., 1967. Laboratory consolidation of carbonate sediments. In: A.F. Richards (Editor), *Marine Geotechnique*. University of Illinois Press, Urbana, Ill., 326 pp.

Robertson Jr., J.O., Rieke III, H.H. and Chilingar, G.V., 1965. Viscosity measurements of aqueous clay suspensions as a tool for determining mineralogic types of clays. *Sedimentology,* 4(3): 181–182.

Sergeev, E.M., 1970. Lithology and engineering geology. In: A.V. Sidorenko et al. (Editors), *Status and Problems of Soviet Lithology,* I. Nauka, Moscow, pp.189–199.

Shishkina, O.V., 1968. Methods of investigating marine and ocean mud waters. In: G.V. Bogomolov et al. (Editors), *Pore Solutions and Methods of Their Study*. Nauka i Tekhnika, Minsk, pp.167–176.

Skeels, D.C., 1943. *The Density of Shales as a Function of Depth, a Compilation of Data.* Standard Oil Co. (N.J.), Geophys. Res. Bull., no.2: 14 pp., unpubl.

Skempton, A.W., 1945. Notes on the compressibility of clays. *Q. J. Geol. Soc. Lond.*, 100: 119–135.

Stetyukha, E.I., 1964. *Formulas of Correlations Between Physical Properties of Rocks and Depth of Burial.* Nedra, Moscow.

Storer, D., 1959. Costipamento dei sedimenti argillosi nel Bacino Padano. In: *I Giacementi Gassiferi dell'Europa Occidentale,* 2. Acad. Nazl. dei Lincei, Roma, pp.519–536.

Strakhov, N.M., 1954. Diagenesis of sediments and its significance for sedimentary ore formation. *Izv. Akad. Nauk S.S.S.R., Ser. Geol.,* 5: 12–49.

Tchillingarian, G., 1952. Possible utilization of electrophoretic phenomenon for separation of fine sediments into grades. *J. Sediment. Petrol.,* 22(1): 29–32.

Teodorovich, G.I. and Chernov, A.A., 1968. Character of changes with depth in productive deposits of Apsheron oil-gas-bearing region. *Sov. Geol.,* 1968(4): 83–93.

Terzaghi, K., 1925a. *Erdbaumeckanik auf bodenphysikalischer Grundlage.* Deuticke, Leipzig, 399 pp.

Terzaghi, K., 1925b. Principles of soil mechanics II, compressive strength of clays. *Eng. News Rec.,* 95: 796–800.

Terzaghi, R.D., 1940. Compaction of lime mud as a cause of secondary structure. *J. Sediment. Petrol.,* 10(1): 78–90.

Tkhostov, B.A., Vezirova, A.D., Vendel'shteyn, B.Yu and Dobrynin, V.M., 1970. In: M.F. Mirchink (Editor), *Petroleum in Fractured Reservoirs.* Nedra, Leningrad, 220 pp.

Van Olphen, H., 1954. Interlayer forces in bentonite. *Proc. Natl. Conf. Clays Clay Miner., 2nd, 1953 – Natl. Res. Counc. Publ.,* 327: 418–438.

Van Olphen, H., 1963a. *An Introduction to Clay Colloid Chemistry.* Wiley, New York, N.Y., 301 pp.

Van Olphen, H., 1963b. Compaction of clay sediments in the range of molecular particle distances. In: W.F. Bradley (Editor), *Clays and Clay Minerals,* II: 178–187.

Vassoevich, N.B., 1955. About origin of oil. In: *Geologic Symposium, Questions on Origin of Oil.* Gostoptekhizdat, Leningrad, 1955(83): 9–98.

Vassoevich, N.B., 1958a. About the origin of oil. Materials on petroleum geology. *Int. Geol. Congr. 20th Sess., Rep. Sov. Geologists,* Moscow, 1: 113–127.

Vassoevich, N.B., 1958b. Probleme der Erdölgenese. *Z. Angew. Geol.,* 4: 512–515.

Vassoevich, N.B., 1960. Experiment in constructing typical gravitational compaction curve of clayey sediments. *Nov. Neft. Tekh., Geol. Ser. (News Pet. Tech., Geol.),* 1960(4): 11–15.

Von Engelhardt, W., 1960. *Der Porenraum der Sedimente.* Springer, Berlin, 207 pp.

Von Engelhardt, W. and Gaida, K.H., 1963. Concentration changes of pore solutions during the compaction of clay sediments. *J. Sediment. Petrol.,* 33(4): 919–930.

Warner, D.L., 1964. *An Analysis of the Influence of Physical-Chemical Factors Upon the Consolidation of Fine-Grained Elastic Sediments.* Thesis, Univ. California, Berkeley, Calif., 136 pp.

Weller, J.M., 1959. Compaction of sediments. *Bull. Am. Assoc. Pet. Geologists.,* 43(2): 273–310.

Woollard, G.P., 1962. Geologic effects in gravity values. In: *The Relation of Gravity Anomalies to Surface Elevation, Crustal Structure, and Geology*, pt.3. Wisconsin Univ. Geophys. Polar Res. Center, Res. Rep., Ser. 62–9: 1–78.

Chapter 3

MECHANICS OF COMPACTION AND COMPACTION MODELS

INTRODUCTION

Until recently, sedimentologists did not pay much attention to the stresses in thick sedimentary sequences. Knowledge of the vertical and lateral stress patterns in a depositional basin is of the utmost importance, and a detailed investigation of the stress state in a basin may aid in anticipating the location of structures favorable for hydrocarbon entrapment. The history and development of depositional basins and the expulsion of interstitial fluids during basin subsidence should be better understood. For a more thorough quantitative understanding of compaction mechanics, the relationship between the total overburden stress, effective stress, and pore stress (pressure) in fine-grained clastics is analyzed in this chapter.

A study of the published literature indicates that there is a general lack of a totally acceptable model for compaction of very fine-grained clastic sediments. There are several apparently conflicting theories in the technical literature as to the genesis of shales, manner in which these rocks are compacted, and the mechanism by which the pore fluid is released during burial. Skempton (1953), for example, believes that overburden load alone is not an effective agent in dewatering deeply buried shales. New concepts have been slow in coming forth following the early work of Athy (1930) and Hedberg (1936). Several of the new concepts and compaction models are described here.

THEORY OF CONSOLIDATION

The theory of consolidation of a saturated clay described by Terzaghi and Peck (1948) is well established in soil mechanics; however, it appears that geologic application of this theory to compaction of fine-grained clastic sediments was first elucidated by Hubbert and Rubey (1959). In nature, the overburden load is divided between the sediment matrix and the interstitial fluid of the pores, so that the total vertical stress at any point consists of the sum of two components: the intergranular stress and the pore-fluid stress. The term "effective pressure" is used to designate the difference between the total overburden pressure and the pore pressure ($p_e = p_t - p_p$). If the vertical permeability of a sediment allows the pore fluid to move out, then the pressure distribution in pore fluids is the same as that of a continuous column of ground water extending to the water-table surface (hydrostatic pressure). The mineral grains contained in the vertical column of water contribute a weight equal to their weight minus the weight of water displaced by the grains (buoyant force). Whether or not buoyant force is effective in clay sediments having very low permeabilities is debatable, however. Inasmuch as the mineral grains support a load equal to the weight of all the overlying water and grains minus the weight transferred to the water by the grains (buoyant force), then the force responsible for compaction is equal to:

$$F_e = F_t - F_p \tag{3-1}$$

where F_e = the effective grain-to-grain force on the horizontal surface A, F_t= total overburden load, and F_p = buoyant force.

The total overburden force, F_t, is equal to:

$$F_t = F_o + W_s + W_f \tag{3-2}$$

where F_o = outside (external) force exerted on the body of sediment under consideration; W_s = weight of solids [$= \gamma_s(1 - \phi)V_b$] ;γ_s is the specific weight of solids, ϕ is fractional volumetric porosity, and V_b is the volume of the sediment under study); and W_f = weight of interstitial fluids (= $\gamma_f \phi V_b$; γ_f is the specific weight of interstitial fluids).

If the buoyant force F_p is equal to the weight of the fluid displaced by the grains [= $\gamma_f V_b(1 - \phi)$] ; and inasmuch as $V_b = A \times D$ (where A is total cross-sectional surface area and D is depth; Z is also used in this book to designate depth) and $\gamma_f \times D$ is equal to pore pressure p_p (fundamental equation of fluid statics, see p.4,5), then $F_p = p_p \times A(1 - \phi)$.

The above expression derived by the writers for the buoyant force is in close agreement with Terzaghi's (1936) view that the force of uplift due to pore pressure is proportional to surface porosity (see Laubscher, 1960). Surface or boundary porosity is the ratio of the pore area to the gross area, along the surface A; it can be also shown that surface porosity on a plane surface is the same as volumetric porosity. In their classic paper, however, Hubbert and Rubey (1959; also see 1960) showed to the satisfaction of the writers that the pore pressure p_p is common to both the water and the clay; and acts over the whole of any surface passed through the porous solid, with the surface porosity being in no way involved. This conclusion, based on the elaborate mathematical analysis by Hubbert and Rubey (1959, 1960), is supported by the experimental results obtained by the writers for sands (see p.5, 6). On assuming that all pores are filled with water, at a depth D (in ft.), the total overburden pressure, p_t (in lb./sq. ft.), resulting from the weight of overlying water and solids can be expressed by the following equation:

$$p_t = [\gamma_s \times (1 - \phi) + \gamma_w \times \phi] D \tag{3-3}$$

where γ_s = specific weight of grains (in lb./cu. ft.); ϕ = porosity, fractional; and γ_w = specific weight of water (in lb./cu. ft.). Inasmuch as the effective pressure, p_e (grain-to-grain stress) is equal to the difference between the total overburden pressure and the pore pressure $[p_e = p_t - p_p]$, and pore pressure at a depth D is equal to $\gamma_w \times D$, then:

$$p_e = [\gamma_s \times (1-\phi) + \gamma_w \phi - \gamma_w] D \tag{3-4}$$

or:

$$p_e = D[(1-\phi)(\gamma_s - \gamma_w)]$$

This equation is considered valid by the writers. Brandt (1955), however, introduced an 85% correction factor (n) into p_p (= $\gamma_w D$) term to take into account the "fact that the internal fluid pressure does not wholly react against the external pressure". His factor, n, is structure dependent and, therefore, is not the same for all sediments; thus, $p_e = p_t - n\,p_p$. Brandt's conclusion, however, is subject to debate as discussed previously. Hamilton (1959) stated that "in normal situations the pressure induced at the level in a rock section does not include the weight of the overlying water as effective pressure on the mineral grains which would cause consolidation".

The writers of the present book were not able to measure the pore pressure in compacting clays at pressures above 100 p.s.i. on using the sophisticated compaction apparatus of Sawabini et al. (1971) (see Chapter 8, p.383), which measures the pore pressures in sands. An important question is: Does pore pressure, as it is understood to exist in sandstones, really exist in shales? Hopefully, the future research work will answer this question.

Accumulation of additional sediments upon older sediments will cause a gradual change in the vertical stress throughout the sediment column. The matrix pressure is redistributed by the grains squeezing closer together so that they bear more load. Many authors (Hubbert and Rubey, 1959; Hottman and Johnson, 1965; Powers, 1967; and others) stated that in thick shale sequences having low permeability, compaction is a slow process and the additional load must be supported by the fluid. This creates abnormally high pore-fluid pressures, which must be balanced by a corresponding decrease in the shale matrix pressure, because the total weight of overlying rock and water to be supported is practically the same.

The above-described concept of the shale-compaction process can be best explained by a mechanical model which is composed of a perforated, round metal plate and the enclosing cylinder which contains a metal spring and water (Fig.49). In this analogy the spring represents the compressible clay particles, the water represents the fluid in the pore space, and the size of the perforations in the metal plate determines the permeability. Here a well-saturated clay can be thought of, and treated mathematically, as a two-phase continuum. The hydrated clay can be envisioned as clean clay plates in mechanical contact with each other with the fluid wetting the clay-particle surfaces and filling the

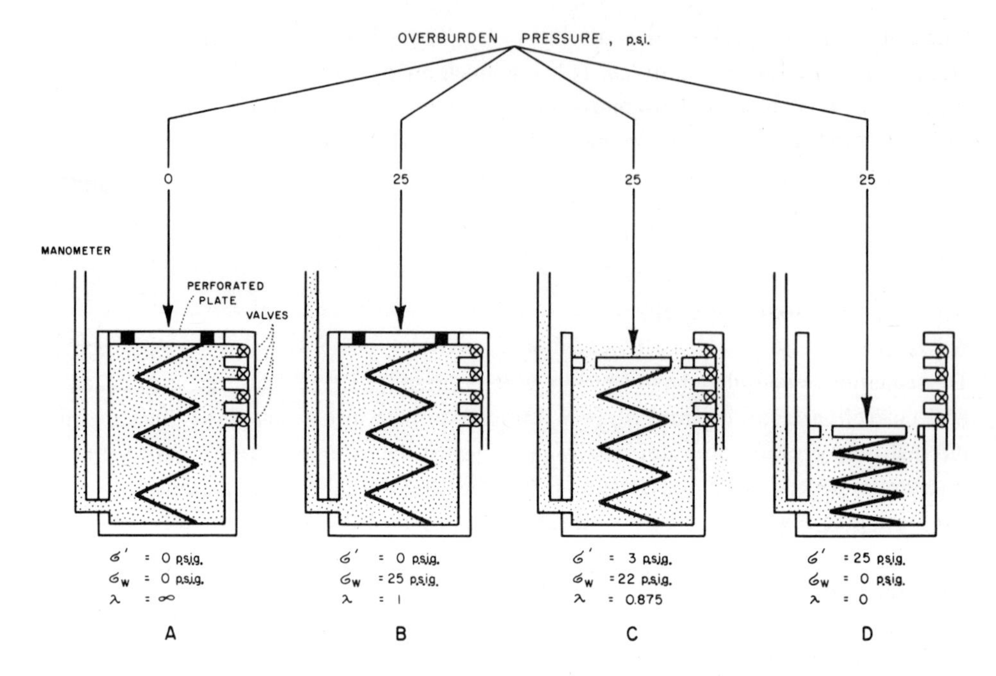

Fig.49. Compaction analogy using a spring and perforated plate. σ' is the effective (intergranular) stress, σ_W is the pore-water stress, and λ is the ratio of the pore-water stress to the overburden stress on the system. (σ' and σ_W are in p.s.i.g., i.e., in pounds per square inch, gauge.)
A. Tightly fitted, frictionless metal plate seals the water in the cylinder. There is no overburden load on the system and perforations in the plate are sealed.
B. A 25-p.s.i. load imposed on the system is entirely carried by the water. Perforations in the plate are closed.
C. Fluid is allowed to flow out through the perforation in response to the overburden load of 25 p.s.i. The plate descends as the fluid escapes. Now the load is distributed partly to the spring and partly to the water.
D. The overburden load is now carried entirely by the spring, and the system is in equilibrium, with no water outflow.

pore space between the particles. If the mechanical model were sealed in such a manner that no fluid could escape through the plate, then the total applied pressure to the system would be carried by the fluid and none by the spring (Fig.49B). The compressibility of the spring is assumed to be so great that the strains produced in the fluid and in the cylinder walls are negligible in comparison (Taylor, 1948, p.223). Fig.49C shows that if the fluid is allowed to escape through the perforations, then the overburden pressure is carried both by the spring and the fluid. As the fluid escapes, the plate sinks lower and lower, compressing the metal spring. The length of time required for the spring to pass from one state of compaction to the next depends on how rapidly the water escapes; this is determined by the size of the perforations in the plate. Equilibrium is reached at a point where none of the overburden stress is borne by the fluid (Fig.49D); however, any additional applied loads cause the plate to compact the spring still further, expelling additional fluid. In this manner the clay layers are thought to be compacted under the weight of the overlying sediments.

In the spring analogy of compaction the following relationship must exist at any particular time (static equilibrium):

$$F_t = F_s + F_w \tag{3-5}$$

where F_t is the total force applied to the system, F_s is the force carried by the spring, and F_w is the force applied to the fluid. If these forces are divided by the total cross-sectional area, A, of the enclosing cylinder, then:

$$\begin{aligned} p_t \text{ or } \sigma &= F_t/A \\ p_e \text{ or } \sigma' &= F_s/A \\ p_p \text{ or } \sigma_w &= F_w/A \end{aligned} \tag{3-6}$$

where p_t or σ is the total stress applied to the system, p_e or σ' is the effective stress, and p_p or σ_w is the pore-water pressure. Thus, eq.3-5 can be written as:

$$\sigma = \sigma' + \sigma_w \tag{3-7}$$

According to the latter equation, the total stress σ, normal to any plane in the clay mass consists of two components: (1) the pore fluid pressure, σ_w, and (2) the effective stress component, σ' which is "effectively" carried by the structure of the clay particles.

The spring analogy fails to agree with actual compaction of a clay in that the pressure conditions are not the same throughout the thickness of the clay mass as they are in the cylinder. In compacting a saturated clay at a given pressure, the water pressure at its surface is atmospheric (zero gage) whereas at short distances inside the clay sample the water pressure is equal to σ-σ'. Fig.50 illustrates a void space surrounded by a shale matrix. In this figure, the total weight of the overburden, which acts downward, and the vertical and horizontal portions of the effective stress are shown. High fluid-pressure gradient at the clay's surface is caused by the rapid expulsion of the fluid from the pores near the surface. Under a constant overburden pressure, the water pressure decreases with time, whereas the intergranular pressure increases.

Hottman and Johnson (1965) stated that a useful expression is the ratio of the fluid stress to the total stress, λ, as presented in Figs.49 and 51.

$$\lambda = \sigma_w/\sigma = p_p/p_t \tag{3-8}$$

Initially, when stress is applied to the closed system, λ has a value of 1 and the system is overpressured (see Chapter 7). At final compaction equilibrium, when the load is carried

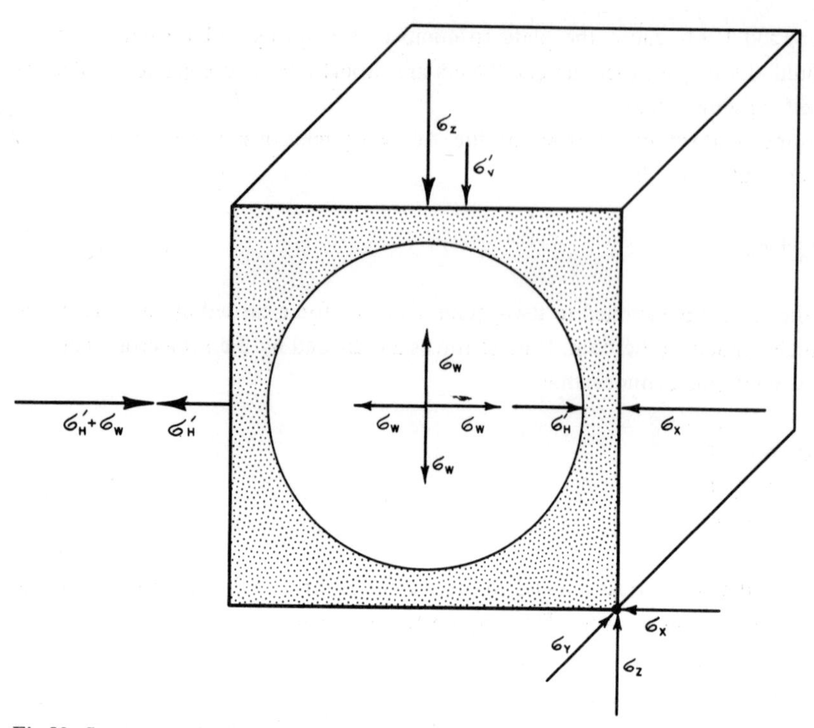

Fig.50. Stress state in shale. Diagrammatic sketch of the stress state in a shale body underground, where σ'_V is the effective (intergranular) stress in the vertical direction, σ'_H is the horizontal effective stress, σ_W is the pore water stress, and σ_Z is the total vertical stress component. The total horizontal stress component in the x-direction σ_X is equal to $\sigma'_H + \sigma_W$.

entirely by the grains (spring), λ is equal to zero. Hottman and Johnson (1965) stated that at the final compaction equilibrium, when the applied load is supported jointly by the grains (spring) and the water pressure, which is simply hydrostatic, the value of λ is equal to approximately 0.465 (Fig.51). The normal pressure gradient on the U.S. Gulf Coast is approximately 0.465 p.s.i./ft. of depth, whereas overburden load gradient is commonly considered to be equal to 1.0 p.s.i./ft. of depth. The hydrostatic pressure gradient depends on the specific weight of water, i.e., salinity.

A similar ratio of the "effective" intergranular stress to the total stress can also be used, which may be denoted by the symbol χ:

$$\chi = \sigma'/\sigma \qquad (3\text{-}9)$$

STATE OF STRESS IN COMPACTING SHALES

The foregoing discussion leads to the problem of the effect of overburden load on the elastic deformation of argillaceous sediments. A basic assumption can be made that the

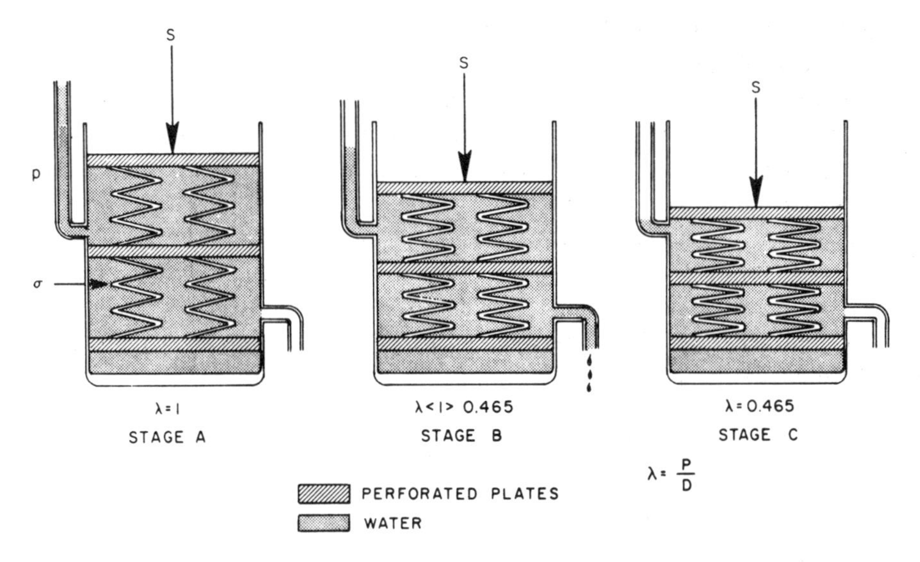

Fig.51. Schematic representation of clay compaction. σ = grain-to-grain bearing strength, S = axial component of total stress (overburden pressure), and p = fluid pressure. $\sigma = S - p$. Stage A – overpressured system. Stage B – water is allowed to escape; springs carry part of the applied load. Stage C – compaction equilibrium; load is supported jointly by the springs and the water pressure, which is simply hydrostatic. (After Terzaghi and Peck, 1948; in: Hottman and Johnson, 1965, p.718.)

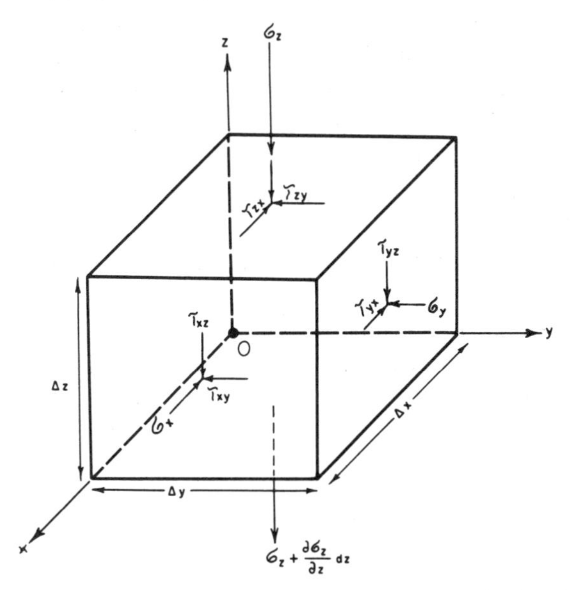

Fig.52. Stress notation in a cubic argillaceous rock slice. Stress notation of the normal component of stress, σ_z, on the plane normal to the z-axis; τ_{zx}, and τ_{zy} refer to the shear stress components in the plane normal to the z-axis and acting in the x- and y-directions, respectively. Point O is in equilibrium with respect to the forces in the x- and y-directions. $\sigma_z + (\delta\sigma_z/\delta z)\mathrm{d}z$ is the incremental change in the vertical stress through the free body.

overburden weight per unit area is equal to the vertical normal component of the stress. There is the possibility, however, that the normal stress at a point in a shale body undergoing compaction at some depth is equal to the overburden weight per unit area plus contributions from the vertical shear components of stress (τ). A total stress field in such a sedimentary body can be specified in terms of its normal and tangential stress components across a given plane surface (Fig.52).

In Fig.52 a slice of an argillaceous sedimentary body that has differential thickness, width, and length is presented. In this elemental volume, as the edge dimensions approach zero, the components of the resultant couple on any cube face become zero. As the edge dimensions approach zero, the stress components, σ and τ, approach a finite value (Rogers, 1964, p.25):

$$F_x = \{\sigma_x \quad \tau_{xy} \quad \tau_{xz}\}\, \Delta y\, \Delta z \tag{3-10}$$

$$F_y = \{\tau_{yx} \quad \sigma_y \quad \tau_{yz}\}\, \Delta x\, \Delta z \tag{3-11}$$

$$F_z = \{\tau_{zx} \quad \tau_{zy} \quad \sigma_z\}\, \Delta x\, \Delta y \tag{3-12}$$

where F_x, F_y, and F_z are the forces in the x-, y-, and z- directions. It should be noted that the pressure (load per unit area) has the dimensions of stress (pounds per square foot, p.s.f.; or pounds per square inch, p.s.i.). Body forces are measured in units of force per unit volume, whereas surface forces are measured in units of force per unit area. Examples of these are specific weight and pressure, respectively. In the above eq.3-10, 3-11, and 3-12, in the case of normal stress, the subscript refers to the direction (axis) normal to the plane on which the stress acts. For the shear stresses, the first subscript denotes the axis perpendicular to the plane in which the stress acts, whereas the second subscript denotes the direction in which the stress acts. It is important to note here that one has to be consistent in considering either (1) all forces acting on the system or (2) all forces acting outward from the system.

If F_t is the normal component of the total force exerted on the element, and F_{tt} is the tangential component of the force, then for any change in F_t or F_{tt}, owing to additional overburden weight, there will be a corresponding change in the shear and tangential stresses:

$$\Delta\sigma = \Delta F_t / A \tag{3-13}$$

$$\Delta\tau = \Delta F_{tt} / A \tag{3-14}$$

Resolution of the total stress field

The stress tensor for a porous, homogeneous, isotropic shale body can be written in the conventional way:

$$S \equiv \begin{vmatrix} \sigma_x & \tau_{xy} & \tau_{xz} \\ \tau_{yx} & \sigma_y & \tau_{yz} \\ \tau_{zx} & \tau_{zy} & \sigma_z \end{vmatrix} \tag{3-15}$$

where S signifies the symmetrical tensor of the total stress; σ_i and τ_{ij} represent the normal and shear forces, respectively, acting on the faces of a unit of an argillaceous sediment.

Next, one can take moments about point O (Fig.52). The tangential stress, τ_{xy}, multiplied by the area in which it acts, gives the force $\tau_{xy}\mathrm{d}z\mathrm{d}y$, and this times $\mathrm{d}x$ gives a clockwise moment about O. The stress τ_{yx} times the area gives $\tau_{yx}\mathrm{d}x\mathrm{d}z$, and the latter times $\mathrm{d}y$ results in a counterclockwise moment $\tau_{yx}\mathrm{d}x\mathrm{d}z\mathrm{d}y$. At equilibrium, the two moments balance each other:

$$\tau_{xy}\mathrm{d}z\,\mathrm{d}y\,\mathrm{d}x = \tau_{yx}\mathrm{d}x\,\mathrm{d}z\,\mathrm{d}y \tag{3-16}$$

or:

$$\tau_{xy} = \tau_{yx} \tag{3-17}$$

Then it follows that:

$$\tau_{xz} = \tau_{zx} \tag{3-18}$$

and:

$$\tau_{yz} = \tau_{zy} \tag{3-19}$$

The total stress array for a point in a cylindrical body under compaction, such as the one studied in the laboratory by numerous investigators (see Chapter 8), can be expressed in cylindrical coordinates r, θ, and z:

$$S \equiv \begin{vmatrix} \sigma_r & \tau_{r\theta} & \tau_{rz} \\ \tau_{\theta r} & \sigma_\theta & \tau_{\theta z} \\ \tau_{zr} & \tau_{z\theta} & \sigma_z \end{vmatrix} \tag{3-20}$$

The total stress tensor can be decomposed into two distinct parts for a body in equilibrium: (1) hydrostatic stress and (2) deviatoric stress.

Hydrostatic stress state. The component attributable to the interstitial fluid is the hydrostatic stress (pressure), σ_w, which can be regarded as being continuous throughout the medium. The normal and shear stress components are given by:

$$P \equiv \begin{vmatrix} \sigma_{wx} & \tau_{wxy} & \tau_{wxz} \\ \tau_{wyx} & \sigma_{wy} & \tau_{wyz} \\ \tau_{wzx} & \tau_{wzy} & \sigma_{wz} \end{vmatrix} \tag{3-21}$$

where P is the hydrostatic tensor. It can be safely assumed that under hydrostatic conditions no shearing stresses exist in the interstitial fluid. As a matter of fact, by definition a fluid is a substance which cannot sustain tangential or shear forces when in static equilibrium. This may not hold true for the adsorbed water because of its probable quasi-crystalline nature. Hubbert and Rubey (1959, p.138) stated that if a viscous liquid occupies the pore space, there are then microscopic shear stresses which are expended locally against the fluid–solid boundaries. Thus, their only macroscopic effect is to transmit to the solid skeleton by viscous coupling whatever net impelling force may be applied to the interstitial fluid.

In any stress system with the principal stresses, σ_x, σ_y, and σ_z, one can define the local mean value of the quantity for the hydrostatic stress, $\bar{\sigma}_w$, as:

$$\bar{\sigma}_w = \tfrac{1}{3}(\sigma_{wx} + \sigma_{wy} + \sigma_{wz}) \tag{3-22}$$

Now, the hydrostatic stress tensor, P, can be represented by:

$$P \equiv \begin{vmatrix} \bar{\sigma}_w & 0 & 0 \\ 0 & \bar{\sigma}_w & 0 \\ 0 & 0 & \bar{\sigma}_w \end{vmatrix} \tag{3-23}$$

and:

$$P = \tfrac{1}{3}(3\bar{\sigma}_w) = \bar{\sigma}_w$$

The above expression represents the hydrostatic pressure of a fluid whether it is flowing or is stationary in the porous system of the shale. Note that $\sigma_{wx} = \sigma_{wy} = \sigma_{wz} = \bar{\sigma}_w$, and that the hydrostatic portion of the total stress system causes only volume changes in the deformed material.

Deviatoric stress state. The second component is known as the stress deviator from the hydrostatic state. It is expressed as the difference between the total stress and the hydrostatic stress which resists deformation:

$$D \equiv \begin{vmatrix} (\sigma_x - \sigma_{wx}) & \tau_{xy} & \tau_{xz} \\ \tau_{yx} & (\sigma_y - \sigma_{wy}) & \tau_{yz} \\ \tau_{zx} & \tau_{zy} & (\sigma_z - \sigma_{wz}) \end{vmatrix} \tag{3-24}$$

where D is the deviatoric part of the total stress tensor. The effect of the deviator stress is to produce a distortion which is elastic or plastic in nature and is introduced into the shale body.

Total stress tensor. If the sediment body is not in equilibrium, the second component will not be a symmetric tensor for $\tau_{xy} \neq \tau_{yx}$. This asymmetric tensor can be subdivided into symmetric and skew-symmetric parts (Ramsay, 1967, p.282). The hydrostatic stress component is the same as in eq.3–22. The second symmetrical part is the deviatoric stress component which can be expressed as:

$$D \equiv \begin{vmatrix} (\sigma_x - \bar{\sigma}_w) & 1/2(\tau_{xy}+\tau_{yx}) & 1/2(\tau_{xz}+\tau_{zx}) \\ 1/2(\tau_{xy}+\tau_{yx}) & (\sigma_y - \bar{\sigma}_w) & 1/2(\tau_{yz}+\tau_{zy}) \\ 1/2(\tau_{xz}+\tau_{zx}) & 1/2(\tau_{yz}+\tau_{zy}) & (\sigma_z - \bar{\sigma}_w) \end{vmatrix} \qquad (3\text{-}25)$$

The skew-symmetric part is termed the disequilibrium component, which causes the shale to undergo a rotation in space and is expressed as:

$$R \equiv \begin{vmatrix} 0 & 1/2(\tau_{xy}-\tau_{yx}) & 1/2(\tau_{xz}-\tau_{zx}) \\ 1/2(\tau_{yx}-\tau_{xy}) & 0 & 1/2(\tau_{yz}-\tau_{zy}) \\ 1/2(\tau_{zx}-\tau_{xz}) & 1/2(\tau_{zy}-\tau_{yz}) & 0 \end{vmatrix} \qquad (3\text{-}26)$$

where R is the disequilibrium component. Such a stress state would be expected if tectonic forces were acting on the shale mass in a basin within a geosyncline. The total stress tensor for a shale body not in equilibrium is expressed as the sum of the above described parts:

$$S = P + D + R \qquad (3\text{-}27)$$

(total stress = hydrostatic stress + deviatoric stress + disequilibrium component)

Each one of the three components making up the state of stress is directly related to the respective component of the strain tensor. The hydrostatic part of the stress system causes changes in volume, the deviatoric stress components cause distortion, and the disequilibrium components cause the material to undergo rotation in space (Ramsay, 1967).

Lo (1969) showed mathematically that the pore pressure induced by shear may be expressed as a sole function of major principal strain. According to him, the only unambiguous and correct principle of superposition of pore pressure is to consider an isotropic stress system and a deviatoric stress system, namely,

$$\begin{bmatrix} \Delta\sigma_1 & 0 & 0 \\ 0 & \Delta\sigma_2 & 0 \\ 0 & 0 & \Delta\sigma_3 \end{bmatrix} = \begin{bmatrix} \Delta\sigma_3 & 0 & 0 \\ 0 & \Delta\sigma_3 & 0 \\ 0 & 0 & \Delta\sigma_3 \end{bmatrix} + \begin{bmatrix} \Delta\sigma_1-\Delta\sigma_3 & 0 & 0 \\ 0 & \Delta\sigma_2-\Delta\sigma_3 & 0 \\ 0 & 0 & 0 \end{bmatrix} \qquad (3\text{-}28)$$

where σ_1 is the total major stress, σ_2 is total intermediate stress and σ_3 is total minor stress.

According to Lo (1969), the physical justification for eq.3-28 lies in the fact that under ambient stress, the induced pore pressure corresponds almost exactly to the applied pressure, because the compressibility of pore water and argillaceous sediment grains are much lower than that of the sediment structure. In the experience of the writers, however, most of the pore-pressure equations presented in the literature give almost identical results providing they are properly used.

Variation of the overburden stress

Assuming no lateral variation in the state of stress owing to tectonic stresses, the stress should vary through a sediment body mainly in the vertical direction. The stress components acting in the z direction at a point $(x, y, z + dz)$ can be presented in the Cartesian coordinate system as follows:

$$(\tau_{xz} + d\tau_{xz})(\tau_{yz} + d\tau_{yz})(\sigma_z + d\sigma_z) \tag{3-29}$$

Inasmuch as dz is small, the changes in $d\sigma_z$, $d\tau_{yz}$, and $d\tau_{xz}$ may be considered as linear variations which depend on the rates of change of stress in the sediment body:

$$d\sigma_z = (\partial\sigma_z/\partial z)\, dz \tag{3-30}$$

$$d\tau_{xz} = (\partial\tau_{xz}/\partial z)\, dz \tag{3-31}$$

$$d\tau_{yz} = (\partial\tau_{yz}/\partial z)\, dz \tag{3-32}$$

Considering the balance of forces on the unit volume for the vertical direction z:

$$\sigma_z + d\sigma_z + \tau_{xz} + d\tau_{xz} + \tau_{yz} + d\tau_{yz} + F_z = \sigma_z + \tau_{xz} + \tau_{yz} + ma_z \tag{3-33}$$

where F_z is the body force, m is the mass of the sediment body, and a_z is the acceleration in the z direction. Eq.3-33 can be expressed as:

$$\text{Accelerating force} = ma_z = \left(\sigma_z + \frac{\partial\sigma_z}{\partial z}\, dz\right) dx\, dy - \sigma_z\, dx\, dy$$

$$+ \left(\tau_{xz} + \frac{\partial\tau_{xz}}{\partial x}\, dx\right) dy\, dz - \tau_{xy}\, dy\, dz + \left(\tau_{yz} + \frac{\partial\tau_{yz}}{\partial y}\, dy\right) dx\, dy$$

$$- \tau_{yz}\, dx\, dz + F_z\, dx\, dy\, dz \tag{3-34}$$

The accelerating force is equal to $\partial^2 u/\partial t^2$, where u is the displacement in the z direction and t is time. Dividing through by the volume of the element, $dxdydz$, one obtains:

$$\partial\sigma_z/\partial z + \partial\tau_{xz}/\partial x + \partial\tau_{yz}/\partial y + F_z = (\partial^2 u/\partial t^2)\rho \qquad (3\text{-}35)$$

where ρ is the density (mass/unit volume; e.g., in slugs per cubic foot).

If the element is in static equilibrium, then the accelerating force is equal to zero:

$$\partial\sigma_z/\partial z + \partial\tau_{xz}/\partial x + \partial\tau_{yz}/\partial y + F_z = 0 \qquad (3\text{-}36)$$

where the body force, F_z, is equal to $-\rho g$.[1] The sediment is assumed to be subjected to body forces owing only to gravity.

Evaluation of the shear stresses due to an overburden load. Eq.3–36 is a partial differential equation expressing both the normal and shear stresses at a point on a plane. It can be written as:

$$\partial\sigma_z = \rho g \partial z - (\partial\tau_{xz}/\partial x)\partial z - (\partial\tau_{yz}/\partial y)\partial z \qquad (3\text{-}37)$$

Integrating with respect to z between the limits of z equal to zero and Z:

$$\int_0^Z d\sigma_z = \rho g \int_0^Z dz - \int_0^Z (\partial\tau_{xz}/\partial x)dz - \int_0^Z (\partial\tau_{yz}/\partial y)\,dz \qquad (3\text{-}38)$$

$$\sigma_z = \rho g Z - \int_0^Z (\partial\tau_{xz}/\partial x)dz - \int_0^Z (\partial\tau_{yz}/\partial y)dz \qquad (3\text{-}39)$$

Eq.3-39 shows that the vertical stress, σ_z, at a point of depth Z, equals the overburden weight per unit area of the sediment less the contributions from the vertical-shear components. Four possible cases are discussed here (see Howard, 1966); the vertical normal component of stress at a point is equal to or nearly identical with the overburden weight of the sediments in three cases:

(1) The vertical-shear stresses are either non-existent or constant in the sediment mass. If $\tau_{xz} = \tau_{yz} =$ constant, then:

$$\partial\tau_{xz}/\partial x = \partial\tau_{yz}/\partial y = 0 \qquad (3\text{-}40)$$

[1] Note that the minus sign is placed here in order to make the compressive forces positive.

Inasmuch as the magnitudes of τ_{xz} and τ_{yz} do not change with depth and the derivative of a constant is equal to zero, the net contribution of shear stresses is zero. Therefore, eq.3-39 becomes:

$$\sigma_z = \rho g Z \tag{3-41}$$

(2) Integrals of the vertical components of shear are approximately equal to zero:

$$\int_0^Z (\partial\tau_{xz}/\partial x)\mathrm{d}z = \int_0^Z (\partial\tau_{yz}/\partial y)\mathrm{d}z \approx 0 \tag{3-42}$$

Sanford (1959) noted that in some types of geologic structures some differences must have existed between the vertical normal component of stress at a point and the weight of the overburden. Namely, the integral of the gradients of the vertical components of shear were non-zero. Over geologic time, however, such stresses may disappear through creep. Eq.3-39 then becomes the same as eq.3-41.

The first two cases are acceptable geological possibilities.

(3) Integrals of the vertical components of shear are equal in value and opposite in direction on two planes perpendicular to each other:

$$\int_0^Z (\partial\tau_{xz}/\partial x)\mathrm{d}z = -\int_0^Z (\partial\tau_{yz}/\partial y)\mathrm{d}z \tag{3-43}$$

Eq.3-39 becomes identical to eq.3-41. This is a restrictive case, geologically speaking. The possibility of it occurring in nature is slight.

(4) The two integrals of the vertical components of shear are equal:

$$\int_0^Z (\partial\tau_{xz}/\partial x)\mathrm{d}z = \int_0^Z (\partial\tau_{yz}/\partial y)\mathrm{d}z \tag{3-44}$$

Therefore, eq.3-39 becomes:

$$\sigma_z = \rho g Z - 2\int_0^Z (\partial\tau_{xz}/\partial x)\mathrm{d}z \tag{3-45}$$

At the present time, the above-described cases are being investigated by geologists.

Bombolakis (1967, p.37) discussed a similar problem with respect to an overthrust block, but did not attempt to show a solution to the problem.

Finite-element methods used in engineering mechanics are well suited to problems of geologic deformation involving discontinuous anisotropic materials, which are heterogeneously distributed and have irregular topographic and internal boundary configurations (Voight and Dahl, 1970). In the continuum approach, the surface of the basin undergoing compaction and/or subsidence is replaced mathematically by an idealized material that deforms in accordance with principles of continuum mechanics. Despite the powerful computational capability of finite-element technology, very little work has been done in basin-wide compaction studies.

Estimation of the magnitude and direction of stress

So far only variations in the stress pattern have been discussed with respect to the vertical overburden load. Berry (1969) has pointed out that the pore-fluid pressures of thick Franciscan and Great Valley of California geosynclinal sediments reach near-lithostatic (or geostatic) values. The origin of these anomalous near-lithostatic fluid pressures is attributed to compression between the granitic Sierran-Klamath and Salinas blocks which act like the jaws of a vise closing in on the argillaceous sediments in the Great Valley (Berry, 1969, p.14). High fluid pressures are thus related to the current tectonic compaction.

It has also been argued that creep (rock flow) will occur in sediments for a non-zero differential stress acting over geologic time. Hubbert and Willis (1957) demonstrated that the following hydrostatic relationship:

$$\sigma_x \approx \sigma_y \approx \sigma_z \approx \rho D \tag{3-46}$$

cannot be the case in tectonically active regions where normal or thrust faulting is prevalent.

In tectonically relaxed sedimentary basins, such as the Gulf Coast Tertiary basin, which are characterized by normal faulting, the minimum stress direction is approximately horizontal and the stress magnitude is approximately equal to 1/3 of the effective pressure of the overburden. On the other hand, in depositional basins under tectonic compression, such as the Great Valley of California, which are characterized by thrust faulting and folding, the minimum stress direction is nearly vertical and the stress magnitude is equal to the effective overburden pressure. The ratio of the horizontal to vertical stress in the latter case is between 2 and 3.

Another theoretical approach to estimating in-situ stresses is to assume that a plane strain condition exists in the horizontal plane at depth (Price, 1959). Under this assumption, the following expression relates the stresses in the x and y directions to the vertical overburden stress for rocks in compression:

$$\sigma_x = \sigma_y = \sigma_h = \frac{\nu}{1-\nu} \sigma_z \tag{3-47}$$

where σ_h is horizontal stress in general, ν is Poisson's ratio, and $\sigma_z = \rho D$. This condition requires the sediment to become isotropically elastic before being subjected to gravity forces. Poisson's ratio values in this case are normally assumed to be equal to 0.25. If the latter value is used in eq.3-47, horizontal stress will be equal to 1/3 of the vertical stress. The question arises as to whether 0.25 is a realistic Poisson's ratio value or not.

According to Birch et al. (1942), the Poisson's ratio ν for consolidated sedimentary rocks ranges from 0.18 to 0.27, which give rise to compressive stress between 0.22 and 0.37 p.s.i./ft. of depth. On the other hand, according to Harrison et al. (1954), the unconsolidated sands and shales found in the Gulf Coast region of Texas and Louisiana can be considered to be in a plastic state of stress and to possess horizontal stresses in excess of 0.37 p.s.i./ft. of depth.

Recently, Geyer and Myung (1970) have used the 3-D velocity log as a tool for determining the elastic moduli of rocks. The 3-D velocity log permits the determination of the velocities of both shear and compressional waves in the borehole-wall rock. These velocities combined with the bulk density obtained from a density log enable computation of Poisson's ratio. This method may possibly provide reliable in-situ Poisson's ratio values.

A most realistic approach, however, is to measure the stresses directly in the sedimentary column. Scheidegger (1962), Dunlap (1963), Fairhurst (1964), Pulpan and Scheidegger (1965), and Perkins (1967) have used pressure data obtained during hydraulic fracturing of oil wells to calculate the stresses. By determining (1) the maximum pressures required for fracture initiation (formation breakdown), (2) the pressure necessary to extend the fracture, and (3) the formation fluid pressure, realistic values for σ_x and σ_y have been obtained. In some cases, however, one of the horizontal stresses is greater than the overburden stress:

$$\sigma_x > \sigma_z > \sigma_y \tag{3-48}$$

This state of stress is one of potential wrench faulting. Gretener (1965) has shown that the large calculated horizontal stress values may be due to the author's assumptions rather than to the pressure data. To date, the approach of measuring stresses directly in the wellbore has not been highly productive. Increased sophistication in the data handling and rock mechanics, however, may ultimately resolve the problems of sedimentary column stresses (L. Kovisars, personal communication, 1971).

PRESSURE–DEPTH–DENSITY RELATIONSHIPS

It has been shown (see above, eq.3-41) that the overburden pressure, for all practical purposes, is the pressure exerted at any depth by the weight of the overlying sediments. By changing the density term in this equation to that of the bulk density of a fluid-saturated shale, the pressure–depth relationships in the field can be determined.

Hubbert and Rubey (1959, p.129) stated that within depths of 1 or 2 km, the pressure of the water as a function of the depth, D, can be closely approximated by the equation:

$$p = \rho_w g D \tag{3-49}$$

where p is the hydrostatic pressure of a column of water extending from the surface of the ground to a depth of D; ρ_w is the density of the water; and g is the acceleration of gravity.

The specific weight of any fluid, γ_f, can be expressed as:

$$\gamma_f = \rho g \tag{3-50}$$

or:

$$\rho = \gamma_f / g \tag{3-51}$$

Thus, pressure at a depth D is equal to:

$$p = \gamma_f \times D \tag{3-52}$$

The specific gravity, SG, is equal to:

$$SG = \gamma_f / \gamma_w \tag{3-53}$$

and:

$$\gamma_f = SG \times \gamma_w \tag{3-54}$$

where γ_w is the specific weight of water. On combining eq.3-52 and 3-54:

$$p = SG \times \gamma_w \times D \tag{3-55}$$

If γ_w is in pounds per cubic foot and D is in feet, then p is in pounds per square foot. The pressure gradient (p/D) for pure water (γ_w = 62.4 lb./cu. ft.) is equal to 0.433 p.s.i./ft. $\left(= \dfrac{62.4 \text{ lb./cu. ft.}}{144 \text{ sq. inch/sq. ft.}}\right)$

Many actual fluid-pressure measurements have been made in the subsurface. Along the

TABLE XI
Pressure gradient and specific gravity for various fluids [1]

Type of fluid [2]	*Specific gravity*	*Specific weight (lb./cu. ft.)*	*Pressure gradient (p.s.i./ft.)*
Crude oil (40° API)	0.8285	51.69	0.359
Fresh water (TDS = 0 mg/l)	1.0000	62.4	0.433
Sea water (TDS = 35,000 mg/l)	1.0256	63.99	0.444
Brine (TDS = 100,000 mg/l)	1.0730	66.95	0.465
Brine (TDS = 200,000 mg/l)	1.1285	70.42	0.489

[1] Levorsen (1958, p.663) pointed out that the pressure gradient averages approximately 0.0043 p.s.i./ft. per specific gravity increase of 0.01.
[2] TDS = total dissolved solids.

Gulf Coast, the fluid pressure increases at the rate of about 0.465 p.s.i. for each foot of depth (hydrostatic pressure gradient for a brine having a specific weight, γ_w, of 67 lb./cu. ft.). The corresponding increase in the shale matrix pressure is 0.535 p.s.i./ft., if one assumes a total overburden pressure gradient of 1 p.s.i./ft. Frederick (1967) presented many examples of the relationship between bottom-hole fluid pressure and depth for areas which are subjected to abnormal pressures. Hubbert and Rubey (1959, p.155) noted a lithostatic pressure as high as 1.06 p.s.i./ft. occurring in the Khaur Field in Pakistan. Levorsen (1958, p.386) reported that the average gradient of oil-field brines is approximately 0.450 p.s.i./ft. Deviations are in part due to the varying salt concentrations in the brines. Table XI gives the specific gravity and pressure gradients of various fluids that could possibly occur in a sand–shale sequence. Fig.53 shows the pressure versus depth relationship for these brines.

A series of curves showing variation in porosity and density of plastic sedimentary rocks with increasing geostatic loading are presented in Fig.54. Ozerskaya (1965) presented the following equation for the variation in porosity with increasing geostatic pressure:

$$\phi = \phi_{max}\, e^{-0.45D} \tag{3-56}$$

where ϕ_{max} = maximum initial porosity of argillaceous sediments, which is commonly considered to be equal to 60%, and D is depth of burial. The 60% value is a good average for deltaic and marine clays; however, ϕ_{max} may be significantly different for continental, lacustrine, deep-ocean, and other types of clays. The formula relating fractional porosity, ϕ, bulk density, ρ_b, and mineralogic density, ρ_s, can be presented as follows:

$$\rho_b = \rho_s(1-\phi) \tag{3-57}$$

Consequently:

$$\rho_b = \rho_s(1-\phi_{max}\, e^{-0.45D}) \tag{3-58}$$

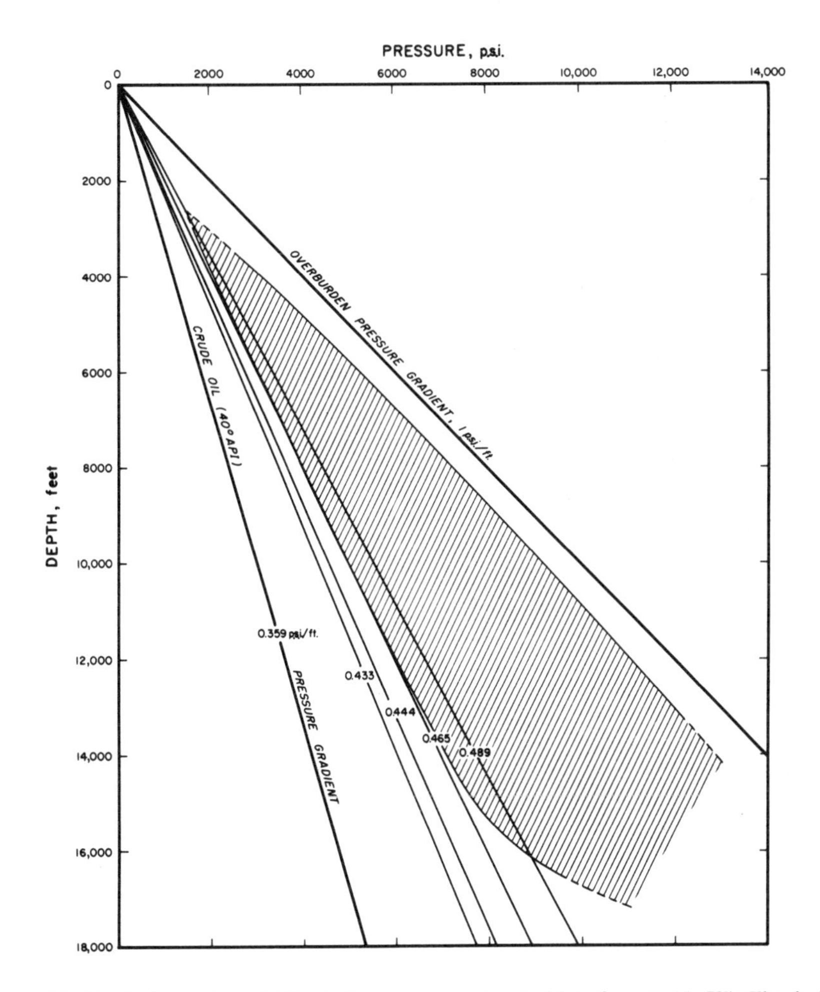

Fig.53. Hydrostatic and lithostatic pressure gradients (data from Table XI). The dashed area indicates the region for reservoirs having abnormally high bottom-hole fluid pressures and which were investigated by the writers.

In preparing Fig.54, the density of mineral grains was assumed to be equal to 2.7 g/cm^3.

It is assumed that the initial maximum porosity of the clayey sediment was equal to 60%; thus, if the values read from Fig.54 indicate lower initial porosity, there are several possible explanations:

(1) Part of the overburden load was removed by erosion.

(2) Uplift of the region.

(3) Excess compaction was caused by geotectonic forces.

(4) Subsequent cementation and filling of pores.

(5) Presence of sand and carbonate fractions.

(6) Wrong initial porosity assumption.

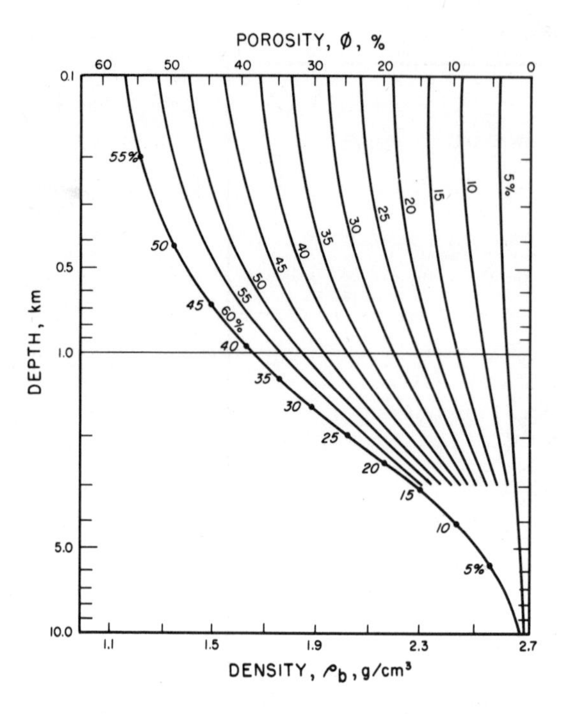

Fig.54. Relationship between bulk density, porosity and depth of burial (geostatic load) for argillaceous sediments. $\phi = \phi_{max}e^{-0.45D}$; $\rho_b = \rho_s(1 - \phi_{max}e^{-0.45D})$; $\rho_s = 2.7$. The numbers designate the values of initial porosity (ϕ_{max}) from 60% to 5% (with 5% intervals). The same numbers shown on the first curve to the left, correspond to curves with different initial porosity, shifted along the depth scale to the 60% curve. (After Ozerskaya, 1965; also see Avchyan and Ozerskaya, 1968, fig.2, p.139.)

Porosity–depth compaction models

As was shown in Chapter 2, the pore volume of clastic sediments and rocks decreases with increasing depth. This decrease in porosity can be used as a convenient measure of the amount of compaction an argillaceous sediment has undergone since deposition. In arriving at a simple compaction model, the problem arises in evaluating the effects of depositional rates and geologic age. Nevertheless, empirical data suggest that the effect of age and depositional rates are commonly predictable.

The effect of temperature is also difficult to evaluate. Experiments by Warner (1964, pp.50–79), however, suggest that the effect of temperatures less than 200°F may not be important other than in accelerating compaction rates.

Most compaction theories utilize clay minerals of idealized size and shape which are influenced by mechanical rearrangement during burial. The following theories, which are presented in chronological order, may enable the reader to better visualize the interrelationship between pressure, porosity reduction and fluid release in shales.

Athy's compaction model. Compaction is a simple process of squeezing out the interstitial fluids and thereby reducing the porosity. In relatively pure shales a definite relationship exists between porosity and depth of burial (Fig.14B). After a sediment has been deposited and buried, the pore volume may be modified by: (1) deformation and granulation of the mineral grains; (2) cementation; (3) solution; (4) recrystallization; and (5) squeezing together of the grains. The continued application of overburden or tectonic stress is the mechanism by which porosity is reduced and bulk density is increased further. Athy (1930) pointed out that the amount of compaction is not directly proportional either to reduction of pore volume or to increase in bulk density because of the above-mentioned processes.

Hedberg's compaction model. Hedberg (1936) stated that because of the numerous processes involved in compaction, it is not possible to express satisfactorily pressure–porosity relationships for clays and shales throughout the entire depth range by any one simple equation. Determinations of the porosities of shale core samples taken from Venezuelan wells (depth of 291 to 6,175 ft.) by Hedberg (1936) have furnished some valuable data (Figs.17 and 55).

From an analysis of this data, Hedberg (1936) proposed a compaction process consisting of three distinct stages (see Figs.55 and 56). The first consists mainly of mechanical rearrangement and dewatering of the clayey mass in the pressure interval from zero to 800 p.s.i. During this time there is a rapid decrease in porosity for small increments of overburden pressure. Expulsion of free water and mechanical particle rearrangement are dominant in the porosity range from 90% to 75%. Some adsorbed water is lost during this stage. Between a porosity of 75% and 35%, adsorbed water is expelled from the sediment. Below a porosity of 35%, the clay particles come in closer contact with each other; this results in greater resistance to further reduction in porosity. According to Hamilton (1959, p.1407), the transition from clay to shale probably occurs at about 35% porosity, because of chemical changes and cementation between the grains which imparts a rigidity to the structure.

In the second stage, between 800 and 6,000 p.s.i. (below 35% porosity), reduction in the sediment volume was attributed by Hedberg to the mechanical deformation of the particles and further expulsion of adsorbed water. There is also some recrystallization of the clay particles during this stage.

In the third and final stage (recrystallization stage) with porosities below 10%, the main mechanism is recrystallization under high pressures. Reduction of the pore volume occurs slowly and only with large pressure increments. The larger crystals may grow at the expense of the smaller ones, and a gradual transition may occur from a shale to a slate and then to phyllite.

Weller's compaction model. Weller (1959) described a compaction process very similar to the one proposed by Hedberg (1936). According to Weller, his composite porosity–depth

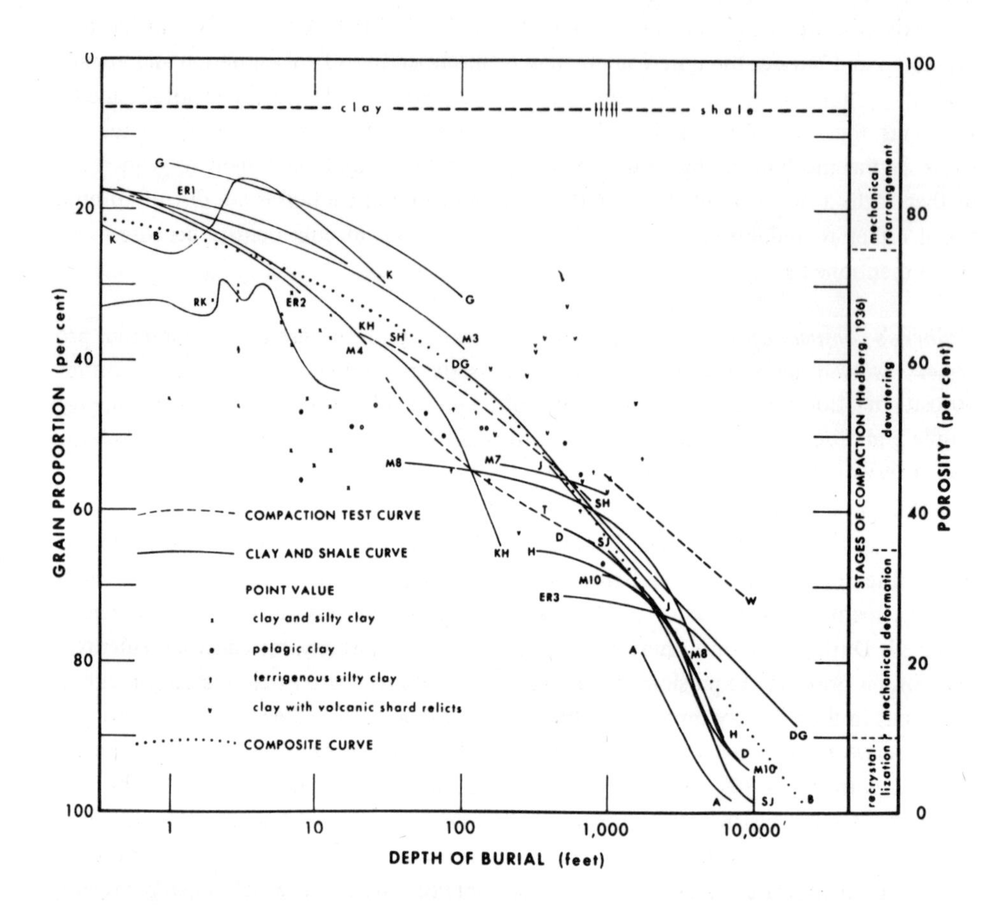

Fig.55. Relationship between depth of burial and grain proportion, showing Hedberg's compaction stages. Composite curve *B* is after Baldwin (1971) – see Fig.42. *A* = composite curve for Oklahoma wells (Athy, 1930, fig.2); *D* = six Venezuelan wells (Dallmus, 1958, figs.17, 18, 22); *DG* = composite curve for Gulf Coast wells (Dickinson, 1953, fig.14); *ER1* = Santa Barbara basin deep-sea core (Emery and Rittenberg, 1952, fig.21C); *ER2* = 13 deep-sea cores off California coast (Emery and Rittenberg, 1952, figs.5–17); *ER3* = five Los Angeles basin wells (Emery and Rittenberg, 1952, fig.30); *G* = seven Lake Mead cores (Gould, 1960, fig.49); *H* = three Venezuelan wells (Hedberg, 1936, fig.2); *J* = generalized curve for JOIDES hole 1 (Beall and Fisher, 1969, fig.3); *K* = fifteen abyssal plain cores (Kermabon et al., 1969, figs.3–8, 17); *KH* = Venezuelan wells (Kidwell and Hunt, 1958, fig.10); *M* = curves *3, 4, 7, 8,* and *10* (Meade, 1966, fig.1A); *RK* = deep-sea core off Nova Scotia (Richards and Keller, 1962, fig.1); *SH* = based on Skempton's compaction test data (Hamilton, 1959, table 1); *SJ* = Skeels' composite for wells (Johnson, 1950, fig.145); *T* = compaction test curve for blue marine clay (Terzaghi, 1925, fig.3, p.743); *W* = curve representing Warner's unpublished compaction test data (Beall and Fisher, 1969, fig.3). (After Baldwin, 1971, fig.2. Courtesy of Society of Economic Paleontologists and Mineralogists).

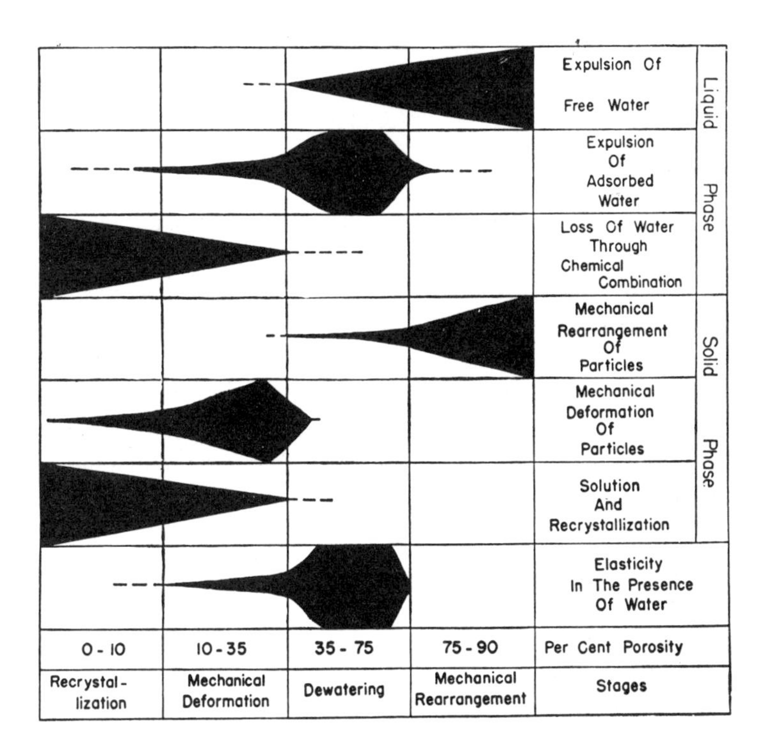

Fig.56. Hedberg's compaction model (processes of consolidation and lithification of clay). (After Hedberg, 1936, fig.7; also see Hamilton, 1959, p.1406.)

curve shown in Figs.17 and 18 represents equilibrium conditions in a continuous column of ordinary mud and shale. This curve, however, is based only on Terzaghi's, Athy's, and Hedberg's data. The porosity–depth relationships can be distorted by the occurrence of carbonates and sands in shales and by abnormally overpressured zones. In addition, application of laboratory soil-compression tests to buried sediments presents some problems.

Weller (1959) proposed a compaction process starting with a mud at the surface having a porosity between 85% and 45%. As the overburden pressure increases owing to sedimentation, the interstitial fluids are expelled from the pore space (porosity range from 45% to 10%). As a result, there is rearrangement of the mineral grains and development of closer packing. Compaction in this stage is related to yielding of the clay minerals between the more resistant grains. Weller theorized that at about 10% porosity, the non-clay mineral grains are in contact with each other, and the clays are being squeezed into the void space. Further compaction (porosity $< 10\%$) requires deformation and crushing of the grains until all porosity is eliminated.

Powers' compaction model. Powers (1967) presented a shale fluid-release theory based on changes in clay minerals and bulk properties with depth in muddy sediments. His theory assumes that mineralogical transformation of montmorillonite to illite occurs during deep burial, with the consequent release of large volumes of bound water from montmorillonite surfaces to interparticle areas where it becomes interstitial water.

In the case of marine montmorillonitic sediments buried to a few hundred feet, a balance is reached between the water retained in the sediment and the water-retaining properties of montmorillonite (Powers, 1967, p.1244). Further increase in overburden stress alone, resulting from deeper burial of the mud is ineffective in squeezing the remaining water out of the plastic sediment. At burial depths greater than 1,500–3,000 ft., most of the water exists as water of hydration and is stacked at least four monomolecular layers thick between the unit layers of montmorillonite. Only a small amount of oriented water occurs between the crystals and particles at depths of about 3,000–6,000 ft. (Fig.57A). At burial depths below about 6,000 ft., montmorillonite is altered to illite and the bound water is desorbed and becomes free pore water (Fig.57B).

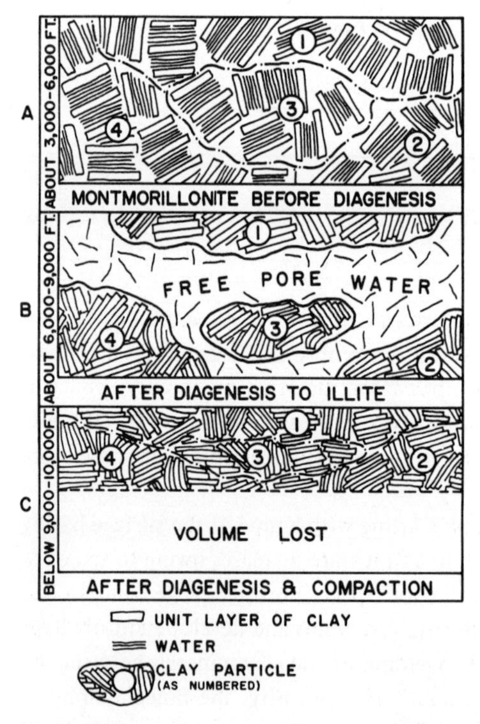

Fig.57. Effect of clay diagenesis on compaction of water from mudrocks, on assuming that same number of particles, crystal aggregates, and unit layers of clay occur in each compaction stage shown. A. No effective porosity or permeability; practically all water is "bound" water. B. Most "bound" water becomes free water; consequently effective porosity and permeability are greatly increased. C. Free water squeezed out; effective porosity, permeability, and original volume are greatly reduced. (After Powers, 1967, fig.1, p.1242. Courtesy of American Association of Petroleum Geologists.)

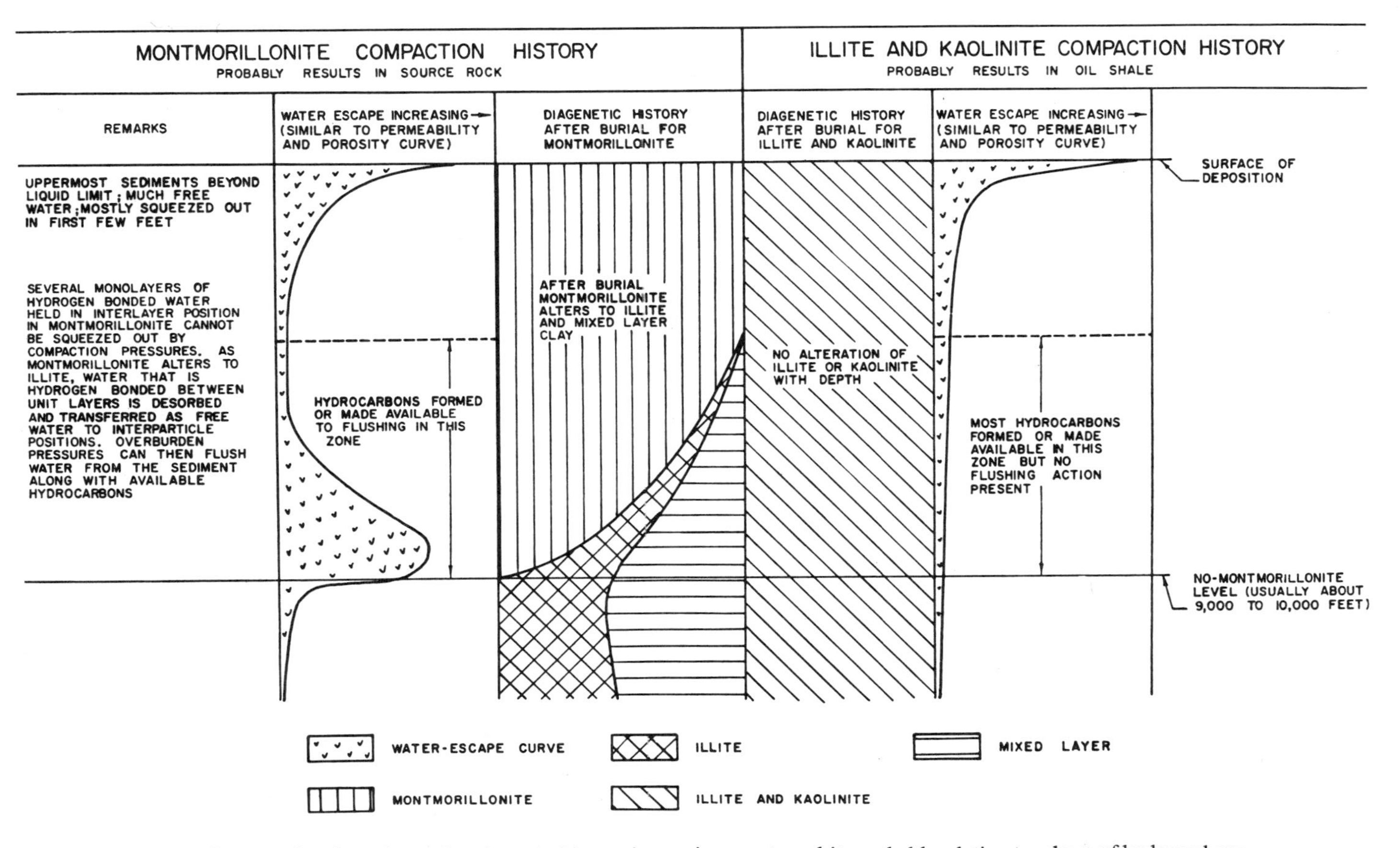

Fig.58. Compaction history of various clays when deposited in marine environments and its probable relation to release of hydrocarbons from mudrocks. (After Powers, 1967, fig.3, p.1245. Courtesy of American Association of Petroleum Geologists.)

This causes a decrease in clay-particle size with a corresponding increase in the effective porosity and permeability at burial depths of 6,000–9,000 ft. Below a depth of 9,000–10,000 ft. the water released from the clay is compacted until a new balance is established corresponding to the water-retaining properties of the illitic alteration product (Fig.57C).

The relationships between water expulsion, type of clay mineral, and depth of burial are illustrated in Fig.58, for both expanding and non-expanding clay deposits (Powers, 1967, p.1245). The water-escape curves are diagnostic of the porosity, permeability, and bulk density of compacting argillaceous sediments. Powers stated that the compaction history of mudrocks depends largely on their original clay composition and the diagenesis which they undergo after burial.

Teodorovich and Chernov's compaction model. Teodorovich and Chernov (1968) recognized the following stages in the compaction of productive Apsheron horizons in Azerbayjan:

(1) The first stage occurs at burial depths of 0 to 8–10 m. There is a rapid compaction during this stage: Porosity of clays decreases from 66% to 40%, whereas that of sandstones–siltstones decreases from 56% to 40%. Large amounts of water are squeezed out during this stage (sedimentogenesis and early diagenesis).

(2) There is a rapid decrease in compaction rate at the interval from 8–10 m to 1,200–1,400 m during the second stage. During this stage porosity of shales and sandstones–siltstones decreases to 20–21%.

(3) The third stage (burial to a depth of 1,400–6,000 m) is characterized by slow compaction. The absolute porosity of sandstones–siltstones at a depth of 6,000 m decreases to approximately 15–16%, whereas that of shales to 7–8%.

Burst's compaction model. A compaction model based on a three-stage dehydration sequence and the transformation of montmorillonite clay to mixed-layer varieties was proposed by Burst (1969). The initial dehydration stage is essentially completed in the first few thousand feet of burial as the interstitial water content is reduced to approximately 30% (20–25% interlayer water and 5–10% residual pore water) (Fig.59). During the second stage, the argillaceous sediment is in a state of quasi-equilibrium as it continues to adsorb geothermal heat. Pressure is relatively ineffective as a dehydrating agent because of the increased density of the interlayer water packet. As soon as the heat accumulation is sufficient to mobilize the interlayer water, one of the two remaining interlayers of bound water (statistically averaged) is discharged into the bulk system. Burst (1969, p.80) stated that the amount of water in movement should constitute 10–15% of the compacted bulk volume. During the third stage, the final water increment, having approximately capillary water density, gradually is forced out of the clay mineral lattice and voids as sediment temperature increases. Fig.60 illustrates the variation of the interlayer water density during compaction. Burst's dehydration–compaction

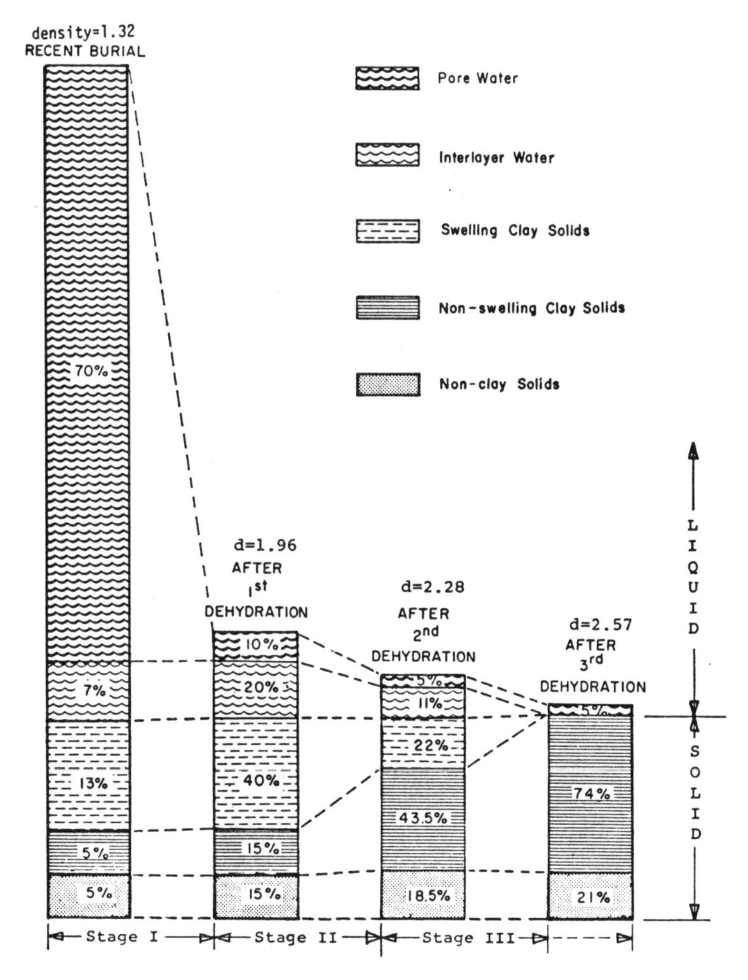

Fig.59. Marine shale bulk composition during dehydration. (After Burst, 1969, fig.6, p.81. Courtesy of American Association of Petroleum Geologists.)

model has not been substantiated by the experiments in the laboratory. Further discussion on this subject is presented in Chapter 7.

Beall's compaction model. Dr. A. Beall (personal communication, 1970), proposed a simple model for consolidation of clastic muds, based on the data from offshore well core samples, Louisiana, the JOIDES Deep Sea Drilling Project, and from high-pressure experiments on marine muds. The initial stage of compaction (down to a depth of approx. 3,300 ft.) primarily involves expulsion of fluids by mechanical processes as in the other earlier theories. Approximately 50% of total consolidation is reached at a very shallow depth. The average calculated pore throat diameters during the first stage are around 6Å.

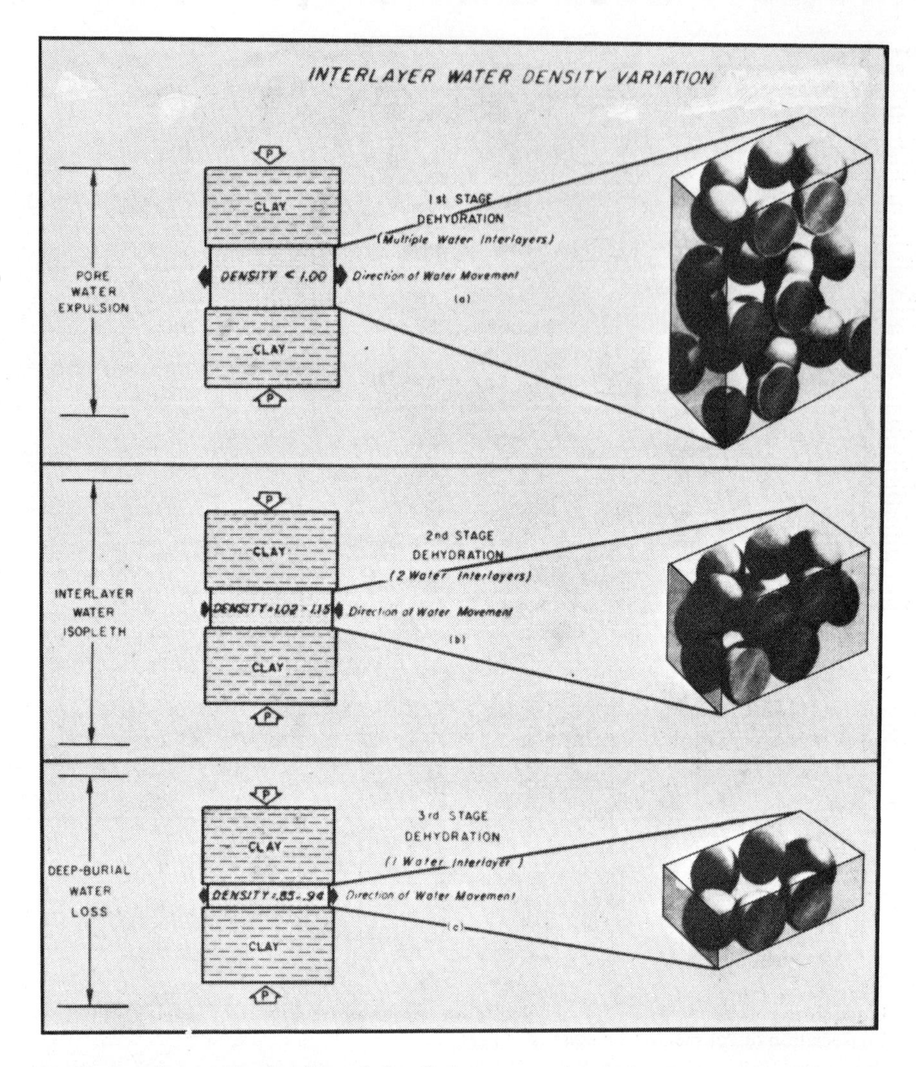

Fig.60. Interlayer water density variation during compaction. (After Burst, 1969, fig.8, p.83. Courtesy of American Association of Petroleum Geologists.)

During the second stage (at depths of 3,300 to approx. 8,000 ft.) approx. 75% of total compaction is complete, and pore throat widths in the muds approach 1 Å. The fluid pressures remain hydrostatic. During the third stage of compaction there is an extremely slow decrease of porosity with depth, and pore throat diameters are generally less than 1 Å. According to Beall, NaCl filtration could probably take place during the third stage, resulting in the expulsion of progressively less saline fluids to associated permeable sands, if the latter are present. In Beall's model, overburden pressure between 8,000 and 12,500 p.s.i. would be required to initiate NaCl filtration in marine muds. In the absence of permeable sands, the excess fluid pressure may be generated during the third stage.

Overton and Zanier (1970) proposed a similar model with four zones having different water types:

(1) Depths less than 3,000 ft. – fresh water.

(2) Depths of 3,000–10,000 ft., depending on the temperature – exponentially increasing salinities.

(3) Depths greater than 10,000 ft. – decreasing salinities to the depth of greatest pressure gradients.

(4) Depths greater than 15,000 ft. – increasing salinities with decreasing water fractions. Physical–chemical changes in shales occur in this indistinct zone.

Overton and Zanier (1970) stated that in the Gulf Coast, sands and shales are difficult to distinguish on SP (self-potential electric log) curves at depths less than 3,000 ft., due to similarity of the waters in them. Water expelled from this interval is loose water, which constitutes 30 to 70% of the rocks. At critical compaction depth (depth around and usually less than 3,200 ft.), shales and sands become readily distinguishable on the SP curve. In zone 2, fresher water is held in the more ordered or crystalline layer next to the clay, whereas saline water is forced into an equilibrium position in an outer layer (large pores in the shale and in the nearest sand) (Overton and Zanier, 1970). As the crystallinity of water increases, ions are expelled into a less-ordered or more fluid layer. Below a depth of 10,000 ft., shales remineralize (Overton and Zanier, 1970) and associated sandstones contain fresher waters. The beginning of this zone (zone 3) is readily apparent on the SP curve. The water freshening is probably due to the expulsion of (1) the last layers of dense, fresh water from shales into sands and/or (2) water of hydration resulting from montmorillonite-to-illite alteration.

DIFFERENTIAL COMPACTION

Differential compaction will occur wherever a compactible unit changes laterally in thickness or compactibility. Differential compaction has been observed where (1) an erosional unconformity of considerable topographical relief has been covered with a compactible sediment (Athy, 1930), (2) shale bodies contain sand lenses (Carver, 1968), and (3) carbonate reefs are embedded in shales (Gussow, 1955). The upper surface of the compactible horizon, which is horizontal at the time of deposition, will deform under loading, and structure will be induced in the overlying strata. Structures formed in this manner will have a maximum amplitude at the upper surface of the compactible unit and decrease systematically upwards, becoming nil at the depositional surface (Labute and Gretener, 1969, p.304). This is not true in the case of structures formed by lateral tectonic compression. If the overburden stress exceeds the load required for maximum possible compaction, all structures will disappear at the critical level required for total compaction (see Gretener and Labute, 1969). No structure will be observed in the strata above this level.

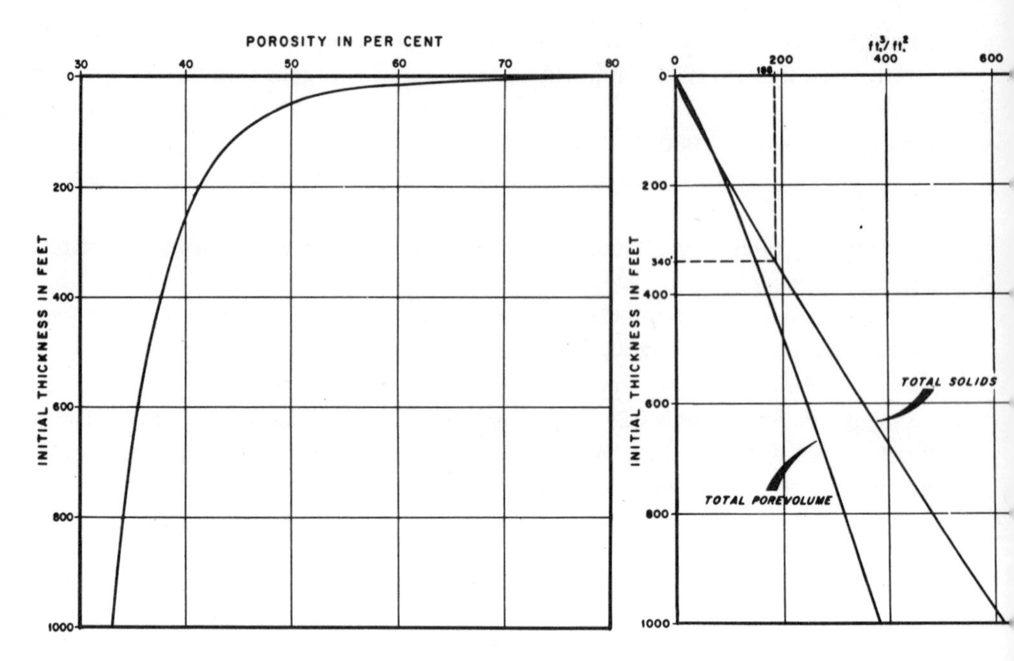

Fig.61. Assumed porosity–depth relationship for a clay rock. For a shale bed 100 ft. thick, the porosity will decrease from 80% at the surface to 45% at the base. For a bed 500 ft. thick, the corresponding values are 80% and 36%. (After Gretener and Labute, 1969, fig.1, p.298.)

Fig.62. Relationship between the initial thickness and total pore volume and total solids, based on porosity–depth relationship presented in Fig.61. (After Gretener and Labute, 1969, fig.2, p.300.)

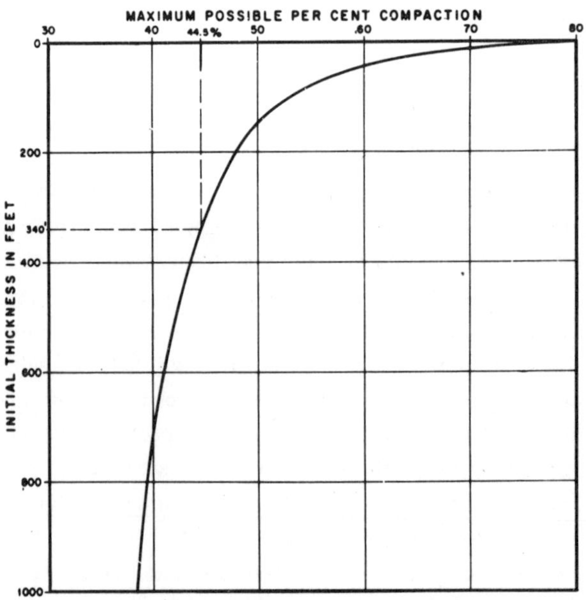

Fig.63. Maximum possible compaction in percent versus initial thickness based on data presented in

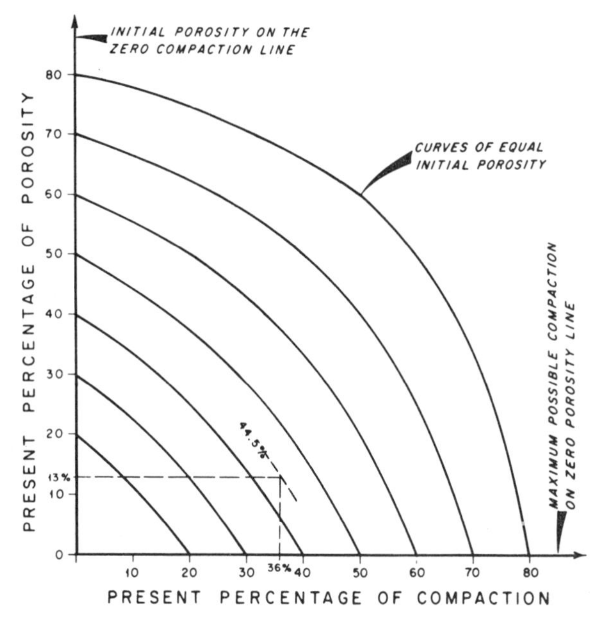

Fig.64. Present compaction in percent as a function of the initial and the present porosities. (After Gretener and Labute, 1969, fig.4, p.301.)

Labute and Gretener (1969, p.310) studied the compactibility of the Leduc Reef and Ireton shale. Although the Leduc carbonate reefs were rigid with respect to the Ireton shales, they showed physical signs of volume reduction. According to Pettijohn (1957, p.215), stylolites, which are commonly observed in the Leduc reefs, may be indicative of volume reduction. The most compactible portion of the Ireton Formation is situated in the central zone (about 500 ft. thick) that is characterized by a high illite-clay content (≈35%). The thickness of this zone ranges from 80 ft. over the reef to 570 ft. off the reef (Labute and Gretener, 1969, p.314).

According to Gretener and Labute (1969), the amount of compaction a certain stratigraphic unit has undergone may be determined. Their approach to this problem is based on the assumption that at the time deposition ends, the porosity within a stratigraphic unit composed of clay will decrease from top to bottom as shown in Fig.61, for example. Graphic integration of this curve will yield the total pore volume and the total solid volume in cubic feet per square foot of horizontal surface area for any given thickness. Fig.62 shows the results of such an integration. The pore and solids curves in Fig.62 allow construction of a third curve relating the maximum possible compaction to initial thickness of sediment (Fig.63). The latter figure illustrates that thick units will compact less on a percentage basis than thin units. The curve in Fig.63 is based on the correct assumption that sediments will compact under their own weight. Gretener and Labute (1969) constructed a series of curves showing relationship between present compaction of a unit (in %) and the initial and present porosities (Fig.64). For example, if the present stratigraphic thickness for a unit is 215 ft. and its porosity is equal to 13%, the unit contains 187 cu.ft./sq.ft. of solids [215 × (1–0.13)=187]. According to Fig.62, the value of 187 cu.ft./sq.ft. of solids corresponds to an initial thickness of 340 ft.; and the maximum possible compaction is equal to 44.5% (Fig.63). Inasmuch as the present porosity is equal to 13%, the present percentage of compaction is equal to 36% (Fig.64).

ESTIMATION OF MAXIMUM EFFECTIVE PRESSURE THAT HAS EVER EXISTED IN AN ARGILLACEOUS SEDIMENT

Numerous argillaceous sediments have been subjected in the past to loads greater than those existing in the same sediments at present; this may result in subsequent expansion (rebound) due to stress release. Casagrande (1936) developed a method for determination of the maximum intergranular pressure to which a given undisturbed sample has been subjected during its past history. An excellent description of this method was presented by Taylor (1948).

A curve, based on test data, of the logarithm of pressure versus the void ratio for a typical undisturbed sample is presented in Fig.65A, showing compression, rebound, and recompression curves (branches). The similarity in the curvature of the compression (original) curve and that of recompression curve at lower pressures, indicates a previous compaction in nature up to a pressure approximately equal to that at point *B*. The method of determining the maximum past pressure involves the following steps (Fig.65B): (1) Location of the point *O* of maximum curvature; this can be done quite accurately by visual inspection. (2) Drawing three lines through this point: (a) horizontal line *OC*, (b) tangent *OB*, and (c) bisector *OD* of the angle formed by the lines *OB* and *OC*. (3) Extension upwards of the straight-line portion of the curve until its intersection with line *OD*; this gives the desired point *E*. The above empirical method if applied to the recom-

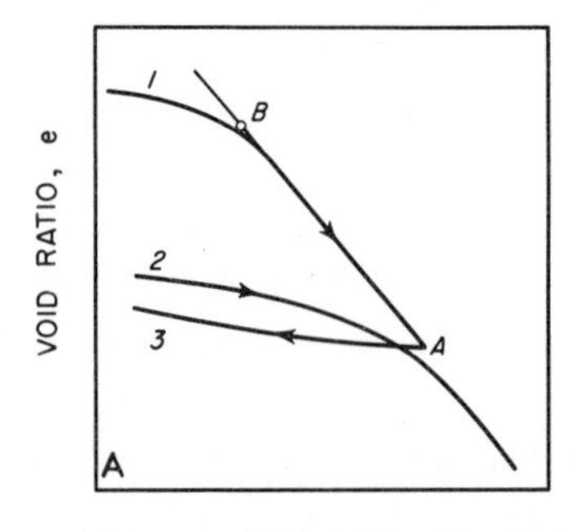

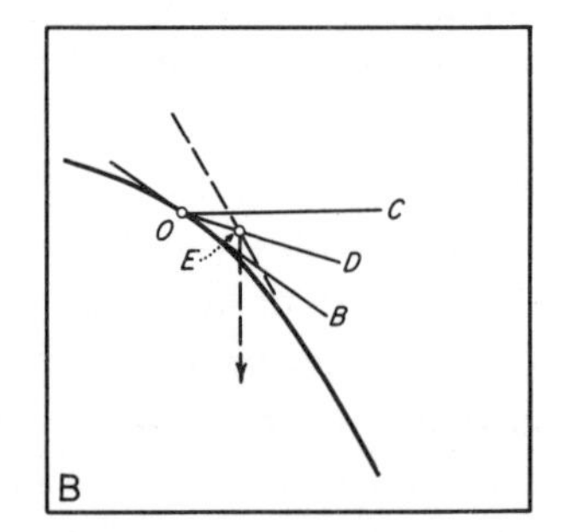

LOG PRESSURE

Fig.65. Method of determining the maximum past intergranular pressure from void ratio versus logarithm of pressure curves. *1* = compression curve; *2* = recompression curve; and *3* = rebound curve. (After Taylor, 1948, fig.12-3, p.278.)

pression curve (Fig.65, curve *2*) would give the point *A*; or point *B* if used on the compression (upper) curve (Fig.65, curve *1*).

If the structure of the sample has been disturbed to an appreciable degree during or after sampling, the pressure obtained by this method tends to be too low (Taylor, 1948, p.279).

Van der Knaap and Van der Vlis (1967) presented an interesting example of determining the maximum effective pressure that has ever existed in a sand and a clay sample from a depth of 3,100 and 3,104 ft., respectively. The samples were obtained from a post-Eocene formation of the Bolivar Coast, Venezuela, with the rubber-sleeve core barrel. These authors plotted void ratio versus effective pressure, which is the difference between the total pressure and the pore pressure (Fig.66). After coring, the stresses were first reduced to atmospheric pressure and then increased in the compression cell to values that have surpassed those prevailing in situ.

The effect of reloading can be observed in Fig.66 for the clay sample. The straight-line portion of the curve at low pressures represents the reloading stage in the laboratory, whereas the straight-line part at high pressures represents a continuation of the virgin compression curve. The sample had never before been subjected to these latter high pressures. According to Van der Knaap and Van der Vlis (1967), the intersection point of the two straight lines for the clay in Fig.66 is indicative of the maximum effective pressure that has ever existed in the material ($p_e \approx 1,750$ p.s.i.).

Van der Knaap and Van der Vlis (1967) artificially subjected the sand sample to unloading and loading cycles. On increasing the pressure to 2,900 p.s.i. (point *A*), there is

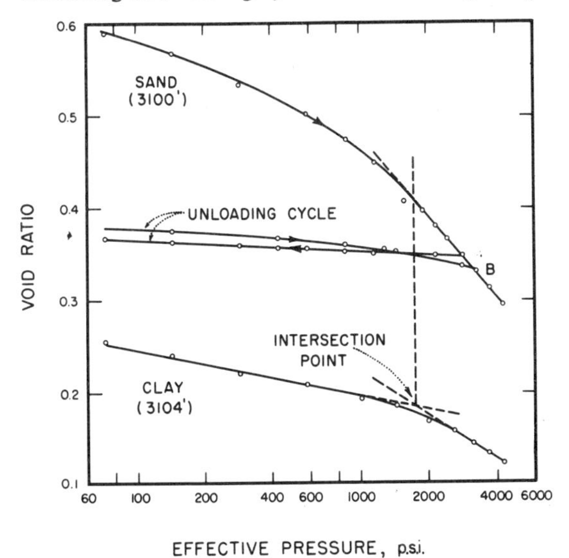

Fig.66. Relationship between the effective pressure (difference between total pressure and pore pressure) and void ratio for adjacent clay and sand samples from a post-Eocene formation of the Bolivar Coast, Venezuela. (After Van der Knaap and Van der Vlis, 1967, fig.4, p.89.)

a marked reduction in the void ratio. As the pressure is then reduced to zero, there is only a small increase in the void ratio, indicating a considerable permanent deformation of the sample. On increasing the load again (to point *B*), the void-ratio values are for the most part slightly higher than those found during the unloading stage; this suggests some hysteresis. There is an abrupt change in the slope as the pressure is raised beyond the point *B*, and the sample behaves as if the initial loading experiment had been continued beyond the point *A* without the intermediate unloading cycle. A marked change in slope is an indication that the virgin compression stage has been reached.

Inasmuch as the clay and sand samples are approximately from the same depth, the transition from reloading to continued virgin loading should take place at identical values of effective pressure. The latter corresponds to the pressure at the maximum depth of burial during geologic history. According to Van der Knaap and Van der Vlis (1967), probably the preservation of their clay sample during and after coring has been better than that of the sand.

One can attempt to compare the pressure obtained on using the present-day burial depth with that determined on using the above-described Casagrande method. If erosion has removed part of the overburden, the latter value is higher. In the case of overpressured conditions (fluids), the pressure obtained on using the Casagrande method would be lower than that obtained using maximum depth of burial that had ever existed. Considerable erosion, however, may complicate the picture. One should also examine porosity, because overpressured formations have higher porosity than that usually encountered at that particular depth. Thus, a systematic analysis may reveal cases of erosion or overpressure. Pressure in pounds per square foot can be calculated by using eq.3-3 and 3-4.

As pointed out by Komornik et al. (1970), by determining the preconsolidation pressure in profiles of clay deposits, the cause of overconsolidation may be identified as being due to the (1) erosion of the overburden, (2) temporary lowering of the water level, or (3) desiccation. For example, in Late Quaternary clays from Israel (Haifa Bay and Ashdod Harbor) the preconsolidation pattern indicated severe desiccation under subaerial conditions, with subsequent lesser desiccation due presumably to the onset of the Flandrian transgression (Komornik et al., 1970). The possibility of cementation causing apparent overconsolidation can be determined by petrographic examination.

REFERENCES

Athy, L.F., 1930. Density, porosity and compaction of sedimentary rocks. *Bull. Am. Assoc. Pet. Geologists*, 14(1): 1–24.

Avchyan, G.M. and Ozerskaya, M.L., 1968. Regularity in consolidation of sedimentary rocks with depth. *Izv. Akad. Nauk, Ser. Geol.*, 1968(2): 137–141.

Baldwin, B., 1971. Ways of deciphering compacted sediments. *J. Sediment. Petrol.*, 41(1): 293–301.

Beall Jr., A.O. and Fisher, A.G., 1969. Sedimentology. In: M. Ewing et al. (Editors), *Initial Reports of the Deep-Sea Drilling Project.* U.S. Govt. Printing Office, Washington, D.C., 1: 672 pp.

Berry, F.A., 1969. Origin and tectonic significance of high fluid pressures in the California coast ranges. *J. Pet. Tech.*, 21(1): 13–14.

Birch, A.F., Schairer, J.F. and Spicer, H.C., 1942. *Handbook of Physical Constants. Geol. Soc. Am., Spec. Pap.* 36: 325 pp.

Bombolakis, E.G., 1967. Analysis of stress at a point. In: R.E. Riecker (Editor), *Rock Mechanics Seminar.* Terrestrial Science Laboratory, Air Force Cambridge Research Laboratories, Bedford, Mass., 1: 31–58.

Brandt, H., 1955. A study of the speed of sound in porous granular media. *Trans. Am. Soc. Mech. Engrs.,* 77: 479–485.

Burst, J.F., 1969. Diagenesis of Gulf Coast clayey sediments and its possible relation to petroleum migration. *Bull. Am. Assoc. Pet. Geologists,* 53(1): 73–93.

Carver, R.E., 1968. Differential compaction as a cause of regional contemporaneous faults. *Bull. Am. Assoc. Pet. Geologists,* 52(3): 414–419.

Casagrande, A., 1936. *Proc. Int. Conf. Soil Mech. Found. Eng.,* 1st, 3: 60–64.

Dallmus, K.F., 1958. Mechanics of basin evolution and its relation to the habitat of oil in the basin. In: L.G. Weeks (Editor), *Habitat of Oil. Am. Assoc. Pet. Geologists, Mem.,* 36: 2071–2124.

Dickinson, G., 1953. Geological aspects of abnormal reservoir pressures in Gulf Coast Louisiana. *Bull. Am. Assoc. Pet. Geologists,* 37(2): 410–432.

Dunlap, J.R., 1963. Factors controlling the orientation and direction of hydraulic fractures. *J. Inst. Pet.,* 49(477): 282–294.

Emery, K.O. and Rittenberg, S.C., 1952. Early diagenesis of California Basin sediments in relation to origin of oil. *Bull. Am. Assoc. Pet. Geologists,* 36(5): 735–806.

Fairhurst, C., 1964. Measurement of in situ rock stresses, with particular reference to hydraulic fracturing. *Rock Mech. Eng. Geol.,* 29(3–4): 129–147.

Ferguson, L., 1963. Estimation of the compaction factor of a shale from distorted Brachiopod shells. *J. Sediment. Petrol.,* 33(3): 796–798.

Frederick, W.S., 1967. Planning a must in abnormally pressured areas. *World Oil,* 164(3): 74–77.

Geyer, R.L. and Myung, J.I., 1970. The 3-D velocity log; a tool for in situ determination of the elastic moduli of rocks. *Ann. Symp. Rock. Mech., 12th, Univ. Missouri, Rolla, Mo.,* pp.71–107.

Gould, H.R., 1960. Character of the accumulated sediment. In: W.O. Smith, C.P. Vetter, G.B. Cummings and others: *Comprehensive Survey of Sedimentation in Lake Mead, 1948–49. U.S., Geol. Surv. Prof. Pap.,* 295: 254 pp.

Gretener, P.E., 1965. Can the state of stress be determined from hydraulic fracturing data? *J. Geophys. Res.,* 70(24): 6205–6212.

Gretener, P.E. and Labute, G.J., 1969. Compaction – a discussion. *Bull. Can. Pet. Geol.,* 17(3): 296–303.

Gussow, W.C., 1955. Time of migration of oil and gas. *Bull. Am. Assoc. Pet. Geologists,* 39(5): 547–574.

Hamilton, E.L., 1959. Thickness and consolidation of deep-sea sediments. *Bull. Geol. Soc. Am.,* 70: 1399–1424.

Harrison, E., Kieschnick Jr., W.J. and McGuire, W.J., 1954. The mechanics of fracture induction and extension. *Trans. Am. Inst. Min. Metall. Engrs.,* 201: 254–255.

Hedberg, H.D., 1936. Gravitational compaction of clays and shales. *Am. J. Sci.,* 5th Ser., 31(184): 241–287.

Hottman, C.E., and Johnson, R.K., 1965. Estimation of formation pressures from log-derived shale properties. *J. Pet. Tech.,* 17(6): 717–722.

Howard, J.H., 1966. Vertical normal stress in the earth and the weight of the overburden. *Geol. Soc. Am. Bull.,* 77(6): 657–660.

Hubbert, M.K. and Rubey, W.W., 1959. Role of fluid pressure in mechanics of overthrust faulting. I. Mechanics of fluid-filled porous solids and its application to overthrust faulting. *Bull. Geol. Soc. Am.,* 70(2): 115–166.

Hubbert, M.K. and Rubey, W.W., 1960. Role of fluid pressure in mechanics of overthrust faulting. *Bull. Geol. Soc. Am.,* 71(5): 617–628.

Hubbert, M.K. and Willis, D.G., 1957. Mechanics of hydraulic fracturing. *Trans. Am. Inst. Min. Metall. Engrs.,* 210: 153–168.

Johnson, F.W., 1950. Shale density analysis. In: L.W. LeRoy (Editor), *Subsurface Geologic Methods.* Colorado School of Mines, Golden, Colo., 2nd ed., 1166 pp.

Kermabon, A., Gehin, C., Blavier, P. and Tonarelli, B., 1969. Acoustic and other physical properties of deep-sea sediments in the Tyrrhenian abyssal plain. *Mar. Geol.,* 7: 129–145.

Kidwell, A.L. and Hunt, J.M., 1958. Migration of oil in recent sediments of Pedernales, Venezuela. In: L.G. Weeks (Editor), *Habitat of Oil.* Am. Assoc. Pet. Geologists, Tulsa, Okla., 1384 pp.

Komornik, A., Rohlich, V. and Wiseman, G., 1970. Overconsolidation by desiccation of coastal Late Quaternary clays in Israel. *Sedimentology,* 14: 125–140.

Labute, G.J. and Gretener, P.E., 1969. Differential compaction around a Leduc reef – Wizard Lake area, Alberta. *Bull. Can. Pet. Geol.,* 17(3): 304–325.

Laubscher, H.P., 1960. Role of fluid pressure in mechanics of overthrust faulting: discussion. *Bull. Geol. Soc. Am.,* 71: 611–615.

Levorsen, A.T., 1958. *Geology of Petroleum.* Freeman, San Francisco, Calif., 703 pp.

Lo, K.Y., 1969. The pore pressure–strain relationship of normally consolidated undisturbed clays. I. Theoretical considerations. *Can. Geotech. J.,* 6(4): 383–394.

Meade, R.H., 1966. Factors influencing the early stages of compaction of clays and sands – review. *J. Sediment. Petrol.,* 36: 1085–1101.

Overton, H.L. and Zanier, A.M., 1970. Hydratable shales and the salinity high enigma. *Soc. Pet. Eng., Am. Inst. Min. Metall. Engrs., 45th Ann. Fall Meet., Houston, Texas,* Paper no.2989: 9 pp.

Ozerskaya, M.L., 1965. Influence of structural factors on density and elastic properties of sedimentary rocks. *Izv. Akad. Nauk S.S.S.R., Ser. Phys. Earth,* 1965(1): 103–108.

Perkins, T.K., 1967. Application of rock mechanics in hydraulic fracturing theories. *Proc. 7th World Pet. Congr.,* 3: 75–84.

Pettijohn, F.J., 1957. *Sedimentary Rocks.* Harper, New York, N.Y., 2nd ed., 718 pp.

Powers, M.C., 1967. Fluid-release mechanisms in compacting marine mudrocks and their importance in oil exploration. *Bull. Am. Assoc. Pet. Geologists,* 51(7): 1240–1254.

Price, N.J., 1959. Mechanics of jointing in rocks. *Geol. Mag.,* 96: 149–167.

Pulpan, H. and Scheidegger, A.E., 1965. Calculation of tectonic stresses from hydraulic well-fracturing data. *J. Inst. Pet.,* 51(497): 169–176.

Ramsay, J.G., 1967. *Folding and Fracturing of Rocks.* McGraw-Hill, New York, N.Y., 568 pp.

Richards, A.F. and Keller, G.H., 1962. Water content variability in a silty core from off Nova Scotia. *Limnology and Oceanography,* 7: 426–427.

Rogers, G.L., 1964. *Mechanics of Solids.* Wiley, New York, N.Y., 250 pp.

Sanford, A.R., 1959. Analytical and experimental study of simple geologic structures. *Bull. Geol. Soc. Am.,* 70(1): 19–52.

Sawabini, C.T., Chilingar, G.V. and Allen, D.R., 1971. Design and operation of a triaxial, high-temperature, high-pressure compaction apparatus. *J. Sediment. Petrol.,* 41(3): 871–881.

Scheidegger, A.E., 1962. Stresses in the earth's crust as determined from hydraulic fracturing data. *Geol. Bauwes.,* 27(2): 45–53.

Skempton, A.W., 1953. Soil mechanics in relation to geology. *Proc. Yorks. Geol. Soc.,* 29: 33–62.

Taylor, D.W., 1948. *Fundamentals of Soil Mechanics.* Wiley, New York, N.Y., 700 pp.

Teodorovich, G.I. and Chernov, A.A., 1968. Character of changes with depth in productive deposits of Apsheron oil-gas-bearing region. *Sov. Geol.,* 4: 83–93.

Terzaghi, K., 1925. Principles of soil mechanics, I: Phenomena of cohesion of clay. *Eng. News Rec.,* 95: 742–746.

Terzaghi, K., 1936. Simple tests determine hydrostatic uplift. *Eng. News Rec.,* 116: 872–875.

Terzaghi, K. and Peck, R.B., 1948. *Soil Mechanics in Engineering Practice.* Wiley, New York, N.Y., 566 pp.

Van der Knaap, W. and Van der Vlis, A.C., 1967. On the cause of subsidence in oil-producing areas. *Proc. 7th World Petrol. Congr.,* 3: 85–95.

Voight, B. and Dahl, H.D., 1970. Numerical continuum approaches to analysis of non-linear rock deformation. *Can. J. Earth Sci.,* 7(3): 814–830.

Warner, D.L., 1964. *An Analysis of the Influence of Physical-Chemical Factors Upon the Consolidation of Fine-Grained Clastic Sediments.* Thesis, Univ. California, Berkeley, Calif., 136 pp.

Weller, F.A., 1959. Compaction of sediments. *Bull. Am. Assoc. Pet. Geologists,* 43(2): 273–310.

Chapter 4

EFFECT OF COMPACTION ON SOME PROPERTIES OF ARGILLACEOUS SEDIMENTS

INTRODUCTION

Sediment constituents exist in three physical states: solid, liquid and gas. The solid phase is represented by minerals and organic materials, the liquid phase by brines and hydrocarbons, and the gaseous phase by natural gases and air. The variability of these three states gives rise to nonuniformity of argillaceous sediments and variation in their characteristics and properties. In addition to the nonhomogeneity, specific interactions among the three phases may occur during compaction. Thus, it is difficult to predict precisely the response of the sediment to a set of external constraints.

Sediment is composed of minerals of various grain sizes with some organic molecules strongly bonded to the minerals and some physically admixed organic matter. The arrangement of the mineral particles influences the properties of the sediment upon compaction. Yet, many physical measurements are made on argillaceous sediment samples that are remolded or disturbed, i.e., not in their natural state. Consequently, when the results of these tests are evaluated, it should be remembered that the in-situ arrangement of particles in the sediment has a definite effect on properties. Particle-size analysis only gives the proportion of different-sized particles present and not their arrangement with respect to each other.

There are three fundamental clay-microstructure arrangements:

(1) *Random structures* having high void ratio arise in argillaceous sediments because of high interparticle repulsion. Dispersion of flocculated clay structure may give rise to random structures.

(2) *Oriented structures* can either be partially oriented or fully oriented. They are thought to be caused primarily by compaction effects and low interaction tendency of the clay particles.

(3) *Flocculated structures* are caused by a decrease in interparticle repulsion. (See Yong and Warkentin, 1966, pp.108–111.)

Inasmuch as there is a definite relationship between the properties of argillaceous sediments and their texture and pore structure, the first section of the present chapter is devoted to this subject. The permeability and rheological behavior of argillaceous sediments is discussed next, followed by elastic and strength properties. Compressibility is also discussed in detail.

Pore-size distribution in clays

Diamond (1970) determined the pore-size distribution and total porosity of a number of oven-dried, naturally-occurring subsoils and clays prepared in various ways by using a mercury injection porosimeter. Knowledge of the mathematical distribution of pore sizes in argillaceous sediments may assist in interpreting the behavior of the material undergoing compaction.

There are two main methods available for determining the pore-size distribution in a porous medium: (1) mercury injection, whereby mercury, a non-wetting fluid, is forced under pressure into the pores; and (2) capillary condensation, which involves the interpretation of adsorption or desorption isotherms in order to calculate that portion of the sorption that is due to capillary condensation in the solid. The latter method, however, is not fully applicable to clays because the maximum pore diameter that can be measured is of the order of 1,000 Å, much smaller than many of the pores present in clays.

Diamond (1970, p.10) presented experimental pore-size distribution data in the form of cumulative pore-diameter curves (Figs.67 and 68). The pore-volume parameter is ex-

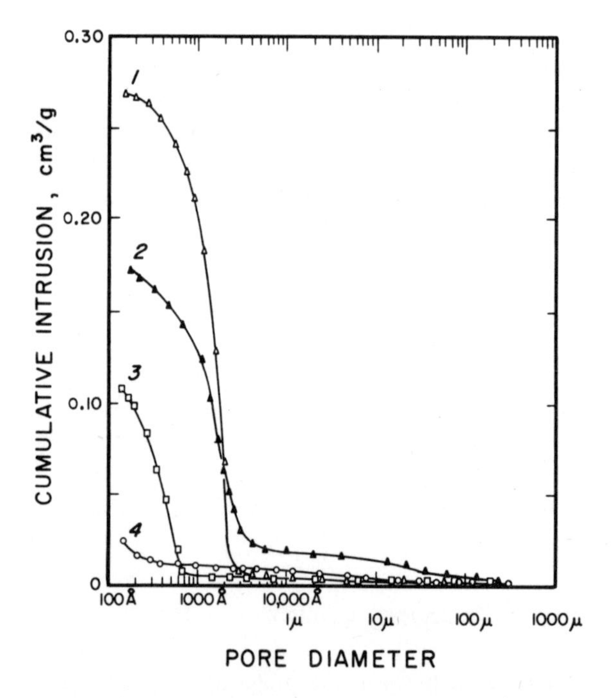

Fig.67. Pore-size distribution curves for American Petroleum Institute reference clay samples. *1* = Macon kaolinite (API no.4); *2* = Garfield nontronite (API no.33a); *3* = Clay Spur montmorillonite (API no.26); and *4* = Fithian illite (API no.35). Cumulative intrusion is expressed as cm^3 of pore space per gram of oven-dried clay. (After Diamond, 1970, p.13, fig.4. Courtesy of *Journal of the Clay Minerals Society.*)

pressed as cubic centimeters of pore space per gram of oven-dried clay. Reproducibility of the pore-size distribution curve was found to depend largely on the homogeneity of the sample. Replicate runs on homogeneous material traced the identical pore-size distribution curve with precision (see Diamond, 1970, p.12, fig.3). Where the clays and soils had a heterogeneous nature, Diamond (1970, p.11) presented only the runs which he judged most representative of a set of different replicates. Fig.67 depicts the pore-size distribution curves for various API standard clays. Fithian illite (API no.35) exhibits an extremely tight microstructure (Fig.67, curve *1*) which is attributed by Diamond (1970, p.11) to small particle size and to the geological consolidation processes.

Clay spur montmorillonite (Fig.67, curve *2*, API no.26) shows comparatively little intrusion by the mercury until reaching a pressure which is sufficient to cause intrusion of pores having 650-Å diameter. A majority of the pores (>90%) apparently have diameters ranging from 140 Å (lower limit of the assay) to 650 Å. According to Diamond (1970, p.13), the strong upward trend of the montmorillonite curve at the 140 Å point probably indicates that most of the residual unintruded pore space lies in diameter classes just beyond this lower limit of 140 Å. The Garfield nontronite (Fig.67, curve *3*, API no.33a) intrusion curve "breaks" around 5,000 Å, rather than at 650 Å, which indicates a wider range of pore diameters than for montmorillonite. Diamond (1970) pointed out that again most of the residual undetected pore space probably occurs as voids just beyond the 140 Å lower limit. For both the montmorillonite and nontronite clays, after oven drying, approximately three-quarters of the total pore space (accessible pores larger in equivalent

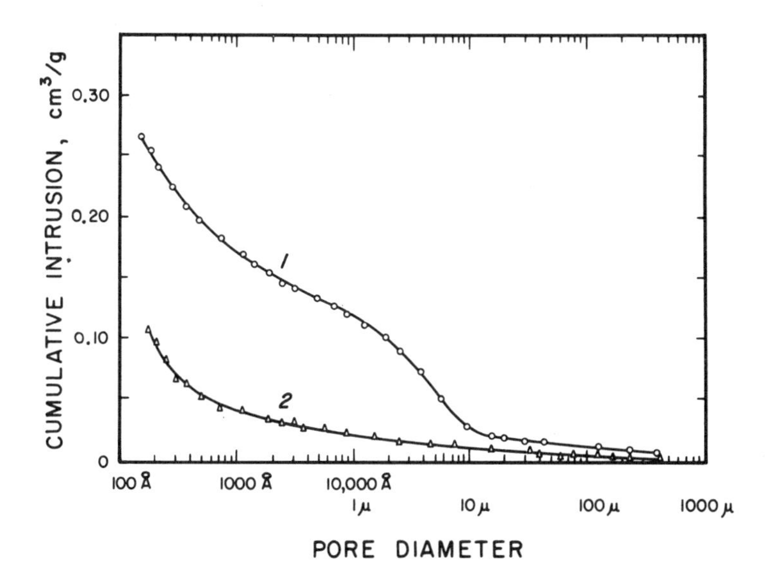

Fig. 68. Pore-size distribution curves for Bedford halloysite (API no.22). *1* = after oven drying; *2* = as received (moisture content = 13% by dry weight). (After Diamond, 1970, p.14, fig.5. Courtesy of *Journal of the Clay Minerals Society.*)

diameter than about 140 Å) could be intruded by mercury at the pressure range available (Diamond, 1970, p.12).

The curve for Macon kaolinite (API no.4) demonstrates that essentially all of the pore space present was intruded by mercury. Diamond (1970, p.13) pointed out that the majority of the pores in the kaolinite are in diameter classes within a narrow range of about 300 Å to about 2,000 Å, or else are accessible only through pores of such equivalent diameters (Fig.67).

Fig.68 shows results for the Bedford halloysite (API no.22), which exhibits a different pattern with respect to that recorded for the other clays. Diamond (1970, p.13) speculated that the difference between the "oven-dried" and "natural" states is due to the partial unrolling of the halloysite tubes during drying.

Pore-size distribution in subsoils is presented in Fig.69. These pore-size distribution curves are quite different from those shown in Fig.67. The Indiana subsoils differ from the API clays by having a higher total pore volume in general, with significantly larger diameter pores. Essentially all the pore space present in the oven-dried samples is available to mercury intrusion at pressures below 15,000 p.s.i. Diamond (1970, p. 15) pointed out that the content of pores less than 1,000 Å in equivalent diameter is negligible in the subsoils, despite the reasonably high clay content of these soils. This is in marked contrast to the API reference clays, which mainly contain pores having equivalent diameters less than about 3,000 Å; oven-dried halloysite is an exception.

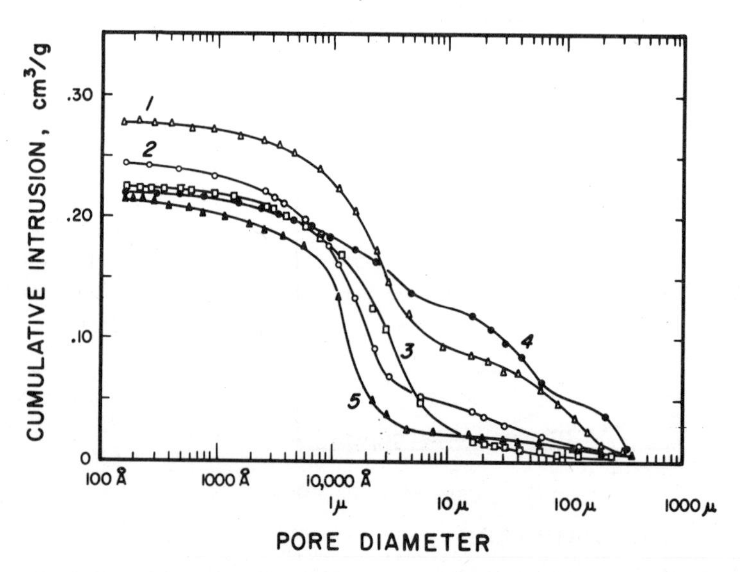

Fig.69. Pore-size distribution curves for samples of selected Indiana subsoils. *1* = Brookston silty clay loam; *2* = Crosby silt loam; *3* = Miami silt loam; *4* = Genessee silt loam; *5* = Fox silt loam. (After Diamond, 1970, p.15, fig.6. Courtesy of *Journal of the Clay Minerals Society.*)

MICROSTRUCTURAL CHANGES IN ARGILLACEOUS SEDIMENTS UNDERGOING COMPACTION

Overburden pressure seemingly causes preferred orientation of the mineral grains. Sections cut in clay samples normal to the pressure direction commonly show a higher concentration of elongated mineral axes than sections cut parallel to the pressure direction. The importance of clay-particles rearrangement is that it changes the permeability, porosity, and other properties of the argillaceous sediments.

Casagrande (1932a), Mitchell (1956), Tan (1957), Lambe (1958), Rosenqvist (1958), Quigley (1961), Martin (1962), Von Engelhardt and Gaida (1963), Pusch (1966, 1970), Meade (1968), and others have investigated the particle arrangement of clays in Recent argillaceous sediments and soils and have presented various idealized images of the observed microstructures. Some studies on the texture of shales and fine-grained sediments have been performed by Bates and Comer (1955), Droste et al. (1958), White (1961), and Gipsen (1965).

Experiments by O'Brien (1963, pp.20–40) on kaolinite and illite pastes and by Martin (1965) with kaolinite slurries indicate that, given sufficient initial water content, reorientation of the clay-mineral grains occurs at pressures less than 1,422 p.s.i. (100 kg/cm^2). Martin (1965) remarked that a pressure of 14.22 p.s.i. (1 kg/cm^2) produced preferred grain orientation, which did not change upon further compaction up to 455 p.s.i. (32 kg/cm^2). Rosenqvist (1959) observed that there is better particle orientation in fresh-water clays than in marine clays.

Rosenqvist (1958) made replicas of freeze-dried clay samples and concluded that the clay-particles arrangement of virgin marine clays is of the type suggested by Tan (1957) (Fig.70). Pusch (1966) investigated the microstructure of Swedish quick clay by applying an ultra-thin sectioning technique and transmission electron microscopy. Quick clay is an ultra-sensitive clay deposited and consolidated in a flocculated state, with subsequent increase in the double-layer potential either by leaching of electrolytes or by addition of certain ions (Rosenqvist, 1966, p.445). Pusch's experimental procedure involved the stepwise replacement of the pore water in the clays with acrylate plastic. Ultra-thin sections 500-Å thick were cut using a precision LKB Ultrotome microtome. These sections were photographed using a Siemens Elmiskop I (80 kV voltage; 50 μ platinum–iridium aperture) electron microscope and the micrographs obtained were used for a study of the clay microstructure.

Pusch (1966) observed that the microstructure of the leached marine quick clay was characterized by a linkage of groups or chains of small particles in and between denser flocs or aggregates, or between larger particles (Fig.71). There was no preferential orientation either of the large or small clay particles in the quick clay of marine origin. The arrangement of the large clay flocs linked together in chains by small particles is almost identical to the hypothetical floc structure postulated by Pusch in 1962 (pp.53, 55). Rosenqvist (1958), on the other hand, showed that marine-clay particle arrangement was

Fig.70. Schematic diagram of Tan's card-house structure in clays. (After Tan, 1957, fig.3, p.87; also: Tan, 1966, fig.1, p.257. Courtesy of Springer Verlag.)

of the type suggested by Tan in 1957, i.e., a card-house type of particle arrangement.

The dimensions of the pore space and particles in the quick clay were defined in accordance with Pusch's earlier concepts used for clay particles (Pusch, 1962) (Fig.72). Histograms of the maximum pore dimensions (a_p) in two samples of the Swedish quick clay are presented in Fig.73. The median, skewness and sorting coefficient of the pores are presented in Table XII. Pusch (1966, p.437) stated that the frequency of smaller pores is probably not representative in the samples tested, because the thickness of the clay sections did not reveal a certain number of very small pores. Furthermore, pores having an a_p value smaller than 0.03 μ could not be identified with certainty. He also observed that the relative pore areas in marine clays are considerably larger than those in

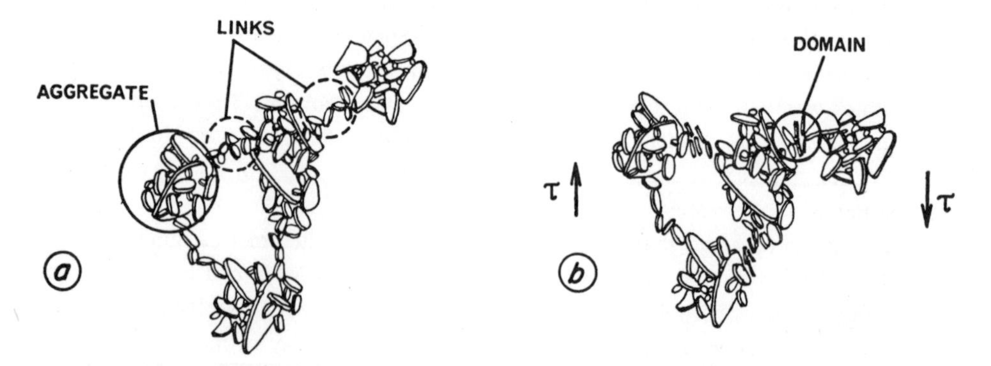

Fig.71. Schematic diagram of microstructure in quick clay and the failure process. a. Natural microstructural pattern. b. Breakdown of particle links resulting in domain formation. (After Pusch, 1970, p.7, fig.8. Courtesy of *Canadian Geotechnical Journal.*)

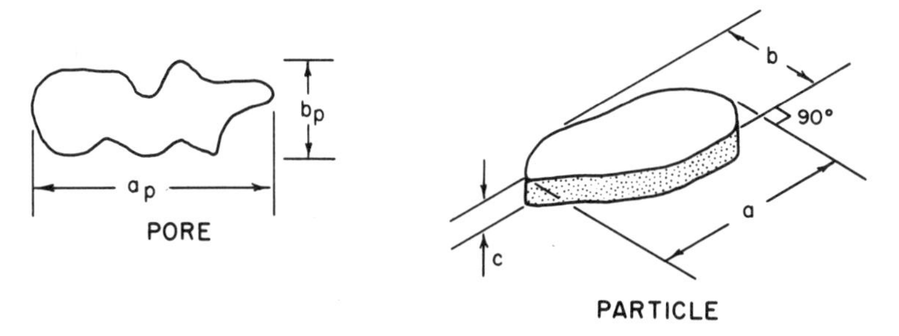

Fig.72. Schematic picture of a pore and clay particle where *a* is the maximum diameter; *b* is the intermediate diameter; and *c* is the thickness of the clay particle. (After Pusch, 1966, 1970.)

fresh-water clays with similar water contents and bulk densities. This is in close accord with the findings of the authors that the permeability of mud cake deposited from a flocculated system is considerably higher than that deposited from a deflocculated clay-plus-water system.

Pusch (1970) studied microstructural changes in a soft marine quick clay from Molndal, Sweden, undergoing unconfined compression in the laboratory. Small specimens for

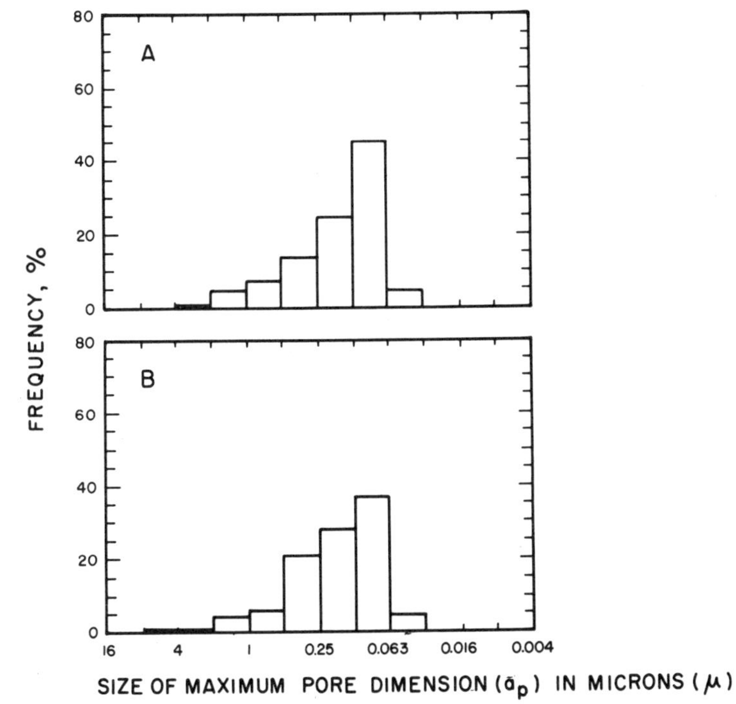

Fig.73. Histogram of the maximum pore dimensions. A. Sample 4643:1. B. Sample 4643:2 (see Table XII). (After Pusch, 1966, p.442, fig.7 and 8. Courtesy of *Engineering Geology*.)

TABLE XII
Pore parameters of two samples from a Swedish quick clay [1]
(After Pusch, 1966, table 2, p.438. Courtesy of *Engineering Geology*)

	Sample 4643 (Strandbacken)	
	micrograph 16252	*micrograph 16257*
a_p (μ)		
M	0.13	0.17
Q_1	0.09	0.10
Q_3	0.25	0.33
a_p/b_p		
M	1.92	1.99
Q_1	1.58	1.54
Q_3	2.40	2.60
S_k of a_p	1.33	1.14
S_k of a_p/b_p	1.03	1.01
log S_o of a_p	0.221	0.259
log S_o of a_p/b_p	0.091	0.114
Pore area (% of total area)	47.8	48.2

[1] Explanation of symbols: M = median value; Q_1 = lower quartile; Q_3 = upper quartile; S_k = skewness = $(Q_1Q_3)/M^2$ (Trask, 1930, 1932); S_o = sorting coefficient = Q_3/Q_1. Log S_o values are directly comparable.

microstructural analysis were cut from samples which had been loaded axially to failure. The natural microstructural pattern of the clay was characterized by a network of small aggregates connected by links of particles. Links between the particles broke down successively on increasing shear deformation and formed domain-like (oriented) groups of particles. The shear mechanism seems to involve movement of the aggregates as units in conjunction with deformation of their links. Pusch (1970, p.6) concluded that at high shear stresses the motion of the aggregates caused large deformations of the connecting links; this resulted in a parallel orientation of the linking particles (Fig.71). The residual strength may correspond to the state where the links are broken due to large shear deformation. A reduction in strength (by 30–50% of the peak shear-strength value) may be caused by dilatancy when the rigid aggregates are approached. Remoulding, which causes successive breakdown of the aggregates, results in further decrease in shear strength. The rigidity of the aggregates seems to govern the shear process. High viscosity of the pore water at low water contents of 15–30% (Rosenqvist, 1959), together with the strong bonding of the closely located individual particles in the aggregates, gives the clay mass considerable rigidity during compaction.

Meade (1968, p.D29) studied the fabric of clays in loosely consolidated sandstones and noted that there was very little indication of progressive development of preferred grain orientation with increasing depth of burial. He concluded that the lack or poor development of the montmorillonite particle orientation was due to the mode of deposition and the water content of the sediment at burial. Raitburd (1960, p.113) reported completely random orientation of illite and kaolinite particles in a marine clay from the Black Sea. Kaarsberg (1959), on the other hand, established that the preferred orientation of illite increased with the depth of burial in flat-lying shales at depths greater than 2,700 ft. White (1961), p.564), however, presented evidence that this did not occur in Paleozoic shales from Illinois. He stated that fissility does not vary systematically with burial depth.

As pointed out by Diamond (1970, p.17), field compaction of soils for engineering purposes involves the application of as many passes of heavy equipment as are required to attain a specific bulk density, after adjusting the loosened soil to an appropriate moisture content. For a given level of compactive effort, the optimal moisture content for a given soil can be determined by repeated impact of a weight falling on the soil from a specified height (standardized test). Compaction at a moisture content that is significantly different from the optimum gives rise to less dense soils, with inferior ability to support loads (Diamond, 1970, p.17).

Pore structure of compacted clays and soils may be of value in defining the details of the compaction process and in determining the degree of compaction. Experimental compaction studies of Edgar Plastic kaolin (a pure commercial clay) by Diamond (1970, p.17) showed that two different microstructure configurations could occur depending on the initial moisture content of the kaolinite. If compaction was carried on at low moisture contents, or with insufficient compactive effort for the moisture content present, a microstructure is produced which does not shrink appreciably on oven drying and in which a significant number of relatively large-diameter pores, ranging up to several microns, are present (Diamond, 1970, p.19). When the kaolinite is compacted at or above the optimum moisture content, very significant shrinkage takes place on oven drying, and the resultant pores have equivalent diameters of less than 2,000 Å. According to Diamond (1970, p.19), these results suggest that the former kaolinite samples have retained much of the structure of the individual, more-or-less spherical aggregates of particles formed in the mixing process, whereas the more effective compaction in the latter case has brought the clay particles into a more nearly homogeneous, quasi-parallel arrangement.

The problem of particle orientation is a complex one. There are many variables which have to be considered, including type of mineral(s), grain shape and size, initial fluid content, electrolyte content of the interstitial fluid, the type of clay-plus-water system from which clay is initially deposited (deflocculated, flocculated, etc.), and the magnitude of overburden pressure.

X-ray diffraction techniques used in compaction studies

Several X-ray diffraction techniques used in investigating the relationship between overburden pressure and changes in the clay fabric have been reported by Buessem and Nagy (1954), Silverman and Bates (1960), Meade (1961), Quigley (1961), Von Engelhardt and Gaida (1963), Meade (1964), Fayed (1966), and others. The diffraction technique probably constitutes the best method for quantitatively measuring particle parallelism in argillaceous sediments.

Von Engelhardt and Gaida (1963) investigated the rearrangement of the clay particles with pressure by cutting laboratory-compacted clay samples at an angle and using a standard goniometer. The intensity of basal plane X-rays reflected from the flat clay surface gives an indication of the number of clay particles with their (001) plane parallel to the sample's surface. As the clay platelets become more highly oriented in a given plane, the X-ray intensity of the basal reflections increases. A high degree of clay-particle parallelism will produce strong (001) reflections, whereas a randomly oriented fabric will give weak (001) reflections. If there is random orientation, the intensity of basal reflections should be the same for all sections cut at different angles through the sample. The degree of particle parallelism can be inferred from measurements of the amplitude of the (001) peaks obtained on the horizontal, smoothly-ground surfaces. Quigley and Thompson (1966, p.64) presented a detailed description of another method of clay sample preparation for such determinations.

Quigley (1961) suggested that increases in the concentration of kaolinite in the surface being X-rayed cause an increase in the peak strength of the (001) line and (hko) reflections independent of parallelism. Martin (1962) expressed fabric (arrangement of the clay particles) in terms of the ratio of the intensity of the (001) peak to (hko) peak so that the concentration effect was eliminated. Von Engelhardt and Gaida (1963, p.925) showed that montmorillonite and kaolinite clay minerals have a tendency to orient themselves normal to the applied pressure direction with increasing overburden pressure (Fig.74). They also discovered that at the same pressure this orientation is improved in the presence of interstitial solutions having low electrolyte concentrations. Probably, the increase in peak heights, at high pressures, is mainly due to densification of the clay-mineral crystal lattice.

In 1968, however, Von Engelhardt (personal communication) stated that his earlier work (Von Engelhardt and Gaida, 1963) was in error: "In the earlier work with Gaida, the sections were prepared through the compressed clay samples at different angles to the

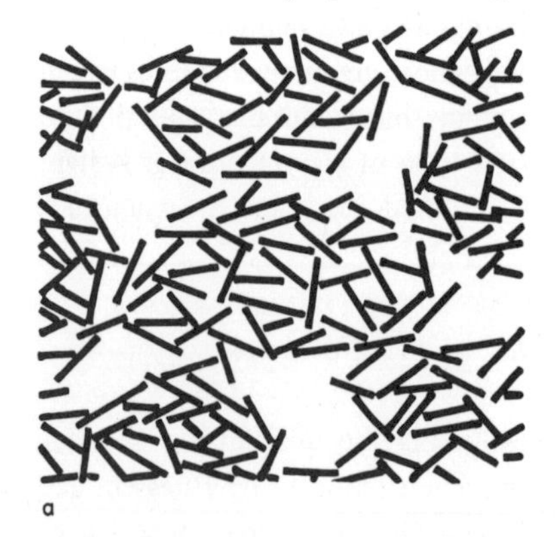

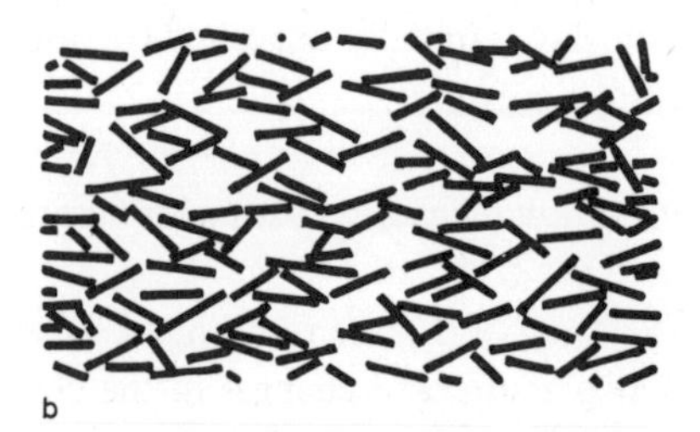

Fig.74. Particle orientation of compacted clay. a. Pore solution of high salinity. b. Pore solution of low salinity. (After Von Engelhardt and Gaida, 1963, p.927, fig.11. Courtesy of *Journal of Sedimentary Petrology*.)

direction of the pressure. The intensity of the (001) reflection was determined for each section. This was not an accurate method. In the meantime we have made new measurements with a texture goniometer, which allows one to incline the plane surface of the compressed sample towards the incidence plane of the X-ray beam. So, we were able to measure for one sample the intensity of (001) peak in dependence on the angle to the pressure direction. The new measurements demonstrate that the results presented in our former publication with Gaida (1963) have to be corrected: based on our experiments there is no dependence of the degree of orientation (textural changes) on the applied pressure (cylinder and piston)." There was no further elaboration by Von Engelhardt on this point.

Quigley and Thompson (1966) studied the fabric of the Leda Clay (composed mainly of illite and chlorite) with respect to axially applied pressure. They concluded that anisotropic consolidation of the Leda Clay produced reorientation of the clay platelets into the plane perpendicular to the direction of the consolidating pressure. The maximum pressure attained in the consolidometer by these investigators was approx. 900 p.s.i. It is probable that reorientation of the fabric of the clays studied by them takes place under 1,000 p.s.i., whereas above 1,000 p.s.i. the increased peak heights can be attributed to an increase in the crystallinity of the clays.

Fayed (1966) obtained results from studying thin sections, which indicated that the basal planes of the micaceous minerals in slate are mostly oriented in a direction parallel to the cleavage plane and the possibility of another orientation was considered to be slight.

Crystal structural changes in clay minerals at different temperatures and at elevated pressures

The effect of stress at low temperature on the structure of various clay minerals and clay-sized particles has been studied by several investigators (e.g., Chilingar et al., 1963a; Rieke et al., 1964; Demirel et al., 1970; and Rieke, 1970). Comparative studies between structures of the various clay types exposed to different overburden pressures, based on DTA scans, X-ray diffraction patterns, and data from the scanning electron microscope, could result in better understanding of the compaction mechanism and the structural behavior of clay minerals before mineral conversion takes place. Internal structural changes in clays result in variations in their mechanical and physical properties.

Crystal-structural analysis normally consists of finding the diffraction spectra of a structure which match the observed diffraction set. Although single "crystals" of a given clay mineral cannot be studied individually in most high-pressure apparatuses, a representative diffraction spectrum can be easily obtained from the phyllosilicate mass which would reflect a change in the structure of the clay. The X-ray pattern demonstrates an averaging effect of many clay particles, most of which are imperfect crystals. Changes in the crystal structure may be indicative of a phase change in the clay mineral. This may be characterized by changes in the unit-cell dimensions, disappearance or appearance of peaks, and size and shape changes in the established peaks on the diffraction pattern.

An excellent example of this is the transition in the polymorphic forms of natural dioctahedral micas under the influence of low-grade metamorphism. Maxwell and Hower (1967) investigated the crystallographic response of illite to increasing stratigraphic depth. They observed the transformation of 1Md to 1M to 2M polymorphic forms of illite (all dioctahedral micas) in the Precambrian Belt Series (Montana) with depth and increasing temperature. The 2M illite polymorph is stable above the temperatures of 200 to 350°C at 15,000 p.s.i. Below this temperature range, the 1M polymorph is apparently the stable form with metastable 1Md being the initial phase at all temperatures, but persisting at lower temperatures. The percentage of 2M illite relative to total illite was determined by obtaining the intensity ratio of an unique 2M peak (2.80 Å) to the standard peak (2.58 Å), which is found in all the dioctahedral mica polymorphs (Maxwell and Hower, 1967, p.847).

Dickite (API no.15). A comparative study among the various X-ray powder diffraction peaks was made by Chilingar et al. (1963a) on dickite clay, a kaolin-family mineral, after exposure to different high pressures at room temperature. Samples of dickite clay, saturated with distilled water, were subjected to overburden pressures of 0 (initial sample), 15,000, 120,000 and 386,000 p.s.i.

There appeared to be no symmetrical trend in size, shape, or height of the peaks, and the X-ray diffraction pattern tends to remain uniform at all pressures. Intensities of the basal reflections of the various samples are approximately equal and show that the preferred orientation of the clay mineral on the prepared slides used in this analysis was approximately the same in all cases. No lateral shifts are discernible in the "d" spacings of the peaks, indicating that the unit-cell volume remained constant with increasing pressure. Schmidt and Heckroodt (1959, p.317) reported that dickite develops a 14 Å reflection after being heated at 600–650°C for one hour. They also stated that the reflections at 7.2 (7.25), 3.57 (3.60), and 2.33 Å did not completely disappear at this temperature. Chilingar et al. (1963a) showed that a 14 Å reflection did not appear nor did the other above-mentioned peaks diminish with increasing overburden pressure at low temperature. Consequently, these dickite peaks are not pressure sensitive. Dickite no.15a is from San Juanito, Chihuahua, Mexico, and was probably formed in situ by the alteration of a porphyritic feldspathic rock, chiefly by hydrothermal action. Although dickite is primarily known as a hydrothermal clay, considerable amounts of the mineral have been observed disseminated in some sandstones in South Africa and Australia.

Halloysite (API no.12). Halloysite is found in various structural forms. The basal spacing of the dehydrated form ($2H_2O$) is about 7.2 Å, whereas the basal spacing of the completely hydrated form ($4H_2O$) is approximately 10.1 Å (Grim, 1968). The difference between the two basal spacings is 2.9 Å, which is the approximate thickness of a single molecular layer of water. The X-ray diffraction pattern of halloysite clay shows that it is in a partially hydrated form (metastable state). The metahalloysite is characterized by a basal-plane spacing ranging from 7.36 to 7.9 Å. According to Grim (1968, p.72), this corresponds to a hydration of about 0.5–1.5 H_2O. The halloysite structure is not fully understood by clay mineralogists, and some investigators feel that it does not belong to the kaolin family of clay minerals.

The X-ray powder diffraction patterns were obtained from halloysite samples, exposed up to pressures of 500,000 p.s.i. (Rieke et al., 1964). The most significant change in the diffraction pattern of halloysite occurred between the pressures of 0 and 1,000 p.s.i. where the 7.45 Å (001) peak becomes sharply defined. An increase in overburden pressure between 1,000 and 15,000 p.s.i. (see Chilingar et al., 1963a, p.1046) does not alter the halloysite X-ray pattern. The 4.40 Å peak is the one with the highest intensity and is highly characteristic of halloysite. The results indicate that the intensity of the 4.40 Å peak (110) remains dominant with respect to the other peaks at all pressures. Drying halloysite at 110°C creates strong reflections at about 7.2 and 3.6 Å (Grim, 1968). This sharpening also showed up on the patterns of Rieke et al. (1964) and Chilingar et al. (1963a) at 7.45 and 3.65 Å with increasing pressure, whereas the heights of 2.55, 2.46, and 2.40 Å peaks are different at different pressures. The intensities of the latter three peaks are sharpest at 400,000 p.s.i. Evidence for no prior pressure exposure of the halloysite clay is that the 7.45 and 3.65 Å peaks become well-defined with increasing pressure. The strong orientation method as described by Aoyagi (1967) was used by Chilingar et al. (1963a) in determining all X-ray powder diffraction patterns, and there was apparently better particle orientation in some slides than in others. This was evidenced by the variation of the height and width of the (001) peak.

Hectorite (Stevensite) (API no.34). The X-ray powder diffraction patterns were determined for samples of hectorite clay from Hector, California (a mixture of stevensite and 50–58% by weight of $CaCO_3$), exposed to autoclave pressures up to 200,000 p.s.i. (Rieke et al., 1964). No significant changes in the diffraction patterns were observed; however, the 2.57 Å line tends to become very weak with increasing pressure (Fig.75).

Davis (1964) studied the phase changes in calcite (calcite I, II, and III; and calcite to aragonite) under increasing pressure. He utilized the change in the calcite line (2.285 Å) (113 peak) as an indicator of the phase transition of calcite I to calcite II. The loss of the (113) peak from the diffraction pattern was taken as an indication of the completed transition (this transition takes place at 15.5 ± 1 kb at 25°C, equivalent to 224,780 ± 14,500 p.s.i.). Rieke (1970) analyzed the pressure effects on the calcium-carbonate impurity in the hectorite clay. Results are presented in Table XIII which show that the phase transition of calcite I to calcite II was being approached as noted by the decrease in the length of the crystallographic axis. The (113) peak also diminished considerably with increasing pressure (Fig.75).

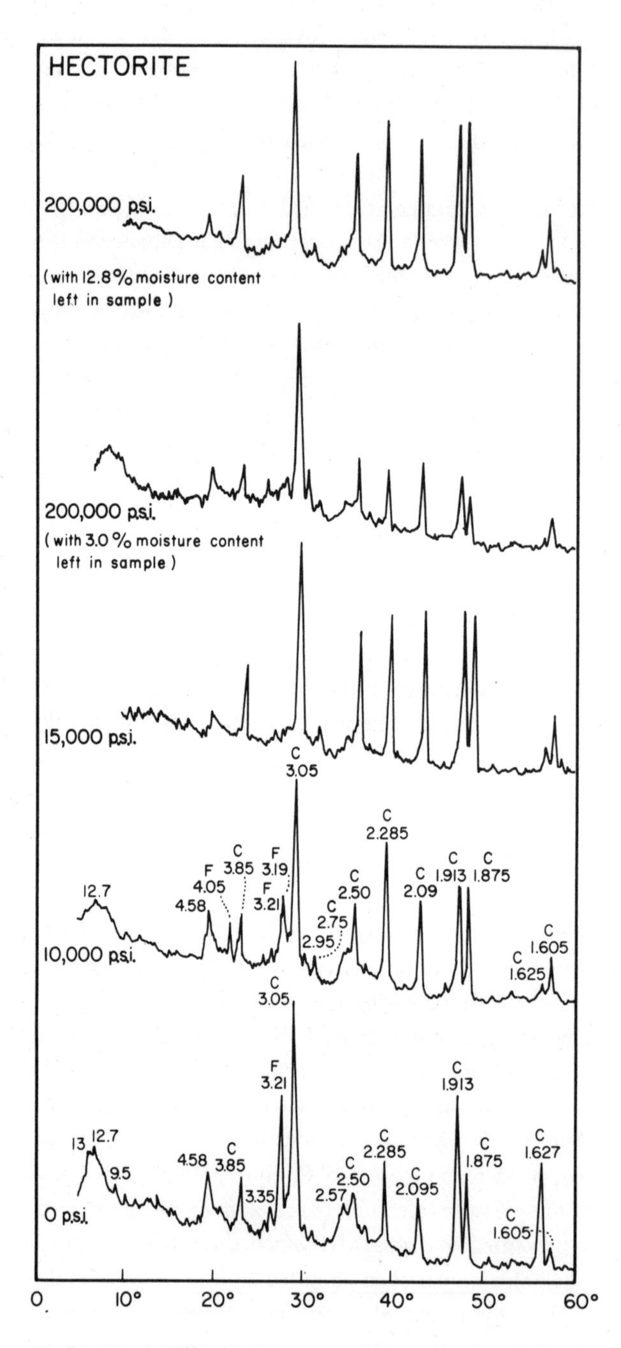

Fig.75. X-ray diffraction patterns of hectorite clay (original sample and samples exposed to 10,000, 15,000 and 200,000 p.s.i. pressure). *F* = feldspar. (After Rieke et al., 1964, fig.5, p.30.)

TABLE XIII
X-ray crystallographic data and grain density for the various phases of calcium carbonate (After Rieke, 1970, p.377)

Phase	*Crystal system*	*Crystallographic axes (A)*			*Z* *	*X-ray density* [1] *(g/cm*^{3}*)*	*Source*
		a_o	b_o	c_o			
Calcite I (0 p.s.i.)	hexagonal–rhombohedral	4.989	–	17.064	6	2.710	Robie et al. (1967)
Calcite I (0 p.s.i.)	hexagonal–rhombohedral	5.035	–	17.167	6	2.646	Rieke (1970)
Calcite I (200,000 p.s.i.)	hexagonal–rhombohedral	5.020	–	17.094	6	2.673	Rieke (1970)
Calcite II (224,809 p.s.i.)	hexagonal–rhombohedral	4.980	–	16.830	6	2.758	Davis (1964)
Calcite III (271,222 p.s.i.)	orthorhombic	8.900	8.900	7.140	10	3.171	Davis (1964)
Aragonite (0 p.s.i.)	orthorhombic	4.949	7.968	5.741	4	2.930	Davis (1964)

* g-formula weights.
[1] Room temperature: 25 ± 5°C.

Montmorillonite. Demirel et al. (1970) studied the variation of the "d" spacing of a calcium montmorillonite under high pressure. A specially built high-pressure apparatus was mounted on an X-ray diffractometer in order that diffraction patterns could be continuously obtained with each increasing pressure increment. The calcium montmorillonite sample was laterally confined and vertically compacted between a beryllium window and a porous nickel plug saturated with water.

Removal of water corresponding to a complete change of d_{001} spacing from 19.5 Å to 15.7 Å occurred at a pressure of 300 kg/cm^2. Demirel et al. (1970) pointed out that at pressures lower than this, both spacings coexisted. A single peak corresponding to 19.5 Å exists at 0 pressure. After the interstitial water is squeezed out and the pressure is released, the expansion from 15.7 Å to 19.5 Å takes place at a pressure lower than 50 kg/cm^2, indicating a pronounced spacing hysteresis. At a pressure of 1,700 kg/cm^2, two peaks corresponding to about 15 Å and 12 Å coexist.

Table XIV presents a summary of X-ray powder diffraction data for a sodium montmorillonite hydrated in ASTM (D-141-52) substitute sea water and compacted under elevated temperature and pressure. Quartz was added to the initial clay sample as an internal reference pattern. All "*d*" spacings were measured with respect to the 4.27 Å quartz peak. Shifts in the *d* spacing of air-dried sodium montmorillonite can be related to pressure–temperature effects and/or chemical pretreatment of clay.

TABLE XIV
X-ray powder diffraction data for a sodium montmorillonite hydrated in A.S.T.M. substitute sea water and compacted at elevated temperatures and pressures

	Run				
	Initial sample	*1*	*2*	*3*	*4*
Pretreatment	none	none	none	NaOH+KCl	HCl+KCl
Temperature (°F)	70	270	330	300	300
Pressure (p.s.i.g.)	0	4000	1000	3000	3000
pH	n.a.[1]	7	7	12.9	0.5
Time of run, days	n.a.	7	10	7	7
Initial moisture content (% dry wt)	n.a.	71	71	68	70
	d (Å)	*d* (Å)	*d* (Å)	*d* (Å)	*d* (Å)
Air dried					
001 *	12.1	12.0	13.0	11.5	10.7
002	6.00	––	6.40	––	––
003	––	––	––	––	––
110, 020	4.48	4.48	4.48	––	4.49
004	3.12	3.14	3.16	3.19	3.19
006	––	––	––	––	––
007	2.06	––	––	––	––
060	–	–	–	–	1.50
Glycolated					
001	17	17	19	17.5	18
002	8.5	8.5	9	9.3	8.4
003	5.65	5.65	5.8	5.55	5.66

[1] n.a. = not applicable;
* = hkl indices; –– = peak not present; – = region not scanned.

Silica. Quartz is a very common mineral in most shales and, therefore, its stability and origin is briefly discussed here. As far as could be determined from a literature survey, there are no published data on the effect of high pressure at low temperatures (< 100°C) on the X-ray diffraction pattern of quartz. All previous work on silica was performed at very high temperatures and pressures. Two well-documented polymorphs of silica, stishovite and coesite, are formed at pressures of 75–125 kb and temperature of 800°C, and 40 kb and 700°C, respectively. The α-quartz is stable below 573°C at atmospheric pressure, whereas β-quartz forms above 573°C; however, the latter reverts to α-quartz when the melt is cooled below 573°C.

Mizutani (1966) investigated the transformation of amorphous silica to a crystalline state of low-cristobalite and low-quartz. Silica is considered to be transformed by a solid

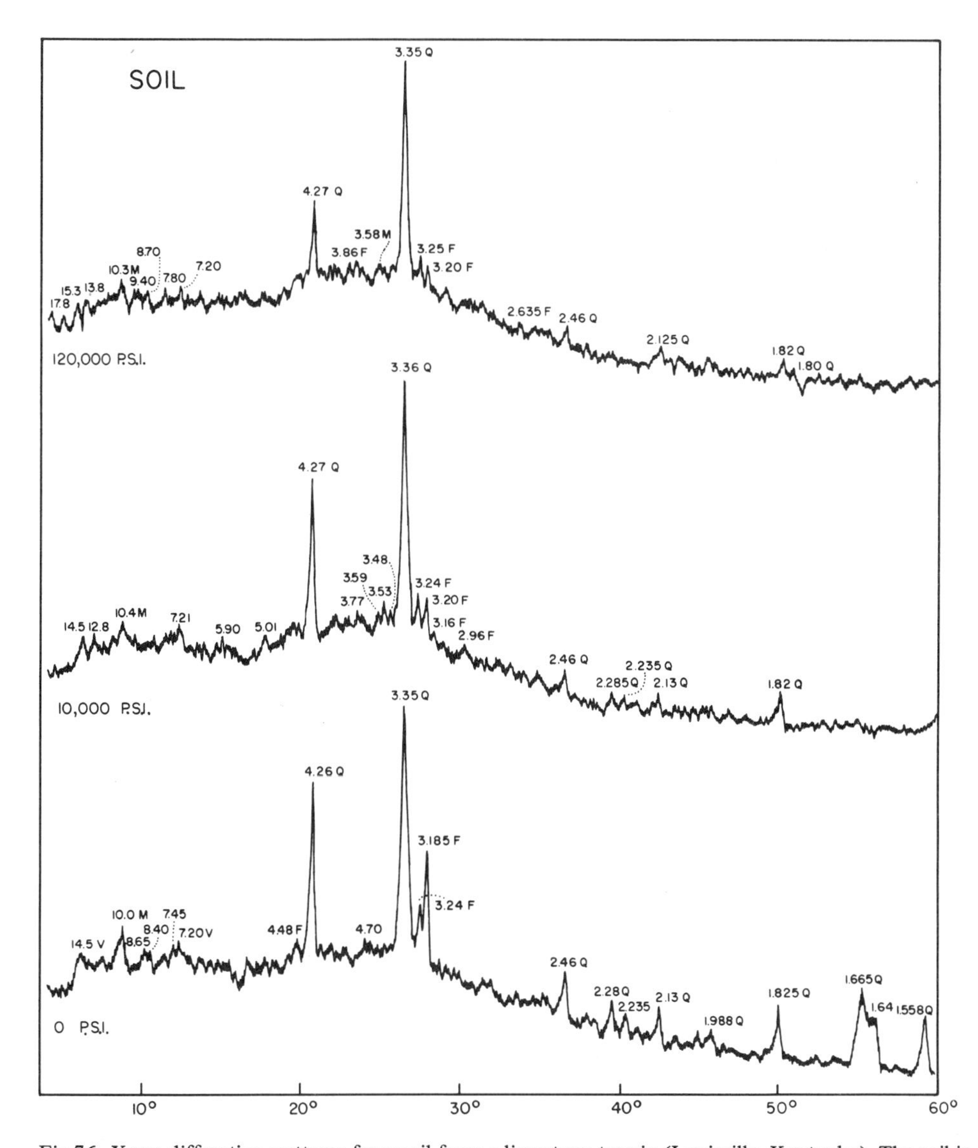

Fig.76. X-ray diffraction patterns for a soil from a limestone terrain (Louisville, Kentucky). The soil is a naturally occurring mixture composed of vermiculite (*V*), illite (mica; *M*), chlorite (*C*), feldspar (*F*), and quartz (*Q*). It was exposed to overburden pressures of 10,000 p.s.i. and 120,000 p.s.i. (After Rieke, 1970, fig.7-7, p.386.)

reaction and by precipitation of dissolved silica from solutions, inasmuch as the crystallization of quartz involves both a solid reaction and precipitation. The rate of transformation depends on the reaction temperature which ranged up to 300°C in Mizutani's laboratory experiments. Apparently, the younger sedimentary rocks contain amorphous silica or cristobalite, whereas in the older rocks (Mesozoic and older) these siliceous materials have been transformed into quartz. The X-ray powder diffraction patterns illustrating variation in the various diffraction peaks of quartz were presented by Mizutani (1966, pp.63 and 78).

Rieke (1970) observed that minor diffraction peaks of clay-sized quartz particles such as at 2.46, 2.28, 2.235, 2.13, 1.988, 1.925, 1.655, 1.64, and 1.558 Å tend to decrease and disappear with increasing pressure (Fig.76). It should be noted, however, that not all the minor quartz peaks were reduced with increasing pressure. The heights of feldspar peaks decreased also with increasing pressure (Fig.76). This diminution effect could be easily attributed to sample preparation (slides).

As shown in this section, X-ray powder diffraction analysis possibly can be used as a tool in the investigation of the stress and temperature history of shales. The response of various minerals to increasing pressure is one of transformation and phase change in their crystal structure. The X-ray analysis made on compacted clay samples may be perfected in the future to estimate the approximate maximum overburden pressure (and temperature) experienced by a shale. Considerable laboratory and field work, however, remains to be accomplished on various clays and minerals under different evironmental conditions. It is suggested by the writers that initially individual clay minerals in the pure state should be investigated thoroughly at various temperatures and pressures.

PERMEABILITY

Credit for the early work on permeability, designated by the symbol k, is usually given to Henry Darcy (1856). The American Petroleum Institute Code no.27, entitled "Standard Procedure for Determining Permeability of Porous Media", discussed many special cases of measuring permeability.

The conventional laboratory technique for determining the permeability or the hydraulic conductivity of saturated clays consists of: (1) placing the clay in a test cell similar to that used by Macey (1940) or McKelvey and Milne (1962) (Figs.197 and 198, pp.367–368), which is connected in series with a capillary tube containing an air–water meniscus or an air bubble; (2) establishing fluid flow under a differential pressure across the sample; and (3) determining the induced flow rate from the rate at which the air bubble or meniscus moves through the capillary tube. A plot of the pressure drop versus the flow rate is a straight line, the slope of which is equal to $q/\Delta p$. A detailed experimental procedure is presented by Young et al. (1964).

Mesri (1969) described special sedimentation tubes in which slurries can be consolidated to pressures ranging from 90 to 125 p.s.f. Samples are then transferred to 2.5-inch diameter consolidation rings and pressure is raised in increments to 64,000 p.s.f., for example, using a pressure ratio of two. The coefficients of permeability can be calculated by fitting Terzaghi's (1943) theory of consolidation to the observed laboratory time-settlement observations and extracting the coefficient of permeability from the calculated coefficient of consolidation.

The method employed by Chilingar et al. (1963a) consisted of measuring the water loss of a clay under constant pressure. Basic assumptions involved were: (1) pressure drop across the length of the sample is the effective pressure on the sample, (2) the moisture content of the clay at a constant overburden pressure varies with time before equilibrium is reached, (3) the water moves in equal amounts towards the upper and lower vents (Fig.208, p.380), (4) the gravitational forces are neglected at the laboratory scale, and (5) Darcy's law holds true.

The simplest expression for Darcy's law of flow through porous media neglects the effect of gravity and refers to a porous medium of constant cross section. All forms of Darcy's law are restricted to the nonturbulent or viscous region of flow. With these limitations in mind, Darcy's equation may be expressed as:

$$q = \frac{kA\,\Delta p}{\mu L} \tag{4-1}$$

where q = rate of flow of one homogeneous single-phase liquid (cm^3/sec); k = permeability in darcys (d); A = cross-sectional area of porous medium (cm^2); Δp = pressure drop along length of sample (atm); μ = viscosity of liquid in centiposes (cp); and L = length of sample (cm).

Work by Darcy and others showed that the rate of flow is directly proportional to the area and the pressure drop, and inversely proportional to the viscosity of flowing fluid and the length. The proportionality constant k, has been called the permeability and its unit has been given the name "darcy" when the units of other variables are as indicated above. A more common unit is the millidarcy (md) which is equal to 0.001 darcy.

Some investigators report permeability as the hydraulic conductivity (K), which was defined by Hubbert (1940) as:

$$K \equiv (k\rho_w g)/\mu \tag{4-2}$$

where K is the hydraulic conductivity (cm/sec), ρ_w is the density of the fluid (g/cm^3), g is the gravity acceleration (980 cm/sec^2), k is the permeability (cm^2), and μ is the viscosity of the fluid (cp).

As pointed out by Katz and Ibrahim (1971), the terms of hydraulic conductivity (K) and permeability (k) are usually used interchangeably in the literature. The units of hydraulic conductivity, K, are in cm/sec, and the following approximate expression for the relationship between K and k (in md) can be written as:

$$K = 9.6 \cdot 10^{-7}\, k \tag{4-3a}$$

or:

$$k = 1.033 \cdot 10^{6}\, K \tag{4-3b}$$

TABLE XV
Selected data on permeabilities of various clays and shales

Type of material	*Permeability* (md)	*Porosity* (%)	*Load* [1] (p.s.i.)	*Investigator*
Wyoming bentonite [2]	$3.3 \cdot 10^{-3}$	– *	7	Lutz and Kemper (1959)
Utah bentonite [2]	$8.44 \cdot 10^{-1}$	–	7	Lutz and Kemper (1959)
Utah bentonite [3]	$5.3 \cdot 10^{-2}$	–	7	Lutz and Kemper (1959)
Bentonite	$2.1 \cdot 10^{-6}$	39	–	Kemper (1961)
Wyoming bentonite	$4 \cdot 10^{-6}$	34	40,000	McKelvey and Milne (1962)
Wyoming bentonite	$5 \cdot 10^{-6}$	41	40,000	McKelvey and Milne (1962)
Wyoming bentonite	$1.35 \cdot 10^{-3}$	90.6	–	Warner (1964)
Wyoming bentonite	$1.45 \cdot 10^{-3}$	84.7	–	Warner (1964)
Wyoming bentonite	$6.26 \cdot 10^{-6}$	66.7	–	Warner (1964)
Wyoming bentonite	$2.9 \cdot 10^{-7}$	56.5	–	Warner (1964)
Kaolinite	1.27	59.6	–	Warner (1964)
Kaolinite	1.03	54.2	–	Warner (1964)
Kaolinite	$8.16 \cdot 10^{-1}$	52.4	–	Warner (1964)
Kaolinite	$2.08 \cdot 10^{-1}$	51.9	–	Warner (1964)
Kaolinite	23.9	58.8	7	Olsen (1966)
Kaolinite	20.4	57.5	14	Olsen (1966)
Kaolinite	$5.68 \cdot 10^{-1}$	36.5	1,425	Olsen (1965)
Kaolinite	$2.98 \cdot 10^{-2}$	22.5	5,000	Olsen (1966)
Halloysite [2]	1.75	–	7	Lutz and Kemper (1959)
Halloysite [3]	$9.69 \cdot 10^{-1}$	–	7	Lutz and Kemper (1959)
Illite	$2.9 \cdot 10^{-3}$	66.7	–	Warner (1964)
Illite	$4.05 \cdot 10^{-4}$	59.4	–	Warner (1964)
Illite	$7.04 \cdot 10^{-5}$	47.6	–	Warner (1964)
Shale (Calif.) [4]	$4 \cdot 10^{-4}$	–	–	Gondouin and Scala (1958)
Shale (Okla.) [4]	$8 \cdot 10^{-4}$	–	–	Gondouin and Scala (1958)
Shale (N. Mex.) [4]	$2 \cdot 10^{-6}$	–	–	Gondouin and Scala (1958)
Shale (Texas) [4]	$7 \cdot 10^{-6}$	–	–	Gondouin and Scala (1958)
Shale (Gulf Coast)	$9 \cdot 10^{-8}$	–	–	Rubey and Hubbert (1959)
Shale (Gulf Coast)	$1.5 \cdot 10^{-8}$	–	–	Rubey and Hubbert (1959)
Shale (Pierre)	$1.1 \cdot 10^{-4}$	21	–	Kemper (1961)
Shale (Muddy)	$<5 \cdot 10^{-2}$	4.7	–	Handin et al. (1963)
Shale (Gulf Coast)	$1 \cdot 10^{-8}$	–	–	Dickey (1970)
Shale (Gulf Coast)	$2.5 \cdot 10^{-9}$	–	10,000	Katz and Ibrahim (1971)

[1] Total stress on the clay–water system.
[2] A calcium water solution was used as the permeant.
[3] A sodium water solution was used as the permeant.
[4] 0.2 *N* sodium chloride solution was used as the permeant.
[5] Air permeability.
* Data not available.

Katz and Ibrahim (1971, p.116) calculated the permeability of a compacting shale under hypothetical conditions (Table XV).

In regard to the question whether Darcy's law is generally obeyed in saturated clays and clayey sediments or not, one must consider the evidence for and against Darcy flow behavior. Two main criteria must be met for Darcy's equation to be valid for clays: (1) the interstitial fluid in the clay pores must exhibit Newtonian behavior (viscosity of the fluid must remain constant with respect to hydraulic gradients imposed on the clay) and (2) the clay particles must be arranged in a rigid manner so that seepage forces do not modify the pore geometry. Evidence often cited for non-Darcy flow is the quasi-crystalline nature of water near the clay surfaces. In the clay that is not highly saturated, the removal of the bound water could probably be non-Darcy in nature. The influence of the quasi-crystalline structure of water in clay pores should increase with decreasing clay porosity. Seepage force can cause clay consolidation so that the porosity is decreased; plugging by finer clay particles and swelling of the clay can also occur.

Olsen (1966, p.292) showed that in the laboratory Darcy's law is obeyed in saturated kaolinite, when the sample porosity varies from 0.558 to 0.225 and the induced hydraulic gradients range from about 0.2 cm/cm to 40 cm/cm. The hydraulic gradient is equal to $\Delta h/l$, where Δh is the head difference between two permeant reservoirs and l is the length of the kaolinite clay sample. Olsen (1966, p.293) explained that the deviations which Hansbo (1960) observed in saturated illitic clays could be attributed to a contamination error (i.e., inaccurate flow-rate measurements due to the movement of liquid around air bubbles in capillary tubes) in the conventional measuring technique. Olsen concluded that Hansbo's data appears to be evidence for, rather than against, Darcy's law. In montmorillonite, the influence of quasi-crystalline water could be substantial for flow in the extremely small pores of the clay (Miller and Low, 1963). Deviations from Darcy's law can occur in any permeability test owing to lack of confinement of the clay sample and to large applied pressure gradients. These conditions introduce fabric changes in the clay. Darcy's law is obeyed when such clays in the laboratory are confined properly and tested at low pressure gradients (Olsen, 1966, p.294).

Jacquin (1965a, b) made an excellent and comprehensive study of fluid flow through argillaceous sandstones and compacted clays. He concluded that generally Darcy's law is not obeyed in compacted clays. The writers of this book, who conducted many experiments on the flow of fluids through clays, agree with this conclusion. Various electrokinetic processes are initiated upon commencement of flow: electroosmosis, reverse osmosis, ion filtration, electrophoresis, etc. In addition, structural changes occur and there is movement of solids. Yet, one can derive empirical equations relating permeability, porosity, and surface area.

Permeability is governed by the (1) manner of packing of grains during deposition, (2) compaction history of sediment, (3) grain-size distribution, (4) shape of grains, and (5) grain size (see Chilingarian et al., 1972).

Considerable evidence is available to show that particle arrangement in clays dominate hydraulic flow properties (Morgenstern, 1969, p.461). Olsen (1966) has shown that deviations between actual permeability behavior and that predicted by the Kozeny–Carman equation (see Langnes et al., 1972) can be explained by the presence of unequal pore sizes that may originate during the formation of initial fabric of the clay.

Specific storage

Specific storage is defined as the volume of water taken into storage, or discharged per unit volume, per unit change in head. Jacob (1940) showed that the specific storage is related to the elastic properties of the sediments and water:

$$S_s = \rho_w g \left(\frac{1}{E_s} + \frac{\phi}{E_w} \right) \tag{4-4}$$

where S_s is specific storage, $\rho_w g$ is weight of unit volume of interstitial fluid, E_s is modulus of compression of the matrix confined in situ, ϕ is porosity, and E_w is the bulk modulus of elasticity of the interstitial fluid (see p.318).

Jacob (1940) defined E_s as:

$$E_s = \sigma_z / \epsilon_z \tag{4-5}$$

where σ_z is the vertical stress and ϵ_z is the vertical strain. Bredehoeft and Hanshaw (1968, p.1100) remarked that E_s is related to the Lamé constants as follows:

$$E_s = \lambda - 2\mu \tag{4-6}$$

where λ and μ are Lamé constants.

Assuming that Hooke's law applies, the specific storage of a perfectly homogeneous confining bed is a constant. Table XVI gives the range in specific storage values for various types of sediments according to Domenico and Mifflin (1965).

TABLE XVI
Specific storage values for clays and sands
(After Domenico and Mifflin, 1965)

Type of sediment	*Specific storage* (cm^{-1})
Plastic clay	$2.0 \cdot 10^{-4}$ to $2.6 \cdot 10^{-5}$
Stiff clay	$2.6 \cdot 10^{-5}$ to $1.3 \cdot 10^{-5}$
Medium-hard clay	$1.3 \cdot 10^{-5}$ to $9.2 \cdot 10^{-6}$
Loose sand	$1.0 \cdot 10^{-5}$ to $4.9 \cdot 10^{-6}$
Dense sand	$4.9 \cdot 10^{-6}$ to $1.0 \cdot 10^{-6}$
Dense sandy gravel	$1.0 \cdot 10^{-6}$ to $4.9 \cdot 10^{-7}$

Permeability of compacted clays

Chilingar et al. (1963c) and Chilingar (1964) related the porosity and surface area of sandstones to their permeability. Papers by Fatt and Davis (1952), Wyble (1958), McLatchie et al. (1958), and Dobrynin (1962) showed that there is a significant decrease in the permeability of sandstones as the overburden pressure is increased (as much as 40% of the original permeability has been reduced at 20,000 p.s.i.). Considerably less information has been published on the effect of overburden pressure on the permeability of clays and shales. Macey (1940, p.635) showed the variation of aqueous conductivity (cm^3 sec^{-1} cm^{-1} $dyne^{-1} \cdot 10^{-12}$) with moisture content for seven ceramic clays. The permeability increased exponentially with increasing moisture content. The clays were saturated with benzene, nitrobenzene, pyridine and water. The units of aqueous conductivity used by Macey (1940) reduce to the reciprocal of dynamic viscosity (μ^{-1}). Dynamic viscosity (M/LT) is expressed in c.g.s. units as dyne-sec/cm^2.

Michaels and Lin (1954) showed that the Kozeny–Carman equation is obeyed over a considerable void-ratio range for various fluids, such as methanol, water, nitrogen, and cyclohexane; however, deviations occurred for all fluids because none of the permeability versus $e^3/1 + e$ curves (straight lines, cartesian plot, permeability in cm^2) passed through the origin. Waidelich (1958a, p.31) plotted void ratio (0.6–2.0 range) versus permeability for clays and discovered that the relationship was an irregular one, i.e., permeability irregularly increased and decreased with increasing void ratio. The general trend, however, was that of increasing permeability with increasing void ratio. Warner (1964) noted similar relationships for kaolinite, illite, and bentonite (Table XV). Bredehoeft et al. (1963) assumed a value of $1 \cdot 10^{-6}$ darcy for confining shale layers in reservoirs. This value was based upon flow rates across the Ordovician Maquoketa Shale (Illinois) as calculated by Walton (1962). Permeabilities ranging from $1 \cdot 10^{-7}$ to $1 \cdot 10^{-10}$ darcys were measured by Young et al. (1964) for argillaceous sandstones and siltstones. Extrapolating their results, they estimated that shales, such as the Joli Fou in western Canada, would have permeabilities as low as 10^{-3} darcys. E.C. Robertson (personal communication, 1966) correlated experimentally permeability with porosity for montmorillonite (API no.30a), illite (API no.35) and pyrophyllite (API no.49). His data show a very wide difference in permeability for a given porosity between the various clays, e.g., at a porosity of 40%, montmorillonite had a permeability of approximately $1 \cdot 10^{-6}$ darcys, whereas that of pyrophyllite was $1 \cdot 10^{-4}$ darcys. Permeability increased with increasing porosity. F.A.F. Berry (personal communication, 1965) calculated the permeability of Miocene shale (Venezuela) to be $6.95 \cdot 10^{-12}$ darcys and that of some recent marine muds to be $3.86 \cdot 10^{-13}$ darcys.

The dependence of shale permeability on the concentration of the electrolyte in interstitial solutions was discussed by Von Engelhardt and Gaida (1963). They reported that clays with interstitial solutions having high salt concentration have higher permeability than those in which the solutions have a low electrolyte concentration.

Permeability probably is also directionally controlled due to the flat plate-like structure of most clay minerals. For a well-compacted clay the permeability perpendicular to the normal applied stress probably should be greater than the vertical permeability. It is possible that this would not be as pronounced in undercompacted shales. Von Engelhardt and Gaida (1963) found that kaolinite and montmorillonite clays had a tendency to orient their basal planes normal to the pressure direction and this orientation increased with increasing overburden pressure. Increased orientation and decreasing porosity with pres-

TABLE XVII
Permeability of various clays at high overburden pressures using distilled water; calculated from water loss (After Rieke et al., 1964, table 4, p. 32)

Type of clay and location	*Overburden pressure* (p.s.i.)	*Permeability* (darcys)	*Time interval* (days)
Halloysite (API no.12)	400,000	$2 \cdot 10^{-13}$	1 (6th through 7th day)
Dickite (API no.15)	15,000	$7 \cdot 10^{-13}$	5 (2nd through 7th day)
Dickite (API no.15)	200,000	$7 \cdot 10^{-14}$	3 (7th through 10th day)
P-95 dry Lake clay (Buckhorn Lake, California)	15,000	$8 \cdot 10^{-12}$	4 (4th through 8th day)
Marine mud (AHF 6943), off Southern California	15,000	$3 \cdot 10^{-11}$	2 (5th through 7th day)

sure indicate that directional permeability in clays can be rather sensitive to pressure; this problem should be thoroughly investigated.

Table XV presents permeability data which were compiled from various sources. The majority of the data fall in the range of 23.9 md to $1.5 \cdot 10^{-9}$ md. Comparison of such diverse data is difficult because of different pressure gradients imposed across the samples, varying water salinities and viscosities, different temperatures, and whether the permeability was taken parallel or perpendicular to the bedding plane.

Data obtained by Chilingar et al. (1963c) and Rieke et al. (1964) on the permeability of clays at various high overburden pressures are presented in Table XVII. Emery and Rittenberg (1952) noted that the computed permeability (computed from annual water loss) of natural marine cores from the southern California sea-floor basins ranged from $5 \cdot 10^{-8}$ (at the top of the core) to $4 \cdot 10^{-10}$ darcys at a depth of 125 inches. Terzaghi and Peck (1948, p.48) showed that such permeabilities are characteristic of practically impervious soils. Emery and Rittenberg (1952, p.762) also measured the permeabilities of four plugs of different lengths from one core section in the laboratory. The measured permeabilities differed by a factor of about two – $1.06 \cdot 10^{-3}$ to $2.30 \cdot 10^{-3}$ darcys. Thus, Emery and Rittenberg (1952, p.763) concluded that compaction is not limited by permeability and that the existing permeability is sufficiently great to allow the egress of water that is displaced by grain deformation or repacking. Apparently, compaction is limited by resistance to grain deformation rather than by the low permeability. This is borne out by the data obtained by Chilingar et al. (1963a).

Mesri and Olson (1971) calculated coefficients of permeability for smectite, illite and kaolinite, using Terzaghi's (1943) theory of one-dimensional consolidation (Fig.77). The clays were homoionized to either the sodium or calcium form and the pore-water electrolyte concentration was varied. According to Mesri and Olson, the reduction in the coefficient of permeability, at constant void ratio, from kaolinite to illite to smectite is largely the result of a reduction in the size of individual flow channels and an increase in the

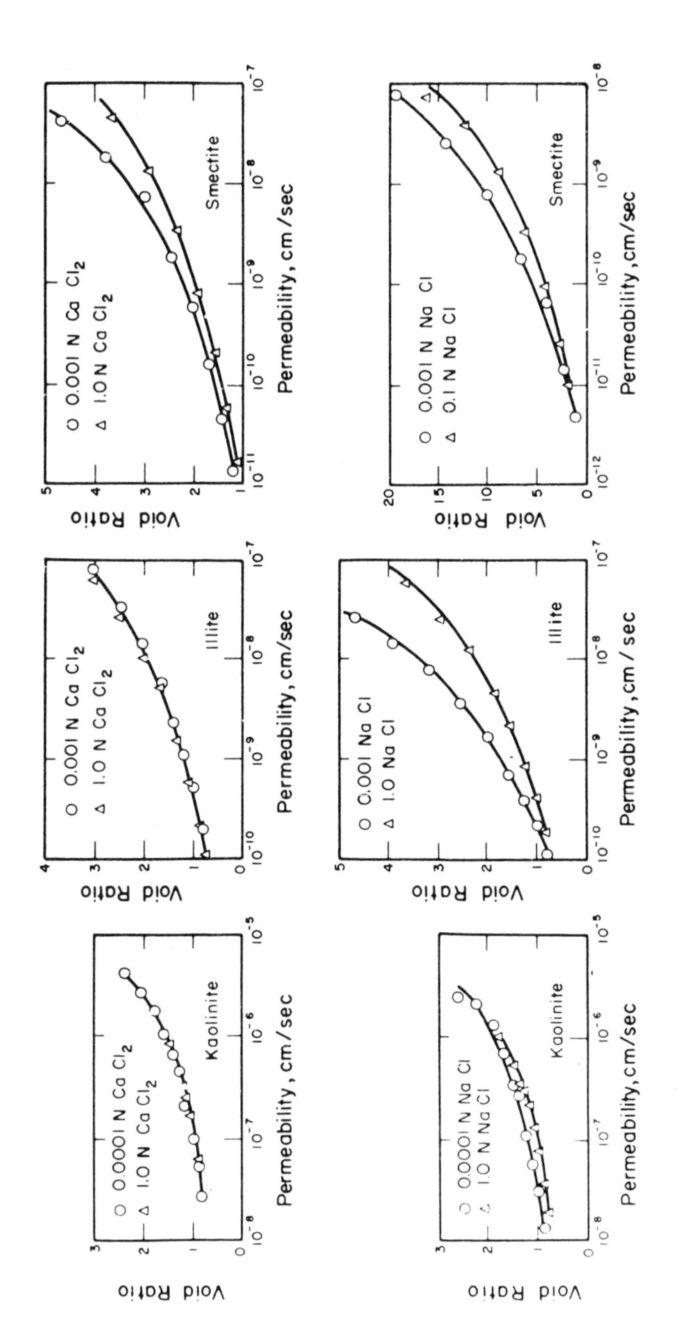

Fig. 77. Coefficients of permeability for kaolinite, illite, and smectite using various electrolyte solutions. (After Mesri and Olson, 1971, fig. 6, p.156. Courtesy of *Journal of the Clay Minerals Society.*)

tortuosity of the flow paths (i.e., mechanical effects). Dispersion or aggregation caused by physicochemical variables also exert great influence on the coefficient of permeability. Dispersion reduces the fluid flow, because it results in channels that are all of nearly the same size and tend to be non-equidimensional. On the other hand, aggregation gives rise to (a) many tiny flow channels, through which there is little flow, and (b) a smaller number of relatively large channels which form the main avenues for the flow. Mud engineers are well aware of this fact because of the necessity to form practically impermeable mud cakes (from drilling fluids) on the walls of wellbores, in order to reduce the amount of water loss to the formation during drilling operations. Flocculated drilling fluids (as a result of drilling through salt, for example) give rise to very permeable mud cakes with resultant high water loss to the formation, whereas deflocculated (dispersed) systems result in low-permeability mud cakes and consequent low water loss to the formation.

Relationship between permeability and porosity

In most sedimentary rocks porosity is related to permeability (e.g., see Chilingar, 1964, p.71; Griffiths, 1958). Yet, this relationship is not well understood and a considerable amount of research work still remains to be done in this field. Grain-size distribution, shape of grains, grain orientation, grain packing, and surface area appear to be important controlling variables (see Chilingar, 1964, and Chilingarian et al., 1972).

Terzaghi (1925b) proposed the following relationship between porosity (ϕ) and permeability (k):

$$k = \lambda(e - 0.15)^3 (1 + e) \tag{4-7a}$$

where λ is a constant and e is void ratio $[e = \phi/(1-\phi)]$. The experimental results show that in the porosity range of 20–80%, this relationship can be expressed in the following form:

$$k = \lambda \times \phi^n \tag{4-7b}$$

where constant n is around 5. The Kozeny–Carman equation, which is commonly used to correlate permeability and porosity, can be presented as follows:

$$k = \lambda \frac{\phi^3}{(1-\phi)^2 s^2} \tag{4-8}$$

where k is permeability, ϕ is porosity, s is specific surface area, and λ is a constant.

Based on extensive experimental results, Nishida and Nakagawa (1969) developed an equation which enables approximate determination of the coefficient of permeability K (in cm/sec) from void ratio, e, and plastic index, PI (in %):

$$\log_{10}K = [e/(0.01\ PI + 0.05)] - 10 \quad (4\text{-}9)$$

In studying the sealing properties of caprocks, Khanin (1968) found that slightly silty clays having a porosity of 1.4–3.4% and permeability to gas of $1 \cdot 10^{-5}$–$6 \cdot 10^{-5}$ md have uniform pore structure, i.e., pore diameter ranges from 0.08 to 0.10 micron (Fig.78). With increasing silt content, in silty clays the pore structure is less uniform – pore diameters range in size from 0.014 to 0.50 or even up to 6.4 microns, with predominance of the 0.016–0.12 micron fraction (Fig.79). The permeability of the latter clays is higher. Although the percentage of larger pores is small, their contribution to the permeability constitutes 40–70% (Figs.78 and 79).

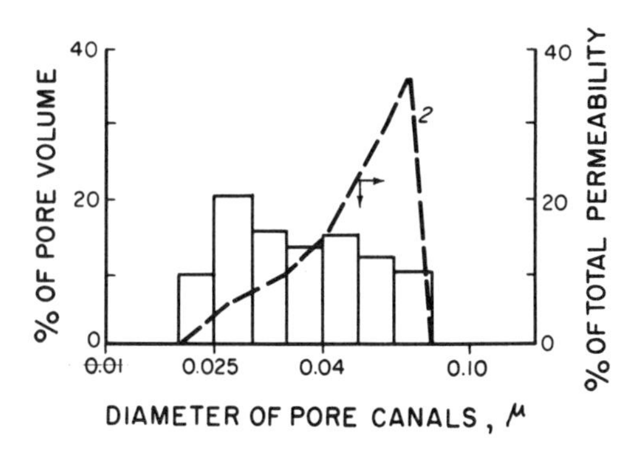

Fig.78. Distribution of diameter of pore canals in slightly silty clay. Core depth of 3,869–3,879 m. *1* = distribution of pore canals; *2* = percentage contribution of pore canals to permeability. (After Khanin, 1968, p.18.)

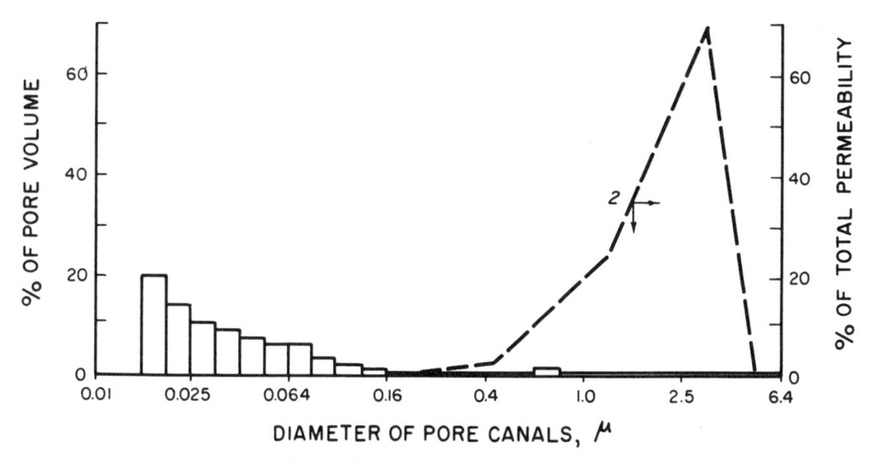

Fig.79. Distribution of diameters of pore canals in silty clay. Core depth of 3,606–3,612 m. *1* = distribution of pore canals of different sizes; *2* = percentage contribution of pore canals of different sizes to total permeability. (After Khanin, 1968, p.19.)

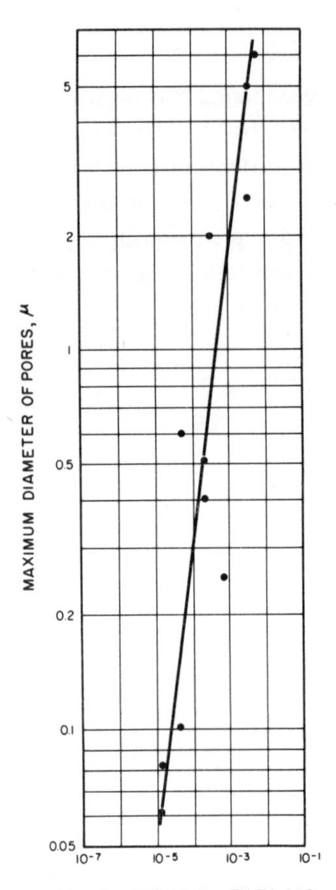

Fig.80. Relationship between permeability and maximum diameter of pores in shales. (After Khanin, 1968, p.19.)

Khanin showed a relationship between the permeability and the maximum diameter of pores in clayey rocks (Fig.80). A mercury injection technique with pressures up to 1,000 atm was used by Khanin (1968) in his experimental work. Permeabilities to gas of dry clay samples were determined at $\Delta p = 10$ atm, after hydrostatic compaction to 90–400 atm. Additional data on the subject has been presented by Khanin (1969) in his excellent book.

Pusch (1966) studied quick clay which was originally deposited in sea water, but subsequently leached in situ. Undisturbed clay samples were taken from 6 m depth at Lilla Edet about 45 km north of Gothenberg. Pusch found that in comparison with eighteen micrographs of clays deposited in fresh or brackish water (sensitivity, $S_t = 10$–20), the relative pore area of the clay from Lilla Edet is about three times greater. The permeability in the horizontal and vertical direction were found to be equal

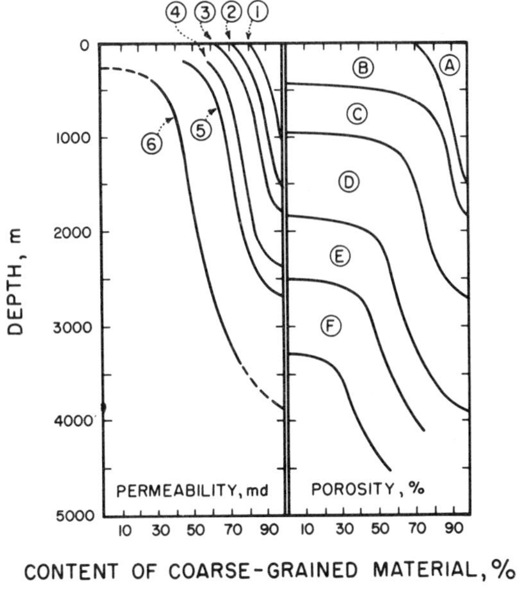

Fig.81. Interrelationship among porosity (effective), permeability (to gas) in md, depth of burial, and lithologic composition of Upper Paleozoic and Mesozoic rocks of the Pre-Caucasus and Pre-Caspian Depression. Permeabilities: area above curve *1* – > 1,000 md; area between curves *1* and *2* – > 500 md; area between curves *2* and *3* – > 100 md; area between curves *3* and *4* – > 10 md; area between curves *4* and *5* > 1 md. Porosities: *A* – > 30%; *B* – 25–30%; *C* – 15–25%; *D* –10–20%; *E* – 5–15%; *F* – 4–10%; area below lower curve – < 5%. (Prepared by B.K. Proshlyakov, 1968, in: Lapinskaya and Proshlyakov, 1970, fig.3, p.117.)

to $7 \cdot 10^{-8}$ cm/sec, whereas those for the fresh- and brackish-water clays are only about 1/4 as high. The permeability was determined from oedometer tests, in which the influence on the percolation process of a few very large pores is much less than in direct permeability tests.

Interrelationship among porosity, permeability, depth of burial and lithologic composition of Upper Paleozoic and Mesozoic rocks of the Pre-Caucasus and Pre-Caspian Depression is presented in Fig.81 (also see Fig.20, p.45). At a particular depth, one can find different porosities and permeabilities, which decrease with decreasing content of detrital material (i.e., sand and silt) and increasing content of fine-grained carbonate-clay fraction. The left portions of the graphs in Fig.81 represent analyses of clays, argillites and marls (Lapinskaya and Proshlyakov, 1970, p.118). At a depth in excess of 4,000 m the formations become practically impermeable. It should be remembered, however, that in the presence of oil and gas in the pores, the formations do not compact as much as they do in the presence of water, and thus retain their porosity. The leaching of calcite present in these rocks complicated the picture.

Robertson (1967) presented a relationship between the permeability and porosity of

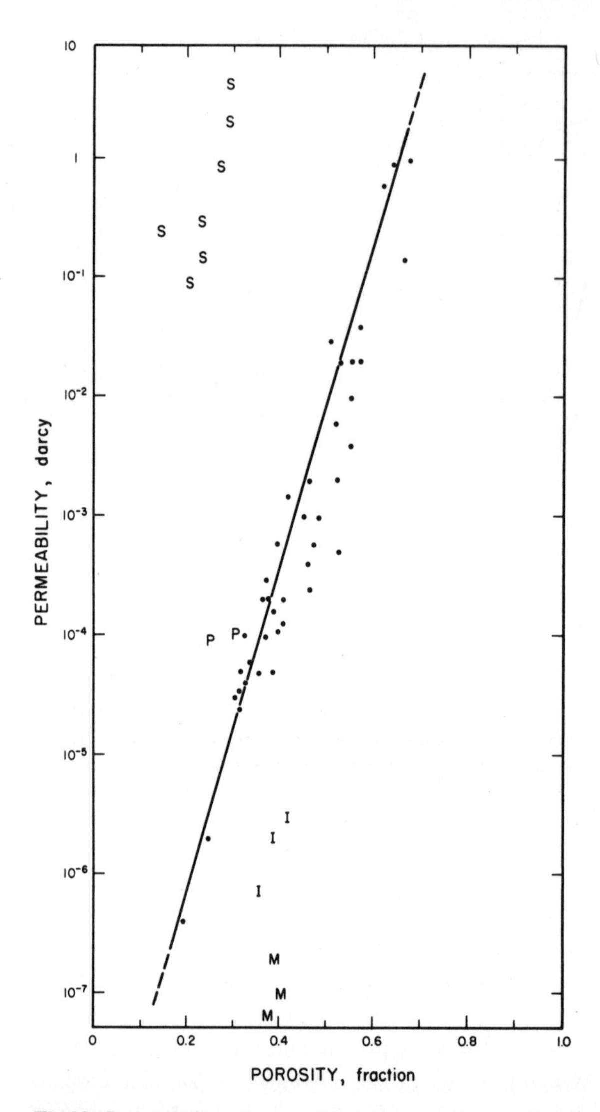

Fig.82. Permeability of aragonite sediment after compaction as a function of porosity. *M* = montmorillonite; *I* = illite; *P* = pyrophyllite, and *S* = sandstones. (After Robertson, 1967, fig.5. Courtesy of University of Illinois Press.)

compacted aragonitic sediment (Fig.82), which behaves similarly to clays on compaction. There is an experimental reduction in permeability with increasing degree of compaction. For comparison purposes, Robertson also tested montmorillonite no.30a (*M*), illite no.35 (*I*) and pyrophyllite no.49 (*P*) (Clay Mineral Standards, American Petroleum Institute Research Project no.49). The samples were saturated with water and then compacted under pressures of 30, 60, and 300 bars. Permeabilities of several sandstones (Fraser, 1935; Fatt, 1953) are also plotted in Fig.82 (points marked *S*).

RHEOLOGY OF ARGILLACEOUS SEDIMENTS

Rheology is the study of the deformation versus time relationship of materials. Certain properties can be ascribed to practically all types of naturally occurring materials. Argillaceous sediment can exhibit several types of behavior during compaction: elastic, perfectly plastic, elasticoplastic, viscoelastic, and viscoplastic.

The rheological behavior of a sedimentary deposit changes with a change in fluid content; this is especially pronounced in clayey sediments. During deposition of clays there is a zone on the sea floor where the sediment acts as a thick suspension with the flow properties of liquids. As the water content is reduced through subsequent burial, the flow of paste-like argillaceous sediments becomes non-Newtonian. As compaction proceeds, the properties change as the water content decreases further. The physical condition of a sediment at a given saturation is termed the *consistency*. Consistency is the resistance to flow of the sediment and, therefore, an indication of its rheological behavior. It is related to the force of attraction between the individual particles or aggregates of these particles. Different argillaceous sediments have different consistency at different saturations and the specification of this condition gives some information about the type of sedimentary material.

A change in the moisture content of the clay results in a material having different mechanical properties. Two properties known as the *liquid* and *plastic limits* were introduced by Atterberg (1911) to provide an empirical but quantitative measure of the degree of plasticity of clays (Skempton, 1970). Terzaghi (1925a; 1926) emphasized their merits for classification of soils. Casagrande (1932b) modified the original tests, whereby they have become internationally standardized in all soil-mechanics laboratories. These consistency limits are not absolute.

The *liquid limit* (*LL*) of a clay is the water content at which, in the remoulded state, it passes from the plastic to an almost fluid condition. The *plastic limit* (*PL*) is the water content at which the remoulded clay passes from the plastic to a friable or brittle condition (Skempton, 1970). The *plasticity index* (*PI*) is defined as the difference between these two limits: $PI = LL - PL$.

The *shrinkage limit* can be defined as the water content at which volume change is no longer linearly proportional to change in water content. Fig.83 illustrates the possible consistency states in argillaceous sediments. Beginning with the liquid state, on losing water during compaction the argillaceous sediment will go through the various states of consistency.

Skempton (1970, p.379) pointed out that the Atterberg limits have two significant advantages. Both the liquid and plastic limits reflect the amount and the type of clay minerals present in an argillaceous sediment (Skempton, 1953) and are, therefore, functions of such properties as total surface area and cation-exchange capacity (Farrar and Coleman, 1967). In addition, they are expressed as water contents, and the natural water content of a clay can be compared directly with its Atterberg limits by a ratio defined as

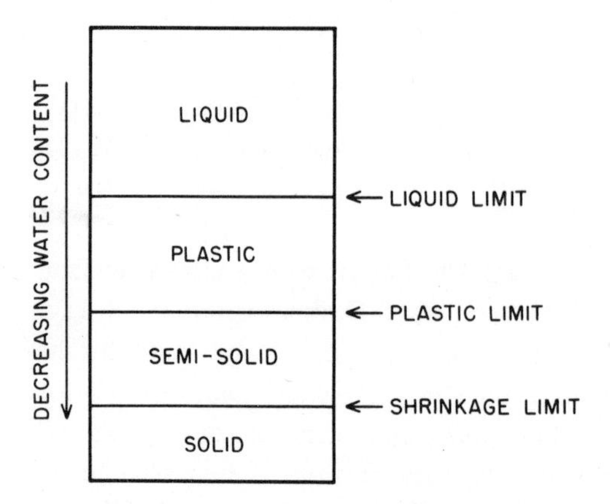

Fig.83. Consistency states possible in an argillaceous sediment.

the *liquidity index* (*LI*):$LI = (w-PL)/(LL - PL)$, where w is the water content of a clay, which is usually defined as the ratio of the weight of water to the dry weight of solids, and is expressed as a percentage.

In his excellent paper, Skempton (1970) has shown that for a wide variety of normally compacted argillaceous sediments, the liquidity index lies inside a narrow range of values, at any given effective pressure, although the corresponding natural water contents of the sediments may vary between very wide limits. Replotting the points in Fig.7 (p.13) in terms of liquidity index, rather than void ratio, would clearly show this. An examination of the position of the various clays in the narrow band of liquidity index versus effective pressure plot reveals that those with a high sensitivity lie towards the upper part of the zone (low effective pressures), whereas those clays with a low sensitivity have a relatively low liquidity index (Skempton, 1970, p.404). The importance of the liquidity index was made clear by Skempton (1970, p.380) by giving a simple example. If two clay samples from different depths have the same natural water content, without a knowledge of the consistency limits, it is possible to infer that the greater pressure acting on the deeper clay sample had caused no additional compaction. As pointed out by Skempton (1970, p.380), however, the liquid limit of the lower sample probably would be greater than that of the upper sample and the liquidity index would decrease with depth.

Rheological models

The mechanical properties of an argillaceous sediment are characterized by the internal friction and the cohesion. Rolling and sliding friction, together with the resistance owing to the geometry and relative position of the clay particles, are usually considered as internal friction. True cohesion is that part of the clay resistance which is due to the force of attraction existing between the clay-mineral particles. If Terzaghi's assumption is accepted that each particle is surrounded by a water shell, which is firmly attached to it

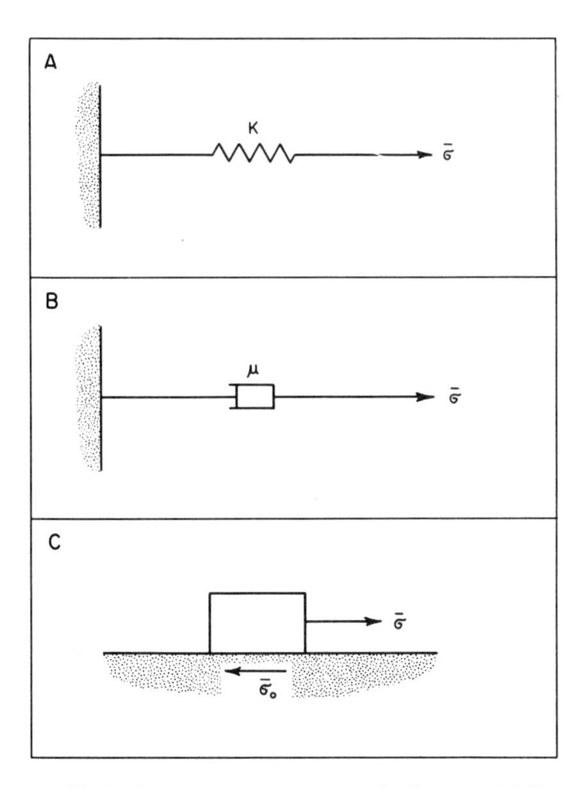

Fig.84. Basic mechanical models. A. Hookean model illustrating the perfectly elastic state; K is spring constant (Hookean modulus) and $\bar{\sigma}$ is stress. B. Newtonian model illustrating the perfectly viscous fluid; μ is viscosity. C. Yield stress model illustrating a minimum pressure or stress gradient which would initiate a strain; $\bar{\sigma}_0$ is frictional resistance which is a constant at maximum value. Boundary conditions are $\bar{\sigma} < \bar{\sigma}_0$ for $\bar{\epsilon} = 0$; $\bar{\sigma} < \bar{\sigma}_0$ for finite movement or strain $\bar{\epsilon}$.

by electrochemical forces, then the applied forces on clays must be transmitted through the water layers separating the individual clay plates (Saada, 1962).

Mechanical models, often referred to as rheological models, can be used to describe the deformation–time behavior of sediments under stress. Only simplified elementary rheological models are presented in this book. These models consist of elementary units of springs and dashpots in series or in parallel. The spring element is generally termed a *Hookean model* (Fig.84A), whereas the dashpot element simulating a purely viscous phenomenon, is a *Newtonian model* (Fig.84B). The spring element under compression represents the elastic characteristics of the sediment matrix. As shown earlier in Chapter 3, the application of an overburden load to a fully saturated sediment–fluid system will initially impose stresses on the interstitial fluid phase within the sediment mass. Dissipation of the applied pressure can be simulated by dashpot action (Fig.84B). In order to simulate the transfer of the load between the pore fluid and the matrix, the dashpot must be combined with a spring in some form of coupling.

The crux of using these models lies in the choice of spring constant. The spring constant in the Hookean model describes the elastic nature of the spring. The rate of pore pressure dissipation simulated by the dashpot arrangement, coupled with the other elements or combinations of elements, will dictate the choice of the coefficient of viscosity. Variables which determine the spring constant and the coefficient of viscosity of a sediment–fluid system include: (1) initial structure of the sediment, (2) type of the clay minerals, (3) the constraints associated with overburden loading, (4) porosity, (5) fluid content, (6) temperature, and (7) geologic time. The greater the activity of the clay (i.e., the more plastic the clay), the greater will be the restrictions placed upon the movement of fluid within the sediment–fluid system. It should be remembered that in the final analysis, the choice of material constants, which are necessary for the rheological models, depends on the understanding of the sediment behavior.

During sedimentation and compaction, the stress–strain behavior of argillaceous sediments does not fall into a simple category defined by the elementary rheological models shown in Fig.84. Inasmuch as the stress–strain relationship with respect to time is complex, it will be necessary to apply a combination of the basic models to the problem. Argillaceous sediments are three-phase systems and the deformation–time characteristics are not simple to model.

A typical deformation–time relationship under a single constant load in an argillaceous sediment may be represented by a Burger model (Fig.85). Whereas it is possible to specify the material constants of elasticity and viscosity to approximate actual physical response under one specific load, it is not possible to apply the same constants for all the ranges of loads that the sediments may be exposed to. The present state of knowledge, coupled with lack of complete data, does not allow exact definition of the relationship required for a variable model, in which the material constants vary with microstructure, stress level, and rate of loading.

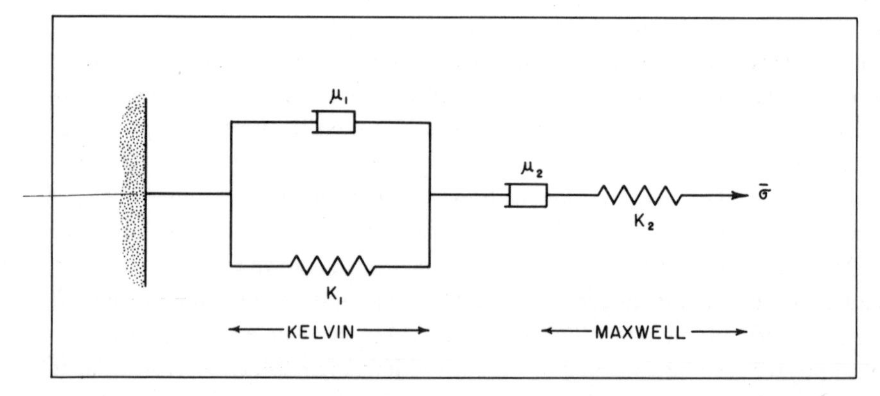

Fig.85. Burger model, coupling various elements in series so that the resultant time–deformation behavior is no longer linear. K is spring constant (Hookean modulus) and μ is viscosity.

Yong and Warkentin (1966, p.230) pointed out that (1) pore-fluid pressure dissipation in primary consolidation may be represented by a spring and dashpot model (Schiffman, 1962), (2) the elastic response of the soil structure can be represented by a spring, and (3) water viscosity influencing the viscosity of the model can be represented by a dashpot. Leonards and Altschaeffl (1964) presented rheological models used for comparison of pore-pressure dissipation rates, which may be considered in terms of the degree of consolidation. The Maxwell model is represented by a series coupling of Hookean and Newtonian models (Figs.84A and B combined in series, also see Fig.85). Mathematically this model is expressed as:

$$\bar{\sigma} = K\bar{\epsilon}_c \exp(-Kt/\mu) \tag{4-10}$$

where $\bar{\sigma}$ is applied stress in p.s.i., K is an elastic constant in p.s.i., $\bar{\epsilon}$ is strain in inch/inch, μ is viscosity in units of stress × time, and t is time. The subscript "c" indicates that the variable is constant. In the above case, $\bar{\epsilon}_c$ is constant at time $t = 0$. Yong and Warkentin (1966, p.87) pointed out that for $\bar{\epsilon} = 0$ at $t = 0$, at constant stress σ_c, the relationship between $\bar{\epsilon}$ and time is a straight line:

$$\bar{\epsilon} = \frac{\bar{\sigma}_c}{K} + \frac{\bar{\sigma}_c t}{\mu} \tag{4-11}$$

This signifies an instantaneous elastic deformation as in the classical Hookean rheological model, followed by a time-dependent deformation as given in the classical Newtonian model.

Schiffman (1962, p.79) states that the physical and mathematical system described by him requires that pore-pressure dissipation be represented by a Maxwell model. The mechanics of the system must conform to known or reasonable approximations of physical reality.

Non-linear time–deformation behavior can be described by using a Burger model which combines in series the Kelvin and Maxwell models (Fig. 85).

Yong and Warkentin (1966, p.90) presented a solution for strain at a constant stress $\bar{\sigma}_c$, acting between time $t = 0 = t_o$ and $t = t_1$:

$$\bar{\epsilon} = \bar{\sigma}_c \left\{ \frac{1}{K_2} + \frac{t}{\mu_2} + \frac{1}{K_1} \left[1 - \exp(-K_1 t/\mu_1)\right] \right\} \tag{4-12}$$

where $\bar{\epsilon}$ is strain, $\bar{\sigma}_c$ is constant stress, t is time, and K and μ are constants. In the above equation, the elastic deformation is given by the first term of the right-hand side (Hookean element in the Maxwell model), whereas the Newtonian element in the Maxwell model provides for the permanent deformation. The Kelvin model relates the after-effects of stress survival (Yong and Warkentin, 1966, p.90). The spring constant, K_2, represents the instantaneous elastic deformation and recovery of the sediment.

It should be stressed that much research work still remains to be done in applying these rheological models to the mechanical behavior of the sediments during compaction. Their immediate use lies in the area of application to the modeling of laboratory compaction of sediments.

ELASTIC AND STRENGTH PROPERTIES OF ARGILLACEOUS SEDIMENTS

Many recent papers in the geological, rock mechanics, and soil mechanics literature are devoted to elastic and viscoelastic models describing the deformational behavior of fluid-saturated rocks and sediments (e.g., De Josselin de Jong, 1968; Hata et al., 1969; Raghavan et al., 1969; Kohlhaas and Miller, 1969). Many publications, however, do not seem to present realistic values for the mechanical and elastic properties so that the models can be appropriately applied. The reason for this is that very little is known about the in-situ elastic and strength properties of unconsolidated and consolidated sediments and especially the properties of very fine-grained clastics. Recently, however, some geotechnical data based on the laboratory and in-place studies of sea-floor sediments have been published (McClelland, 1956; Richards, 1961; Moore, 1962; Richards and Hamilton, 1967; Inderbitzen, 1969). Geophysical investigations employing reflection and refraction seismology, borehole seismology, and resistivity methods have also supplied considerable amounts of information about the mass and elastic properties of marine sediments (Laughton, 1957; Smith, 1962; Lawrence, 1964; Kermabon et al., 1969; Hamilton, 1970a, b, 1971a, b). These studies provide basic data and attempt to establish relationships among the various sediment properties, so that these properties can be estimated and/or predicted by using results of simple tests. Geophysical studies can provide the sedimentologist with approximate values for the sediment properties and their variations with increasing burial depth.

Sedimentologists have shown that rapidly-deposited marine sediments are not normally compacted near the water–sediment interface, because the overburden stress is largely borne by the interstitial fluids rather than by the sedimentary particles (Richards and Hamilton, 1967). These fluid-saturated sediments have low compressive and shear strengths and are of an elasticoplastic nature. The precise physical nature of these sediments can best be described in rheological terms.

Knowledge of the elastic or elasticoplastic nature of argillaceous sediments is required for the solution of many problems associated with compaction and deformation processes. In the following sections certain terms are defined and a summary on elastic and strength properties of rocks is presented.

Elastic properties of unconsolidated and indurated sediments

In contrast to the consolidation tests, the results of which are usually plotted in the form of void ratio-versus-log pressure curves for unconsolidated sediments, the stress–strain characteristics of sediments define the elastic nature of the material. The stress–strain behavior of any sediment depends on a number of factors including mass properties, fluid saturation, particle arrangements, type of strain conditions (i.e., uniaxial, triaxial, or polyaxial), stress history, confining pressure, shear stress, and duration of the loading.

Judd (1959) stated that a major research problem is how to correlate data from in-situ field tests with data from laboratory tests on rock specimens. It is also important to determine which test is more diagnostic. It is very difficult to sample an unconsolidated sediment adequately for testing purposes without destroying or altering those properties which the test is designed to determine. Laboratory values of the various elastic moduli are determined either statically or dynamically. Static (destructive) testing involves the mechanical application of stress to the sample, whereas in dynamic (non-destructive) testing the sample is vibrated and the velocity of the shear and compressional waves through the medium are measured. Table XVIII presents equations showing the interrelationships between various static and dynamic parameters for various elastic moduli. The values of the elastic moduli calculated from static measurements of cores are sometimes significantly different from the values determined by dynamic testing of the same sediment sample. Zisman (1933) and Ide (1936) found that, with few exceptions, the dynamically determined moduli were from 4 to 20% larger than statically determined

TABLE XVIII
Some important equations relating the elastic constants in an isotropic sedimentary body

Elastic constants	*Dimensions* [1]	*Relationship*
Modulus of elasticity, E (Young's modulus)	F/L^2	$E = 2G(1 + \nu)$ E = slope of stress-strain curve
Modulus of rigidity, G (shear modulus)	F/L^2	$G = \frac{E}{2(1+\nu)}$ $G = \frac{3}{4}(\rho v_p^2 - \kappa)$ *
Bulk modulus, κ * (incompressibility)	F/L^2	$\kappa = \frac{E}{3(1-2\nu)}$ $\kappa = \rho(v_p^2 - \frac{4}{3}v_s^2)$ **
Lamé constant, λ	F/L^2	$\lambda = \kappa - \frac{2}{3}G$
Poisson's ratio, ν ν = lateral strain : longitudinal strain	dimensionless	$\nu = \frac{E - 2G}{2G}$ $\nu = \frac{3\kappa - \rho v_p^2}{3\kappa + \rho v_p^2}$ **

[1] F = force; L = length.
* Reciprocal of κ is called compressibility; ρ is density.
** Dynamic equations utilizing the compression-wave velocity, v_p, and shear-wave velocity, v_s.

moduli. Judd (1959, p.73) remarked that there is doubt whether laboratory-determined dynamic moduli can be correlated with static moduli or not, because the static test imposes a relatively long-time continuous load. This factor is important in the evaluation of the results, because static tests would be more nearly analogous to actual sediment loads; and, secondly, continuous loading more closely measures the elasticoplastic properties of the sediment. Rapid application of a load, as in dynamic testing, appears to negate the possibility of obtaining data on the plastic strength of the sediments.

Lawrence (1964) listed the following reasons why laboratory-determined dynamic constants are subject to inaccuracies as opposed to dynamic values obtained from borehole seismology: (1) Confining stress from which the rock sample has been relieved is difficult to simulate. (2) Only the more competent rock core samples are recovered from drilling operations. (3) Core size is frequently too small to adequately represent the heterogeneous nature of the material. (4) The physical properties of the sample may change with time. (5) Elastic properties of unconsolidated sediments are difficult to determine because the elastic limits of the sample are easily exceeded in the testing process. (6) The effects of anisotropy are difficult to determine by laboratory techniques.

Several reasons were offered by Geyer and Myung (1970) to account for the differences between static measurements and in-situ measurements of the same rock. (1) Most static measurements are made at room temperature and atmospheric pressure. (2) Micro- or macro-fractures in the sample lead to lower calculated values for the elastic moduli than would be obtained for the same samples at higher pressures, when cracks and voids tend to close. (3) Sample size is insufficient. Both methods leave much to be desired; however, some of the objections can be minimized by restoring the natural state of the sample as closely as possible.

The most widely used indirect method of determining values of elastic moduli is based on the study of the behavior of elastic waves transmitted dynamically through the medium. Fig.86 diagrammatically illustrates the two main wave-path arrangements that are used in seismic work. Reflection seismology makes use of waves following the OAR_1 path, in which a portion of a transmitted wave encountering a series of discontinuities at depth is reflected at each boundary (Smith, 1962, p.34). Most smaller-scale seismic investigations are carried out using this technique. Refraction seismic shooting illustrated by OBR_2 makes use of the fact that transmitted waves obey Snell's law in crossing the discontinuities between layers of lower and higher velocity material (Smith, 1962, p.34). Inasmuch as elasticity usually increases at a more rapid rate with increasing depth than does density, the deeper layers have higher transmission velocities that cause curved paths to return to the surface some distance from the shot point. Smith (1962) reported that the majority of seismic work done at sea to date has been the refraction technique in order to obtain the greater depths of penetration necessary to reach the Mohorovičić discontinuity. Acoustic investigations are popular for near-shore studies (Fig.86).

Recognizing the shortcomings of laboratory results, borehole seismology techniques, such as the 3-D velocity-logging tool, have been developed to obtain the values of various

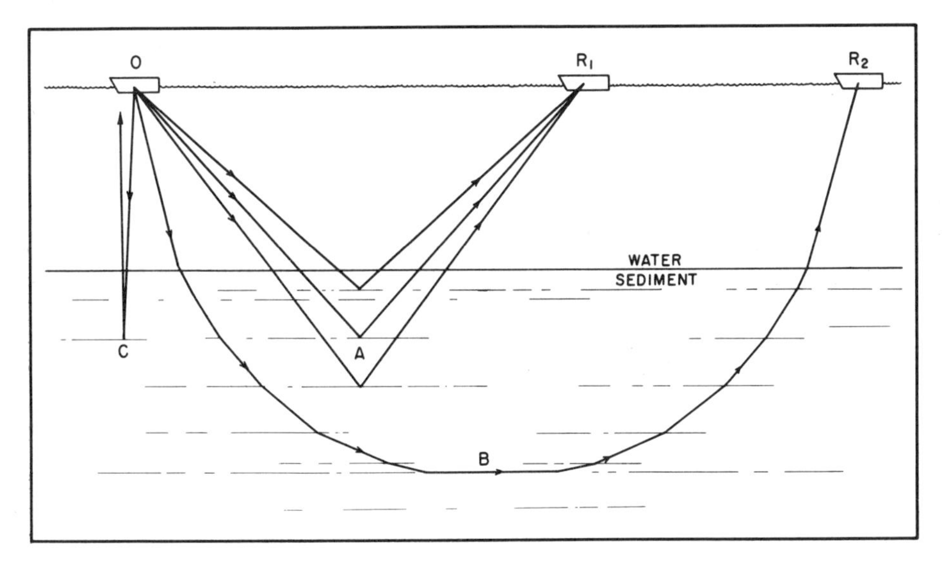

Fig.86. Wave paths followed in seismic reflection (*A*), seismic refraction (*B*), and acoustic (*C*) studies of sediments. (After Smith, 1962, fig.2, p.35.)

elastic moduli under more natural conditions. Geyer and Myung (1970) evaluated the effectiveness of this method and showed that in-situ elastic-moduli values, when compared to static values, may not be related by a simple proportionality factor. This was attributed to the difference in the downhole environment.

Elastic moduli. There are a number of problems involved in treating the unconsolidated and semiconsolidated argillaceous sediments as elastic bodies. Sediments are not continua, but are granular, initially unaggregated, solid-mineral particles interspaced with fluids and organic molecules. The sediment's macroscopic behavior depends upon the nature of particle contacts and the deformation of the mineral grains at these contacts. In addition, sediments are for the most part heterogeneously deposited, with their characteristic properties varying considerably from point to point in vertical and horizontal directions. The conditions of their deposition are generally such that the layers are anisotropic. Instead of studying sediments as a discrete system consisting of an assemblage of mineral grains, it is more expedient to consider sediments as a continuum and to use the methods and results of continuum mechanics for studying their mechanical properties (see Chapter 3). The simplest continuum model for argillaceous-sediment behavior is linear elasticity.

In order to attempt to predict the distribution of stress and strain in a sedimentary layer one assumes that the sediment is perfectly elastic, i.e., all the deformation is recoverable and, upon removal of the overburden load, the sedimentary layer would return to its initial state. Buried confined sediment layers, whether they are unconsolidated or indurated, are assumed to follow Hooke's law of deformation for very small loading increments. In 1676, Hooke discovered experimentally that stress is proportional to strain as long as the stress does not exceed or closely approach the elastic limit. The amount of change under load in the shape or size of a body per unit of the original dimension is termed strain (Swartz, 1962). It may be a change in a lineal direction, in the angle between lineal dimensions, or a change in volume. The types of stress–strain curves obtained for argillaceous rocks are rather similar to those obtained during the testing of metals. Metallurgical terminology has been accepted by geologists in describing the deformation of rocks. In nature, millions of years may be required for strains in rocks to change significantly. Consequently, a constant strain rate may be assumed to hold for geologic conditions during compaction instead of a highly variable strain rate. Mann and Semple (1970) presented a computer program by which geotechnical data of this type could be reduced and graphically illustrated; this has reduced the number of man-hours required in plotting, calculating, and interpreting such data.

Historically, the elastic coefficients of solids were defined in terms of the relationship between the components of an applied stress system and the components of the resulting strain system. Drabble (1970) presented a very advanced thermodynamic approach in order to predict the mechanical properties and elastic behavior under pressure in terms of the basic composition and structure of the solid.

Perhaps the most significant drawback from a practical approach, in using the classical theory of elasticity to represent unconsolidated sediment behavior, is that it cannot account for the dilatancy effect. When unconsolidated or weakly indurated sediments are subjected to a pure shear-stress increment, a volumetric strain, which can be either an expansion or a contraction, is usually observed. Several other important aspects of sediment behavior not accounted for are the non-linear shear stress–strain curve and the partially inelastic behavior. The sedimentologist should be fully aware of all these problems; however, a detailed mathematical examination is beyond the scope of this book.

Modulus of elasticity (Young's modulus). A common method of studying the elastic properties of a sediment in a laboratory is by applying an axial stress to a right circular cylinder cut from the rock or a circular slice cut or molded from a loose sediment. For most argillaceous sediments, the resulting stress–strain curve when plotted may take an approximately linear form ending abruptly in failure, for example, at point x, as shown in Fig.87.

Swartz (1962) pointed out that each type of stress will be related to its resultant strain by its own modulus of elasticity. The ratio of change in stress, σ (force per unit area), to a change in strain, ϵ (change in length per unit length), is termed the modulus of elasticity E ($E = \Delta\sigma/\Delta\epsilon$); this applies to both the linear and curvilinear portion of a stress–strain plot. This one-dimensional definition of linear elasticity, in which uniaxial strain is proportional to uniaxial stress, can be generalized to three dimensions by assuming that each component of strain is a linear function of the components of stress. Boundary conditions can be expressed as:

(1) uniaxial stress: $\sigma_z \neq 0, \sigma_x = \sigma_y = 0$
uniaxial strain: $\epsilon_z \neq 0, \epsilon_x = \epsilon_y = 0$

(2) biaxial stress: $\sigma_z \neq 0, \sigma_x \neq 0, \sigma_y = 0$
biaxial strain: $\epsilon_z \neq 0, \epsilon_x \neq 0, \epsilon_y = 0$

Underwood (1967, p.105) reported that published data on results of modulus determinations from laboratory tests often do not indicate whether the modulus of elasticity was taken from the initial tangent modulus through the origin, secant modulus or chord modulus. Sedimentologists are referred to an excellent paper by Swartz (1962, p.962) for a detailed treatment of the subject.

Simons (1963) presented an experimental technique for measuring the dynamic modulus of elasticity of shales under stresses similar to those existing in situ. Fig.88 exhibits the variation in modulus of elasticity values for two different shales at confining pressures of 0, 4,000 and 8,000 p.s.i. and a variable rate of loading. The differences in values between the Yegua Shale (Louisiana) and the Canyon Shale (Texas) reflect differences in the degree of induration, prior stress–strain history, depth of burial, and age. Bureau of Mines Young's modulus data on shales derived from static tests ranged from $1.7 \cdot 10^6$ p.s.i. to $7.6 \cdot 10^6$ p.s.i. (tangent modulus of elasticity at the midstrength point) (Obert and Duvall, 1967). Rieke (1965) presented both compressional and tensional static modulus of elasticity values for various slates and related them to other physical properties of the rocks such as bulk specific gravity, index of alteration, and compressive strength. The modulus of elasticity of shales increases significantly with increasing confining pressure according to the data presented by Smith et al. (1969) (Fig.89).

Elastic properties of a homogeneous, isotropic, linearly elastic body following Hooke's

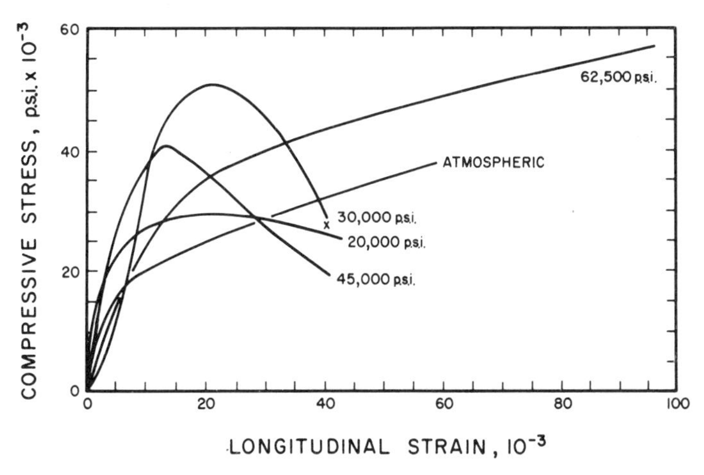

Fig.87. Stress–strain curves for various shale samples with hydrostatic confining pressure as a parameter (triaxial test); x denotes failure point. Compressive stress measured by a load cell. (After Smith et al., 1969, fig.9, p.6.)

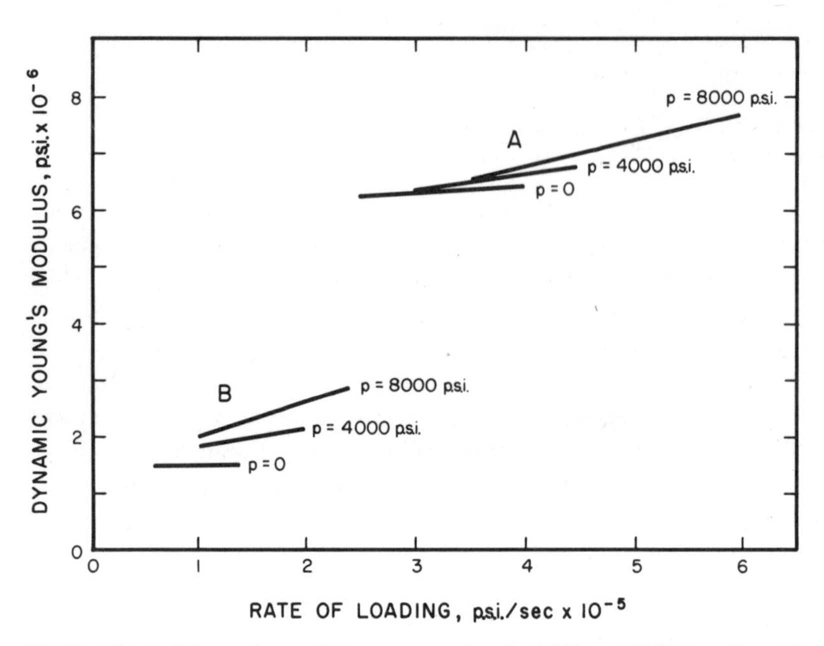

Fig.88. Effect of increasing confining pressure (p = 0; 4,000; and 8,000 p.s.i.) and the rate of loading on dynamic Young's modulus. A. For the Canyon Shale (Pennsylvanian; depth = 10,849 ft.) B. For the Yegua Shale (Eocene; depth = 6,711 ft.). All measurements were made at room temperature. (After Simons, 1963, fig.11.)

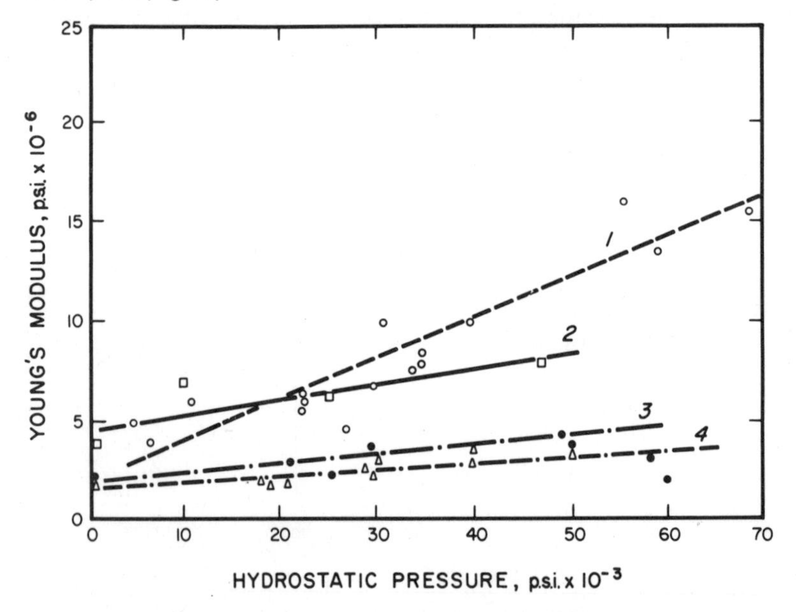

Fig.89. Variation of Young's modulus (secant modulus measured at the point where the strain is equal to 10^{-3}) with confining pressure. *1* = granite; *2* = marble; *3* = shale; *4* = sandstone. (After Smith et al., 1969, fig.12, p.7.)

TABLE XIX
Average measured and computed elastic constants for North Pacific sediments. Laboratory measurements at a temperature of 23°C and a pressure of 1 atm [1]
(After Hamilton, 1971a, table 1, p.587. Courtesy of *Journal of Geophysical Research*)

Sediment type	*Measured*			*Computed*			
	ϕ (%)	ρ (g/cm^3)	v_p (m/sec)	κ (dynes/cm^2 $\cdot 10^{10}$)	ν	G (dynes/cm^2 $\cdot 10^{10}$)	v_s (m/sec)
Continental terrace (shelf and slope)							
Sand							
Coarse	38.6	2.03	1836	6.6859	0.491	0.1289	250
Fine	43.9	1.98	1742	5.6877	0.469	0.3212	382
Very fine	47.4	1.91	1711	5.1182	0.453	0.5035	503
Silty sand	52.8	1.83	1677	4.6812	0.457	0.3926	457
Sandy silt	68.3	1.56	1552	3.4152	0.461	0.2809	379
Sand–silt–clay	67.5	1.58	1578	3.5781	0.463	0.2731	409
Clayey silt	75.0	1.43	1535	3.1720	0.478	0.1427	364
Silty clay	76.0	1.42	1519	3.1476	0.480	0.1323	287
Abyssal plain (turbidite)							
Clayey silt	78.6	1.38	1535	3.0561	0.477	0.1435	312
Silty clay	85.8	1.24	1521	2.7772	0.486	0.0773	240
Clay	85.8	1.26	1505	2.7805	0.491	0.0483	196
Abyssal hill (pelagic)							
Clayey silt	76.4	1.41	1531	3.1213	0.478	0.1498	312
Silty clay	79.4	1.37	1507	3.0316	0.487	0.0795	232
Clay	77.5	1.42	1491	3.0781	0.491	0.0514	195

[1] ϕ = porosity; ρ = bulk saturated density; v_p = compressional-wave velocity; κ = bulk modulus; ν= Poisson's ratio; G = rigidity (shear modulus); and v_s = shear-wave velocity.

law can be computed by knowing the values of two elastic parameters. To define any material elastically, two elastic constants out of five listed in Table XVIII are required. As an example, if the coefficient of rigidity, G, which determines the change of form under pure shear, and the bulk modulus, κ, which controls the change in volume under pure compression, are known, other elastic parameters can be easily calculated. Each of the constants presented in Table XVIII determines the changes in volume and form of a homogeneous body in a purely elastic process. Equations relating these various parameters are also presented in Table XVIII.

Hamilton (1971a, p.580) measured the elastic properties of various recent marine sediments in the laboratory using dynamic tests (see Table XIX). Although the values of his measured and computed properties are mostly for recent marine sediments from various environments, they should be close to those of similar sediments. Hamilton's (1971a)

TABLE XX
Bulk modulus of compression for various sediments
(After Domenico and Mifflin, 1965, table 1, p.566)

Material	E (lb./ft.^{2})
Plastic clay	$1 \cdot 10^4$ to $8 \cdot 10^4$
Stiff clay	$8 \cdot 10^4$ to $1.6 \cdot 10^5$
Medium-hard clay	$1.6 \cdot 10^5$ to $3 \cdot 10^5$
Loose sand	$2 \cdot 10^5$ to $4 \cdot 10^5$
Dense sand	$1 \cdot 10^6$ to $1.6 \cdot 10^6$
Dense sandy gravel	$2 \cdot 10^6$ to $4 \cdot 10^6$
Rock, fissured, jointed	$3 \cdot 10^6$ to $6.25 \cdot 10^7$
Rock, sound	Greater than $6.25 \cdot 10^7$

study is based on the application of the Hookean elasticity equation to results from dynamic tests.

Domenico and Mifflin (1965) presented some values for the bulk modulus of elasticity (?-termed bulk modulus of compression by them) for various sediments (Table XX).

Bulk modulus. The concept of interstitial fluid storage and its flow in argillaceous sediments implies that a mathematical solution would require an understanding of the changes in the physical strength and elastic properties. As shown previously in Chapter 2, volume reduction in saturated sediments and rocks can take place only when there is a loss of fluid from the pores. With expulsion of fluid from the pores of an elemental clay volume, the volume of clay decreases, resulting in a decrease in thickness. A change in vertical dimension is related to the compressibility of the element. Compressibility by definition is the reciprocal of the bulk modulus of compression. A separate section in this chapter discusses thoroughly the compressibilities of clays.

The best values for dynamic bulk moduli of water-saturated sediments are obtained when compressional- and shear-wave velocities and densities are known (Hamilton, 1971a, p.581). Urick (1947) proved experimentally that an aggregate theory expressed as:

$$c_{sw} = \phi c_w + (1 - \phi) c_s \tag{4-13}$$

where c_{sw} is the compressibility of the suspension, ϕ is the porosity, c_w is the water compressibility, and c_s is compressibility of the solids, could be applied to kaolinite clay suspensions in distilled water. A deflocculant was used by Urick to prevent formation of flocculated clay-mineral aggregate structures, thereby creating a true suspension without rigidity. Hamilton (1971a, p.583) pointed out that the frame bulk modulus has to be considered in computations of the bulk modulus of the system. With increasing porosity, the dynamic frame bulk modulus decreases. When porosity is zero, the bulk modulus of the frame equals the bulk modulus of the solid material.

Rigidity. The elastic resistance of a body to shearing forces is called rigidity. Hamilton (1971a,p.594) showed that dynamic rigidity increases sharply in marine clastics with decreasing grain size and increasing porosity to maximum values in very fine sands, silty sands, and coarse silts (Table XIX). Maximum values of dynamic rigidity occur between mean grain sizes of 0.09 to 0.04 mm, and porosities between 55 and 60%. Maximum rigidities occur between bulk densities of 1.7 and 1.8 g/cm^3.

In argillaceous sediments, dynamic rigidity is related to cohesion between fine clay-size particles. Cohesion is the resistance to shear stress and is caused by physicochemical forces between the particles. Cohen (1968) demonstrated an increase in dynamic rigidity in flocculated structures formed by both bentonite and kaolinite with time. Hamilton (1971a, p.595) pointed out that slow deposition rates and longer time may lead to increase in sediment cohesion owing to compaction and cementation. Humphries and Wahls (1968) showed that the dynamic modulus of rigidity increases with increasing effective stress and decreasing void ratio in clays. Kaolinite shear-modulus values were on the order of 6–10 times higher than those for bentonite. The dynamic rigidity of most sands and clays increases with increasing effective stress.

Poisson's ratio. Poisson's ratio is defined as the ratio of the unit transverse strain to the unit longitudinal strain (Table XVIII). The Poisson's ratio of a liquid (having zero rigidity) is 0.50. Judd (1959) stated that for rough computations, Poisson's ratio for rocks generally is regarded as 0.25. Ranges of Poisson's ratio for some principal marine sediment types are given in Table XXI (Hamilton, 1971a). Barkan (1962, p.13) noted that Poisson's ratio is always less for sands than for clays; however, Hamilton (1971a, p.600) indicated that this is not always true. Poisson's ratio varies with porosity and density in the same manner as does dynamic rigidity.

TABLE XXI
Poisson's ratio for some principal marine sediment types
(After Hamilton, 1971a, table 3, p.599, Courtesy of *Journal of Geophysical Research.*)

Sediment	*Poisson's ratio*			*No. of samples*
	max.	*min.*	*avg.*	
Continental terrace				
Sands (all grades)	0.496	0.416	0.470	14
Clayey silt	0.499	0.447	0.478	40
Abyssal plain (turbidite)				
Silt-clays	0.496	0.466	0.484	54
Abyssal hill (pelagic)				
Silt-clays	0.499	0.467	0.487	41

Eaton (1969) showed by using a rearranged form of the Hubbert–Willis equation that Poisson's ratio values could be computed with respect to increasing burial depth in the Gulf Coast sedimentary sequence (see previous discussion in Chapter 3, p. 102):

$$\frac{\nu}{1-\nu} = \frac{(p_{\mathrm{w}}/D)-(p_{\mathrm{p}}/D)}{(p_{\mathrm{t}}-p_{\mathrm{p}})/D} = \frac{p_{\mathrm{w}}-p_{\mathrm{p}}}{p_{\mathrm{t}}-p_{\mathrm{p}}} \tag{4-14}$$

where p_{w} is the wellbore pressure in p.s.i., D is depth in feet, p_t is the overburden pressure in p.s.i., p_{p} is pore pressure in p.s.i., and ν is Poisson's ratio.

Some computed values for Poisson's ratio for the Gulf Coast sedimentary rocks ranged from 0.25 at a depth of 1,000 ft. to approx. 0.5 at 20,000 ft., which would indicate that the sedimentary layers at a depth of 20,000 ft. approach the state of being incompressible in a plastic failure environment (Eaton, 1969). In many instances these values are not reasonable for rocks at these depths. Adams (1951, p.66) stated that with an increasing overburden pressure, Poisson's ratio should increase, but not strikingly. Data from Geyer and Myung (1970) showed very little increase in Poisson's ratio with depth. There are some shortcomings involved in using the Hubbert–Willis equation because of the following assumptions: (1) existence of uniaxial strain conditions; (2) rapid increase in Poisson's ratio with depth, reaching an upper limit of 0.5; and (3) existence of overburden pressure gradient of 1 p.s.i./ft. throughout the entire sedimentary column.

Lamé constant. The quantity $[\nu E/(1+\nu)(1-2\nu)]$ is known as Lamé's constant, λ (Table XVIII). Although Lamé's constant does not seem to have physical significance, it facilitates the calculation of other elastic moduli.

Effect of pore fluids on elastic properties. Gregory (1967) noted that liquid pore fluids in rocks and sediments cause increases in the dynamic bulk modulus and dynamic Poisson's ratio, but reduce the dynamic modulus of elasticity and dynamic modulus of rigidity. Modulus of elasticity decreases with increasing fluid saturation; this is in conformance with elastic theory. Liquid saturation tends to reduce the rate of change of elastic moduli as a function of axial or uniform loading. The rate of change in the elastic moduli is also reduced as the stress level increases. The presence of pore fluids in rock samples caused the P-wave velocity to increase and the S-wave velocity to decrease at all pressure levels. Gregory (1967) investigated wave propagation in sandstone and limestone under uniform and axial compaction.

Strength properties of argillaceous sediments

There is considerable confusion in the literature concerning the term "strength" of a sediment or a rock. Ultimate strength of a sedimentary rock is defined as the stress at the time of sudden failure under an applied load (see Fig.87). In unconsolidated sediments, the strength or bearing capacity of the material is defined in terms of the amount of load per unit area that causes the sediment to deform, i.e., to become plastic under stress.

Shear strength of rocks is not measured directly, but is computed for a given normal stress from the principal stresses at failure by Mohr's theory. The maximum shear stress is one-half the difference of the maximum axial stress and the hydrostatic pressure. The triaxial compression test permits a determination of the shearing strength of unconsolidated and indurated sediments.

Shear strength. Coulomb's formula which expresses the shearing strength, τ, in p.s.i. is:

$$\tau = C + (\sigma_z - \sigma_w) \tan \theta \tag{4-15}$$

where C is the cohesion in p.s.i., σ_z is the vertical compressive stress in p.s.i., σ_w is the pore stress in p.s.i., and θ is the internal angle of friction. Hamilton (1971a, p.593) stated that shear strength in sediment composed of silt and clay, in the absence of normal stress, is defined by the cohesion alone. In sands and most clays, shear strength and dynamic rigidity increase with increasing effective pressure.

Underwood (1967, p.104) reported that weakly compacted shales have cohesion values as low as 2 p.s.i. and a friction angle as low as 5°, whereas some calcareous and quartz-zone shales of Cambrian age have cohesion values ranging from 1,160 p.s.i. to 3,390 p.s.i. and friction angle varying from 45° to 64°.

Compressive strength. Underwood (1967, p.103) stated that the compressive strength of

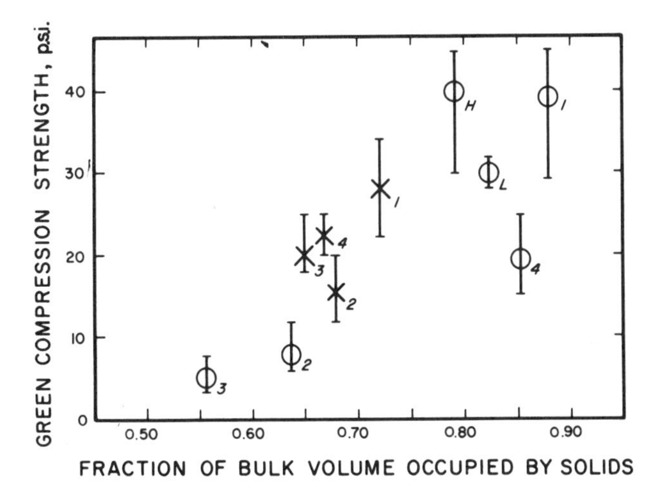

Fig.90. Average green compressive strengths of natural bentonites. Vertical lines indicate the deviation from the average values. Circles are Na-bentonites, whereas crosses are Ca-bentonites. Sample numbers are also presented, as given by Bradley. (After Bradley, 1956, fig.1, p.43.) Green strength – load at failure of unconfined sample as conducted on a synthetic molding sand testing machine (Bradley, personal communication, 1972.)

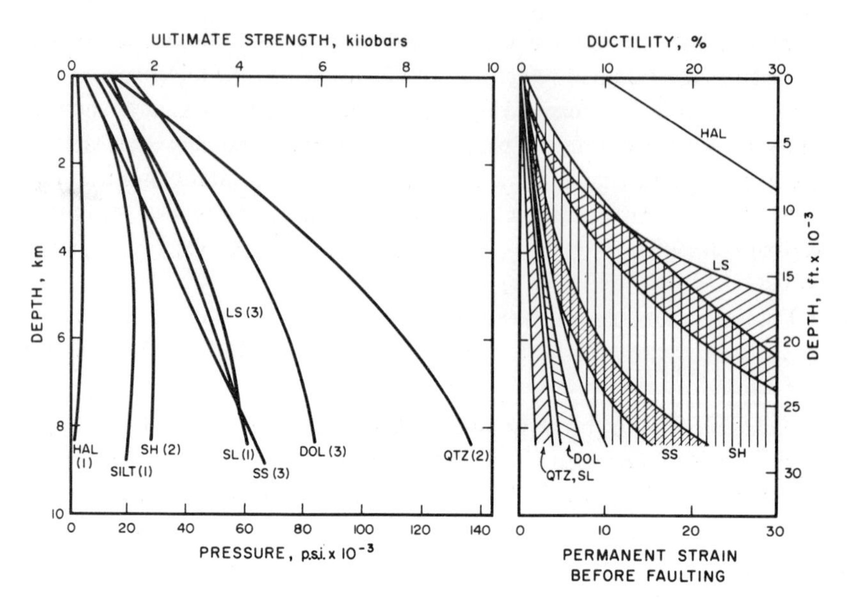

Fig.91. Ultimate compressive strengths and ductilities of dry rocks as functions of burial depth at a temperature gradient of 30°C/km. (After Handin et al., 1963, fig.26, p.748. Courtesy of American Association of Petroleum Geologists.)

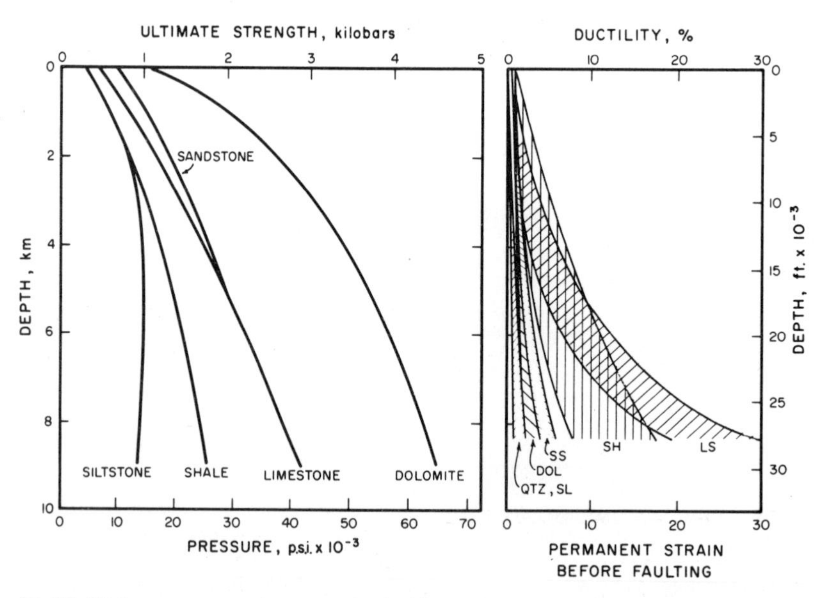

Fig.92. Ultimate compressive strengths and ductilities of water-saturated rocks as functions of burial depth at a temperature gradient of 30°C/km. (After Handin et al., 1963, fig.27, p.749. Courtesy of American Association of Petroleum Geologists.)

shale ranges from less than 25 p.s.i. for weaker compacted shales to more than 15,000 p.s.i. for well-cemented shales.

Bradley (1956) cored several commercial bentonite beds and determined their green compressive strengths. Fig.90 shows that the compressive strength of the various bentonites are roughly proportional to the fraction of the bulk volume of the rock which is actually occupied by solids. The strengths are of about the same order of magnitude as the load of overburden which deposits bear in the field. Maximum green compressive strengths of synthetic molding sands, which have solid fractions of about 0.9, are of the order of 30–40 p.s.i. (Bradley, 1956). The initial apparent relative viscosities of dilute thixotropic suspensions relate to shearing stresses of the order of only 10^{-3} times the strength of these concentrated systems.

The effects of pore pressure on ultimate strength and ductility of some common sedimentary rock types is given in Figs.91 and 92. Comparison of data in Figs.91 and 92 reveals that pore pressure tends to reduce both the ultimate compressive strength and ductility at all depths, inasmuch as these parameters depend principally upon effective confining pressure, and formation pressure lowers the effective overburden pressure in porous rocks (Handin et al., 1963, p.751). Handin and Hager (1958) showed that the deformational behaviors of all homogeneous sedimentary rocks are essentially similar. One can use Figs.91 and 92 with a degree of confidence to estimate the strengths and

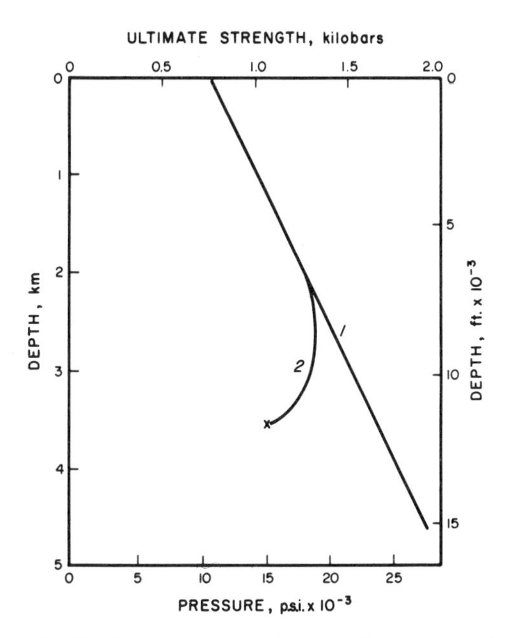

Fig.93. Ultimate compressive strength of water-saturated Berea Sandstone as a function of depth for a "normal" formation (pore) pressure gradient ($p_p \approx 0.5$ of hydrostatic confining pressure) and for an abnormal gradient measured in Gulf Coast, Louisiana, U.S.A. *1* = strength for normal pressure; *2* = strength for abnormal pressure. (After Handin et al., 1963, fig.28, p.752.)

ductilities of these rock types at any realistically simulated depths to about 30,000 ft. (Handin et al., 1963, p.751).

Abnormal formation pressures (see Chapter 7) strongly affect the compressive strength of a rock. Fig.93 illustrates the variation of the ultimate compressive strength of water-saturated Berea Sandstone under normal and abnormal pressures (Handin et al., 1963, p.752). Shaley sediments under such conditions probably would show even a more pronounced reduction in strength.

Relationship between strength and texture (orientation) of clays

The compressive strength of clays under uniaxial compression increases with increasing compaction pressure until a maximum value is reached at an "optimal loading pressure". Thereafter, compressive strength does not change to any appreciable degree. This phenomenon is possibly due to variation in the texture of compacting clay. Shibakova and Shalimova (1967) investigated the optimal loading pressure for kaolinite, hydromica, and montmorillonite clays. The moisture content of the initial samples corresponded to the lower plasticity limit. After compaction, the samples were tested at air-dry condition. The optimal loading pressures for montmorillonite, hydromica, and kaolinite were found to be equal to 20–25 kg/cm^2, 60 kg/cm^2, and 80 kg/cm^2, respectively. The coefficients of degree of orientation, C_o, and relative orderliness, U, were determined with the polarizing microscope. The coefficient of orientation, C_o, shows the degree of parallelism (expressed in %) of clay particles in any part of a sample. This coefficient is equal to $[1-(I_{min}/I_{max})] \times 100$, where I_{min} is the light flow in the position of extinction and I_{max} is the light flow in the position of illumination (between crossed nicols). The coefficient of degree of orderliness, U, is equal to area of thin section occupied by microsections (aggregates), in which clay particles are oriented, divided by the total area of thin section.

For kaolinite clay, the coefficient of degree of orderliness, U, gradually increases from 50% to 80% on increasing loading from 5 to 80 kg/cm^2 (Fig.94). At low pressures ($\approx$ 5 kg/cm^2), sections cut parallel and perpendicular to the loading direction have similar textures. On increasing the loading pressure, the U coefficient increases in the sections parallel to the loading direction.

In the case of hydromica, there were no non-oriented clay masses; however, the individual packets (microblocks, aggregates or domains) are oriented at random with respect to each other. Although the increase in loading pressure does not result in reorientation of microblocks, it does cause increase in the degree of orientation of particles within individual packets, i.e., the coefficient C_o increases from 50% to 72% (Fig.94).

The montmorillonite clay tested was composed of primary, rounded, tabular aggregates (0.05–1 mm in size) and individual clay particles. At a pressure of 5 kg/cm^2, the aggregates predominate. On compaction, the coefficient of orientation within the aggregates remains more or less constant (65–90%). Up to 60 kg/cm^2, relative proportions of the aggregates and the clay mass remain constant (U = 70%). On the other hand, the

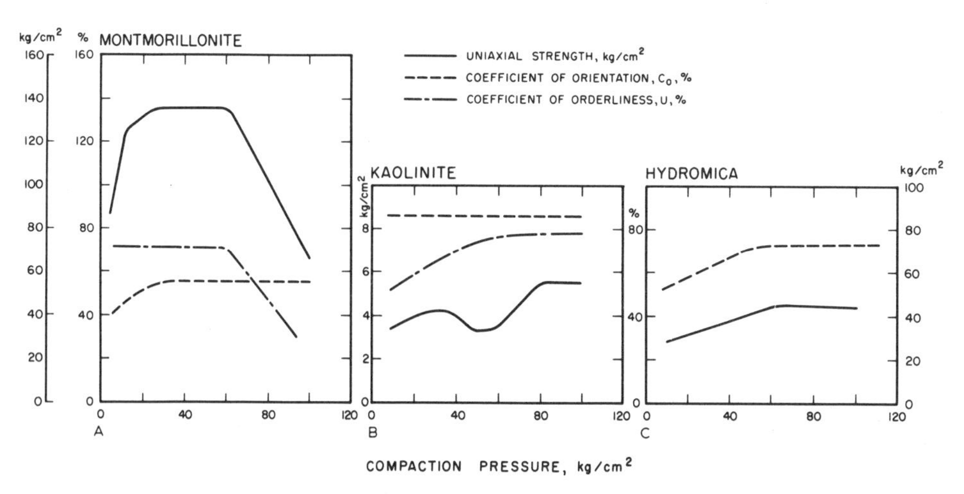

Fig.94. Variation in strength and textural coefficients of clays on compaction. A. Montmorillonite clay from the Askanskiy deposit of Georgia (95% clay). B. Kaolinite clay of Novoselitskiy deposit of Ukraine (70% of clay particles are less than 0.001 mm in size). C. Cambrian clay (hydromica) of Leningrad (40% of clay particles are less than 0.001 mm and 20–30% are 0.01–0.05 mm in size.) (After Shibakova and Shalimova, 1967, fig.1, p.91.)

degree of orientation of clay particles outside the aggregates first increases (C_0 = 38% at 5 kg/cm^2 and C_0 = 53% at 25 kg/cm^2) and then remains practically the same. At a loading pressure of 80–100 kg/cm^2, which corresponds to a marked decrease in strength, the aggregates are destroyed and U decreases to 30%. Thus, it appears that the strength of montmorillonite clay depends on the strength of primary aggregates which have a high degree of particle orientation. The reasons for the destruction of aggregates at pressures above 60 kg/cm^2 are not known.

In general, the strength of each clay type is determined by its texture and increases with increasing degree of orientation of individual clay particles.

COMPRESSIBILITY OF ARGILLACEOUS SEDIMENTS

Many properties of clays change with decreasing porosity under high overburden pressure. The magnitude of these changes is related to the pore compressibility of the clays, whereas at high pressures it is also related to the compressibility of the mineral grains themselves. The studies on compressibilities of sandstones and carbonates are not numerous, but include some important papers by Carpenter and Spencer (1940), Fatt (1953), Hall (1953), Harville (1967), Harville and Hawkins (1967), Fatt (1958a,b), McLatchie et al. (1958), Van der Knaap (1959), Dobrynin (1962), Knutson and Bohor (1963), and Van der Knaap and Van der Vlis (1967).

Even fewer studies have been published on the compressibility characteristics of shales and clays; they include contributions by Macey (1940), Skempton (1953), Paulding et al. (1965), Van der Knaap and Van der Vlis (1967), Rieke et al. (1969), and Chilingar et al. (1969).

As pointed out by Cebell and Chilingarian (1972), data on the compressibility of sediments can be employed in the solution of a variety of problems, and may shed light upon the migration of petroleum from shales into reservoir rocks. Fluid withdrawal from poorly indurated sands and shales usually results in ground subsidence at the surface. A knowledge of the compressibilities of the formations involved will aid in estimating the amount of subsidence and, whenever necessary, in formulating a program of subsidence control. In many cases, petroleum reservoir sands and associated shales undergo compaction contemporaneously with production of oil, gas, and water. This compaction often provides additional reservoir energy, in which case it becomes important in predicting ultimate recovery of oil and gas from the reservoir. In addition, a better understanding of the nature of clay, or shale, compressibility should precede studies on oil migration from argillaceous sediments.

Sedimentologists are aware of the effects of differential compaction which can provide closure above bar sands, channel sands, reefs, etc., in a predominantly shale section. A reasonably accurate estimate of the shale compressibility would enable the construction of a regional, pre-compaction shale isopach. This may produce a more regular, blanket type of isopach which could aid in delimiting anomalies when compared with the normal isopach (Cebell and Chilingarian, 1972).

The important types of water associated with clays are: (1) pore water, (2) interlayer water, and (3) hydroxyl-lattice water (Grim, 1968). Pore water, in turn, includes liquid free water present in the mid-regions of pores between clay particles and the adsorbed, non-liquid, "quasi-crystalline" water bonded to and partially penetrating the exterior surfaces of clay particles (Grim, 1968). The latter water molecules are bonded to oxygen atoms in the clay mineral and, possibly, to the hydroxyls and exchangeable cations as well. Much has been written concerning the distinctly different character of adsorbed water (Grim, 1968). Cebell and Chilingarian (1972) defined the clay solids as "including all the water normally associated with clays, but excluding the liquid pore water".

There are four major kinds of compressibility parameters that should be considered: (1) bulk compressibility, c_b, (2) pore compressibility, c_p, (3) grain or matrix compressibility, c_s, and (4) fluid compressibility, c_f.

Relationships among remaining moisture content, void ratio and compressibility

Bulk compressibility, c_b, is defined as: (change in bulk volume)/(unit bulk volume)/(unit change in pressure). Two different methods can be used to determine compressibility. In the first method, the remaining moisture content, M (percent of dry weight) is determined for clay suspensions following compaction. The remaining moisture content is then converted to void ratio, e, as shown below, from which the theoretical compressibility, c_{be}, is computed and plotted versus compaction pressure. This method assumes that all of the compaction occurs within the voids, leaving the clay solids unaltered, and that the clay solids have a fixed specific gravity. Based on experimental work, Chilingar and Knight (1960) and Chilingarian and Rieke (1968) presented formulas relating remaining moisture content, M (in percentage of dry weight), to compaction pressure, p, in p.s.i. Above a certain pressure (e.g., about 1,000 p.s.i. for montmorillonite clay), the experimental points fit a straight line. For example, for halloysite clay the relationship can be expressed as $M = 0.401 - 0.066 \log p$. Furthermore, the related plots of void ratio, e, versus $\log p$ and c_{be} versus $\log p$ are also very close to being straight lines.

The other method of determining compressibility involves the actual measurement of height reduction of a clay compact having a constant cross-sectional area on increasing the compaction pressure. Bulk compressibility determined from the latter method was designated c_{bh} by Cebell and Chilingarian (1972).

The void ratio, e, of a sediment saturated with water is defined as:

$$e = \frac{\text{volume of water}}{\text{volume of solids}} = \frac{V_w}{V_s} \tag{4-16}$$

or:

$$e = \frac{(\text{weight water/specific weight of water})}{(\text{weight solids/specific weight of solids})} = \frac{(W_w/\gamma_w)}{(W_s/\gamma_s)} \tag{4-17}$$

The remaining fractional moisture content, M', is equal to:

$$M' = \frac{\text{weight of water}}{\text{weight of solids}} = \frac{W_w}{W_s} \tag{4-18}$$

Thus:

$$e = \frac{W_w}{W_s} \times \frac{\gamma_s}{\gamma_w} = M' \times \text{sp. gr. solids} \tag{4-19}$$

where sp. gr. is equal to γ_s/γ_w and is assumed by many investigators to remain constant during compaction.

The bulk compressibility, c_b, is defined mathematically as:

$$c_b = -\frac{1}{V_b} \lim_{\Delta p \to 0} \frac{\Delta V_b}{\Delta p}$$

or: (4-20)

$$c_b = -\frac{1}{V_b} \frac{dV_b}{dp}$$

where V_b is the bulk volume and p is the effective compaction pressure. (Effective pressure = total overburden pressure − pore pressure. Pore pressure was atmospheric in the experiments conducted by many investigators.) Bulk volume, V_b, is equal to the sum of the solids volume, V_s, and the volume of the voids, V_v:

$$V_b = V_v + V_s \tag{4-21}$$

and:

$$dV_b = dV_v + dV_s \tag{4-22}$$

When the assumption is made that the change in volume of the solids, dV_s, is equal to zero, eq. 4-20 becomes:

$$c_b = -\frac{1}{V_v + V_s}\frac{dV_v}{dp} \tag{4-23}$$

Multiplying numerator and denominator by V_s and defining $e = V_v/V_s$, eq. 4-23 can be expressed as:

$$c_{be} = -\frac{1}{e+1}\frac{de}{dp} \tag{4-24}$$

where c_{be} is the compressibility as determined from the above-described moisture content and void-ratio versus pressure relationships, and on assuming a constant specific gravity for clay. Inasmuch as the void ratio, e, versus log of pressure, p, plot is commonly a straight, or nearly straight, line on semi-log paper, namely:

$$e = a - n \log p \tag{4-25}$$

where a is the intercept on the y-axis and n is the essentially constant slope of the straight line, then:

$$\frac{de}{dp} = \frac{-n}{2.303\,p} \tag{4-26}$$

Thus:

$$c_{be} = -\frac{1}{e+1}\left(\frac{n}{2.303\,p}\right) \tag{4-27}$$

Rearranging (Dr. Lyman L. Handy, personal communication, 1970):

$$\log c_{be} = -0.3623 - \log p - \log\left(\frac{a+1}{n} - \log p\right) \tag{4-28}$$

The term $(-\mathrm{d}e/\mathrm{d}p)$ is known in soil mechanics as the coefficient of compressibility, a_v, and is the stress/strain ratio of the clay. Inasmuch as fractional porosity, ϕ, is equal to V_v/V_b, and $\mathrm{d}\phi = \mathrm{d}V_v/V_b$, the bulk compressibility in terms of porosity from eq.4-20 is equal to:

$$c_b = -\mathrm{d}\phi/\mathrm{d}p \tag{4-29}$$

The second method of determining the bulk compressibility is a more direct one. The cross-sectional area of the clay compact remains constant, whereas changes in thickness, h, of the clay compact are measured directly with increasing compaction pressure. The bulk compressibility determined from height versus pressure relationships is designated as c_{bh}. As stated previously, eq.4-20 for bulk compressibility is as follows:

$$c_b = -\frac{1}{V_b}\frac{\mathrm{d}V_b}{\mathrm{d}p} \tag{4-20}$$

Thus:

$$c_{bh} = -\frac{1}{Ah}\frac{A\mathrm{d}h}{\mathrm{d}p} = -\frac{1}{h}\frac{\mathrm{d}h}{\mathrm{d}p} \tag{4-30}$$

or:

$$c_{bh} = -\frac{1}{h}\frac{n'}{2.303\,p} \tag{4-31}$$

where n' is the slope of the plot of thickness h versus log of pressure p at any selected pressure. The slope n' is not constant as was the slope n of the void ratio versus log of pressure plot (straight line).

By definition pore compressibility is expressed mathematically as:

$$c_p = -\frac{1}{V_v}\left(\frac{\mathrm{d}V_v}{\mathrm{d}p}\right) \tag{4-32}$$

where c_p is the pore compressibility, V_v is the pore volume, $\mathrm{d}V_v$ is change in pore volume, and $\mathrm{d}p$ is change in the external pressure. It can be easily shown that the bulk compressibility divided by the porosity yields the pore compressibility. Starting with eq.4-23, dividing both sides by ϕ:

$$\frac{c_b}{\phi} = -\frac{1}{V_b\phi}\frac{\mathrm{d}V_v}{\mathrm{d}p} \tag{4-33}$$

and substituting V_v for $(V_b \times \phi)$:

$$\frac{c_b}{\phi} = c_p \tag{4-34}$$

By definition the mineral-grain compressibility, also known as the rock-matrix compressibility, is mathematically expressed as:

$$c_s = -\frac{1}{V_s}\left(\frac{dV_s}{dp}\right) \tag{4-35}$$

Dobrynin (1962, p.361) showed that the bulk compressibility can be expressed in terms of c_s, ϕ, and c_p, if $\Delta V_s \neq 0$:

$$c_b = \phi c_p + (1 - \phi)c_s \tag{4-36}$$

Van der Knaap (1959, p.183) pointed out that fluids in the pores can influence the elastic properties of the porous medium. By definition, the isothermal compressibility of a fluid is:

$$c_f = -\frac{1}{V_f}\left(\frac{dV_f}{dp}\right)_T \tag{4-37}$$

where c_f is fluid compressibility, V_f is the fluid volume, and T is the temperature at which c_f is determined.

A common assumption is that if porous rock or clay is completely saturated with fluids, all additional applied pressure is completely borne by the pore fluids and that the initial grain to grain contact stress remains constant. At high pressures this may not be valid and the relative compressibilities of the pore fluid and grains must be taken into account. Volume decrease of the material at very high pressures may be the result of continued elastic deformation and fracturing of the mineral grains and the compressibility of the pore fluids. The proper total compressibility expression for a clay, shale, or reservoir rock may then contain the compressibility terms for oil, gas, water, and pores. Such an expression for the total compressibility is generally written in terms of separate phase compressibilities by volumetric phase saturation weighting (Craft and Hawkins, 1959, p.273):

$$c_t = S_o c_o + S_g c_g + S_w c_w + c_p \tag{4-38}$$

where c_t is "total system" compressibility, c_o is oil compressibility, c_g is gas compressibility, and c_p is pore compressibility; S_o, S_g, and S_w are the oil, gas, and water saturations (percentage of total effective pore volume) of the formation, respectively. Ramey (1964, p.447) presented generalized plots which speed the process of obtaining estimates of fluid and total system compressibilities.

The "effective" shale compressibility can be defined as the sum of the shale pore and water compressibilities. This is especially valid for the Tertiary basins where there is a large influx of water into the producing overpressured reservoirs from the surrounding argillaceous sediments. It should be noted that certain shales have a large hydrocarbon content and this should be taken into consideration through the addition of proper variables.

Some data on compressibilities of fluids have been presented by Long and Chierici (1961) and Langnes et al. (1972).

Experimental data on compressibility

Although many consolidation (compressibility) studies on clays and shales have been performed in soil-mechanics laboratories for more than 50 years, these tests have been limited largely to a low-pressure range ($\leqslant$ 1,000 p.s.i.). During the same period, high-pressure confining tests on reservoir rocks have exceeded 15,000 p.s.i. Most investigators used mainly well-consolidated sandstones or limestones in their laboratory experiments. Knutson and Bohor (1963), however, worked with reservoir rocks typical of the Texas–Louisiana Gulf Coast region (orthoquartzites to calcareous subgraywackes), whereas Van der Knaap and Van der Vlis (1967) worked with unconsolidated clays and sands from the Bolivar Coast in Venezuela.

Carpenter and Spencer (1940) measured the "pseudo bulk compressibility" [$\beta = -(1/V_b)(\partial V_v/\partial p_t)$, where β is pseudo bulk compressibility in p.s.i.$^{-1}$, V_b is the original bulk volume in cm^3, ∂V_v is the change in void volume in cm^3, and ∂p_t is the change in the applied pressure in p.s.i.] of various consolidated sandstones in an attempt to investigate whether or not fluid withdrawal from Gulf Coast reservoirs and the resulting volume reduction could account for subsidence. Compressibilities calculated from Carpenter and Spencer data on the Woodbine Sandstone are presented in Fig.95. Carpenter and Spencer (1940) did not keep either the external pressure or the internal pore pressure constant in their series of experiments.

Hall (1953, p.310) utilized the reduction in pore space, at a constant overburden pressure, to obtain the "effective" compressibility values [$= -(1/V_v)(\partial V_v/\partial p_i)_{p_t}$] of sandstones and limestones. In his equation, p_i is the internal fluid pressure in the voids of the rock and p_t is the constant overburden pressure. He introduced another compressibility term, "formation compaction" [$= -(1/V_v)(\partial V_v/\partial p_t)_{p_i}$], which is part of the total rock compressibility. Hall's values for "effective" reservoir rock compressibilities ranged from $1 \cdot 10^{-7}$ to $3.4 \cdot 10^{-6}$ p.s.i.$^{-1}$ by gradually reducing static fluid pressure in the core (0 to 26% porosity range) at a constant external hydrostatic pressure p_t of 3,000 p.s.i. Hall showed that as porosity increases, the "effective" rock compressibility decreases. Hall (1953, p.310) applied a constant external hydrostatic pressure to his lucite-jacketed cores and allowed the pore pressure to be reduced in steps.

Fatt (1958a,b) and McLatchie et al. (1958) studied the relationship between compres-

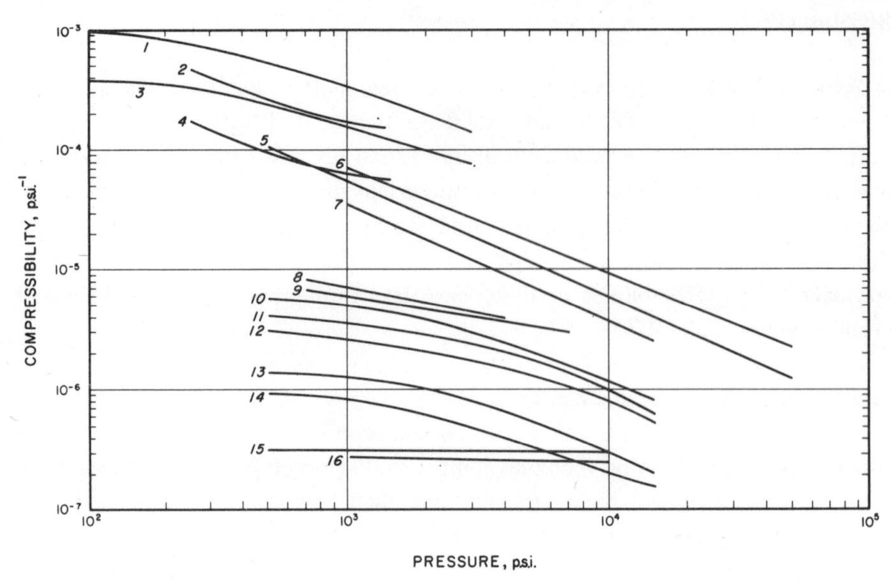

Fig.95. Relationship between compressibility (p.s.i.$^{-1}$) and applied pressure (p.s.i.) for unconsolidated sands, illite clay, limestone, sandstones, and shale. (After Sawabini, 1971, p.118.)

No.	*Investigator*	*Rock type*	*Type of applied pressure*	*Compressibility*
1	Sawabini and Chilingarian (1971)	California unconsolidated sands	triaxial (?)[4]	pore
2	Kohlhaas and Miller (1969)	California unconsolidated sands	uniaxial (?)	pore
3	Sawabini and Chilingarian (1971)	California unconsolidated sands	triaxial (?)[4]	bulk
4	Kohlhaas and Miller (1969)	California unconsolidated sands	uniaxial (?)	bulk
5	Chilingar et al. (1969)	illite clay (API no.35) (wet) [1]	uniaxial	bulk
6	Chilingar et al. (1969)	illite clay (API no.35) (dry)	uniaxial	bulk
7	Knutson and Bohor (1963)	Repetto Fm. (Grubb zone) (wet) [2]	net confining[5]	pore
8	Knutson and Bohor (1963)	Lansing–Kansas City Limestone (wet) [2]	net confining	pore
9	Carpenter and Spencer (1940)	Woodbine Sandstone (wet)[1]	net confining	pseudo bulk
10	Fatt (1958b)	feldspathic graywacke (no.10) (wet) [3]	net confining	bulk
11	Fatt (1958b)	graywacke (no.7) (wet) [3]	net confining	bulk
12	Fatt (1958b)	feldspathic graywacke (no.11) (wet) [3]	net confining	bulk
13	Fatt (1958b)	lithic graywacke (no.12) (wet) [3]	net confining	bulk
14	Fatt (1958b)	feldspathic quartzite (no.20) (wet) [3]	net confining	bulk
15	Podio et al. (1968)	Green River Shale	net confining	bulk
16	Podio et al. (1968)	Green River Shale (wet) [1]	net confining	bulk

[1] Saturated with distilled water.
[2] Saturated with formation water.
[3] Saturated with kerosene.
[4] Actually approaches hydrostatic.
[5] Net confining pressure (hydrostatic) = $p_e = \sigma - 0.85\, p_p$.

sibility and rock composition. They reported that unconsolidated sediments, which are poorly sorted and contain clay, have higher compressibilities than do consolidated and well-sorted sands. Fatt (1958b, p.1924) found that his measured bulk compressibilities in sandstones are a function of rock composition for a given grain shape and sorting. If sandstones are divided into two classes (one with well-sorted, well-rounded grains and the other with poorly sorted, angular grains), then for each class the compressibility is a linear function of the amount of intergranular material (Fatt, 1958b, p.1937). The procedure used by Fatt (1958b) and Hall (1953) was similar to that of Carpenter and Spencer (1940), but in the former case the fluid was expelled under constant external pressure with a reduction in pore pressure rather than an increase in the external stress. This is believed to closely duplicate reservoir conditions. Fatt's (1958b) procedure was to apply a constant external stress to the core and decrease or increase the pore pressure. His apparatus simultaneously measured both the bulk and pore volume compressibilities at room temperature. Volume changes of the core in the pressure cell were measured through the use of a linear potentiometer that could resolve a movement of $1 \cdot 10^{-3}$ inch.

Fatt (1958b) reported measurements of bulk compressibilities on a variety of petroliferous sandstones (0–15,000 p.s.i. net confining pressure range) (Fig.95). His "pseudo" bulk compressibility (of a Boise Sandstone sample) is not the same property measured by Carpenter and Spencer (1940, p.17), and was defined by Fatt as $[-(1/V_b)(\partial V_b/\partial p_i)_{p_t}]$, where p_i is the internal fluid pressure and p_t is the constant external pressure. This volume change $(\partial V_b/\partial p_i)$ is believed to represent essentially the volume change that a reservoir undergoes as the fluid pressure is depleted by production. The net confining pressure is defined as the difference between the applied stress and the pore pressure. It is important to note that Fatt's work included the development of several analytical analog models constructed out of rubber and glass spheres and cubes with which he attempted to explain the mechanical behavior of porous rocks.

Van der Knaap (1959) noted that pore compressibility increases with decreasing porosity. Dobrynin (1962, p.361) suggested that between certain minimum and maximum pressures, the relationship between pore compressibility and logarithm of pressure can be approximated by a straight line.

Geertsma (1957) demonstrated that the bulk and pore-volume compressibilities are functions of the elastic and viscous deformation of the rock matrix and rock bulk material, and the porosity. On the basis of mathematical analysis, he also showed that for many oil sands, which show isotropic elastic behavior, the pore-volume compressibility measured in the laboratory under uniform and constant pressure is about twice the compressibility in the reservoir.

Paulding et al. (1965, p.70) reported the compressibility of a kaolinite clay sample (95% saturated with water) up to 32,000 p.s.i. In this experiment no water egression from the sample was allowed. The bulk compressibility was found to be equal to $8.8 \cdot 10^{-7}$ p.s.i.$^{-1}$ at 32,000 p.s.i.

Thomas (1966, p.4) studied the effects of overburden pressure on shale during re-

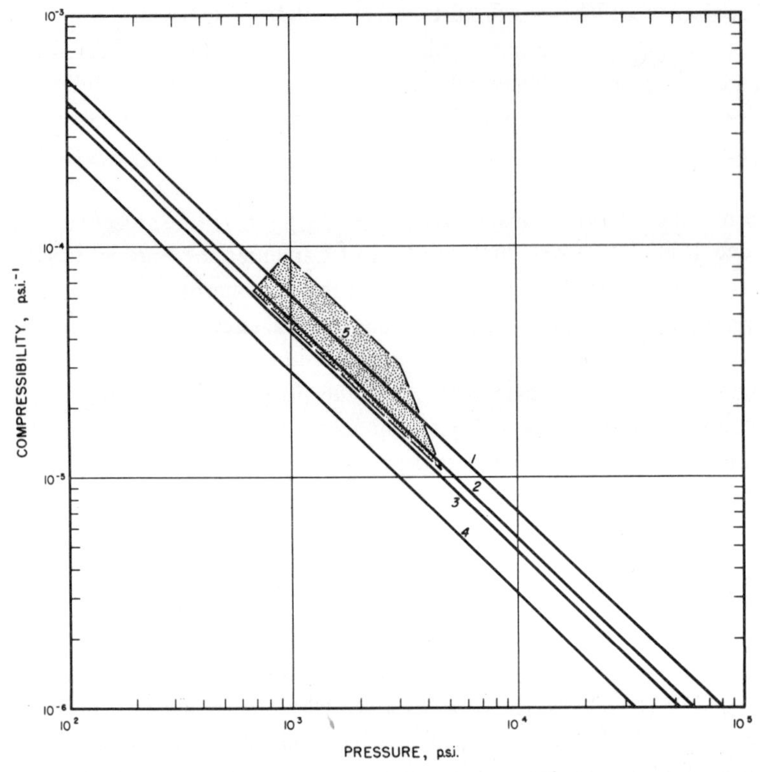

Fig.96. Relationship between bulk compressibility in p.s.i.$^{-1}$ and effective pressure in p.s.i. for various clays. *1* = illite (API no.35); *2* = halloysite (API no.12); *3* = dickite (API no.15); *4* = soil from weathered limestone terrain (Louisville, Kentucky); and *5* = the area where compressibilities of post-Eocene clay samples from the Bolivar Coast, Venezuela, fall (*5* = after Van der Knaap and Van der Vlis, 1967). Samples *1, 2, 3* and *4* were hydrated in distilled water. (After Rieke et al., 1969, fig.5, p.824.)

torting and noted that initially there was an increase in the bulk volume under low overburden pressures. Above 1,000 p.s.i.g. * the bulk volume decreased.

Van der Knaap and Van der Vlis (1967) found that a straight-line relationship exists between the log of the bulk compressibility and the log of the "effective" pressure, which in this case was equal to the direct applied axial load (Figs.96 and 97). Bulk compressibilities of unconsolidated clays and sands decreased with increasing overburden pressure.

Rieke et al. (1969) and Chilingar et al. (1969) presented some data on compressibility of clays (Tables XXII and XXIII; Figs.96 and 97).

Rieke et al. (1969) also showed the relationship between the void ratio, e, and the logarithm of pressure for various clays as presented in Table XXII and Figs.98 and 99.

Figs.100–102 show relationships between bulk compressibility, c_b, and pressure for halloysite, hectorite and illite clays, as determined by Cebell and Chilingarian (1972).

---

* p.s.i.g. = pounds per square inch, gauge.

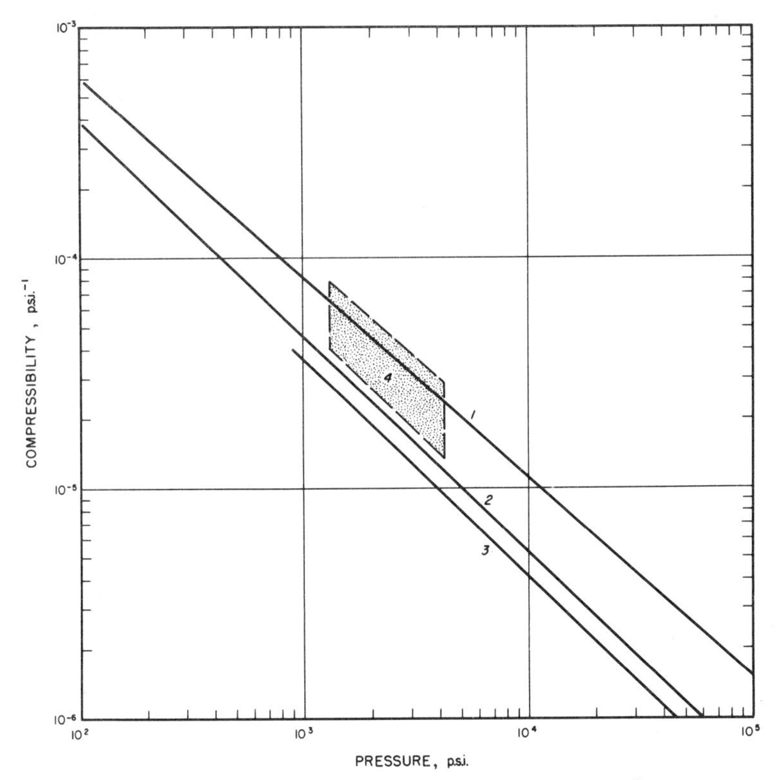

Fig.97. Relationship between the bulk compressibility in p.s.i.$^{-1}$ and effective pressure in p.s.i. *1* = montmorillonite (API no.25); *2* = kaolinite (API no.4); *3* = P-95 dry lake clay (Buckhorn Lake, California); *4* = the area encompassing compressibilities of post-Eocene sand samples from the Bolivar Coast, Venezuela (*4* = after Van der Knaap and Van der Vlis, 1967). Samples *1, 2* and *3* were hydrated in distilled water. (After Rieke et al., 1969, fig.4. p.823.)

Each figure consists of three curves: (1) actual bulk compressibility, c_{bh}, based on direct thickness-versus-pressure measurements; (2) bulk compressibility, c_{be}, determined from remaining moisture content-versus-pressure curves on assuming a constant density of 2.0 g/cm^3; and (3) bulk compressibility determined from remaining moisture content-versus-pressure curves on assuming a density of 3.0 g/cm^3. In Fig.102, curve *4* was also calculated from the moisture-versus-pressure relationship but on assuming different densities at different pressures, based on the observed results (Table II) that solid densities decrease with pressure. The difference between the c_{bh} and c_{be} curves in Fig.100 is probably due to the fact that constant densities were assumed in calculating the c_{be} values. As shown in eq.4-19, void ratio, e, would decrease as the density of solids decreases; and there is a consequent increase in slope n. Thus, examination of eq.4-27 shows that the net result is a relative increase in c_{be} values as density decreases (Fig.102, curve *4*).

According to Cebell and Chilingarian (1972), the actual compressibility, c_{bh}, of halloysite clay starts to exceed the calculated compressibility, c_{be}, at about 5,000 p.s.i. In

TABLE XXII
Void ratio and compressibility equations for various clays [1]
(Modified after Rieke et al., 1969, table 1, p.822)

Type of clay	*Assumed density* (g/cm^3)	*Relationship between void ratio and effective pressure*	*Compressibility equations*
Montmorillonite [2] (API no.25) [3]	2.60	$e = 2.69 - 0.467(\log p)$	$c_b = 3.25 \cdot 10^{-2} p^{-0.874}$
Illite (API no.35)	2.67	$e = 1.335 - 0.23(\log p)$	$c_b = 3.9 \cdot 10^{-2} p^{-0.926}$
Illite, dry powder [4]	–	$h = 0.7405 - 0.0885(\log p)$ *	$c_b = 4.5 \cdot 10^{-2} p^{-0.915}$
Halloysite (API no.12)	2.55	$e = 1.01 - 0.163(\log p)$	$c_b = 3.3 \cdot 10^{-2} p^{-0.946}$
Kaolinite (API no.4)	2.63	$e = 0.885 - 0.153(\log p)$	$c_b = 3.5 \cdot 10^{-2} p^{-0.946}$
Dickite (API no.15)	2.60	$e = 0.682 - 0.128(\log p)$	$c_b = 3.05 \cdot 10^{-2} p^{-0.980}$
Hectorite (API no.34)	2.66	$e = 0.718 - 0.123(\log p)$	$c_b = 2.85 \cdot 10^{-2} p^{-0.954}$
P-95 dry clay (Buckhorn Lake, Calif.)	2.53	$e = 0.70 - 0.116(\log p)$	$c_b = 2.8 \cdot 10^{-2} p^{-0.946}$
Soil from limestone terrain (Louisville, Kentucky)	2.67	$e = 0.50 - 0.0816(\log p)$	$c_b = 2.25 \cdot 10^{-2} p^{-0.952}$

[1] Compressibility equations were calculated from void ratio-versus-pressure curves $[c_b = -\{1/(e+1)\}(de/dp)]$.
[2] At pressures above 1,000 p.s.i.
[3] American Petroleum Institute, 1951, Preliminary Reports, Reference Clay Minerals, Research Project 49. These clays were obtained from Ward's Natural Science Establishment, Inc., 3000 Ridge Road East, Rochester, New York.
[4] All other samples were initially saturated with distilled water; compressibility of dry powder was determined from eq.4-31.
* Sample thickness = ½ inch at 1,000 p.s.i.

TABLE XXIII
Compressibility equations calculated from thickness, h, versus overburden pressure, p, measurements for various clays hydrated in distilled water $[c_b = -(1/h)(dh/dp)]$
(After Sheth et al., 1972)

Type of clay	*Compressibility equations obtained using different varieties of the same type of clay*
Dickite	$c_b = 3.9 \cdot 10^{-2} p^{-0.924}$; $c_b = 3 \cdot 10^{-2} p^{-0.788}$
Halloysite	$c_b = 4.2 \cdot 10^{-2} p^{-0.848}$; $c_b = 4.7 \cdot 10^{-2} p^{-0.864}$
Hectorite (containing 62% $CaCO_3$)	$c_b = 3.4 \cdot 10^{-2} p^{-0.787}$; $c_b = 4.4 \cdot 10^{-2} p^{-0.830}$
Illite	$c_b = 3.5 \cdot 10^{-2} p^{-0.815}$; $c_b = 3.7 \cdot 10^{-2} p^{-0.825}$; $c_b = 3.8 \cdot 10^{-2} p^{-0.93}$
Kaolinite	$c_b = 4 \cdot 10^{-2} p^{-0.811}$; $c_b = 3.5 \cdot 10^{-2} p^{-0.784}$
Montmorillonite *	$c_b = 2.8 \cdot 10^{-2} p^{-0.732}$; $c_b = 2.7 \cdot 10^{-2} p^{-0.735}$

* For the best straight line drawn through the experimental points ($p > 1{,}000$ p.s.i.). Actually the curve is concave upwards.

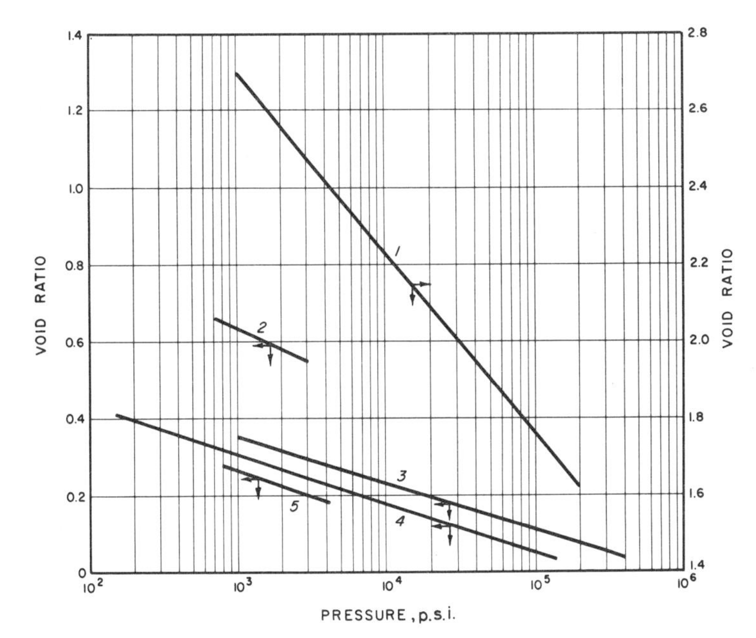

Fig.98. Relationship between void ratio, *e* (volume of pores/volume of solids) and effective pressure in p.s.i. *1* = montmorillonite (API no.25); *2* = clay from Bolivar Coast; *3* = hectorite containing 50% $CaCO_3$ (API no.34); *4* = dickite (API no.15); and *5* = clay from Bolivar Coast, Venezuela (curves *2* and *5* are after Van der Knaap and Van der Vlis, 1967.) (After Rieke et al., 1969, fig.2, p.820.)

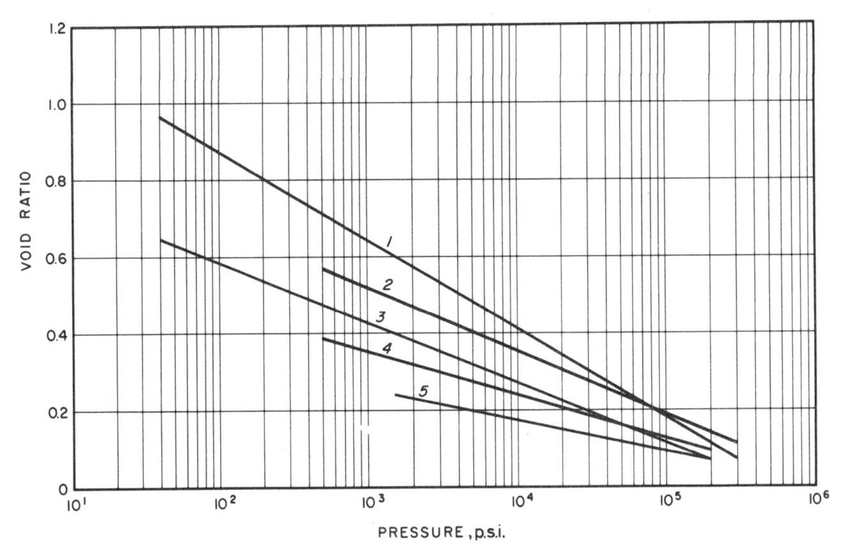

Fig.99. Relationship between void ratio, *e*, and effective pressure in p.s.i. *1* = illite (API no.35); *2* = halloysite (API no.12); *3* = kaolinite (API no.4); *4* = P-95 dry lake clay (Buckhorn Dry Lake, Calif.); *5* = soil from weathered limestone terrain (Louisville, Kentucky). (After Rieke et al., 1969, fig.3, p.821.)

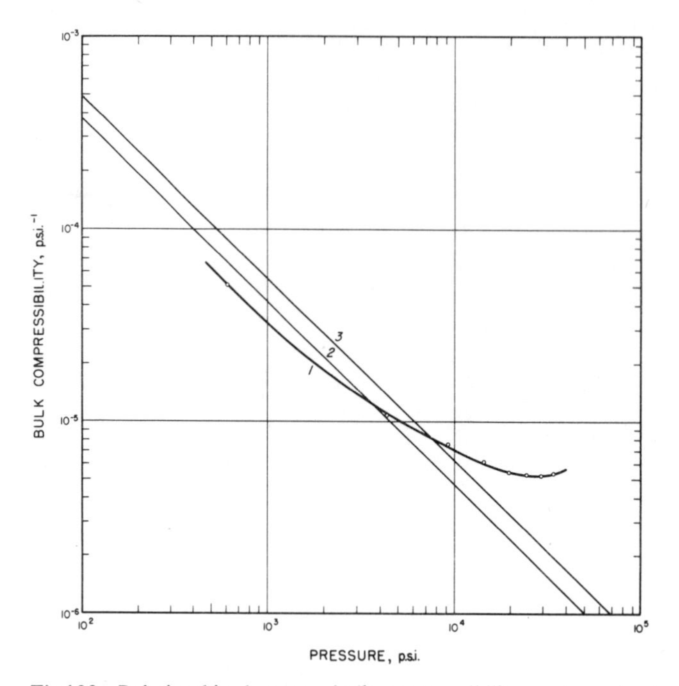

Fig.100. Relationship between bulk compressibility and overburden pressure for halloysite clay. *1* = bulk compressibility (c_{bh}) determined from thickness-versus-pressure measurements; *2* = bulk compressibility (c_{be}) calculated from void ratio-versus-pressure relationship on assuming a constant solids density of 2.0 g/cm^3; *3* = bulk compressibility (c_{be}) calculated from void ratio-versus-pressure relationship on assuming a constant solids density of 3.0 g/cm^3. (After Cebell and Chilingarian, 1972, fig.3, p.800. Courtesy of American Association of Petroleum Geologists.)

the case of hectorite and illite clays, these pressures were 3,000 and 9,000 p.s.i., respectively. The hectorite clay contained 62.3% of $CaCO_3$, as determined by the ascarite adsorption technique (Kolthoff and Sandell, 1938). Cebell and Chilingarian (1972) stated that the wide divergence of curves in the case of illite clay is probably due to the fact that in determining curves *2, 3* and *4* in Fig.102, a mixture of several different illites was inadvertently substituted for the illite no.35.

Actual compressibilities of halloysite, hectorite and illite clays saturated in distilled water range from $5.1 \cdot 10^{-5}$ to $5.35 \cdot 10^{-6}$ p.s.i.$^{-1}$, $4.48 \cdot 10^{-5}$ to $4.08 \cdot 10^{-6}$ p.s.i.$^{-1}$ and $3.35 \cdot 10^{-5}$ to $6.85 \cdot 10^{-6}$ p.s.i.$^{-1}$, respectively, in the 600–34,000 p.s.i. overburden pressure range (Cebell and Chilingarian, 1972). Above a certain pressure, actual compressibilities, determined from thickness-versus-pressure measurements, lie above those determined from remaining moisture content-versus-pressure curves and with assumed constant densities for the solids. Namely, the measured compressibilities are higher than those calculated from void ratio-versus-pressure curves. This could be explained by assuming that the apparent density of halloysite and illite clay solids decreases with increasing pressure (see p.35). This apparent decrease may be caused by increasing the amounts of "driven-in"

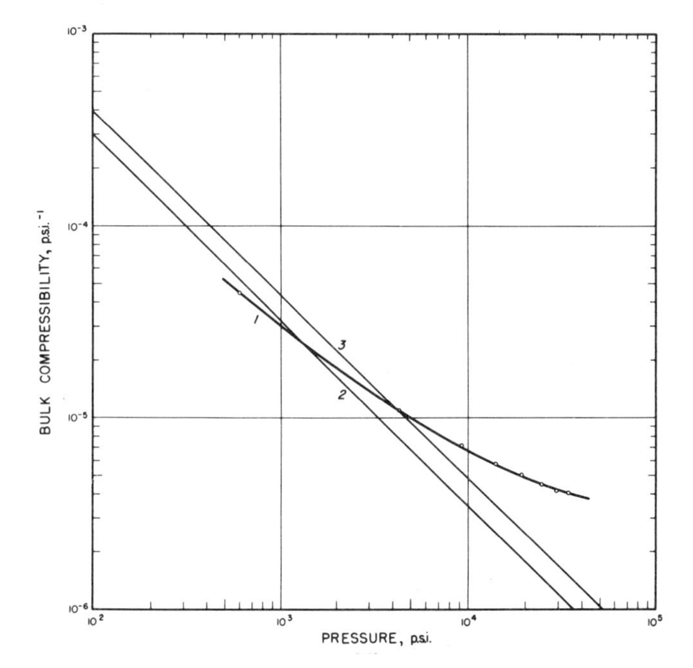

Fig.101. Relationship between bulk compressibility and overburden pressure for hectorite clay. *1* = bulk compressibility (c_{bh}) determined from thickness-versus-pressure measurements; *2* = bulk compressibility (c_{be}) calculated from void ratio-versus-pressure relationship on assuming a constant solids density of 2.0 g/cm^3; *3* = bulk compressibility (c_{be}) calculated from void ratio-versus-pressure relationship on assuming a constant solids density of 3.0 g/cm^3. (After Cebell and Chilingarian, 1972, fig.4, p.801. Courtesy of American Association of Petroleum Geologists.)

interlayer or adsorbed water at higher pressures. There is also a possibility that the adsorbed water remaining at higher pressures has a lower density than ordinary bulk water (see Anderson and Low, 1958).

Podio et al. (1968) computed the dynamic elastic properties of dry and water-saturated Green River Shale samples from compressional- and shear-wave velocity measurements. Bulk compressibility values of both wet and dry samples and their relationship to confining pressure are presented in Fig.95. Values of compressibility for the wet samples are lower than those for dry shale. In all instances, the bulk compressibility decreases with increasing confining pressure.

Compressibility of dry clays. Rieke (1970) determined the bulk compressibility of various loose clay powders (finer than 200 mesh or $< 74\ \mu$ in size). During the compaction, an Ames dial gauge was used to measure precisely piston deflections of 1/10,000 inch (see Fig.208). The dial gauge rests on a hexagonally shaped lever arm that is fastened to the movable bottom piston. It is fastened by using a clamp block, which is clamped to the piston, and the lever arm is screwed into the block. The block and the lever arm are constructed so that they are aligned perpendicular to the travel of the piston.

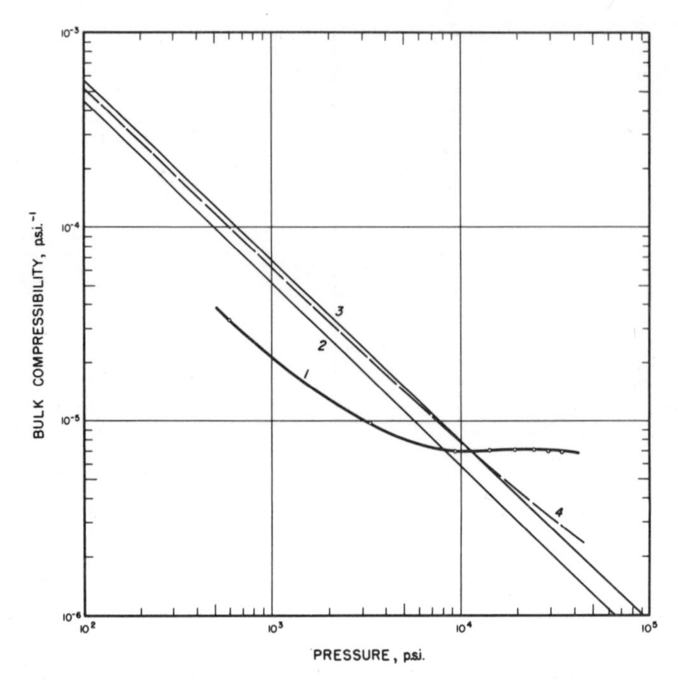

Fig.102. Relationship between bulk compressibility and overburden pressure for illite clay. *1* = bulk compressibility (c_{bh}) determined from thickness-versus-pressure measurements; *2* = bulk compressibility (c_{be}) calculated from void ratio-versus-pressure relationship on assuming a constant solids density of 2.0 g/cm^3; *3* = bulk compressibility (c_{be}) calculated from void ratio-versus-pressure relationship on assuming a constant solids density of 3.0 g/cm^3; and *4* = bulk compressibility (c_{be}) calculated from void ratio-versus-pressure curve corrected for the observed density decrease with compaction. (After Cebell and Chilingarian, 1972, fig.5, p.801. Courtesy of American Association of Petroleum Geologists.)

The linear upward movement of the piston was employed to calculate the change in volume of a dry clay powder as a function of pressure. The linear movement of the pistons during compaction is the algebraic sum of the change in height (thickness) of the clay compact and the elastic compressive strains in the cylinder and pistons. The latter changes were measured experimentally during pressure calibration of the gauge and sample-chamber.

All samples were placed as loose powders in the cylinder. According to Rieke (1970), the crucial problem was one of determining the exact height of the sample as placed in the cylinder. This was accomplished by using the principle that the relationship between the clay powder weight and its length (compact length) is essentially linear at a constant compacting pressure (Fig.103). The initial compacting pressure of 1,000 p.s.i. (gauge pressure = 200 p.s.i.) was selected in most cases. At this pressure, a series of clay powders of varying weights were compacted and then their lengths were measured immediately, through the use of a vernier caliper, after extrusion (removal) from the cylinder. All

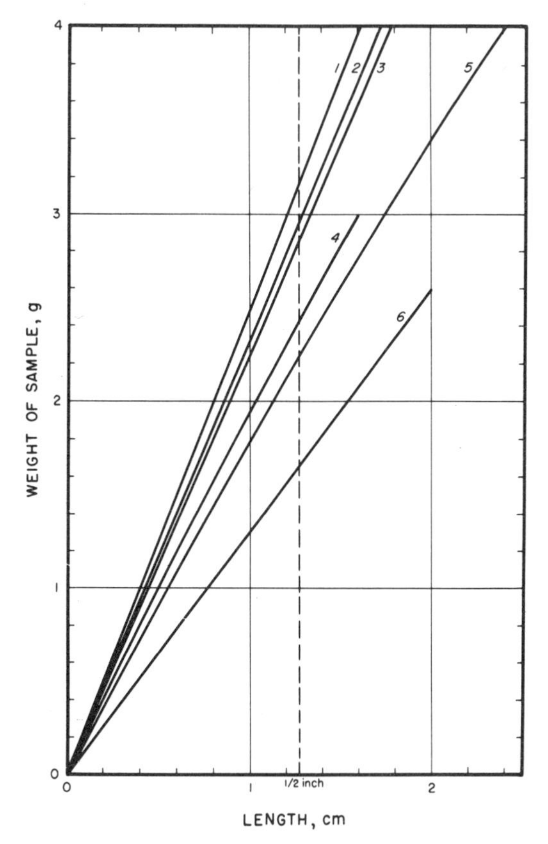

Fig.103. Relation between the weight (g) of some clay powders and their compact length (cm) under a certain constant compaction pressure (p_c). *1* = illite (API no.35; p_c = 11,300 p.s.i.); *2* = montmorillonite (API no.25; p_c = 1,000 p.s.i.); *3* = kaolinite (API no.4; p_c = 6,000 p.s.i.); *4* = dickite (API no.15; p_c = 1,000 p.s.i.); *5* = halloysite (API no.12; p_c = 1,000 p.s.i.), *6* = attapulgite (API no.46; p_c = 1,000 p.s.i.). The constant compaction pressure, p_c, was applied to the powder for 15 min. (After Rieke, 1970, fig.6-4, p.312.)

powders were weighed to 1/10,000 of a gram. The powder sample was then placed in the thick-walled cylinder on top of the bottom plug and piston, which were positioned at a zero dial reading. The upper piston and plug were placed on top of the sample and the unit was capped. Initial height was calculated by adding the total deflection read on the dial gauge when the sample was pressurized to its constant compacting pressure; in most cases this was 1,000 p.s.i. (Fig.103). Experiments were so designed by Rieke (1970) that the sample thickness at the constant compacting pressure would be 0.500 inch (equal to its diameter).

Normally, 1,000 p.s.i. was sufficient to give some competence to the clay powder. In the case of attapulgite and halloysite, 1,000 p.s.i. caused crushing of the elongated mineral grains and consequently some shearing occurred in the clay compact. This was mini-

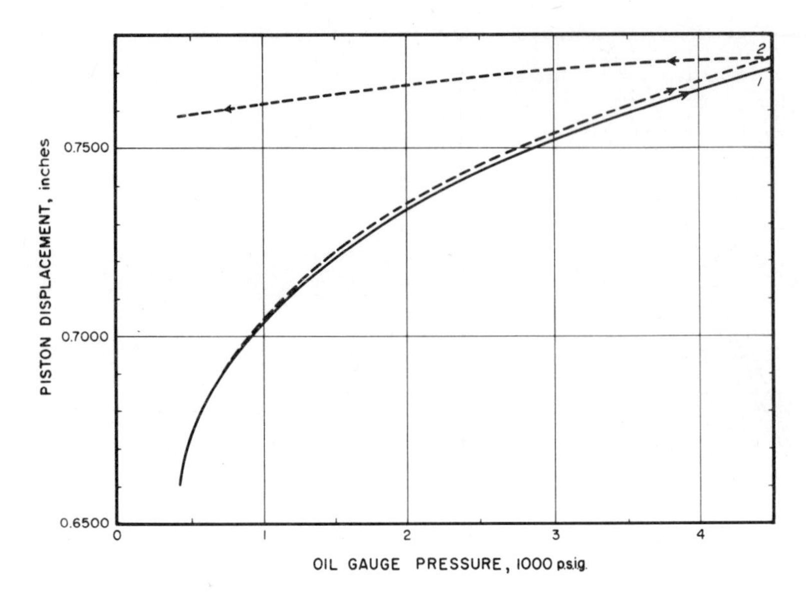

Fig.104. Secondary compression effects in kaolinite clay. *1* = initial reading of the indicator dial gauge; *2* = 30-min reading (dashed line) both on the upstroke and downstroke cycles. (After Rieke, 1970, p.316.)

mized by using lubricants or a lower constant compacting pressure. Creep (secondary compression) was accounted for, in the compressibility experiments, by keeping the desired pressure up until it stabilized and there were no further large deflections recorded by the dial gauge. One-half hour time interval was sufficient between readings at the same pressure level (Fig.104).

The volume of a porous cylindrical rock sample in a thin covering jacket can be changed by changing the external pressure on the outside of the jacket or changing the internal pore pressure. In most consolidation tests of the other investigators cores were jacketed in such a manner in thin copper or tin foil and lucite or rubber tubes, that the external pressures could be effectively applied hydrostatically to the whole core. The jacket also prevented penetration of the pressurizing fluid into the pores. Cleary (1959, p.9) described a triaxial cell in which he measured bulk linear compressibilities of sandstones and employed bonded wire-resistance strain gauges to measure the axial strain.

The compressibility of the dry powders was determined by Rieke (1970) from the change of thickness as a function of the "effective" pressure (Fig.105 and Table XXIV). "Effective" or "net" pressure is defined as the difference between the total pressure and the pore pressure. In these experiments the pore pressure was atmospheric and was considered as a zero reference level. Eq.4-31 was used to calculate the compressibility of these clays from the height-versus-pressure curves; equations representing the relationship between sample thickness (height) and effective pressure are given in Table XXIV.

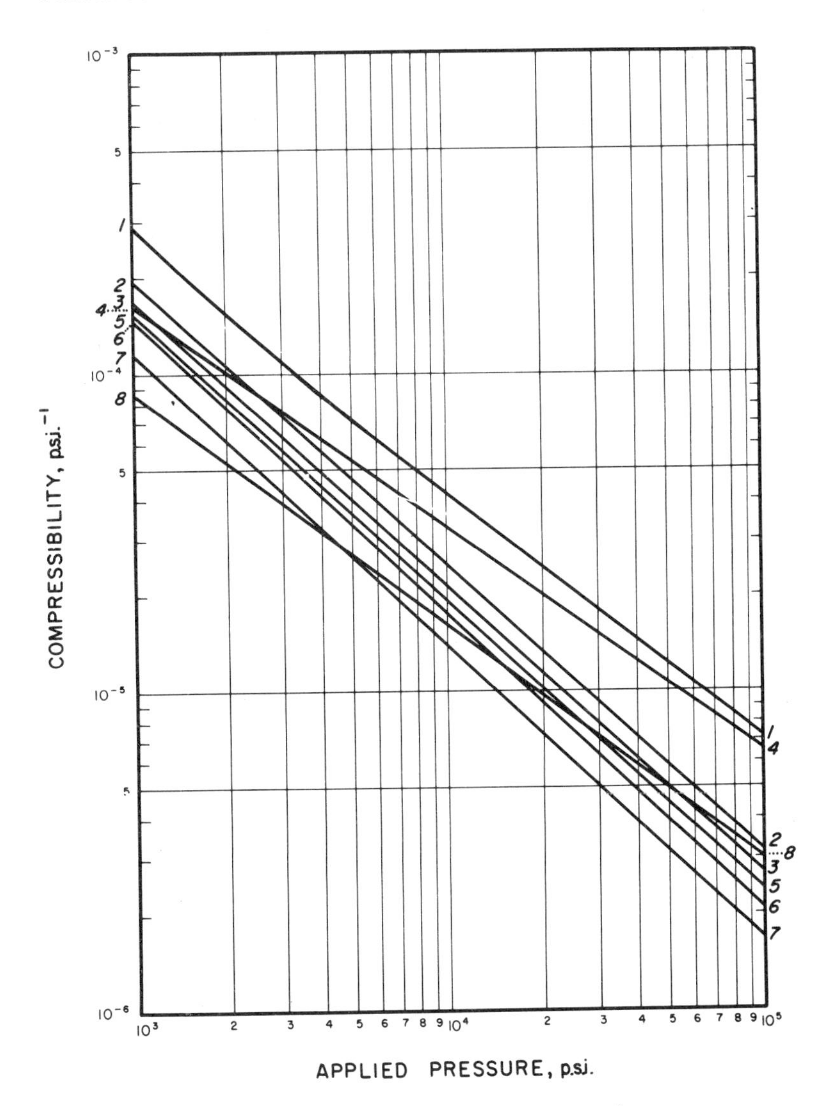

Fig.105. Relationship between bulk compressibility in p.s.i.$^{-1}$ and effective pressure in p.s.i. for dry clay powders. *1* = halloysite (API no.12); *2* = dickite (API no.15); *3* = illite (API no.35); *4* = attapulgite (API no.46); *5* = kaolinite (API no.4); *6* = hectorite containing 50% $CaCO_3$ (API no.34); *7* = montmorillonite (API no.25); *8* = lake clay derived from the Chattanooga Shale and soil derived from limestone terrain (Louisville, Kentucky) (both materials are represented by curve *8*). (After Rieke, 1970, p.321.)

Compressibility can be used as an indication of the extent to which the density of a powder is increased through applied pressure. One of the most widely used compaction parameters in dry powders is the "compression ratio", which is defined as the ratio of the compact density at a given pressure to the apparent density of the loose clay powder.

TABLE XXIV
Relationship between pressure and thickness, and compressibility for various dry clay powders. (After Rieke, 1970, p.329)

Type of clay	*Relationship between sample thickness and effective pressure* [1]	*Compressibility equations*
Attapulgite (API no.46)	$h = 0.3921 + 0.1222 \log p - 0.0307 \log p^2$	$c_b = 0.88 \cdot 10^{-2} p^{-0.699}$
Dickite (API no.15)	$h = 0.7857 - 0.0962 \log p$	$c_b = 3.96 \cdot 10^{-2} p^{-0.8925}$
Halloysite (API no.12)	$h = 0.9416 - 0.1472 \log p$	$c_b = 2.27 \cdot 10^{-2} p^{-0.733}$
Hectorite (API no.34) [2]	$h = 0.7213 - 0.0737 \log p$	$c_b = 4.15 \cdot 10^{-2} p^{-0.9825}$
Illite (API no.35)	$h = 0.6698 - 0.07506 \log p$	$c_b = 3.4 \cdot 10^{-2} p^{-0.982}$
Kaolinite (API no.25)	$h = 0.8599 - 0.929 \log p$	$c_b = 3.58 \cdot 10^{-2} p^{-0.9091}$
Montmorillonite (API no.25)	$h = 0.6495 - 0.05812 \log p$	$c_b = 3.29 \cdot 10^{-2} p^{-0.9348}$
Lake clay (Louisville, Ky.)	$h = 0.6177 - 0.0202 \log p - 0.00653 \log p^2$	$c_b = 0.59 \cdot 10^{-2} p^{-0.7293}$
Soil derived from limestone terrain	$h = 0.06903 - 0.0516 \log p - 0.00303 \log p^2$	$c_b = 0.59 \cdot 10^{-2} p^{-0.7293}$

[1] Initial sample thickness = ½ inch.
[2] Containing 60% $CaCO_3$.

Obviously, this "quantitative" description of the compaction behavior of a clay powder provides only limited information.

Any mathematical reduction of the experimental data into curves relating the compressibility and effective pressure of the clay compact requires the following information: (1) the weight of the clay powder, (2) the thickness of the sample in the cylinder at a given pressure, (3) the cross-sectional area of the cylindrical borehole, and (4) the dial-gauge reading at zero pressure on the compact. According to Rieke (1970, p.323), in his experiments the diameter of the clay compact was larger than that of the borehole by about 1/10,000 inch when the sample was removed from the cavity; however, the two cross-sectional areas were assumed to be equal in calculating volume of compacts.

As pointed out by Schwartz and Weinstein (1965, p.346), the expanding use of powdered materials, whether ceramic or refractory metal, has stimulated interest in the compressibility of granular material. Major research effort was directed toward determining the pressure–porosity relationship and understanding of the compacting and bonding mechanism (Cooper and Eaton, 1962; Shapiro, 1963). Schwartz and Weinstein (1965) concluded that the Coulomb yield criterion may be used to compute loads and stress distribution in the compacting of ceramic powders. The Coulomb yield criterion can be expressed mathematically as follows:

$$\tau_f = C + \sigma_f \tan \theta \qquad (4\text{-}39)$$

where τ_f is magnitude of shear stress at failure, σ_f is normal compressive stress at failure, C is cohesion due to bond forces between the clay particles, and θ is angle of internal friction, the tangent of which is the slope of a line encompassing the failure stress-state plots (shear stress plotted versus compressive stress) on a Mohr diagram.

Cooper and Eaton (1962) determined compaction behavior of four dry ceramic powders ($CaCO_3$, MgO, SiO_2 and Al_2O_3) of widely different hardness, but of essentially the same particle-size fraction (230–325 mesh: 44–62 μ). Softer powders should yield a greater fractional volume compaction at a given pressure. Cooper and Eaton (1962) expressed the fractional volume compaction, V_r^* in terms of applied pressure, p, as follows:

$$V^* \equiv \left(\frac{V_o - V}{V_o - V_\infty}\right) = a_1 \exp\left(-\frac{k_1}{p}\right) + a_2 \exp\left(-\frac{k_2}{p}\right) \tag{4-40}$$

where a_i and k_i are coefficients; V_o is initial total volume when no holes are filled at zero pressure; V is compacted volume at some pressure p; and V_∞ is compacted volume when all pores are filled (zero porosity; volume at theoretical compaction). Cooper and Eaton summarized the values of coefficients a_i and k_i and porosities of powders in a tabular form (Table XXV).

Compaction of dry clay powders is probably governed by two dominant processes: (1) filling of the pore space having the diameter of the same order of magnitude as the clay particles, and (2) filling of the voids which have diameters much smaller than the clay particles. Filling of the large pores may occur as a result of clay particles sliding past one another, which may require elastic deformation or even slight fracturing or plastic flow of the clay. Filling of the small pores occurs by plastic flow or by fragmentation of the particles. If the sum of coefficients $a_1 + a_2 = 1$, then compaction of dry powders can be described in terms of the two separate processes; but if $a_1 + a_2 < 1$, then this is an indication that other processes, i.e., cold working, must become operative before complete compaction (theoretical density) is achieved. Cooper and Eaton (1962, p.101) found the following values of $a_1 + a_2$ for the four powders: alumina (0.85), silica (0.85), magnesia (0.90), and calcite (1.00). They noted that the filling of the smaller voids needed considerably higher pressures than those required to close the larger pores. The

TABLE XXV

Summarized values of coefficients a_i *and* k_i and porosities ϕ_1 and ϕ_∞ for various powders (After Cooper and Eaton, 1962, p.101)

Material	k_1 (p.s.i.)	k_2 (p.s.i.)	a_1	$a_1 + a_2$	ϕ_1	ϕ_∞
Alumina	3,100	50,000	0.50	0.85	0.28	0.09
Silica	2,400	54,000	0.60	0.85	0.21	0.07
Magnesia	2,400	49,000	0.65	0.90	0.21	0.05
Calcite	1,450	42,000	0.68	1.00	0.18	0.00

TABLE XXVI
Comparison of fractional volume compaction of hectorite (API no.34), calcite and silica [1]
(After Rieke, 1970, Table 6-2, p.328)

Pressure (p.s.i.)	*Fractional volume compaction*		
	hectorite	*calcite*	*silica*
6,000	0.871	0.540	0.400
11,300	0.899	0.620	0.500
23,000	0.929	0.645	0.560
36,000	0.951	0.740	0.610
49,000	0.962	0.800	0.650
74,000	0.982	0.860	0.695

[1] The data for the latter two powders are from Cooper and Eaton (1962). Fractional volume was calculated by using eq.4-40.

harder and more brittle the powder, the higher is the ultimate porosity. It appears that only in soft materials, such as clays, can the theoretical density be reached, because plastic flow occurs at lower pressures (Cooper and Eaton, 1962, p.101).

Rieke (1970, p.327) determined fractional volume compaction, V^*, for hectorite clay (API no.34). The hectorite tested contained 50% $CaCO_3$ by weight and, therefore, the results could be compared to Cooper and Eaton's (1962) data on $CaCO_3$. Table XXVI presents data which show that hectorite clay has a higher fractional volume compaction than does $CaCO_3$ or SiO_2. This is to be expected because hectorite has a hardness of 1–1½ on Mohs' scale, whereas those of $CaCO_3$ and SiO_2 are 3 and 7, respectively.

The non-linearity of attapulgite, as expressed by the least-squares equations in Table XXIV, is possibly owing to the needle-like shape of the clay particles which rearrange and close-pack differently than do plate-like clay particles and to the crushing of these grains where rearrangement cannot effectively take place. Sudden pressure releases recorded on the oil-pressure gauge were normally associated with some cracking noise. This was possibly an indication of failure of the compact. Other non-linear relationships occur in the natural soil and lake clay and these are possibly due to some crushing of clay-size quartz and feldspar grains (Table XXIV). The extent of grain crushing can be investigated by using an electron microscope.

Comparison among compressibilities of dry and hydrated clays and sands

Van der Knaap (1959, p.183) found in his experiments that fluids in the pores influenced the bulk and pore compressibility values of sandstones and limestones. From samples studied by Van der Knaap bulk or pore compressibilities were sometimes as much as twice as high when the rocks were saturated with water than when dry. Data on compressibility of clays obtained by Van der Knaap and Van der Vlis (1967) for the

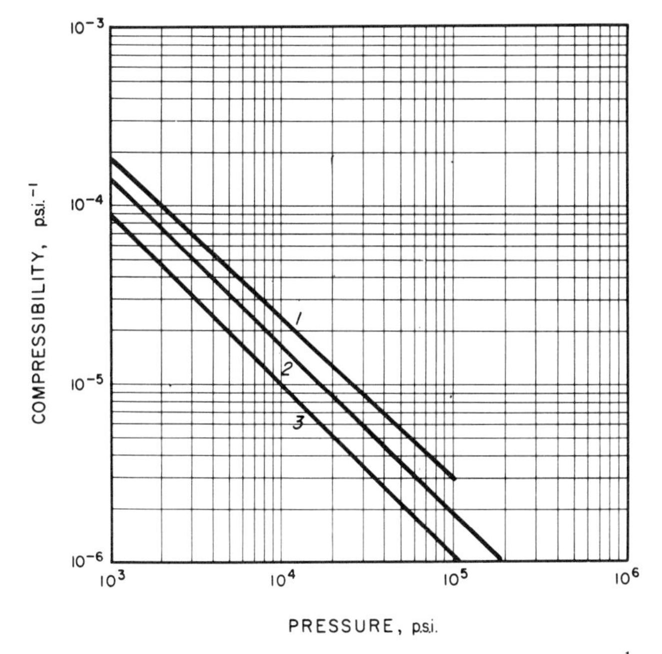

Fig.106. Relationship between bulk compressibility in p.s.i. $^{-1}$ and pressure in p.s.i. *1* = illite dry powder (API no.35), *2* = illite hydrated in distilled water (API no.35), *3* = hectorite hydrated in distilled water (API no.34). (After Rieke et al., 1969, fig.7, p.827.)

Bolivar Coast in Venezuela are shown in Fig.96; whereas their data for sands are presented in Fig.97. These data show the high values of compressibility of formation sand and clay under in-situ conditions. The results obtained by Rieke et al. (1969) and those by Van der Knaap and Van der Vlis (1967) are of comparable magnitude for all hydrated clays.

The compressibility of dry illite powder between the pressures of 1,000 and 100,000 p.s.i. is linear on a log-log plot of compressibility versus effective pressure (Fig.95 and 106), and the curve for dry illite powder lies above that for illite initially saturated with distilled water. The same relationship (higher compressibility values of dry clay powder compared to its hydrated form) holds true for dickite, halloysite, hectorite, kaolinite and soil from weathered limestone terrain. Apparently, dry clays close-pack more than do hydrated and enlarged clay particles. This, in part, may be due to the low permeability of hydrated clays, which prevents expulsion of water. Montmorillonite saturated with distilled water has compressibility values very close to those in a dry state (compare Tables XXII–XXIV; Fig.97 and 105). This may be attributed to the high porosity of montmorillonite saturated with distilled water (see Fig.98; and Fig.30, Chapter 2, p.59) as compared to other clays, possibly due, in part, to a "card house" structure in which the clay platelets are aligned edge to face forming a honeycomb-like structure.

Like sandstone, clay and shale compressibility is dependent on many variables. In the case of hydrated clays the compressibility of the pores is influenced by the following variables: (1) compacting pressure, (2) fluid pressure in the pores, (3) permeability, and (4) degree of fluid saturation. In most competent rocks, the bonding material has a high specific surface which is affected by the nature and amount of fluid in contact with it. Van der Knaap (1959, p.183) found that wetting causes a certain amount of softening of the bonding material between the rock grains; this could explain why wet competent rocks have higher compressibilities than do dry rocks. This, however, does not seem to be the case for clays and shales as shown by the experimental data of Podio et al. (1968) and Chilingar et al. (1969) (see Fig.95).

Terzaghi (1925a) noticed that there was a relationship between the bulk compressibility of the clay, the grain-shape factor and the forces of molecular attraction or repulsion within the adsorbed water layer. No doubt, the time factor and geothermal gradient also affect the bulk compressibility of clays and shales in situ. Geertsma (1957, p.331) theoretically showed that compressibility is independent of the pore shape; he assumed a continuous homogeneous matrix that is isotropic in situ. Gondouin and Scala (1958, p.179) remarked that shale pore compressibilities, as calculated from laboratory data, were at least one order of magnitude larger than the pore compressibilities of sandstones. Geertsma (1957) reported that the laboratory-measured compressibilities, by the "net confining" method, may be greater than those in the reservoir by a factor of two. In the "net confining" method, the sandstones and limestone cores are placed under a simulated hydrostatic overburden pressure, whereas the stresses in a reservoir may not be hydrostatic (i.e., pressure is equal in all directions at a given point).

Relationship between the bulk compressibility and overburden pressure for montmorillonite clay saturated in sea water, on using both uniaxial and triaxial compaction apparatus, is presented in Fig.107. The bulk compressibilities obtained on using triaxial compaction apparatus of Sawabini et al. (1971) range from $4.5 \cdot 10^{-4}$ to $3 \cdot 10^{-5}$ p.s.i.$^{-1}$ in the 500–20,000 p.s.i. pressure range. They are higher by about 300–500% than compressibilities obtained in the uniaxial compaction equipment described by Rieke et al. (1969) (see Fig.208).

Relationship between the compressibility and the applied overburden pressure for various types of rocks, unconsolidated sands, and clays are presented in Fig.95. It seems that compressibilities of clays are very close to those of unconsolidated sands.

Van der Knaap and Van der Vlis (1967) in their classic paper on subsidence concluded that clay and sand layers compact almost to the same extent, the main difference being that the low permeability to water of the clay prevents instantaneous compaction and time effects become important. It should be remembered, however, that based on laboratory experiments of Chilingar and Knight (1960) equilibrium is reached in the laboratory within a short period of time (1–16 days depending on the type of clay).

As stated by Van der Knaap and Van der Vlis (1967, p.94), one can assume that a certain volume of subsidence at the surface of an oil field is caused by an equal volume of reservoir compaction. Subsidence at the surface is spread over a larger area and, conse-

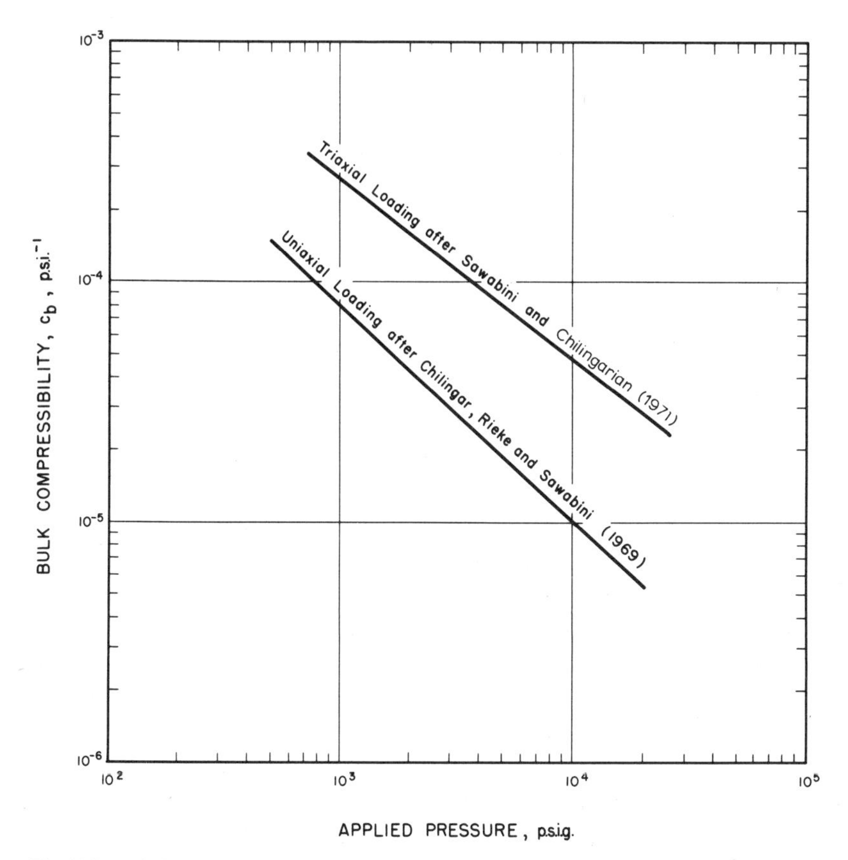

Fig.107. Relationship between the bulk compressibility, c_b, and applied pressure, p_t, for montmorillonite clay (API no.25) saturated in sea water using (1) triaxial and (2) uniaxial loading. (After Chilingar et al., 1969, and Sawabini and Chilingarian, 1971.)

quently, the subsidence at a given point in the field may not be equal in magnitude to the formation compaction directly below this point. Estimates of subsidence caused by compaction can be made if the thickness of the clay layers, net thickness of the producing horizons, rate of pressure decline, and the correct compressibility–pressure relationships are known.

There is a time lag between fluid withdrawal and compaction involved in subsidence. In the case of highly permeable rocks, it is small. Removal of the interstitial fluid places a direct load on the individual mineral grains throughout the reservoir. Even if the direct load involves crushing of the rock matrix, it should proceed rapidly to completion as there is little resistance to adjustment.

On the other hand, in the case of clay lenses or shales, each shale layer can be thought of as a small high-pressure reservoir. If the pressure in the reservoir rock is reduced, the overburden pressure is transferred also to the clay. Inasmuch as shales and clays have low permeability and their pores are mostly filled with fluid, the load is not immediately transferred to the matrix due to the inability of the rock to transmit the fluid rapidly. Subsidence resulting from fluid removal, therefore, may not occur rapidly and time lags of many years may be involved (Van der Knaap and Van der Vlis, 1967).

Effect of temperature on compressibility

Based on relatively few studies, one can conclude that rock compressibility increases with increase in temperature.

Carpenter and Spencer (1940) measured the bulk compressibility for two cores at two different temperatures: 91 and 146°F. One core showed increase in compressibility with temperature, whereas the other showed a decrease. These authors concluded that there is no definite relationship between compressibility and temperature.

Kaul (1965) studied the effect of temperature and type of pore fluids on the compressibility of clays. He concluded that the compressibility of clays depends on the type of fluid present in the pores and increases with the increase in temperature.

Von Gonten and Choudhary (1969) studied the effect of temperature and pressure on the pore compressibility of rocks. They measured the pore compressibility of eight sandstone and one limestone cores at 75 and 400°F. Their results lead to a conclusion that the pore compressibility of rocks increases with the increase in temperature. They found that the pore compressibility at 400°F is about 21% higher than that at room temperature.

Lobree (1968) measured the compressibilities of Berea, Bandera and Boise sandstones at 76 and 400°F. The results of his studies showed that the bulk, formation and rock solid compressibilities of Boise Sandstone, bulk compressibility of Berea Sandstone and bulk and formation compressibilities of Bandera Sandstone increased with increasing temperature. He defined bulk compressibility as the change in bulk volume with increasing external stress at constant pore pressure; formation compressibility, as the change in pore volume with pore pressure at constant external stress; pore compressibility, as the change in pore volume with the external stress at constant pore pressure; and rock solid compressibility, as the change in rock solid volume with applied stress (if tested without jacket, the external stress is equal to the pore pressure) (Gomaa, 1970).

Creep

The following procedure was followed by Rieke (1970) in studying the "creep" of two dry clay powders (montmorillonite and illite) under various axially applied loads:

(1) Both clays were preconsolidated at 1,000 p.s.i. at 65°F until pressure–deflection equilibrium was reached.

(2) A load of 5,000 p.s.i. was applied, and the time and indicator dial gauge readings were recorded. Under this constant stress the following dial deflection readings were taken initially, at total elapsed times of 0.25, 1, 2.25, 4, 6.25, 9, 12.25, 16, 20.25, 25, 36,

49, 64, 81, 100, 121, and 144 min, and then at random intervals up to 6,400 min. This resulted in good spacing of points in the study of short-term secondary compression in dry clays.

(3) At the end of 6,400 min, the pressure was increased to 10,000 p.s.i. and readings were taken as described above. This procedure was also repeated at 20,000 p.s.i. and 40,000 p.s.i.

The acquired data could be presented in two ways: (1) the dial deflections may be plotted versus the square root of time on a cartesian graph paper or the dial deflections versus time on a semilog graph paper, and (2) strain may be plotted versus the square root of time on a cartesian graph paper. The results of uniaxially loaded soils are usually presented in the first manner. From such a plot the primary compression ratio, r_c, can be determined. The primary compression ratio is defined as the ratio of primary compression to total compression according to Terzaghi's theory of consolidation (Taylor, 1948, p.240; Lambe, 1958, p.82):

$$r_c = \frac{\frac{10}{9}(d_s - d_{90})}{d_o - d_f} \tag{4-41}$$

where d_s = corrected zero point (see Taylor, 1948, p.240), d_{90} = dial reading at 90% of consolidation, d_o = initial dial reading, and d_f = final dial reading.

The purpose of the theoretical study of consolidation (Terzaghi's theory, 1925) is to set up an equation from which the pore-pressure and void-ratio values may be determined at any point and at any time in a stratum of consolidating clay of any thickness. Terzaghi's theory of consolidation is expressed as:

$$C_v \frac{\partial^2 p_i}{\partial z^2} = \frac{\partial p_i}{\partial t} \tag{4-42}$$

where C_v = coefficient of consolidation, p_i = internal fluid pressure (hydrostatic excess pressure), z = coordinate distance measured downward from the surface of the clay sample, and t = time. The coefficient of consolidation, C_v, which is discussed in greater detail later (p.205), can be expressed as:

$$C_v = \frac{K(1+e)}{a_v \gamma_w} \tag{4-43}$$

where K = hydraulic conductivity in cm/sec, e = void ratio, a_v = coefficient of compressibility [= $-(de/dp_i)$] in cm^2/g, and γ_w = specific weight of water in g/cm^3.

The larger the value of r_c, the better is the agreement between the computed and the actual rate of subsidence. In most cases, the study of hydrated clays would give more representative r_c values than that of dry ones.

Finite strain. Strain was arbitrarily defined as:

$$\epsilon = \frac{L_f - L_o}{L_o} \qquad (4\text{-}44)$$

where L_f is the final length, L_o is the initial length and ϵ is the finite strain. This definition gives strain a positive value in the case of a tension test and a negative sign in the case of a compression test. Strain also can be defined as:

$$\epsilon = \frac{L_f - L_o}{L_f} \qquad (4\text{-}45)$$

It is preferable to define strain so that it will be a positive quantity in the compression test:

$$\epsilon = \frac{L_o - L_f}{L_o} \qquad (4\text{-}46)$$

Axial strain versus square root of time curves were constructed by Rieke (1970) in order to illustrate primary creep (secondary compression) of dry montmorillonite and illite clays in short-period secondary compression tests (Figs.108 and 109). Secondary compression can be defined as plastic flow under a constant stress.

It is well documented that there are time lags between the occurrence of compaction and the time when the load was applied during the compression of hydrated clays and soils (Terzaghi, 1938; Taylor, 1948). In the case of hydrated clay two phenomena contribute to this time lag: (1) time required for the escape of the pore fluid (referred to as the hydrodynamic lag) and (2) plastic lag. The escape of water is controlled by the permeability which controls the flow of the interstitial fluids. Plastic time lag, owing to plastic action in adsorbed water near grain-to-grain contacts or points of nearest approach to contact, is complicated and is only partially understood. The frictional lag in sands may be thought of as a simple form of plastic lag. Terzaghi's consolidation theory included an understanding of the hydrodynamic lag, but does not recognize the existence of plastic lag. In practice, the Terzaghi theory has not always given satisfactory results in predicting settlement in clayey soils. Compressibility and creep data on the dry clays might give some insight into the phenomenon of plastic lag.

Taylor (1948, p.245) observed that intergranular pressure is not a fixed value as assumed by Terzaghi, but rather depends on the rate of compression. Intergranular pressure is considered to consist of two parts, static case pressure and plastic resistance to compression. Taylor (1948, p.246) stated that plastic resistance is dependent on the speed of compression, but it also could be a function of the type of cementation or bonding.

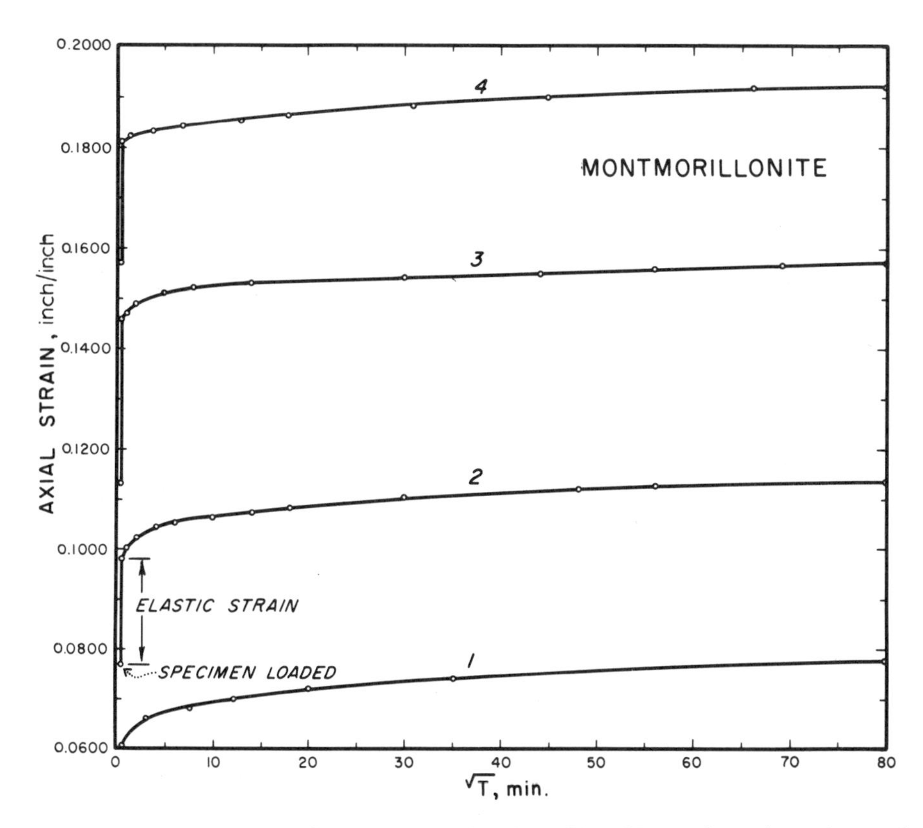

Fig.108. Time–strain curves for dry montmorillonite (API no.25) powder under various static pressures (p_c). *1* = p_c = 5,000 p.s.i.; *2* = p_c = 10,000 p.s.i.; *3* = p_c = 20,000 p.s.i.; *4* = p_c = 40,000 p.s.i. (After Rieke, 1970, fig.6, p.349.)

From Figs.108 and 109 it can be noted that: (1) Curves for both dry clay powders show similar secondary compression patterns. (2) Dry illite is more compressible than dry montmorillonite. (3) The axial strain values of both clays are close to each other. (4) Both clays only exhibit primary creep, even though pressures up to 40,000 p.s.i. were employed by the writers. Secondary creep, which corresponds to pseudoviscous flow, was not evident nor was tertiary creep observed. (5) Apparently not much time is needed to reach essentially the end of deformation of the powders in the laboratory (Rieke, 1970, p.354).

Changes in compression index with compaction

The compression index, C_c, is defined as follows (Taylor, 1948, p.217):

$$C_c = \frac{e_o - e}{\log_{10}(p/p_o)} \tag{4-47}$$

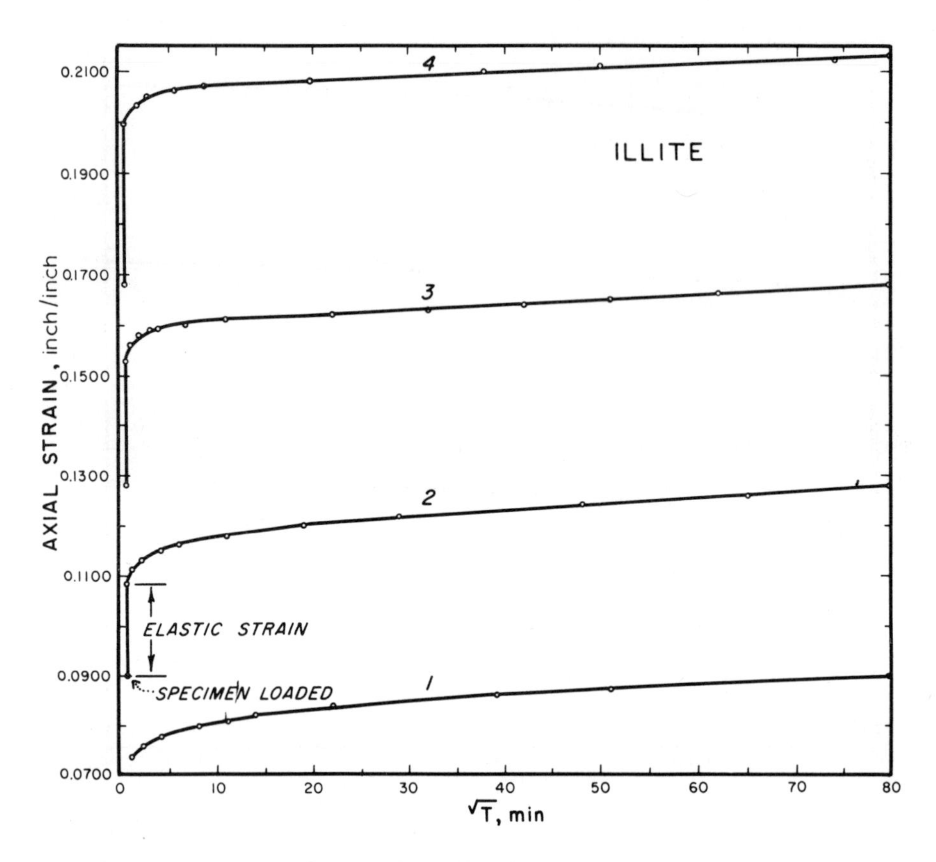

Fig.109. Time–strain curves for dry illite (API no.35) powder under various static pressures (p_c). *1* = p_c = 5,000 p.s.i.; *2* = p_c = 10,000 p.s.i.; *3* = p_c = 20,000 p.s.i.; and *4* = p_c = 40,000 p.s.i. (After Rieke, 1970, fig.6-13, p.351.)

where e_o is void ratio at pressure p_o. The arbitrarily chosen value for p_o is commonly 1 ton/sq. ft. or 1 kg/cm^2, although the straight line often has to be projected backward to reach this pressure.

According to Richards and Hamilton (1967, p.100), the compression index can be easily determined on a semi-log plot by selecting a value of p which is equal to $10p_o$ and then computing $C_c = e_o - e$. Following a report by Skempton (1944), Terzaghi and Peck (1948, p.66) noted that C_c is related to liquid limit, LL, of a sediment as follows:

$$C_c = 0.009\,(LL - 10\%) \tag{4-48}$$

Thus, a crude estimate of sediment compressibility can be made without a consolidation test.

According to Skempton (1944) (also see Richards and Hamilton, 1967), in the pres-

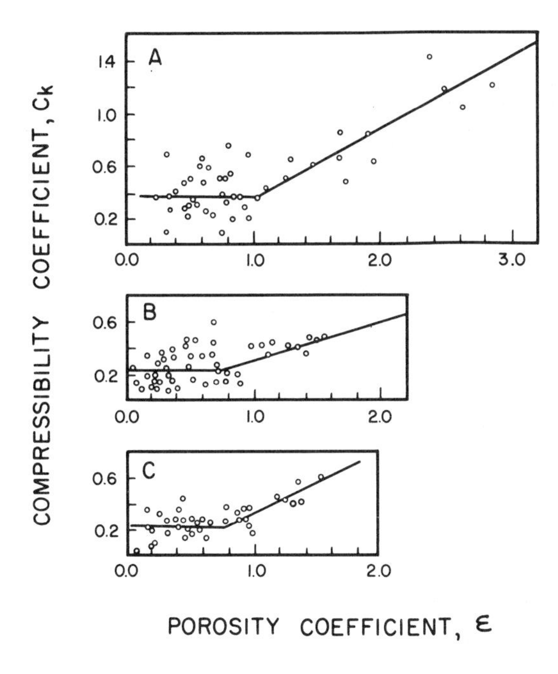

Fig.110. Relationship between compression index C_c (coefficient of compression, C_k, in Russian terminology) and void ratio, e (coefficient of porosity, ϵ, in Russian terminology) in monomineralIic clays. A. Montmorillonite clays. B. Predominantly hydromica and muscovite. C. Kaolinite clays. (Based on results of numerous investigators, in: Sergeev, 1970, fig.3, p.198.)

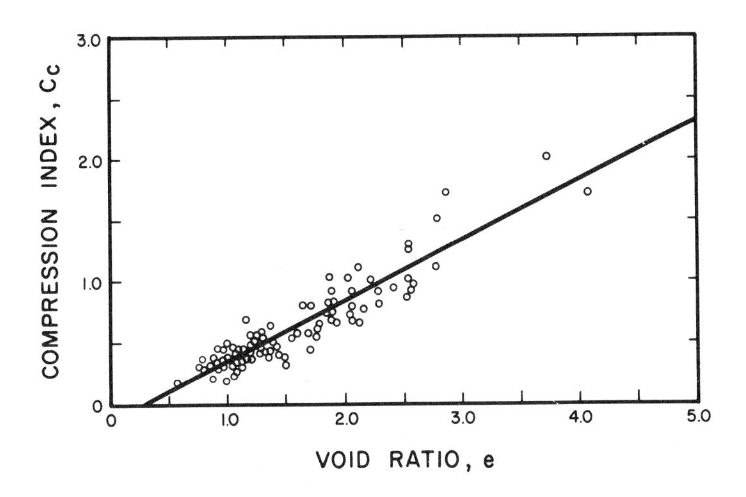

Fig.111. Relationship between compression index and void ratio for Late Quarternary clays in Haifa Bay and Ashdod Harbor in Israel. (After Komornik et al., 1970, fig.12, p.137. Courtesy of *Sedimentology*.)

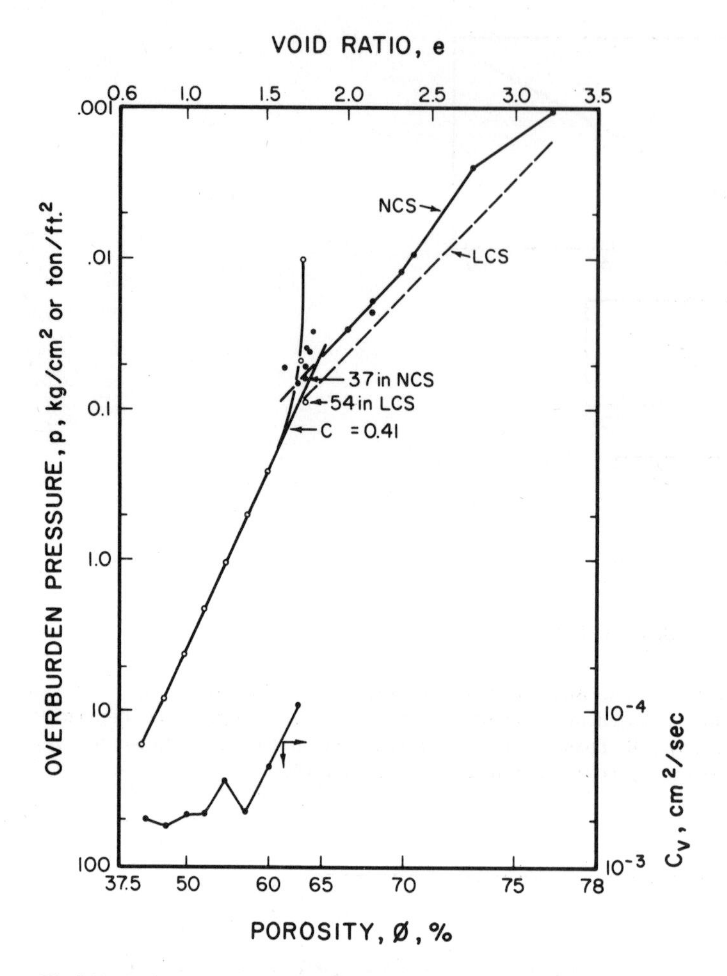

Fig.112. Relationship among void ratio, e, porosity, ϕ, coefficient of consolidation, C_v, and overburden pressure, p. Determined from a laboratory test (open circles) on a distributed sample, 36–38 inch. (37 inch. avg.) below the top of gravity core A-31, assuming no core shortening (*NCS*), and 54 inch. average, assuming linear core shortening (*LCS*). (After Richards and Hamilton, 1967, fig.3a, p.105.)

sure range of 1.0–10.0 kg/cm^2 the C_c varies from 0.20 for silty clay to 0.83 for highly colloidal clay.

Sergeev (1970) presented graphs relating coefficient of compression, C_c, to void ratio, e, for various monominerallic clays (Fig.110). As previously shown by Zhiangirov (1968), clays containing only strongly adsorbed water deform as hard elastic bodies. On removing free water as a result of compaction, when the moisture content of clay becomes equal to or less than the maximum value of hygroscopicity, the strength cannot be appreciably altered on further compaction.

Komornik et al. (1970) showed linear correlation between the compression index and void ratio for Late Quarternary clays in Haifa Bay and Ashdod Harbor in Israel (Fig.111). They defined compression index C_c as a "decrease of void ratio of a sediment due to a tenfold increase of effective unit load from p to $10p$, i.e., the slope of the void ratio versus log pressure curve".

Coefficient of consolidation

The coefficient of consolidation, C_v, relates decrease in volume of a sediment with pressure and time and, thus, enables prediction of rate of settlement. C_v (in cm^2/sec) is derived from the curve of volume decrease with time for each load increment of the consolidation test and can be expressed by the following equation:

$$C_v = \frac{K}{\gamma_w M_v} \tag{4-49}$$

where K is coefficient of permeability in cm/sec, γ_w is density of pore water in g/cm^3, and M_v is coefficient of volume compressibility in cm^2/g:

$$M_v = \frac{e_1 - e_2}{(p_2 - p_1)(1 + e_1)} = \frac{a_v}{1 + e_1} \tag{4-50}$$

Pressure, p, is in kg/cm^2, void ratio, e, is dimensionless, and coefficient of compressibility, a_v, is in cm^2/g. It is important to note here that the coefficient of consolidation may change with confining stress (see Fig.112).

In studying deep-sea, fine-grained sediments, having median grain-size diameter of less than $4\,\mu$, up to a depth of approx. 100 inches below the sediment–water interface, Richards and Hamilton (1967, p.100) determined that, with an exception of one core sample, all values for C_v fall within a range of 3.2 to $6.0 \cdot 10^{-4}$ cm^2/sec. The maximum range of all samples and all pressures was found to be 2.3 to $33.8 \cdot 10^{-4}$ cm^2/sec (Richards and Hamilton, 1967, p.100). Based on the extensive experience of Terzaghi (1955), however, the full range is from 0.01 to $100 \cdot 10^{-4}$ cm^2/sec.

VARIATION OF PHYSICAL PROPERTIES WITH DEPTH IN RECENT SEDIMENTS OF YORK RIVER, SOUTHEASTERN VIRGINIA, USA (A CASE STUDY)

Professor Richard W. Faas of Lafayette College, Easton, Pennsylvania (personal communication, 1971) studied the mass physical properties, and their variation with depth, of some fine-grained sediments from the York River of southeastern Virginia (Fig.113). Cores averaging 45 cm in length were carefully extracted from the sediment using SCUBA

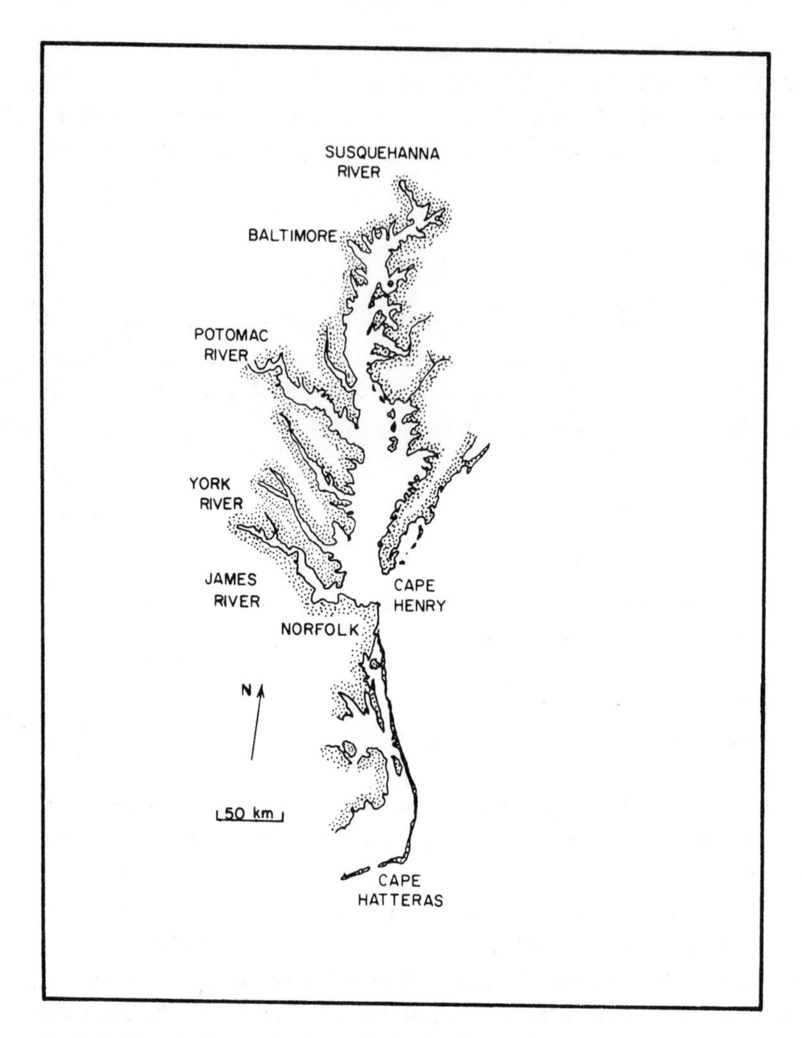

Fig.113. Map of Chesapeake Bay and associated estuarine systems.

techniques and immediately analyzed for bulk (wet) density, moisture content, and interstitial water salinity. A few drops of interstitial water were extracted under low pressures and placed on a A-B Goldberg refractometer. Variations in the index of refraction of the interstitial water were converted into salinity values. Moisture content was obtained using standard techniques. Fig.114 shows the variation in moisture content and interstitial water salinity down the core. There is a considerable variation in moisture content, with the largest change occurring within the upper 15 cm. Salinity shows similar variations with depth, from an initial value of 26 to 31‰. The increase appears to become significantly larger below 40 cm. The sediment core 13-C_1 was taken from a water depth of 30 ft. It appeared homogeneously textures throughout and only slightly bioturbated in

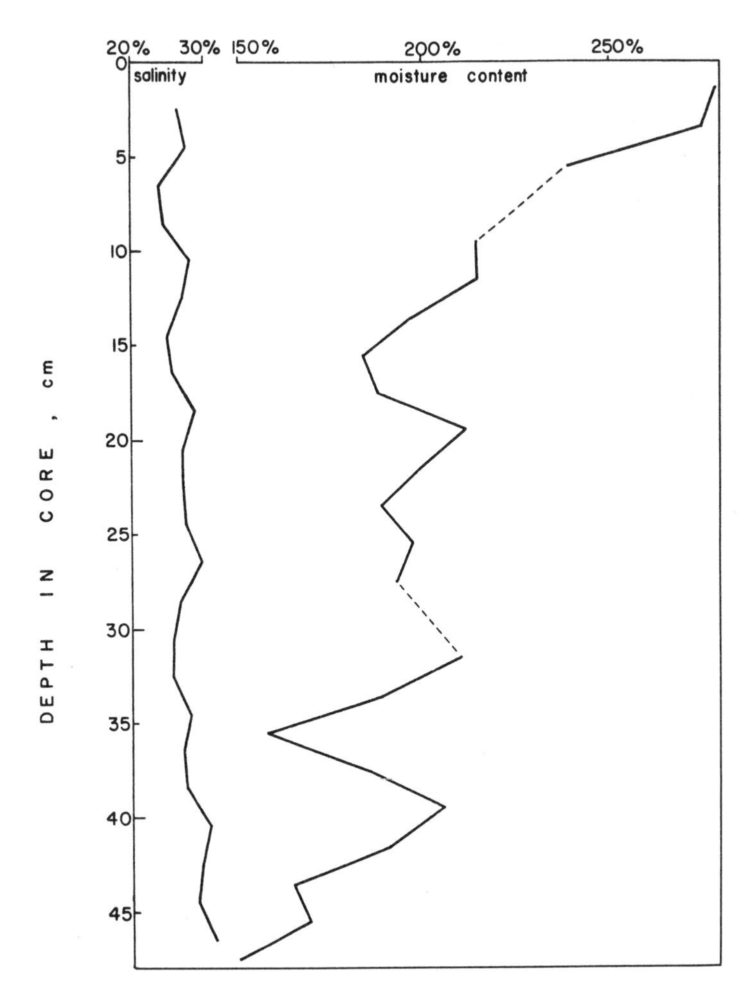

Fig.114. Variation in moisture content and interstitial water salinity with depth in core no.13-C_1. (After R.W. Faas, personal communication, 1971.)

the upper 15 cm. The size analysis showed the presence of 68% of clay-sized particles, 30.5% silt, and 1.5% of sand. This size distribution remained nearly constant for the entire core. The X-ray analyses indicated the presence of illite and chlorite, with illite being the dominant clay mineral.

Experiments were performed to determine the variation in the consolidational characteristics of these fine-grained sediments. Fig.115 shows void ratio versus logarithm of pressure ($e - \log p$) curves obtained from different depths in core 14-D. The position of the break between the reloading curve and the virgin compression curve shifts toward lower pressures with depth, with the lowermost curve showing the least sediment strength. This situation is not what one might expect, inasmuch as the preconsolidation

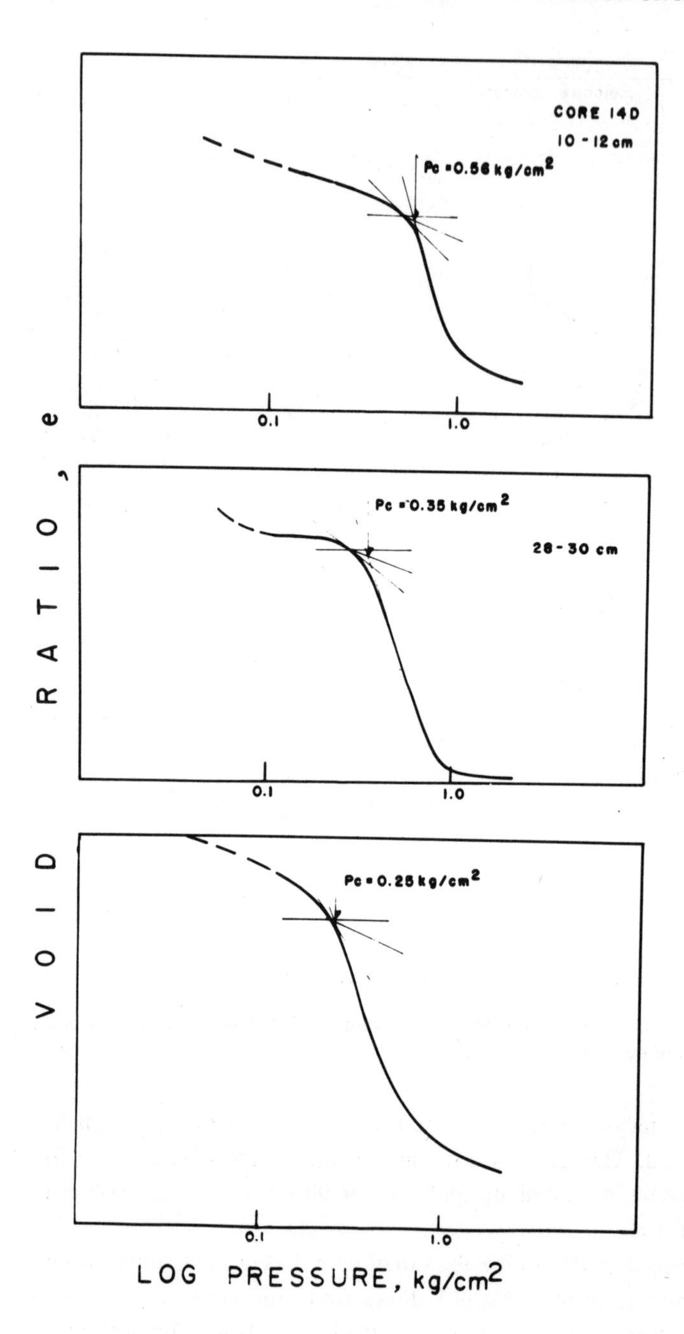

Fig.115. Void ratio versus logarithm of pressure ($e - \log p$) curves at different depths in core no.14-D. (After R.W. Faas, personal communication, 1971.)

pressure (the apparent maximum load the sediment has experienced) is greater than the pressures exerted by the overburden at that point. Successive $e - \log p$ diagrams indicate that the preconsolidation pressure becomes less with depth, finally reaching a level (48–50 cm) where it is hardly greater than the overburden pressure.

The usual explanation for such an anomalous behavior is that either the sediment has been exposed to desiccating conditions at some time in its history, or that it has become cemented through organic or inorganic means, causing an increase in "apparent" strength. The reason for such an anomalous behavior in the sediments studied by R.W. Faas is not known. He suggested, however, that a mechanism of consolidation is operating in these York River sediments in which closely-spaced sedimentary particles, although completely surrounded by water, exert great interparticle attraction due both to proximity and an extremely large amount of surface area. With normal compaction and flocculation, the particles approach each other and form small aggregates. These are still surrounded by water, but the attraction to an adjacent aggregate is diminished. Further compaction brings more aggregates in contact, forming increasingly larger bundles of particles, each of which is still surrounded by water, or a water-rich gel. Attractive forces within each of the larger bundles are greater than the force between two adjacent bundles separated by the water-rich gel. A gradient will exist showing a decrease in moisture content associated with a decrease in water-rich connections. The compressive strength of this sediment will then become normal for that depth. Owing to the excessive interparticle attraction in the upper part of the sediment column, however, sediment responds as if it were "over-consolidated".

The analysis of 15 samples from four cores provided a value for the coefficient of consolidation; it ranged from 1.3 to $73.0 \cdot 10^{-4}$ cm^2/sec for all samples, with an average of $11.9 \cdot 10^{-4}$ cm^2/sec. This value compares favorably with $9.2 \cdot 10^{-4}$ cm^2/sec, the average of a number of these coefficients measured on similar sediment types from Atlantic and Mediterranean deep-water cores (Richards and Hamilton, 1967). The coefficient of consolidation is a complex one, being dependent upon variations in permeability, void ratio, and compressibility. Consequently, the close agreement between consolidation co-

TABLE XXVII
Variation in physical properties of York River fine-grained sediments with depth, core 14B
(After R.W. Faas, personal communication, 1971)

Depth (cm)	*Moisture content* (%)	*Void ratio* *e*	*Bulk (wet) density* (g/cm^3)	*Atterberg limits*			*Salinity* (‰)
				LL	*PL*	*PI*	
8–10	220	5.7	1.35	157	43	114	23.5
14–16	220	5.7	1.33	158	43	115	27.5
30–32	219	5.7	1.38	150	43	107	29
42–45	167	4.4	1.30	160	38	122	32

efficient for sediments from such widely separated environments may be simply fortuitous, yet one is tempted to speculate that similar sediments will show similar responses, regardless of differences in origin (R.W. Faas, personal communication, 1971).

Variation of some physical properties with depth in York River sediments, as represented by examination of core no.14B, is shown in Table XXVII. Salinity increases with depth (also see Fig.114), whereas the Atterberg limits and other water-dependent properties remain unchanged down to a depth of 45 cm, where a sharp change occurs. The behavior of the bulk (wet) density at a depth of 45 cm is hard to explain. At the same level, other cores (14A, 14C, 14D) show high values for bulk (wet) density (R.W. Faas, personal communication, 1971).

In conclusion, it can be stated that with further accumulation of research data, this chapter should be expanded and a separate book be published on the subject.

REFERENCES

Adams, L.H., 1951. Elastic properties of materials of the earth's crust. In: B. Gutenberg (Editor), *Internal Constitution of the Earth*. Dover Publ., New York, N.Y., 439 pp.

Anderson, D.M. and Low, P.F., 1958. Density of water adsorbed by lithium-, sodium-, and potassium-bentonite. *Soil Sci. Soc. Am. Proc.*, 22: 99–103.

Aoyagi, K., 1967. Mineralogical study of sedimentary rocks in the oil fields of Japan by the X-ray diffraction method, and its application to petroleum geology (Pt. I). *Clay Sci.*, 3(1–2): 37–54.

Atterberg, A., 1971. Die Plastizität und Bindigkeit Liefernden Bestandteile der Tone. *Int. Mitt. Bodenkd.*, 1: 4–37.

Barkan, D.D., 1962. *Dynamics of Bases and Foundations*. McGraw-Hill, New York, N.Y., 434 pp.

Bates, T.F. and Comer, J.J., 1955. Further observations on the morphology of chrysotile and halloysite. *Clays Clay Miner., Proc. Natl. Conf. Clays Clay Miner., 3rd* (1954), pp.237–248.

Bradley, W.F., 1956. The green compression strength of natural bentonites. *Clays Clay Miner., Proc. Natl. Conf. Clays Clay Miner., 4th (1955)*, pp.41–44.

Bredehoeft, J.D. and Hanshaw, B.B., 1968. On the maintenance of anomalous fluid pressures: I. Thick sedimentary sequences. *Geol. Soc. Am. Bull.*, 79(9): 1097–1106.

Bredehoeft, J.D., Blyth, C.R., Waite, W.A. and Maxey, G.B., 1963. Possible mechanism for concentration of brines in subsurface formations. *Bull. Am. Assoc. Pet. Geologists*, 47: 257–269.

Buessem, W.R. and Nagy, B., 1954. The mechanism of the deformation of clay. *Clays Clay Miner., Proc. Natl. Conf. Clays Clay Miner., 2 (1953)*, 327: 480–491.

Carpenter, C. and Spencer, G.B., 1940. Measurements of compressibility of consolidated oil-bearing sandstones. *U.S. Bur. Mines, Rep. Invest.* No. 3540, 20 pp.

Casagrande, A., 1932a. The structure of clay and its importance in foundation engineering. *J. Boston Soc. Civil Engrs.*, 19: 168–209.

Casagrande, A., 1932b. Research on the Atterberg limits of soils. *Public Roads*, 13: 121–136.

Cebell, W.A. and Chilingarian, G.V., 1972. Some data on compressibility and density anomalies in halloysite, hectorite and illite clays. *Bull. Am. Assoc. Pet. Geologists*, 56: 796–802.

Chilingar, G.V., 1964. Relationship between porosity, permeability, and grain-size distribution of sands and sandstones. In: L.M.J.U. van Straaten (Editor), *Deltaic and Shallow Marine Deposits*, Elsevier, Amsterdam, 1: 71–75.

Chilingar, G.V. and Knight, L., 1960. Relationship between pressure and moisture content of kaolinite, illite and montmorillonite clays. *Bull. Am. Assoc. Pet. Geologists*, 44(1): 101–106.

Chilingar, G.V., Rieke III, H.H. and Robertson Jr., J.O., 1963a. Relationship between high overburden pressure and moisture content of halloysite and dickite clays. *Geol. Soc. Am. Bull.*, 74: 1041–1048.
Chilingar, G.V., Rieke III, H.H. and Robertson Jr., J.O., 1963b. Degree of hydration of clays. *Sedimentology*, 2(4): 341–342.
Chilingar, G.V., Main, R. and Sinnokrot, A., 1963c. Relationship between porosity, permeability, and surface areas of sediments. *J. Sediment. Petrol.*, 33(3): 759–765.
Chilingar, G.V., Rieke III, H.H. and Sawabini, C.T., 1969. Compressibilities of clays and some means of predicting and preventing subsidence. *Symp. Land Subsidence – Int. Assoc. Sci. Hydrol. and UNESCO, Tokyo, Japan*, 89(2): 377–393.
Chilingarian, G.V. and Rieke III, H.H., 1968. Data on consolidation of fine-grained sediments. *J. Sediment. Petrol.*, 38(3). 811–816.
Chilingarian, G.V., Beeson, C.M. and Ershaghi, I., 1972. Porosity and permeability. In: Rhodes W. Fairbridge (Editor), *Encyclopedia of Earth Sciences*, IV A: 964–970.
Cleary, J.M., 1959. Hydraulic fracture theory. III. Elastic properties of sandstone. *Ill. State Geol. Surv. Circ.*, 281: 44 pp.
Cohen, S.R., 1968. *Measurement of the Viscoelastic Properties of Water-Saturated Clay Sediments.* Thesis, U.S. Naval Postgraduate School.
Cooper Jr., A.R. and Eaton, L.E., 1962. Compaction behavior of several ceramic powders. *J. Am. Ceram. Soc.*, 45(3): 97–101.
Craft, B.C. and Hawkins, M.F., 1959. *Applied Petroleum Reservoir Engineering.* Prentice-Hall, Englewood Cliffs, N.J., 437 pp.
Darcy, H., 1856. *Les Fontaines Publiques de la Ville de Dijon.* Victor Dalmont, Paris.
Davis, B.L., 1964. *High-Pressure X-Ray Investigation of $CaCO_3$-II and $CaCO_3$-III at 25°C and of the Calcite–Aragonite Transition in the 300–500°C Temperature Interval.* Thesis, Univ. of California, Los Angeles, Calif., 133 pp.
De Josselin de Jong, G., 1968. Consolidation models consisting of an assembly of viscous elements or a cavity channel network. *Geotechnique*, 18: 195–228.
Demirel, T., Olson, T.W. and Handy, R.L., 1970. X-ray diffraction analysis of calcium montmorillonite under high pressures and drainage. *Natl. Clay Conf., Ann. Meet., Miami, Fla.* (Abstract), p. 16.
Diamond, S., 1970. Pore size distribution in clays. *Clays Clay Miner.*, 19 (1): 7–23.
Dickey, P.A., 1970. Comment on Chocolate Bayou Field. *J. Pet. Tech.*, 22(5): 560–569.
Dickey, P.A., Shriram, C.R. and Paine, W.R., 1968. Abnormal pressures in deep wells of southwestern Louisiana. *Science*, 160(3828): 609–615.
Dobrynin, V.M., 1962. Effect of overburden pressure on some properties of sandstones. *Trans. Am. Inst. Min. Metall. Engrs.*, 225: 360–366.
Domenico, P.A. and Mifflin, M.D., 1965. Water from low-permeability sediments and land subsidence. *Water Resour. Res.*, 1(4): 563–576.
Drabble, J.R., 1970. Elastic constants under pressure. In: H.L.I.D. Pugh (Editor), *Mechanical Behavior of Materials under Pressure.* Elsevier, Amsterdam, 785 pp.
Droste, J.B., White, G.W. and Vatter, A.E., 1958. Electron micrography of till matrix. *J. Sediment. Petrol.*, 28(3): 345–350.
Eaton, B.A., 1969. Fracture gradients prediction and its application in oilfield operations. *J. Pet. Tech.*, 21(10): 1353–1360.
Emery, K.O. and Rittenberg, S.C., 1952. Early diagenesis of California Basin sediments in relation to origin of oil. *Bull. Am. Assoc. Pet. Geologists*, 36(5): 735–806.
Farrar, D.M. and Coleman, J.D., 1967. The correlation of surface area with other properties of nineteen British clay soils. *J. Soil Sci.*, 18: 118–124.
Fatt, I., 1953. The effect of overburden pressure on relative permeability. *Trans. Am. Inst. Min. Metall. Engrs.*, 198: 325–326.
Fatt, I., 1958a. Pore structure in sandstones by compressible sphere-pack models. *Bull Am. Assoc. Pet. Geologists*, 42(8): 1914–1923.

Fatt, I., 1958b. Compressibility of sandstones at low to moderate pressures. *Bull. Am. Assoc. Pet. Geologists*, 42(8): 1924–1957.

Fatt, I. and Davis, D.H., 1952. The reduction in permeability with overburden pressure. *Trans. Am. Inst. Min. Metall. Engrs.*, 195: 329.

Fayed, L.A., 1966. X-ray diffraction study of orientation of the micaceous minerals in slate. *Clay Min.*, 6(4): 333–340.

Fertl, W.H. and Timko, D.J., 1971. Application of well logs to geopressure problems in the search, drilling and production of hydrocarbons. *Colloq. Assoc. Rech. Tech. Exploit. Pet. Commun.* 4: 15 pp.

Fraser, H.J., 1935. Experimental study of the porosity and permeability of clastic sediments. *J. Geol.*, 43: 910–1010.

Geertsma, J., 1957. The effect of fluid pressure decline on volumetric changes of porous rocks. *J. Pet. Tech.*, 9(12): 331–339.

Geyer, R.L. and Myung, J.I., 1971. The 3-D velocity log; a tool for in situ determination of the elastic moduli of rocks. 12th Ann. Symp. Rock Mech., Univ. Mo. (Rolla). In: G.B. Clark (Editor), *Dynamic Rock Mechanics.* Am. Inst. Min. Metall. Engrs., pp.71–107.

Gipson Jr., M., 1965. Application of the electron microscope to the study of particle orientation and fissility in shale. *J. Sediment. Petrol.*, 35(2): 408–414.

Gomaa, E.M., 1970. Compressibility of rocks and factors affecting them. *Am. Inst. Min. Metall. Engrs., Ann. Student Contest Meeting, April, Univ. Calif., Berkeley,* presented paper.

Gondouin, M. and Scala, C., 1958. Streaming potential and the S.P. log. *J. Pet. Tech.*, 10(8): 170–179.

Gregory, A.R., 1967. Mode conversion technique employed in shear wave velocity studies of rock samples under axial and uniform compaction. *J. Soc. Pet. Engrs.*, 7(2): 136–148.

Griffiths, J.C., 1958. Petrography and porosity of the Cow Run Sand, St. Marys, West Virginia. *J. Sediment. Petrol.*, 28: 15–30.

Grim, R.E., 1968, *Clay Mineralogy*. McGraw-Hill, New York, N.Y., 596 pp.

Hall, H.N., 1953. Compressibility of reservoir rocks. *Trans. Am. Inst. Min. Metall. Engrs.*, 198: 309–311.

Hamilton, E.L., 1970a. Sound velocity and related properties of marine sediments. North Pacific. *J. Geophys. Res.*, 75(23): 4423–4446.

Hamilton, E.L., 1970b. Reflection coefficients and bottom losses at normal incidence computed from Pacific sediment properties. *Geophysics*, 35(6): 995–1004.

Hamilton, E.L., 1971a. Elastic properties of marine sediments. *J. Geophys. Res.*, 76(2): 579–604.

Hamilton, E.L., 1971b. Prediction of in situ acoustic and elastic properties of marine sediments. *Geophysics*, 36(2): 266–284.

Handin, J. and Hager Jr., R.V., 1958. Experimental deformation of sedimentary rocks under confining pressure: tests at high temperature. *Bull. Am. Assoc. Pet. Geologists*, 42: 2892–2934.

Handin, J., Hager Jr., R.V., Friedman, M. and Feather, J.N., 1963. Experimental deformation of seaimentary rocks under confining pressure: pore pressure tests. *Bull. Am. Assoc. Pet. Geologists*, 47(5): 717–755.

Hansbo, S., 1960. Consolidation of clay with reference to the influence of vertical sand. *Swed. Geotech. Inst. Proc.*, 18: 41–46.

Harville, D.W., 1967. *Rock Collapse as a Producing Mechanism in Superpressure Reservoirs.* Thesis, Louisiana State Univ., 94 pp.

Harville, D.W. and Hawkins Jr., M.F., 1967. Rock compressibility and failure as reservoir mechanisms in geopressured gas reservoirs. *Ann. Symp. Abnorm. Subsurf. Pressures, 1st, Louisiana State Univ.*

Hasegawa, H. and Ikeuti, M., 1966. On the tensile strength test of disturbed soils. In: J. Kravtchenko and P.M. Sirielys (Editors), *Rheology and Soil Mechanics*. Springer, New York, N.Y., pp.405–412.

Hata, S., Ohta, H. and Yoshitani, S., 1969. A theoretical approach to stress–strain relations of clays. In: *Land Subsidence. IASH–UNESCO, Publ.* no.89; AIHS 2 : 563–572.

Hubbert, M.K., 1940. The theory of ground-water motion. *J. Geol.*, 48(1): 785–944.

Humphries, W.K. and Wahls, H.E., 1968. Stress history effects on dynamic modulus of clay. *J. Soil Mech. Found. Div., Am. Soc. Civ. Engrs.*, 94(SM2): 371–389.

Ide, J.M., 1936. Comparison of statically and dynamically determined Young's modulus of rocks. *Proc. Natl. Acad. Sci.*, 22: 81.

Inderbitzen, A.L., 1969. *Relationships Between Sedimentation Rate and Shear Strength in Recent Marine Sediments off Southern California.* Thesis, Stanford Univ., 113 pp.

Jacob, C.E., 1940. On the flow of water in an elastic artesian aquifer. *Trans. Am. Geophys. Union*, Pt. 2: 574–586.

Jacob, C.E., 1950. Flow of ground water. In: H. Rouse (Editor), *Engineering Hydraulics*. Wiley, New York, N.Y., pp.321–386.

Jacquin, C., 1965a. *Etude des Ecoulements et des Equilibres de Fluides dans les Sables Argileux.* Thesis, Univ. Paris, France. *Rev. Inst. Franç. Pét.*, 20(4): 49 pp.

Jacquin, C., 1965b. Interactions entre l'argile et les fluides écoulements à travers les argiles compactes; Etude bibliographique. *Rev. Inst. Franç. Pét.*, 20(10): 1475–1501.

Judd, W.R., 1959. Effect of the elastic properties of rocks on civil engineering design. In: P.D. Trask (Editor), *Symposium on Rock Mechanics, Geol. Soc. Am., Eng. Geol. Case Histories*, 3: 76 pp.

Kaarsberg, E.A., 1959. Introductory studies of natural and artificial argillaceous aggregates by sound-propagation and X-ray diffraction methods. *J. Geol.*, 67(4): 447–472.

Katz, D.L. and Ibrahim, M.A., 1971. Threshold displacement pressure considerations for caprocks of abnormal-pressure reservoirs. *Conf. Drill. Rock Mech., 5th, Soc. Pet. Engrs.*, Austin, Texas, Paper no.3222.

Kaul, B.K., 1964. Effect of temperature on the swelling of compressed clays. *Current Sci.*, 33: 170–173.

Kaul, B.K., 1965. Effect of temperature and pore fluids on the compressibility of clay (study on the compressibility of clays). In: *Behavior of Soils under Stress, Indian Inst. Sci. Proc.*, 1(A8): 31 pp.

Kawakami, F. and Ogawa, S., 1966. Yield stress and modulus of elasticity of soil. In: J. Kravtchenko and P.M. Sirielys (Editors), *Rheology and Soil Mechanics.* Springer, New York, N.Y., pp.354–363.

Kemper, W.D., 1961. Movements of water as affected by free energy and pressure gradients. *Soil Sci. Soc. Am. Proc.*, 25: 255–265.

Kermabon, A., Gehin, C. and Blavier, P., 1969. A deep-sea electrical resistivity probe for measuring porosity and density of unconsolidated sediments. *Geophysics*, 34(4): 554–571.

Khanin, A.A., 1968. Evaluation of sealing abilities of clayey cap rocks in gas deposits. *Geol. Nefti i Gaza*, 1968(8): 17–21.

Khanin, A.A., 1969. *Oil and Gas Reservoir Rocks*. Nedra, Moscow, 366 pp.

Knutson, C.F. and Bohor, B.F., 1963. Reservoir rock behavior under moderate confining pressure. In: C. Fairhurst (Editor), *Rock Mechanics.* Pergamon, New York, N.Y., pp.627–658.

Kohlhaas, C.A. and Miller, F.G., 1969. Rock-compaction and pressure-transient analysis with pressure-dependent rock properties. *Soc. Petrol. Engrs., Fall Meet., Denver, Colo.*, Paper 2563: 7 pp.

Kolthoff, I.M. and Sandell, E.B., 1938. *Textbook of Quantitative Inorganic Analysis*. Macmillan, New York, N.Y., 749 pp.

Komornik, A., Rohlich, V. and Wiseman, G., 1970. Overconsolidation by desiccation of coastal Late Quaternary clays in Israel. *Sedimentology*, 14: 125–140.

Lambe, T.W., 1958. The structure of compacted clay. *J. Soil Mech. Found Div., Am. Soc. Civil Engrs.*, 84(SM2, Pt. 1): 1654–1711.

Langnes, G.L., Robertson, J.O. and Chilingar, G.V., 1972. *Secondary Recovery and Carbonate Reservoirs.* Elsevier, Amsterdam, 303 pp.

Lapinskaya, T.A. and Proshlyakov, B.K., 1970. Problem of reservoir rocks during exploration for oil and gas at great depths, pp.115–121. In: A.V. Sidorenko et al. (Editors), *Status and Problems of Soviet Lithology, III*. Nauka, Moscow, 286 pp.

Laughton, A.S., 1957. Sound propagation in compacted ocean sediments. *Geophysics*, 22: 233–260.

Lawrence, H.W., 1964. In situ measurement of the elastic properties of rocks. *Proc. Symposium Rock Mech., 6th, Rolla, Mo.*, pp.381–389.

Leonards, G.A. and Altschaeffl, A.G., 1964. Compressibility of clay. *J. Soil Mech. Found. Div., Am. Soc. Civil Engrs.*, 90(SM5, Pt. 1): 133–155.

Lobree, D.T., 1968. *Measurement of the Compressibilities of Reservoir Type Rocks at Elevated Temperatures.* Thesis, Univ. California, Berkeley, Calif., 78 pp.
Long, G. and Chierici, G., 1961. Salt content changes compressibility of reservoir brines. *Pet. Eng.*, 1961 (July): B25–B31.
Lutz, J.F. and Kemper, W.D., 1959. Intrinsic permeability of clay as affected by clay–water interaction. *Soil Sci.*, 88(2): 83–90.
Macey, H.H., 1940. Clay–water relationships. *Proc. Phys. Soc. (Lond.)*, 52(5): 625–656.
Mann, C.J. and Semple, R.M., 1970. Computer reduction and plotting of marine sediment data. *Mar. Tech. Soc. J.*, 4(1): 53–57.
Martin, R.T., 1962. Adsorbed water on clay: a review. In: E. Ingerson (Editor), *Clays Clay Miner., Proc. Natl. Conf. Clays Clay Miner., 9 (1960)*, pp.28–70.
Martin, R.T., 1965. Quantitative measurements of wet clay fabric (abs.). In: E. Ingerson (Editor), *Clays Clay Miner., Proc. Natl. Conf. Clays Clay Miner., 14 (1964)*, pp.271–287.
Maxwell, D.T. and Hower, J., 1967. High-grade diagenesis and low-grade metamorphism of illite in the Precambrian Belt Series. *Am. Mineralogist*, 52: 843–857.
McClelland, B., 1956. Engineering properties of soils on the continental shelf of the Gulf of Mexico. *Proc. Texas Conf. Soil Mech. Found. Eng., 8th, Austin, Texas*: 28 pp.
McKelvey, J.G. and Milne, I.H., 1962. Flow of salt solutions through compacted clay. In: E. Ingerson (Editor), *Clays Clay Miner., Proc. Natl. Conf. Clays Clay Miner., 9 (1960)*, pp.248–259.
McLatchie, A.S., Hemstock, R.A. and Young, J.W., 1958. The effective compressibility of reservoir rock and its effects on permeability. *Trans. Am. Inst. Min. Metall. Engrs.*, 213: 386–388.
Meade, R.H., 1961. X-ray diffractometer method for measuring preferred orientation in clays. *U.S. Geol. Surv., Profess. Pap.*, 424-B: B273–B276.
Meade, R.H., 1964. Removal of water and rearrangement of particles during the compaction of clayey sediments – review. *U.S. Geol. Surv., Profess. Pap.*, 497-B: 23 pp.
Meade, R.H., 1968. Compaction of sediments underlying areas of land subsidence in Central California. *U.S. Geol. Surv., Profess. Pap.*, 497-D: 39 pp.
Mesri, G., 1969. *Engineering Properties of Montmorillonite.* Thesis, Univ. Illinois, Urbana, Ill.
Mesri, G. and Olson, R.E., 1971. Mechanisms controlling the permeability of clays. *Clays Clay Miner.*, 19(3): 151–158.
Michaels, A.S. and Lin, C.S., 1954. The permeability of kaolinite. *Ind. Eng. Chem.*, 46(6): 1239–1246.
Miller, R.J. and Low, P.F., 1963. Threshold gradient for water flow in clay systems. *Soil Sci. Soc. Am. Proc.*, 27: 605–609.
Millot, G., 1970. *Geology of Clays*. Springer, New York, N.Y., 429 pp.
Mitchell, J.K., 1956. The fabric of natural clays and its relation to engineering properties. *Proc. Highway Res. Board*, 35: 693.
Mitchell, J.K. and Younger, J.S., 1967. Abnormalities in hydraulic flow through fine-grained soils. In: *Permeability and Capillarity of Soils. Am. Soc. Test. Mater., Spec. Tech. Publ.*, 417: 106–139.
Mizutani, S., 1966. Transformation of silica under hydrothermal conditions. *J. Earth Sci.*, Nagoya Univ., 14: 56–88.
Moore, D.G., 1962. Bearing strength and other physical properties of some shallow and deep-sea sediments from the North Pacific. *Geol. Soc. Am. Bull.*, 73: 1163–1166.
Morelock, J., 1969. Shear strength and stability of continental slope deposits of western Gulf of Mexico. *J. Geophys. Res.*, 74(2): 465–482.
Morgenstern, N.R. and Tchalenko, J.S., 1967. Microstructural observations on shear zones from slips in natural clays. *Proc. Geotech. Conf., Oslo*, 1: 147–152.
Morgenstern, N.R. and Tamuly Phukan, A.L., 1969. Non-linear stress–strain relations for a homogeneous sandstone. *J. Rock Mech. Min. Sci.*, 6: 127–142.
Nishida, Y. and Nakagawa, S., 1969. Water permeability and plastic index of soils. In: *Land Subsidence. IASH-UNESCO, Publ.* no.89; *AIHS*, 2: 573–578.
Obert, L. and Duvall, W.T., 1967. *Rock Mechanics and the Design of Structure in Rock*. Wiley, New York, N.Y., 650 pp.

O'Brien, N.R., 1963. *A Study of Fissility in Argillaceous Rocks.* Thesis, Univ. Illinois, Urbana, Ill., 80 pp.

Olsen, H.W., 1966. Darcy's law in saturated kaolinite. *Water Resour. Res.*, 2: 287–294.

Osipov, Yu.B., 1969. Toward question of evaluation of three-dimensional texture of rocks (using example of uniaxial compaction and consolidation of clayey pastes). *Vest. Mosk. Univ., Geol.*, 1969(2): 62–70.

Paulding, B.W., Cornish, R.H., Abbott, R.W. and Finlayson, L.A., 1965. *Behavior of Rock and Soil Under High Pressures.* Tech. Rep. No. AFWL-TR-65-51.

Podio, A.L., Gregory, A.R. and Gray, K.E., 1968. Dynamic properties of dry and water-saturated Green River Shale under stress. *J. Soc. Pet. Engrs.*, 8(4): 389–404.

Pusch, R., 1962. Clay particles. *Statens Rad Byggforskning. Handl.*, 40: 150 pp.

Pusch, R., 1966. Quick-clay microstructure. *Eng. Geol.*, 1(6): 433–443.

Pusch, R., 1970. Microstructural changes in soft quick clay at failure. *Can. Geotech. J.*, 7(1): 1–7.

Quigley, R.M., 1961. *Research on the Physical Properties of Marine Soils.* Mass. Inst. Tech., Soil Eng. Div., Publ., 117.

Quigley, R.M. and Thompson, C.D., 1966. The fabric of anisotropically consolidated sensitive marine clay. *Can. Geotech. J.*, 3(2): 62–73.

Raghavan, R., Scorer, J.D.T. and Miller, F.G., 1969. An investigation by numerical methods of the effect of pressure dependent rock and fluid properties on well flow tests. *Soc. Pet. Engrs. Fall Meet., Denver, Colo.*, Paper 2617.

Raitburd, Ts.M., 1960. Study of microstructure of clays by X-ray methods. In: *Reports to the Meeting of the International Commission for the Study of Clays.* Acad. Sci. U.S.S.R., Dept. Geol., Geol. Sci., Comm. for Study of Clays, pp.108–118.

Ramey Jr., H.J., 1964. Rapid method for estimating reservoir compressibilities. *Trans. Am. Inst. Min. Metall. Engrs.*, 231: 447–454.

Riccio, J.F., 1962. Porosity and permeability studies of some unconsolidated sediments. *Compass*, 39(2): 51–63.

Richards, A.F., 1961. *Investigations of Deep-Sea Sediment Cores, I. Shear Strength, Bearing Capacity, and Consolidation.* U.S. Navy Hydrographic Office, Tech. Rep., 63: 70 pp.

Richards, A.F. and Hamilton, E.L., 1967. Investigation of deep-sea sediment cores. III. Consolidation. In: A.F. Richards (Editor), *Marine Geotechnique*, Univ. Illinois Press, Urbana, Ill., 327 pp.

Rieke III, H.H., 1965. *Rock Mechanics Applied to the Solution of Slope Stability Problems in the Santa Monica Slates.* Thesis, Univ. Southern California, Los Angeles, Calif., 143 pp.

Rieke III, H.H., 1970. *Compaction of Argillaceous Sediments (20–500,000 p.s.i.).* Thesis, Univ. Southern Calif., Los Angeles, Calif., 682 pp.

Rieke III, H.H., Chilingar, G.V. and Robertson Jr., J.O., 1964. High-pressure (up to 500,000 p.s.i.) compaction studies on various clays. *Int. Geol. Congr., 22nd Ses., New Delhi, India*, Pt. 15, Sect. 15: 22–38.

Rieke III, H.H., Ghose, S.K., Fahhad, S.A. and Chilingar, G.V., 1969. Some data on compressibility of various clays. *Proc. Int. Clay Conf.*, 1: 817–828.

Robertson, E.C., 1967. Laboratory consolidation of carbonate sediments. In: A.F. Richards (Editor), *Marine Geotechnique*. Univ. Illinois Press, Urbana, Ill., 327 pp.

Rosenqvist, I.Th., 1958. Remarks to the mechanical properties of soil–water systems. *Geol. Fören. Stockholm Forh.*, 80(4): 435–457.

Rosenqvist, I.Th., 1959. Physicochemical properties of soils; soil–water systems. *J. Soil Mech. Found., Div. Am. Soc. Civil Engrs.*, 85(SM2): 31–53.

Rosenqvist, I.Th., 1966. Norwegian research into the properties of quick clay – a review. *Eng. Geol.*, 1(6): 445–450.

Rubey, W.W. and Hubbert, M.K., 1959. Role of fluid pressure in mechanics of overthrust faulting. II. Overthrust belt in geosynclinal area of western Wyoming in light of fluid-pressure hypothesis. *Bull. Geol. Soc. Am.*, 70(2): 167–206.

Saada, A.S., 1962. A rheological analysis of shear and consolidation of saturated clays. *Highway Res. Board Bull.*, 342 – *Natl. Res. Counc.*, pp.52–76.

Sawabini, C.T., 1971. *Triaxial Compaction of Unconsolidated Sand Core Samples under Producing Conditions at a Constant Overburden Pressure of 3,000 p.s.i.g. and Constant Temperature of 140° F.* Thesis, Univ. Southern California, Los Angeles, Calif., 178 pp.

Sawabini, C.T. and Chilingarian, G.V., 1971. Triaxial compaction of unconsolidated sand cores and clays and chemistry of squeezed-out solutions. *8th Int. Sediment. Congr.* (Abstracts).

Sawabini, C.T., Chilingar, G.V. and Allen, D.R., 1971. Design and operation of a triaxial, high-temperature, high-pressure compaction apparatus. *J. Sediment. Petrol.*, 41(3): 871–881.

Schiffman, R.L., 1962. A rheological analysis of shear and consolidation of saturated clays. In: *Stress Distribution in Earth Masses. Bull. Highway Res. Board*, 342: 76–84.

Schmidt, E.R. and Heckroodt, R.O., 1959. A dickite with an elongated crystal habit and its dehydroxylation. *Mineral. Mag.*, 32(247): 314–323.

Schwartz, E.G. and Weinstein, A.S., 1965. Model for compaction of ceramic powders. *J. Am. Ceram. Soc.*, 48: 346–350.

Sergeev, E.M., 1970. Lithology and engineering geology. In: A.V. Sidorenko et al. (Editors), *Status and Problems of Soviet Lithology, I.* Nauka, Moscow, pp.189–199.

Shamsiev, A.A., Suleymanov, T.Kh. and Kuliev, V.T., 1967. Compressional characteristics of clays at high specific loading (up to 2,500 kg/cm^2). *Dokl. Akad. Nauk Azerb. S.S.R.*, 23(8): 29–32.

Shapiro, I., 1963. *Fundamental Studies of Compressibility of Powders.* Gov. Rep. No. ASD-TDR-63-147, Contract No. AF33-(616)-8006: 92pp.

Sheth, S.B., Al-Zamil, Z. and Chilingarian, G.V., 1972. Some data on compressibility and density of various clays. *Eng. Geol.*, in press.

Shibakova, V.C. and Shalimova, E.M., 1967. Optimal loading pressure and clay texture. *Vestn. Moscov. Univ.*, (4): 90–94.

Silverman, E.N. and Bates, T.F., 1960. X-ray diffraction study of orientation in the Chattanooga Shale. *Am. Mineralogist*, 45(1): 60–68.

Simons, L.H., 1963. Elastic properties of shales. *Conf. Drilling Rock Mech., Soc. Pet. Engrs., 1st, Austin, Texas*, 496 pp.

Skempton, A.W., 1944. Notes on the compressibility of clays. *Q. J. Geol. Soc. Lond.*, 100(1/2): 119–135.

Skempton, A.W., 1953. Soil mechanics in relation to geology. *Proc. Yorks. Geol. Soc.*, 29: 33–62.

Skempton, A.W., 1970. The consolidation of clays by gravitational compaction. *Q. J. Geol. Soc. Lond.*, 125(3): 373–411.

Smith, R.J., 1962. *Deep-Ocean Core Boring and Soil Testing Investigations.* U.S. Naval Civil Eng. Lab. Tech. Note N-445, Port Hueneme, Calif.: 66 pp.

Smith, J.L., DeVries, K.L., Bushnell, D.J. and Brown, W.S., 1969. Fracture data and stress–strain behavior of rocks in triaxial compression. *Exp. Mech.*, 9(8): 348–355.

Swartz, J.H., 1962. Some physical constants for the Marshall Islands Area. *U.S. Geol. Surv., Profess. Pap.*, 260-A: 953–989.

Tan, T.K., 1957. Report on soil properties and their measurements. *Proc. Int. Conf. Soil Mech. Found. Eng., 4th, Lond.*, 3: 87–89.

Tan, T.K., 1966. Determination of the rheological parameters and the hardening coefficients of clay. In: J. Kravtchenko and P.M. Sirielys (Editors), *Rheology and Soil Mechanics.* Springer, New York, N.Y., pp.256–272.

Taylor, D.W., 1948. *Fundamentals of Soil Mechanics.* Wiley, New York, N.Y., 700 pp.

Tazhibaeva, P.T. (Editor-in-Chief), 1970. *Investigation and Utilization of Clays and Clay Minerals. Proc. Symp., Alma-Ata, Kaz. S.S.R., Sept. 16–21, 1968.* Nauka, Kaz. S.S.R., 307 pp.

Terzaghi, K., 1925a. *Erdbaumechanik auf bodenphysikalischer Grundlage.* Deuticke, Leipzig, 399 pp.

Terzaghi, K., 1925b. Principles in soil mechanics. III. Determination of the permeability of clay. *Eng. News Rec.*, 95(21): 832–836.

Terzaghi, K., 1926. Simplified soil tests for subgrades and their physical significance. *Public Roads*, 7: 153–162.

Terzaghi, K., 1938. Settlement of structures in Europe and methods of observation. *Trans. Am. Soc. Civil Engrs.*, 103: 1432–1448.

Terzaghi, K., 1943. *Theoretical Soil Mechanics*. Wiley, New York, N.Y., 510 pp.

Terzaghi, K., 1955. Influence of geological factors on the engineering properties of sediments. *Econ. Geol.*, 50th Ann. Vol.: 557–618.

Terzaghi, K. and Peck, R.B., 1948. *Soil Mechanics in Engineering Practice*. Wiley, New York, N.Y., 566 pp.

Thomas, G.W., 1966. Some effects of overburden pressure on oil shale during underground retorting. *J. Soc. Pet. Engrs.*, 6(1): 1–8.

Trask, P.D., 1930. Mechanical analysis of sediments by centrifuge. *Econ. Geol.*, 25: 581–599.

Trask, P.D., 1932. *Origin and Environment of Source Sediments of Petroleum*. Gulf Publ. Co., Houston, Texas, 67 pp.

Underwood, L.B., 1967. Classification and identification of shales. *J. Soil Mech. Found., Div. Proc. Am. Soc. Civil Engrs.*, 93(6): 97–116.

Urick, R.J., 1947. A sound velocity method for determining compressibility of finely divided substances. *J. Appl. Phys.*, 18: 983–987.

Van der Knaap, W., 1959. Non-linear behavior of elastic porous media. *Trans. Am. Inst. Min. Metall. Engrs.*, 216: 179–186.

Van der Knaap, W. and Van der Vlis, A.C., 1967. On the cause of subsidence in oil-producing areas. *Proc. 7th World Petrol. Congr., Mexico City, Mexico.* Elsevier, Amsterdam, 3: 85–95.

Von Engelhardt, W. and Gaida, K.H., 1963. Concentration changes of pore solutions during the compaction of clay sediments. *J. Sediment. Petrol.*, 33(4): 919–930.

Von Engelhardt, W. and Tunn, W.L.M., 1955. The flow of fluids through sandstones. *Ill. State Geol. Surv., Circ.*, 194: 17 pp.

Von Gonten, W.D. and Choudhary, B.K., 1969. Effect of pressure and temperature on the volume compressibility. *Soc. Pet. Engrs., 44th Ann. Meet., Denver, Colo.*, Pap. no.2526 (preprint).

Waidelich, W.C., 1958a. Influence of liquid and clay mineral type on consolidation of clay–liquid systems. In: Winterkorn, H.F. (Editor), *Water and Its Conduction in Soils. Highway Res. Board, Spec. Rep.*, 40; *Natl. Acad. Sci. – Natl. Res. Counc. Publ.*, 629: 24–42.

Waidelich, W.C., 1958b. *Physicochemical Factors Influencing the Consolidation of Soils.* Thesis, Princeton Univ., Princeton, N.J.

Walton, W.C., 1962. Selected analytical methods for well and aquifer evaluation. *Ill. Water Surv. Bull.*, 49: 81 pp.

Warner, D.L., 1964. *An Analysis of the Influence of Physical-Chemical Factors Upon the Consolidation of Fine-Grained Elastic Sediments.* Thesis, Univ. California, Berkeley, Calif., 136 pp.

White, W.A., 1961. Colloid phenomena and sedimentation of argillaceous rocks. *J. Sediment. Petrol.*, 31(4): 560–570.

Wyble, D.O., 1958. Effect of applied pressure on conductivity, porosity and permeability of sandstones. *Trans. Am. Inst. Min. Metall. Engrs.*, 213: 431–432.

Yong, R.N. and Warkentin, B.P., 1966. *Introduction to Soil Behavior.* McMillan, New York, N.Y., 451 pp.

Young, A., Low, P.F. and McLatchie, A.S., 1964. Permeability studies of argillaceous rocks. *J. Geophys. Res.*, 69(20): 4237–4245.

Zhiangirov, R.S., 1968. Compressibility of clayey sediments and rocks. Problem 12: Engineering geology in government planning. *Int. Geol. Congr., 23d Sess., Rep. Sov. Geologists*, Moscow, pp.102–108.

Zisman, W.A., 1933. Comparison of the statically and seismologically determined elastic constants of rocks. *Proc. Natl. Acad. Sci.*, 19(7): 680.

Chapter 5

CHEMICAL ALTERATIONS AND BEHAVIOR OF INTERSTITIAL FLUIDS DURING COMPACTION

INTRODUCTION

This chapter is mainly concerned with the effect of compaction on the geochemical make-up and origin of subterranean waters. The other diagenetic processes which affect the chemical composition of interstitial waters are also briefly discussed. Chemical properties of subsurface waters are considered in relation to the mode of their expulsion during compaction. In addition, a brief discussion is presented on the interrelationship between water and oil phases during compaction.

To date, the formation of sodium-chloride solutions in rocks at depth is not well understood. In the presence of salt domes or salt beds, brines of high salinity in the surrounding strata owe their origin in large measure to solution of the available salt. Formation waters with high salinity, however, also occur in sedimentary rocks which are not associated with salt deposits. It seems probable that under certain conditions gravitational compaction acts as the main mechanism producing concentrated salt solutions. Some consideration is also given to other mechanisms that in one way or another change the chemistry of fluids during compaction and early diagenesis.

In geosynclines, where the depositional rate is rapid, large quantities of water are continuously extracted from the hydrosphere during sedimentation. Recent muds, for example, may contain up to about 80% water by volume (Degens and Chilingar, 1967, p.478). The interstitial fluids occupy the pore space of most of the buried sediments; upon compaction of the sediments, the connate waters, as the interstitial fluids are sometimes termed, are expelled into the associated sandstones. The speed with which the water is expelled from the original argillaceous sediments depends not only on the overburden pressure and the physical and chemical properties of fluids, but also on the texture, structure, and mineral composition of the sediment.

The composition of fluids contained in the pore space of permeable sediments at depth is known from the many analyses of formation waters obtained from drill holes and producing oil wells. In many sedimentary basins the salinity of fluids increases with increasing depth, especially in those fluids that are associated with basins older than the Tertiary. These fluids differ greatly from sea water in chemical composition. All trapped fluids in the sediments are believed to have undergone chemical alteration which is a function of time, temperature and pressure. In studying these fluids, the basic assumption is made that the salinity and ionic composition of sea water, especially since the Mesozoic, was the same as today. The original water, expelled by compaction, has been gradually replaced by or mingled with meteoric water, especially if the sediments were uplifted or exposed at the earth's surface by tectonic activity or erosional processes. The majority of analyzed fluids from these sediments are derived from permeable strata.

In the past, most investigators have ignored the pore fluids associated with less permeable shales. One reason for this was that the permeability of shales and argillaceous sediments is so low that the exposed shale intervals in oil wells rarely produce fluids in measurable quantities. In addition, the fluid that is produced for the most part is not representative because of contamination. The contamination, which may cause concentration (or dilution) in field samples, may occur either during drilling and production of the oil well, or may be due to associated salt deposits. One way in which data on interstitial fluids in shales can be obtained is by core analysis. The pore solution can be flushed or distilled out of the core, permeability permitting, or the core can be pulverized and then leached with distilled water in order to obtain the soluble salts. The latter technique, however, will not give the true composition of the salts dissolved in the pore solution, because any soluble minerals present in the rock will also be dissolved along with the salts. Von Engelhardt and Gaida (1963) leached interstitial salts from Jurassic claystone and shale core samples. They felt that analysis for the chloride content of the leached solution alone would be sufficient because it is very likely that it would be derived from the pore fluid only. The proportions of calcium, magnesium, and sodium ions may be changed by the base exchange of the clays. They also concluded that the chloride content was of the same magnitude as that in associated permeable strata, which is questioned by other investigators as discussed later. Possibly, the formation-logging techniques (electric, radioactive, etc.) may be improved to such an extent in the future that the chemistry of fluids in shales could be determined directly without using leaching or other procedures.

Changes in concentration of interstitial fluids during the process of compaction, as reported by different investigators, are presented in this chapter and related to field data obtained from various literature sources.

CLASSIFICATION OF INTERSTITIAL SOLUTIONS

Interstitial solutions can be classified as (1) syngenetic (formed at the same time as the enclosing rocks), and (2) epigenetic (owe their origin to subsequent infiltration of meteoric and other waters into already formed rocks). The main processes which alter the chemistry of buried waters are: (1) physical (compaction), (2) chemical (reactions between rock minerals, organic matter and interstitial solutions, etc.), (3) physicochemical (filtration through charged-net clay membranes, adsorption and base exchange, etc.), (4) electrochemical, and (5) biochemical.

As soon as buried sea water comes in contact with minerals and organic matter, many different processes become operative that could result in: (1) concentration of dissolved solids, (2) preferential increase or decrease of various dissolved mineral and organic components, or (3) desalting of the water. The diagenetic alterations are often so pronounced that the present-day chemistry no longer reflects the original chemical composition of the water at the time of deposition. In most cases, however, evidence suggests that the pore waters in petroleum-bearing strata are ultimately derived from the sea. Obviously, subsequent infiltration of meteoric waters may obscure the picture. In the latter case the waters can be called syngenetic–epigenetic.

Many chemical classifications have been proposed: Tolstikhin (1932, 1954), De Sitter (1947), Durov (1948), Stiff (1951), Vassoevich (1954), Krejci-Graf et al. (1957), Gorrell (1958), Rainwater and White (1958), Chave (1960), Eremenko (1960), and Lomonosov and Pinneker (1969). The subject of classification of waters has been reviewed by several investigators, e.g., by Chilingar and Degens (1964) and by Konzewitsch (1967).

Most oil-field waters are highly mineralized (sp. gr. reaches 1.13 and higher). They are commonly characterized by the presence of bicarbonate ion and absence, or very low concentration, of SO_4^{2-} ion. Calcium-chloride (hard) waters commonly have a salinity range of 7–20° Bé.[1] Decrease in sulfate content can be linked to microbiological activities or to inorganic precipitation, because the solubility product of calcium sulfate is commonly exceeded when the salinity of brine exceeds 10%.

COMPOSITION OF INTERSTITIAL SOLUTIONS RELATED TO THAT OF SEA WATER

Degens et al. (1964) analyzed the oxygen-isotope composition of a number of connate waters ranging in age from the Cambrian to the Tertiary, and reported that the $\delta^{18}O$ values of the highly saline oil-field brines do not deviate appreciably from the $\delta^{18}O$ of modern sea water (also see Degens and Chilingar, 1967). Deviation from the mean value in some of the samples into the negative range of $\delta^{18}O$ are always well correlated to a decrease in salinity. This feature can be easily explained by effects of dilution with meteoric waters

---

[1] There are two Baumé scales in use: One is for liquids heavier than water and the other for liquids lighter than water. The expression for "lighter" liquids is:

$$°Bé = \frac{140}{\text{sp.gr. @ } 60°F} - 130$$

and for "heavier" liquids:

$$°Bé = 145 - \frac{145}{\text{sp.gr. @ } 60°F}$$

The specific gravity is the ratio of the weight of a unit volume of fluid to the weight of the same volume of distilled water at 60°F.

during migration of brine, or by subsequent infiltration caused by a change in the geological setting through uplift, denudation, etc.

Graf et al. (1965, 1966) studied the isotopic fractionation by shale micropore systems and the origin of calcium-chloride waters in the Illinois and Michigan basins. A small isotopic fractionation resulting from passage of water through micropores in shales is postulated as an explanation for the regularity remaining between deuterium and ^{18}O contents after both have been normalized. The concentration of deuterium was normalized by Graf et al. (1965) to the value for meteoric precipitation at Chicago, and ^{18}O was normalized to the value at 25°C for equilibrium with pre-Tertiary marine limestones. If the rate of mixing of the brine behind each shale barrier is small relative to the rate of passage through the shale, the heavy fraction left behind from a given unit of brine will be pushed through the next time. Thus, not much more than a single-stage fractionation will be observed for each shale barrier. The expulsion of interstitial fluids by compaction from strata not receiving recharge is slow enough to permit significant diffusional mixing of the fluids remaining. Thus, a large δD micropore effect would build up in that strata, but the $\delta^{18}O$ effect would decay by equilibrium with the wall rock (Graf et al., 1965, p.22). Consequently, one would expect that an increase in deuterium content would be greatest at depth in a basin, although flushing of the fluids could occur with time.

The similarity between the isotope characteristics of brines and modern sea water

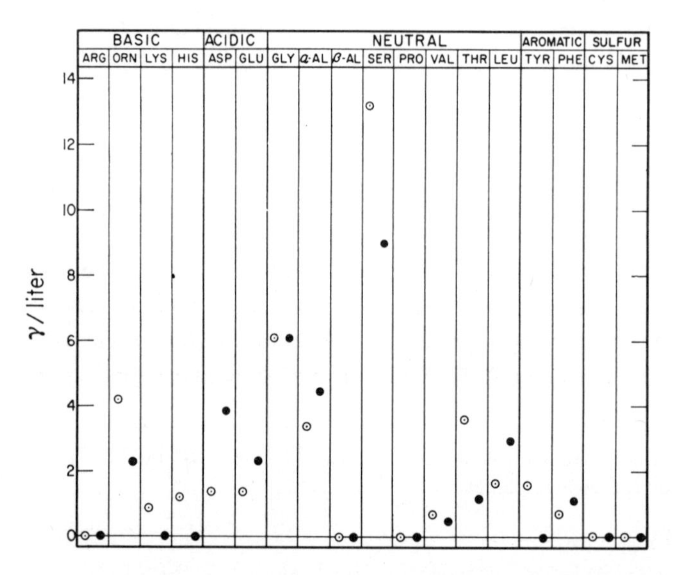

Fig.116. Comparison of amino-acid distribution in petroleum brine waters of Paleozoic age and modern sea water ($1\gamma = 1\mu g$). ⊙ = Ocean water; • = petroleum brine water, adjusted to salinity of mean ocean water (3.5%). Abbreviations: *AL* = alanine; *ARG* = arginine; *ASP* = aspartic acid; *CYS* = cystine; *GLU* = glutamic acid; *GLY* = glycine; *HIS* = histidine; *LEU* = leucine; *LYS* = lysine; *MET* = methionine; *ORN* = ornithine; *PHE* = phenylalanine; *PRO* = proline; *SER* = serine; *THR* = threonine; *TYR* = tyrosine; *VAL* = valine. (After Degens et al., fig.12. p.216, 1964.)

suggests that the concentration of inorganic salts has not been accomplished by syngenetic evaporation in most cases studied. Consequently, other possible reasons were sought by the writers and are discussed later in this chapter. Slight deviations into the positive range of $\delta^{18}O$ values in some samples studied may have been caused by original evaporation in a surface environment, or by isotope equilibration with the surrounding mineral matter for millions of years (Degens and Epstein, 1962).

The concentration of amino acids in oil-field brine waters is a function of salinity (Degens et al., 1964), i.e., the content of amino acids increases with increasing salt concentration. On adjusting the salinity of brine waters to that of present-day oceans and applying the same calculation factors to the original amino-acid values, the similarities between the amino-acid spectra in the Recent sea water and fossil brines is striking (Fig.116).

SOLUBILITY AND MOBILITY OF VARIOUS SALTS AND IONS

The solubilities of various salts in distilled water at 18°C are presented in Table XXVIII. The solubility of NaCl does not change appreciably with increasing temperature: 371 g/l at 60°C, for example, as compared to 358.6 g/l at 18°C.

The solubilities of Na_2SO_4 and Na_2CO_3 increase appreciably with increasing temperature: for example, 453 and 466 g/l, respectively, at 60°C as compared to 168.3 and 193.9 g/l at 18°C. On the other hand, the solubility of $CaSO_4$ at 100°C is equal to 1.5 g/l as compared to 2.01 g/l at 18°C (see Table XXVIII), namely, it decreases with increasing temperature.

The solubility of salt decreases in the presence of other salts having an ion in common, whereas it increases when there is no common ion effect. The solubility of $CaSO_4$, for example, increases four-fold in the presence of NaCl: at a NaCl concentration of 1.3018 mole/l, solubility of $CaSO_4$ is equal to 0.853 g/100 ml of solution. In the presence of Na_2SO_4, the solubility of $CaSO_4$ initially decreases to 1.4 g/l (at 16.3% concentration of Na_2SO_4), and then increases to 2.6 g/l (at 222.6 g/l of Na_2SO_4). Solubility of $CaSO_4$ drops to 0.5 g/l in the presence of $MgSO_4$, whereas $MgCl_2$ increases $CaSO_4$ solubility. On

TABLE XXVIII
Solubilities of various salts in distilled water at 18°C
(After Posokhov, 1966, p.165)

Salt	*Solubility* (g/l)	*Salt*	*Solubility* (g/l)	*Salt*	*Solubility* (g/l)
$MgCl_2$	558.1	Na_2SO_4	168.3	Na_2CO_3	193.9
$CaCl_2$	731.9	$MgSO_4$	354.3	$MgCO_3$	0.022
NaCl	358.6	$CaSO_4$	2.01	$CaCO_3$	0.013

the other hand, the solubility of $CaSO_4$ markedly decreases in the presence of $CaCl_2$: in the presence of 7.3 g/l of $CaCl_2$ at 25°C, the $CaSO_4$ solubility is lowered to 1.3 g/l; in the presence of 280 g/l of $CaCl_2$, the $CaSO_4$ solubility constitutes only 0.2 g/l (Stankevich, 1959). Other interrelationships between solubilities have been discussed by Posokhov (1966).

The mobility of hydrated ions depends on their valence and ionic radius (Posokhov, 1966, p.157):

(a) highest $NO_3^- > Cl^- > SO_4^{2-} > CO_3^{2-}$ lowest
(b) highest $K^+ > Na^+ > Ca^{2+} > Mg^{2+}$ lowest

The above order, however, is only theoretical. It changes with the nature of the medium (pH, adsorption processes, etc.).

In Table XXIX, data on radii of hydrated and non-hydrated ions, their ionic potential, and polarization (= valency/ion radius ratio) are presented. The size of the ions should largely determine the mobility coefficient through the clay–shale membrane and the replacement coefficient (base exchange). With increasing polarization, the replacing power of an ion decreases. The data on ionic potential is of value because elements having low ionic potentials are most likely to remain in true solution.

The experimental data obtained by Larsen (1967) tend to indicate that the size of hydrated ions influences the rate of ion rejection by a reverse osmosis process and the order should be $Mg^{2+} > Ca^{2+} > SO_4^{2-} > Na^+ > Cl^-$. This order, however, was not followed in the natural brines studied by Collins (1970). According to Collins (1970), "the tremen-

TABLE XXIX
Radii, ionic potentials, and polarization of various ions
(See Nightingale, 1959)

Ion and valence	*Nonhydrated radius* (Å)	*Hydrated radius* (Å)	*Ionic potential*	*Polarization (valency/ non-hydrated ion radius)*
Barium (+2)	1.35	4.04	0.68	1.48
Boron (+3)	0.23	–	0.08	–
Bromide (–1)	1.95	3.30	1.95	0.51
Calcium (+2)	0.99	4.12	0.50	2.02
Chloride (–1)	1.81	3.32	1.81	0.55
Iodide (–1)	2.16	3.31	2.16	0.46
Lithium (+1)	0.60	3.82	0.60	1.67
Magnesium (+2)	0.65	4.28	0.33	3.08
Potassium (+1)	1.33	3.31	1.33	0.75
Sodium (+1)	0.95	3.58	0.95	1.05
Strontium (+2)	1.13	4.12	0.57	1.77
Sulfate (–2)	2.90	3.79	1.45	0.69

dous concentration of barium and iodide in the natural brines, along with depletion of sulfate, magnesium, and potassium, indicates that these brines probably were not subjected to a clay–shale reverse-osmosis process; they were severely altered before or after subjection to such a process; the clay–shale reverse osmosis is radically different from experimental desalinization reverse-osmosis processes; or the waters were altered during diagenetic reactions." He concluded that dolomitization probably was important in controlling the amounts of Ca^{2+}, Mg^{2+}, and Sr^{2+} in Louisiana brines; and the enrichment of barium and iodide was due to exchange and leaching reactions.

BRIEF REVIEW OF VARIOUS PROCESSES AFFECTING CHEMICAL COMPOSITION OF INTERSTITIAL WATERS

There are systematic chemical differences (both qualitative and quantitative) between the ancient and modern connate waters. Magnesium, which is abundant in sea water, is present only in minor amounts in oil-field waters and the opposite is true for calcium. Calcium-chloride waters, which are not formed in surface environment, are widespread among oil-field brines. This feature can possibly be linked to the dolomitization process.

Magnesium may also replace various cations in chlorites and clay minerals. Adsorption and exchange phenomena may explain variations in K/Na and Ca/Na ratios in fossil and modern interstitial solutions; the former ratio also appears to be temperature dependent (White, 1965, p.359). It should be remembered, however, that the significance of base exchange decreases with depth and, therefore, could not account for the formation of deeply buried calcium-chloride waters. Graf et al. (1966) noted that the calcium content of the Michigan Basin brines is very high, greater than that of sodium in some fluid samples, and increases proportionately with increase in total solids content. A definite relationship appears to exist between calcium and total solids content in oil-field brines of various geological ages (Figs.117 and 118). Graf et al. (1966) felt that the more obvious geologic processes are inadequate as explanations of the origin of concentrated calcium-chloride brines, which occur in geosynclines free of major orogeny. Two simple models were proposed by Graf et al. (1966) for deriving such a composition: (1) by shale ultrafiltration of the dissolved solids contained in the original sea water, and (2) by mixing various proportions of fresh water and sea water.

Field data from various formations showing relationships between the Na/Ca ratio and the total solids are presented in Fig.119. In this figure brines from formations older than Pliocene lie generally parallel to the *Y*-axis (total solids), whereas brines from younger formations or marine sediments lie more or less parallel to the *X*-axis (Na/Ca ratio). In some cases there is a tendency for the Na/Ca ratio to increase with increasing total solids (generally older formations), whereas in others (generally younger formations) the reverse is true. It is of interest to note that the interstitial waters of the Sespe Formation have a closer affinity with the more "stabilized" brines than the interstitial waters from the

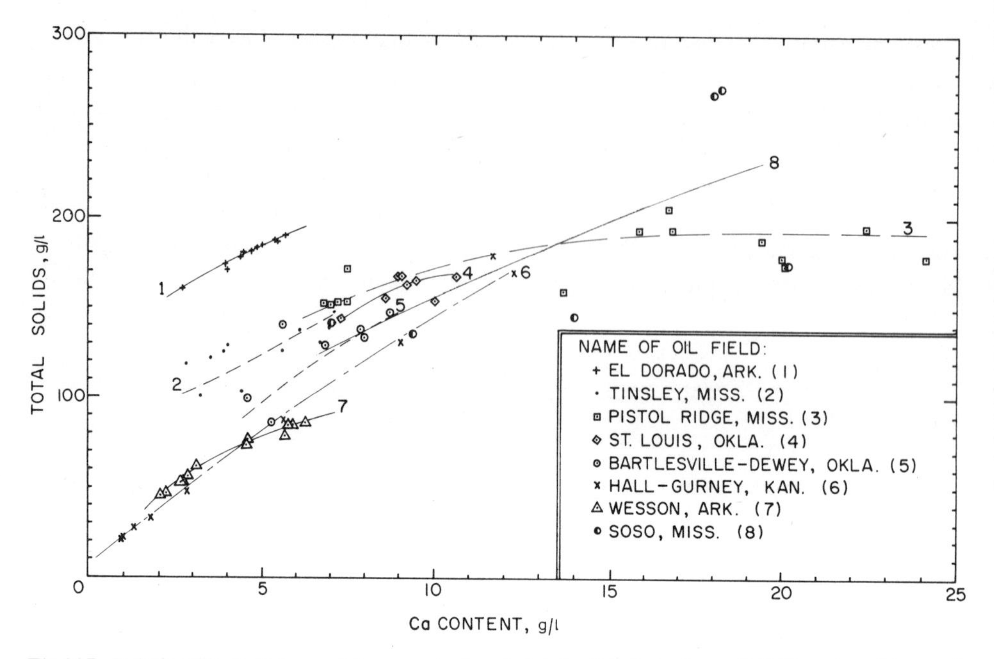

Fig.117. Relationship between calcium content and total solids in various oil-field waters. Name of field, sampled interval, and geologic age: (1) El Dorado East, 2,150–3,172 ft., Upper Cretaceous; (2) Tinsley, 4,797–5,770 ft., Upper Cretaceous; (3) Pistol Ridge, 7,423–10,964 ft., Upper Cretaceous; (4) St. Louis, 2,610–3,174 ft., Ordovician–Pennsylvanian; (5) Bartlesville–Dewey, 1,241–2,557 ft., Cambrian–Pennsylvanian; (6) Hall–Gurney, 2,610–3,293 ft., Ordovician–Pennsylvanian; (7) Wesson, 2,132–3,598 ft., Lower Cretaceous–Upper Cretaceous; and (8) Soso, 6,498–12,045 ft., Lower Cretaceous–Upper Cretaceous. (Data from Rall and Wright, 1953; Wright et al., 1957; Hawkins et al., 1963a, 1963b.)

younger Pico Formation; the stratigraphic distance between the Sespe and Pico Formations in Ventura County, California, is approximately 20,000 ft.

The distribution of Cl^- with depth may be affected by the proximity of salt, which should promote higher salt concentrations in pore fluids. For example, Manheim and Sayles (1970) observed marked increases in interstitial salinity in two drill holes located in the Gulf of Mexico at a water depth of more than 3,500 m. They attributed these increases to diffusion of salt from buried evaporites. In one hole, however, on penetrating the oil-permeated cap rock of a salt dome, they encountered fresh water, which could have formed during oxidation of petroleum hydrocarbons and decomposition of gypsum to form native sulfur.

The diffusion can be expressed by the following formula:

$$\frac{\partial C}{\partial t} = -K\frac{\partial^2 C}{\partial x^2} \tag{5-1}$$

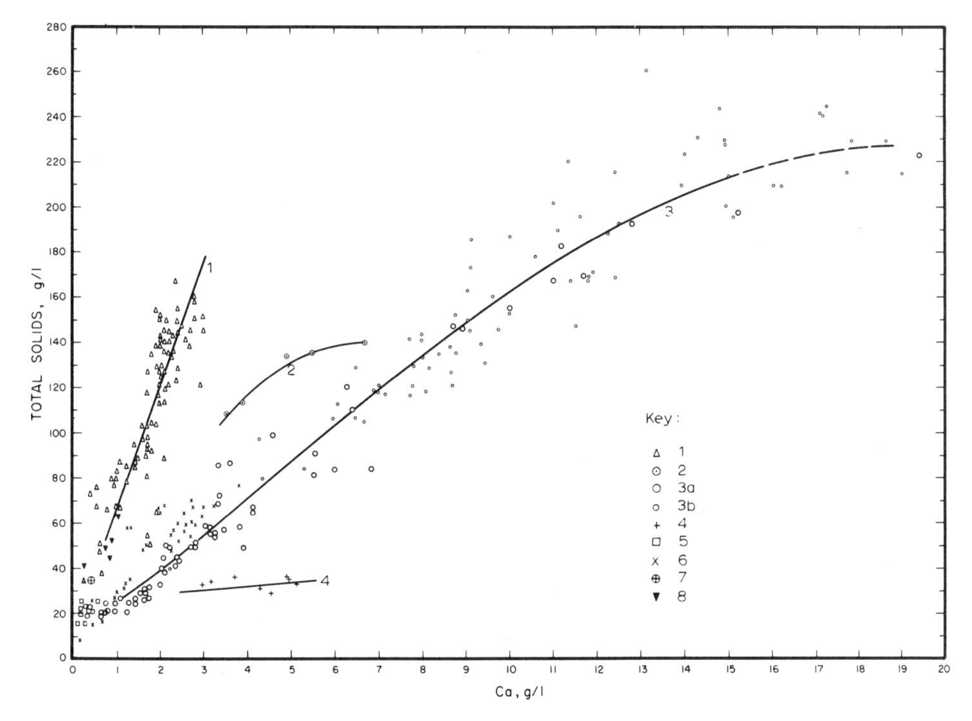

Fig.118. Relationship between calcium and total solids content in various oil-field brines, each curve representing a different formation except curve 3. *1* = Wilcox Formation, Eocene, Gulf Coast; *2* = Ste. Genevieve Limestone, Mississippian, Illinois Basin; *3a* = Arbuckle Limestone, Cambrian–Ordovician, Kansas; *3b* = Bartlesville Sandstone, Pennsylvanian, Oklahoma; *4* = Sespe Formation, Oligocene, California; *5* = Pico Formation, Pliocene, California; *6* = Nacatoch Sandstone, Cretaceous, Texas–Louisiana; *7* = sea water; *8* = laboratory data obtained by the writers during compaction experiments. (Data from Rall and Wright, 1953; Wright et al., 1957; Gullikson et al., 1961a; Hawkins et al., 1963a, 1963b, 1964; and Graf et al., 1966.)

where C = concentration, t = time, K = diffusion constant, and x = distance from the salt.

Posokhov (1966) discussed in detail many of the factors which affect the chemical composition of underground waters: (1) physiogeographic, (2) geologic, (3) hydrogeologic, (4) biologic, (5) physical, and (6) physicochemical. Under physicochemical factors Posokhov discussed: (a) oxidation–reduction conditions of underground waters, (b) solubility of salts, (c) diffusion, (d) osmosis, (e) gravitational differentiation, (f) mixing of different waters, and (g) base exchange.

Clay particle – fluid interface

The interface between the interstitial fluid and clay particles is composed of an adsorbed water layer containing ions. The nature of adsorbed water has long been a subject of speculation and experimentation. Most investigators are in agreement that the water

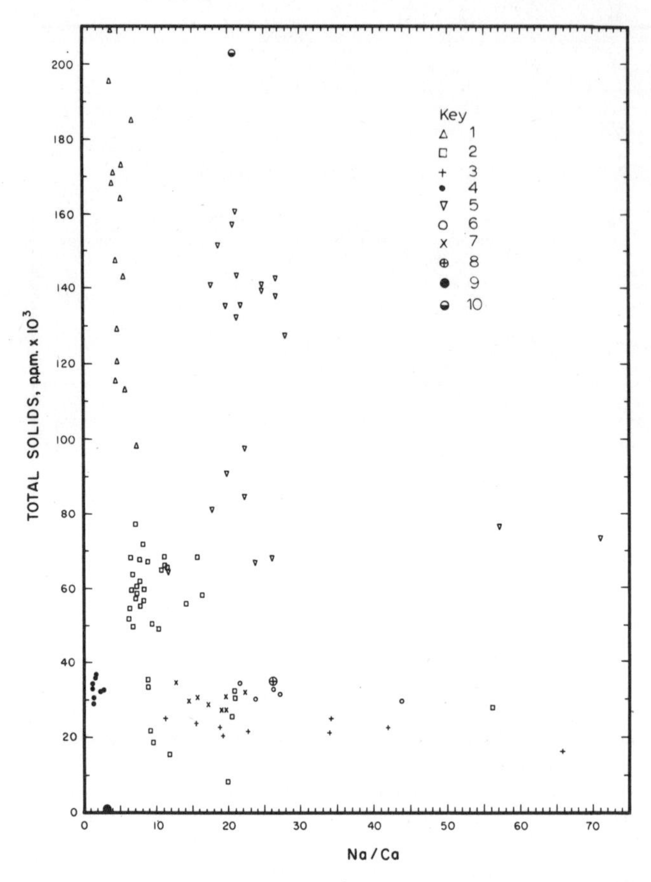

Fig.119. Relation between the Na/Ca weight ratio and total solids in oil-field brines. *1* = Bartlesville Sandstone, Pennsylvanian, Oklahoma; *2* = Nacatoch Formation, Upper Cretaceous, Texas–Louisiana; *3* = Pico Formation, Pliocene, California; *4* = Sespe Formation, Oligocene, California; *5* = Wilcox Formation. Eocene, Gulf Coast; *6* = Recent marine sediments (Siever et al., 1965); *7* = Wilmington Oil Field, California. (Data after Wright et al., 1957; Gullikson et al., 1961a; Hawkins et al., 1963a, 1964.) *8* = sea water; *9* = river water; and *10* = Great Salt Lake (Bentor, 1961).

adsorbed on the surface of the clay particle differs in its physical and chemical properties from normal liquid water (Grim, 1968, p.234). There is little agreement, however, on the manner or magnitude of the deviation between the adsorbed and free water in clays. There are two main viewpoints: (1) existence of a fixed fluid layer, where the water structure is similar to that of ice (quasi-crystalline solid), or (2) a two-dimensional fluid in which each bond between the water molecules in the adsorbed layer is stronger than in normal water.

In the fixed fluid-layer theory, each bond between the clay surface and water molecule is stronger than the bonds between normal water molecules. This bond strength is due to the negatively charged clay surface and the positively charged hydrogen ions in the water molecule. Such an adsorbed, quasi-crystalline water layer resists both normal and shear stresses better than normal water.

According to the two-dimensional fluid theory, the number of bonds that can form between the polar water molecules is considerably less than the number of bonds between water molecules and the clay particle (see Martin, 1960, p.110). There is a greater resistance of the adsorbed water layer to normal stress. It is believed that the maximum bond strength is developed when a clay-water system contains the maximum amount of water that can develop a definite configuration (Grim, 1968, p.248). Grim (1968, p.246) accepted the quasi-crystalline viewpoint. Martin (1962) stated that the published data appear to support both viewpoints, and all of the presented data could be contested for various reasons; however, he favored the two-dimensional-liquid model. In the opinion of the writers perhaps both mechanisms may be operative; however, it appears that the adsorbed water structure is generally similar to that of ice (see Grim, 1968, p.246).

Double-layer theory

It is assumed that when colloidal clay particles are hydrated in water, two solution phases are formed: (1) saturating (hydrating) phase, and (2) the internal phase. The clay lattice possesses a net negative charge on its plate-like flat surface which is a result of isomorphous substitution of ions into the lattice. Positive charges are apparently active at the edges of the plates. The internal phase is strongly influenced by the fixed charge on the clay surface. The distribution of water and ions between the two phases can be thought of as constituting an electrical double layer which surrounds the clay particle. One layer is represented by the negative charges on the surface of the clay, whereas the second layer is composed of oppositely charged cations in the water. The second layer is a consequence of the requirement of electro-neutrality, which states that if charged colloidal particles are present in a system, oppositely charged ions also must be present in an equivalent amount (Mysels, 1959, p.299). Several models, described below, have been developed to explain the electrolyte-ion behavior and distribution within the double layer surrounding a colloidal clay particle.

Helmholtz double-layer model (rigid double layer). The Helmholtz model theory states that all the neutralizing positive ions of the double layer, approximately one molecule thick, are in a single plane as shown in Fig.120B. The electrical charges are uniformly distributed in the layer and on the solid. This theory proposes that the layer of water molecules in contact with the clay platelet is immobile, whereas the remainder of the water molecules in the double layer are mobile.

Gouy double-layer model (diffuse double layer). Gouy (1910) believed that a rigid array of charges of fixed thickness at the particle surface could not exist. He proposed that the positive charges in the Helmholtz double layer would attract charges of opposite sign.

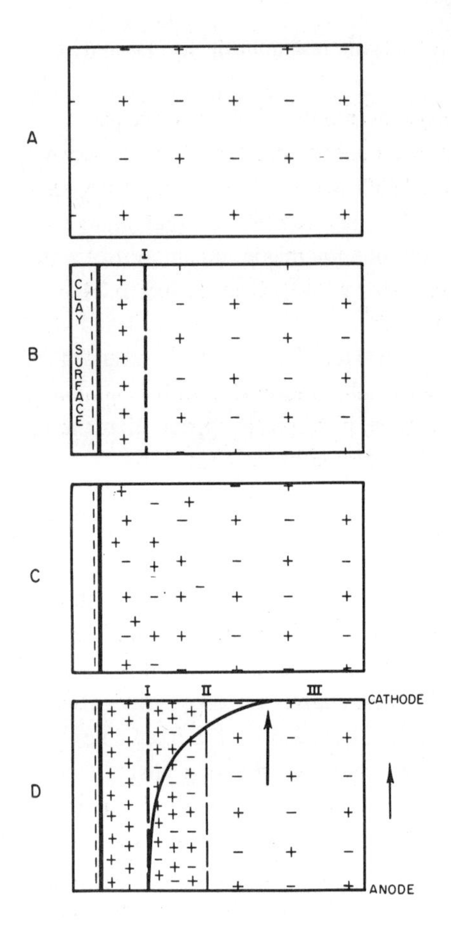

Fig.120. Schematic diagram of the writers' interpretation of the various electric double-layer models. A. Interstitial solution undisturbed by a charged clay surface. B. Helmholtz double layer in a capillary. *I* = fixed part of the double layer. C. Gouy double layer in a capillary. D. Stern double layer in a capillary. *I* = fixed part of the double layer; *II* = mobile part of the double layer; *III* = free interstitial fluid. Arrow points in direction of the interstitial fluid movement on application of direct electric current, whereas the solid curved line indicates velocity profile.

This attraction results in a non-uniform distribution of charges in the liquid near the solid–liquid interface. The ions have a tendency to diffuse away towards the bulk of the solution. This is analogous to the situation in the earth's atmosphere. If thermal agitation is present, it will prevent the formation of the compact double layer. Near the clay's surface, the ionic atmosphere is fairly dense, whereas at greater distances from the surface, its density decreases until the net charge density is zero. All the ions within the outer layer are mixed; however, the cations predominate (Fig.120C).

Stern double-layer model (combination double layer). This theory is a combination of the other two above-described models. According to the Stern theory, a layer of positive ions, that are rigidly adsorbed on the clay's surface, exists immediately adjacent to the negatively charged particles (Fig.120D). This rigidly adsorbed layer is approximately one ion thick. The Gouy electrical diffuse layer lies adjacent to the rigidly adsorbed one. The mobile positive ions in the diffuse layer are shielded from the negative surface and thus can move in an external field.

Chemical equilibrium between the negatively-charged clay particle and liquid must be maintained at all times. The types and concentrations of ions present in the diffuse portion, however, are thought to vary continuously, depending upon the concentration of the external liquid and the pH. If new ions are introduced into the external fluid, a portion of them will replace ions that were previously within the diffuse double layer. The Stern layer concept is important in formulating the theory of "electrodiagenesis".

Electrodiagenesis

The effect of current flow and fluid movements, resulting from natural electric potentials, on chemistry of interstitial solutions is not well understood, and considerable amount of research still remains to be done in this field. For example, many different minerals form as a result of applying direct electric current to marine muds containing various electrolytes: allophane, allophanoid, calcite, gypsum, hematite, hisingerite, hydrogoethite (hydrolepidocrosite), hydrohematite, limonite, magnetite, natron ($Na_2CO_3 \cdot 10H_2O$), nontronite, and trona (Rieke et al., 1966; Serruya et al., 1967). Thus, "electrodiagenesis" may change the chemistry of interstitial solutions.

Amba (1963) observed brine migration in capillary tubes and porous media caused by the tangential movement between the fixed and mobile portions of the Stern double layer, when an electric potential was applied. The cations in the mobile layers drag the trapped water molecules and add viscous movement to the bulk liquid as they migrate toward the cathode (Adamson, 1966).

Movement of fluids as a result of compaction could create electrical potentials which, in turn, may be responsible for selective migration of ions (Serruya et al., 1967). Borovitskiy (1969) reported that on evaporation of water from the soils, the resulting electrical potential may be higher than 250 mV. The magnitude of electrical potentials in sediments should be investigated both in the field and in the laboratory. In the presence of electro-osmotic forces (Adamson et al., 1966), the anions will move towards the anode, whereas the cations will move towards the cathode. The zone near the anode will have lower pH with resulting solution of carbonates and enrichment in aluminium and iron cations. Calcium bicarbonate may precipitate out as $CaCO_3$ at the cathode with evolution of CO_2. If the water–sediment interface forms the anode, then Cl^- will migrate upwards and concentrate in the upper layers (Borovitskiy, 1969).

Role of gravity and temperature gradients in formation of fossil brines

Dissolved ions tend to settle out just as sediment itself tends to settle out of a water column. Calculations by Mangelsdorf et al. (1970) show that gravitational settling of ions in an isothermal sediment column could produce increases in concentrations of ions in interstitial fluids: 1% per 100 m of depth for chloride and 4% per 100 m of depth for strontium.

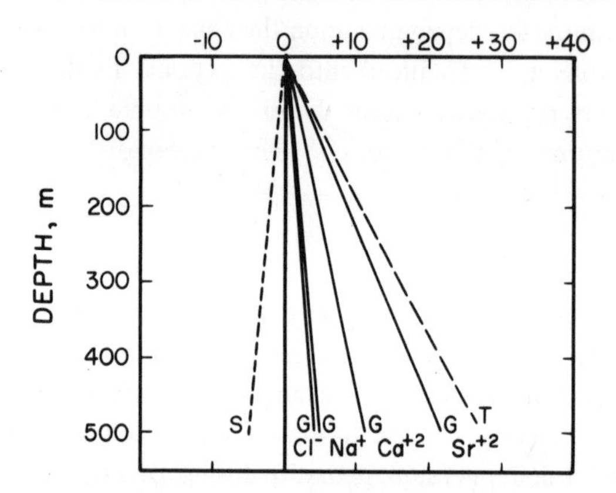

Fig.121. Estimated changes in steady-state concentration of interstitial waters in a sediment column owing to various mechanisms. Starting solution is sea water. G = downward enrichment caused by gravitational settling of specific ions. S = changes due to Soret migration of ions toward cooler pole (thermal gradient = 3.3°C/100 m; 0.01 M NaCl); L = approximate changes owing to the net movement of ions toward hotter pole in presence of cation-exchanging media (thermal pumping). Total enrichment should approximate sum of G and T. (After Mangelsdorf et al., 1970, fig.2, p.624. Courtesy of American Association of Petroleum Geologists.)

Mangelsdorf et al. (1970) pointed out that migration of ions as a result of thermal gradient (Soret effect) could cause minor salt enrichment upward towards the colder region (pole); however, the presence of cation-exchanging particles such as clays would reverse this tendency, namely, the salt will be pumped downward. According to these authors, 5% enrichment in Cl^- per 100 m of depth may occur under steady-state conditions. Estimated changes in equilibrium concentration of pore waters in a sediment column owing to various mechanisms are presented in Fig.121.

FLUID RELEASE MECHANISMS

The gravitational compaction process is considered by many petroleum engineers and geologists to explain the origin of the interstitial fluids and the associated abnormal high pressures in Tertiary geosynclines. Inasmuch as this process is based on principles of soil mechanics, it is simple to understand and, if one chooses, can be easily studied through the use of various mathematical models.

The alternative fluid-release and brine-concentration mechanisms, some of which have not been completely documented and substantiated in the field and/or laboratory, include osmosis, filtration through semipermeable clay membrane, clay-mineral alteration, and chemical cementation.

Semipermeable membrane, osmosis, and reverse osmosis

Although Lindgren (1933, p.133) stated that fine-grained rocks could behave as semipermeable membranes, the review of literature seems to indicate that DeSitter (1947) was the first to suggest that some brines may have been concentrated by the "salt sieving" action of clayey sediments which permitted water to escape, retaining some of the ions of dissolved salts. A semipermeable membrane has pores large enough to permit passage of all small ions, but it is impervious to large ionized colloidal particles.

It is possible that osmosis occurs in sediments. Osmosis can be defined simply as the spontaneous flow of water across a membrane from a dilute solution to a more concentrated solution. Under isothermal conditions, osmotic phenomena will continue to operate across a membrane as long as there is a difference in activity across the membrane. The osmotic-pressure differential at a constant temperature is almost directly proportional to the concentration differential, and for a given concentration differential it increases with the absolute temperature (in: Glasstone, 1946, p.654). The osmotic coefficient of NaCl changes slightly in the temperature range from 25° to 105°C; that of $MgSO_4$ decreases with increasing temperature, whereas that of Na_2SO_4 increases appreciably with increasing temperature. It is important to establish the effect of geothermal temperature upon osmotic pressure in sediments.

Shales separating sands, in which the waters have different salinities, can be osmotic media through which hydraulic gradients can develop in the direction of the more saline sand. In this case the shales cannot be of great thickness in order for this mechanism to operate well. Both field and laboratory data indicate that many cases of anomalous pressures can be caused by the osmotic processes (Hanshaw and Zen, 1965, p.1381). In the case when dissolved components in interstitial fluids are being filtered out by one or more strata rich in clay minerals, the pressure on the influx low-salinity side should be anomalously high under equilibrium conditions. Evaporite beds (source of highly saline brines) are not necessary for establishing the high fluid pressure, but clearly their presence would help.

Bischoff and Ku (1970) studied pore fluids which were extracted from deep-sea sediment cores taken during an east to west traverse from Africa to the Mid-Atlantic Ridge between latitudes 18°N and 20°N. According to them, the constancy of Cl^- in samples of varying clay/$CaCO_3$ ratios negate semipermeable-membrane behavior of compacting clays, at least for the top 10 m of sediment. If any membrane process were active, there should have been large differences in chlorinity between pure clay and calcareous ooze sediments, and this was not the case. The writers, however, would like to point out that a semipermeable membrane effect possibly becomes operative at a certain minimum void ratio, as discussed later.

The effectiveness of clay as a semipermeable membrane-filtering device has been demonstrated by several investigators (Teorell, 1935; Davis, 1955; Hanshaw, 1962; and McKelvey and Milne, 1962). Experiments with argillaceous sediment membranes, containing cation-exchanging clay minerals, and water systems containing NaCl and $CaCl_2$, have shown that water molecules can pass through the membranes, whereas movement of some salts is impeded. The fixed negative charges on the clays apparently repulse the anions as they approach the membrane.

When the porosity in clays is high, apparently the permeability is sufficiently large to allow water and dissolved salts to pass through the strata in proportions approximately equal to their abundance on the "input" side of the clay membrane. On the other hand, a well-compacted shale has low porosity and very low permeability, and the fixed negative charges on adjacent clay-mineral grains are so close together that anions in solution of the "input" side are repelled and cannot escape. White (1965, p.348) stated that the adsorbed cations can still move between adjacent exchange sites and through the membrane. An electrical streaming potential is set up across the membrane and the cations cannot continue to move through the membrane without increasing the electrical imbalance. The uncharged water molecules pass through the membrane in the direction of decreasing water pressure, thus increasing the salt content on the "input" side. McKelvey and Milne (1962) generally found higher upstream- to downstream-salinity ratios (2/1 to 8/1) with compressed bentonite membranes.

White (1965, p.350) noted that the salt solution on the chloride concentrated side of the semipermeable (Donnan) membrane will be acid and on the low-salt side alkaline. This was based on a membrane assumed to be only permeable to Na^+, H^+, and water molecules. The sodium ions diffuse to the low-sodium side, but the chloride ions cannot diffuse through the membrane, whereby an electrical charge across the membrane is set up. Hydrogen ions then diffuse to the high-salt side in response to the potential until equilibrium is attained. Osmotic pressure develops on the high-salt side due to the addition of water. For additional discussion on the subject, the reader is referred to the stimulating paper by White (1965).

Electrochemistry of semipermeable clay membranes. Shale beds in situ may be considered to be ideal membrane electrodes. The best argument for this assumption is the observed constancy of a "shale baseline" on spontaneous-potential logs found in drill holes in every part of the world (Wyllie, 1955). A quantitative theoretical treatment of the electrochemical properties of clays is given by the Teorell–Meyer–Sievers theory of membrane

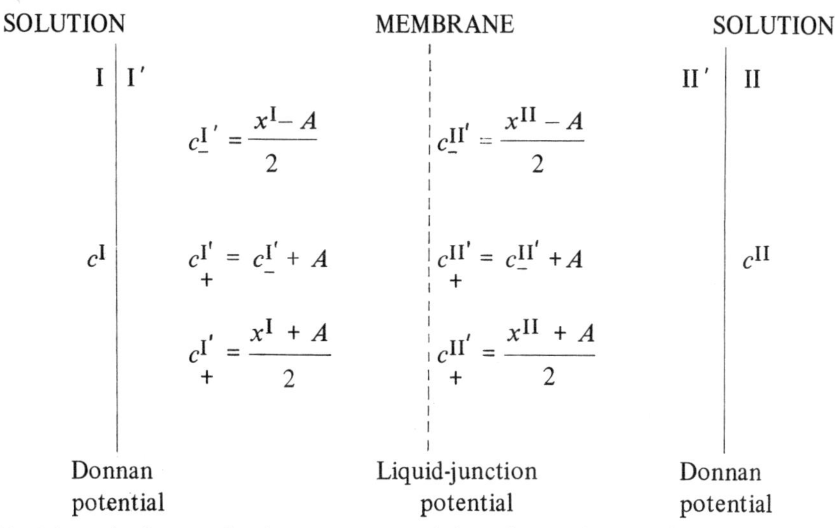

Fig.122. Schematic diagram showing a system consisting of two solutions of electrolytes (*I* and *II*) separated by an intervening membrane. *I'* and *II'* are the peripheral laminae of the membrane.

behavior (Davis, 1955), which, according to calculations based on the spontaneous potential curve of electric logs, approximates the behavior of shales in situ.

Considering a system consisting of two solutions of electrolytes separated by an intervening membrane, which can be represented by the scheme in Fig.122, *I* and *II* lie in the solution phases, whereas *I'* and *II'* are in the peripheral portion of the membrane which is represented only as a cation exchanger. The Teorell–Meyer–Sievers theory, which considers a single symmetrical electrolyte, is based on the following equation derived from Nernst's theory of the diffusion potential:

$$FE_j = RT \int_{I}^{II} \frac{t_+ - t_-}{t_+ + t_-} \, d\ln c \tag{5-2}$$

where F is the Faraday equivalent, E_j is the so-called junction potential, R is the gas constant, T is the absolute temperature, c is the electrolyte concentration, and t is the transference number which represents the relative amount of electricity carried by the ionic species, i, across a given plane. Eq. 5-2 can be expressed in terms of the electrochemical (diffusion) potential E across the membrane:

$$E = \frac{RT}{F} \int_{I}^{II} \frac{t_+ - t_-}{t_+ + t_-} \, d\ln c \tag{5-3}$$

It is determined by the ionic migration velocities, U_i ($U_+ = U_{\text{cation}}$, $U_- = U_{\text{anion}}$), and concentrations, c_i, of the ions at the given phase:

$$t_i = \frac{U_i c_i}{\Sigma_i U_i c_i} \tag{5-4}$$

so that:

$$E = \frac{RT}{F} \int_{\mathrm{I}}^{\mathrm{II}} \frac{U_+ c_+ - U_- c_-}{U_+ c_+ + U_- c_-} \, \mathrm{d} \ln c \tag{5-5}$$

Between the peripheral laminae of the membrane and the adjacent solution phases, i.e., between I and I′ and II and II′ two Donnan equilibria are assumed to exist. Davis (1955, p.49) expressed the thermodynamic equation for the Donnan equilibrium as:

$$(f_\pm^2 c^2)^{\mathrm{I}} = (f_+ c_+ f_- c_-)^{\mathrm{I}'} \quad \text{and} \quad (f_\pm^2 c^2)^{\mathrm{II}} = (f_+ c_+ f_- c_-)^{\mathrm{II}'} \tag{5-6}$$

where f represents activity coefficients. Inasmuch as f_+ and f_- are not measurable, it is assumed that:

$$(f_\pm^2)^{\mathrm{I}} = (f_+ f_-)^{\mathrm{I}'} \quad \text{and} \quad (f_\pm^2)^{\mathrm{II}} = (f_+ f_-)^{\mathrm{II}'} \tag{5-7}$$

so that:

$$(c^2)^{\mathrm{I}} = (c_+ c_-)^{\mathrm{I}'} \quad \text{and} \quad (c^2)^{\mathrm{II}} = (c_+ c_-)^{\mathrm{II}'} \tag{5-8}$$

This mathematical assumption for suspensions of clay in contact with dilute solutions is probably not even approximately valid and, furthermore, one does not know how valid it is for rigid membranes (Davis, 1955, p.49).

If A is the number of fixed unit charges on a negatively-charged membrane, one may state that:

$$c_+ = c_- + A \tag{5-9}$$

where c_+ and c_- refer to the peripheral laminae of the clay membrane at I′ and II′.

The integral of eq.5-5 represents the sum of two external integrals, and an internal integral within and extending across the clay membrane, shown in Fig.122. The Donnan potential I–I′ is expressed as:

$$E_{\mathrm{I-I'}} = \frac{RT}{F} \ln \frac{c^{\mathrm{I}}}{\frac{1}{2}(x^{\mathrm{I}} + A)} \tag{5-10}$$

The liquid potential I′–II′ is expressed as:

$$E_{\mathrm{I'-II'}} = \frac{RT}{F} u \ln \frac{x^{\mathrm{I}} + uA}{x^{\mathrm{II}} + uA} \tag{5-11}$$

The Donnan potential II–II′ is expressed as:

$$E_{\mathrm{II-II'}} = -\frac{RT}{F} \ln \frac{c^{\mathrm{II}}}{\frac{1}{2}(x^{\mathrm{II}} + A)} \tag{5-12}$$

where

$$x = (4c^2 + A^2)^{1/2} \quad \text{and} \quad u = \frac{U_+ - U_-}{U_+ + U_-}$$

The total potential across the clay membrane is thus:

$$E = \frac{RT}{F} \left[\ln \frac{c^{\mathrm{I}}}{c^{\mathrm{II}}} \frac{x^{\mathrm{II}} + A}{x^{\mathrm{I}} + A} + u \ln \frac{x^{\mathrm{I}} + uA}{x^{\mathrm{II}} + uA} \right] \tag{5-13}$$

Eq.5-13 has the mathematical properties that for $A \gg c^{\mathrm{I}}$ or c^{II}, the last term approaches zero, whereas the first term approaches $\ln(c^{\mathrm{I}}/c^{\mathrm{II}})$. Thus, the equation for a "perfect" electrode (Nernst equation) is obtained:

$$E = \frac{RT}{F} \ln \frac{c^{\mathrm{I}}}{c^{\mathrm{II}}} \tag{5-14}$$

If $A \ll c^{\mathrm{I}}$ or c^{II}, then the first term approaches zero, whereas the last term approaches $u \ln(c^{\mathrm{I}}/c^{\mathrm{II}})$. Consequently, eq. 5-13 reduces to the ordinary liquid-junction potential:

$$E = \frac{RT}{F} u \ln \frac{c^{\mathrm{I}}}{c^{\mathrm{II}}} \tag{5-15}$$

These potentials, therefore, represent the upper and lower possible limits. The larger the absolute value of A, the higher the permissible activities (concentrations), which may be used before the developed potential seriously diverges from the Nernst potential. From electric-log calculations, one obtains eq.5-14 and not eq.5-15; therefore, the conclusion is that shales function as perfect membranes. Wyllie (1955, p.291) found that a great majority of shale specimens gave rise to potentials, when separating solutions of high ionic strength, considerably below the theoretical Nernst potential. Wyllie's laboratory data indicate that shale specimens do not act as ideal membranes through which electric current is effectively transmitted only by cations.

McKelvey and Milne (1962) studied the salt-filtering ability of compacted bentonite clay membranes under fluid injection pressures of 5,000–10,000 p.s.i. The primary aim of their investigation was to demonstrate experimentally the salt-filtering phenomenon in the case of clay materials. Preliminary experimental results indicate that compaction pressures of 10,000 p.s.i. would be required to attain porosities (24–41%) at which salt filtering would be significant (McKelvey and Milne, 1962, p.250).

Diagenetic clay mineral alteration

Powers (1967) proposed a new fluid-release mechanism based on the current knowledge of colloidal clay chemistry and clay mineralogy. In the presence of potassium, montmorillonite releases its intracrystalline water and becomes illite in a true mineral alteration.

Montmorillonite adsorbs far more water, primarily as interlayer water, than do the less expanding clays such as illite. During burial, the last few monomolecular layers of oriented water adsorbed on montmorillonite surfaces are not expelled by overburden pressures. Along the Louisiana–Texas Gulf Coast at a depth of about 6,000 ft., montmorillonite comprises approximately 60% of the clay volume. The diagenesis of montmorillonite to illite and mixed layer clays begins at a depth of 8,500 ft. and is fairly complete at a depth of 15,000 ft., where only 20% of the clay volume is montmorillonite (Jones, 1968, p.5). The change of montmorillonite to illite, caused by the adsorption of potassium, gives rise to the release of large amounts of bound water from the interlayer surfaces to interparticle areas, where it becomes normal pore water. A corresponding decrease in the clay particle size occurs along with an increase in the effective porosity, permeability, and free pore-water content. As mentioned before, most investigators agree that water adsorbed on clay surfaces differs in structure and physical properties from bulk liquid water. Unfortunately, there is considerable uncertainty concerning the direction and magnitude of the property deviations from bulk liquid water. According to several investigators (in: Martin, 1962, p.32), the last few layers of oriented water on the Na-montmorillonite surface have abnormally high densities. These densities may exceed 1.4 g/cm^3 (in: Martin, 1962), which is believed by some investigators to be caused by molecular forces between the unit layers of montmorillonite. Possibly, when the oriented water is released, it undergoes a density decrease with a corresponding increase in volume; this may produce the abnormally high fluid pressures observed in Tertiary basins (Burst, 1959). Powers (1967) felt that this process is universal wherever montmorillonite has been buried, and that it is mainly controlled by temperature rather than by overburden pressure or time. This point, however, can be argued on the basis of thermodynamic principles. Jones (1968) stated that the over-all recharging mechanism is keyed to the geothermal temperature regime. The waters released from the shales are less saline, less dense, and less viscous than waters of associated sands in the same geothermal environment. (Along the Gulf Coast the temperature ranges from 200° to 250°F at depths of 10,000–12,000 ft.) The alteration of montmorillonite to illite is also discussed in Chapter 7.

Chemical cementation

Coleman (1967) suggested that dewatering of buried clays is caused by chemical cementation. He studied some superficial deposits (depth of 150 ft.) of the Atchafalaya Basin (Louisiana Gulf Coast) in which there were no apparent differences in the thickness of the clay layers in the upper and lower cores that could be attributed to compaction. These clay layers show an increase in the compressive strength, degree of dewatering, and quantity of cementing agents with depth of burial. Most soil engineers have interpreted these effects to be a result of compaction caused by an overburden load. Ho and Coleman (1969) believed that Meade's explanation (Meade, 1963) of the increase in strength owing to the amounts of exchangeable cations is also an oversimplification of the problem. Their data shows that in addition to gravitational compaction with depth of burial in the Recent sediments of the Atchafalaya Basin, there is a gradual chemical cementation and gradual replacement of the pore-water space by secondary minerals ($CaCO_3$, Fe_2O_3, $FeCO_3$, and Mg and Mn compounds of unknown nature). This could be an important factor in the dewatering of argillaceous sediments.

EFFECT OF COMPACTION ON THE CHEMISTRY OF SOLUTIONS SQUEEZED OUT OF CLAYS AND MUDS – EXPERIMENTAL DATA AND THEORETICAL ANALYSIS

Large amounts of water are squeezed out of the continental and marine sediments during compaction and lithification. The overburden pressures on these sediments may reach magnitudes of 14,000–36,000 p.s.i. in geosynclinal basins. It seems that most of the salts present in the waters, which are trapped during sedimentation, are squeezed out

TABLE XXX
Mineralization and content of various ions in solutions squeezed out at different overburden pressures from montmorillonite clay (no.25, Upton, Wyo.) saturated with sea water
(After Rieke et al., 1964)

Overburden pressure (p.s.i.)	*Percentage of the concentration in solution squeezed-out at 100 p.s.i.*					
	Cl^-	Na^+	Ca^{2+}	Mg^{2+}	SO_4^{2-}	*total mineralization*
100	100	100	100	100	100	100
400	91–95	93–95	75–84	–	84–95	–
1,000	70–83	84 *	67 *	80 *	–	–
3,000	40–82	25(?)–87	50–62	60 *	67–81	–
10,000	36–61	–	–	–	–	–
40,000	36 *	37 *	25 *	–	38 *	–
90,000	–	–	–	–	–	20 *

* Only one trial.

TABLE XXXI
The percentage increase in the resistivity of solutions squeezed out of marine mud with increasing overburden pressure
(After Rieke et al., 1964)

Overburden pressure	*Percentage increase in resistivity as compared to resistivity of solution squeezed out at 500 p.s.i.*
1,000	2.3– 6.5
2,000	3.5–15.2
3,000	10.5–19.6
7,000	16.3–32.9
14,000	18.6–37.0
30,000	23.2–45.6
40,000	25.6–48.0

during the initial stages of compaction. The laboratory results obtained by Buneeva et al. (1947), Kryukov and Komarova (1954), Kryukov and Zhuchkova (1963), and Rieke et al. (1964) showed that mineralization of squeezed-out solutions progressively decreases with increasing overburden pressure.

Data obtained by Rieke et al. (1964) are presented in Tables XXX and XXXI, and Fig.123, which show that the mineralization of solutions squeezed out during the different stages of compaction is a function of overburden pressure. The degree of reduction in mineralization and the content of various ions in solutions squeezed out at higher overburden pressures from montmorillonite clay (API no.25) saturated with sea water is presented in Table XXX and Fig.123.

Rieke et al. (1964) determined the percentage increase in the resistivity of expelled solutions from marine mud with increasing overburden pressure (Table XXXI). The mud was obtained from the Santa Cruz Basin, off the coast of southern California. These results support the finding of Kryukov et al. (1962, p.1365) that the mineralization of squeezed-out solutions changes with pressure.

In some experiments conducted by Rieke et al. (1964), the percentage decrease in concentrations of the principal cations and anions with increasing pressure was about the same. This suggested that (1) the ions being removed represent interstitial electrolyte solution and do not include the adsorbed cations, and (2) the analysis for a single ion in the effluent (for example, Cl^-) might reveal as much as the analysis for all of the ions.

The results of Kryukov and Zhuchkova (1963) which are presented in Figs. 124–126, also demonstrate that the concentration of solutions squeezed out at higher pressures is lower. Apparently, the last portions of water (adsorbed?) squeezed out of sediments are poor in electrolytes. According to Manheim (1966), the threshold pressure for chloride in 0.86 N NaCl solution was about 20,000 p.s.i. (also in Na-bentonite), whereas for fresher waters the influence of pressure on composition was noted at lower pressures. For ordinary sediments, the pressure threshold for influence on the composition of water is

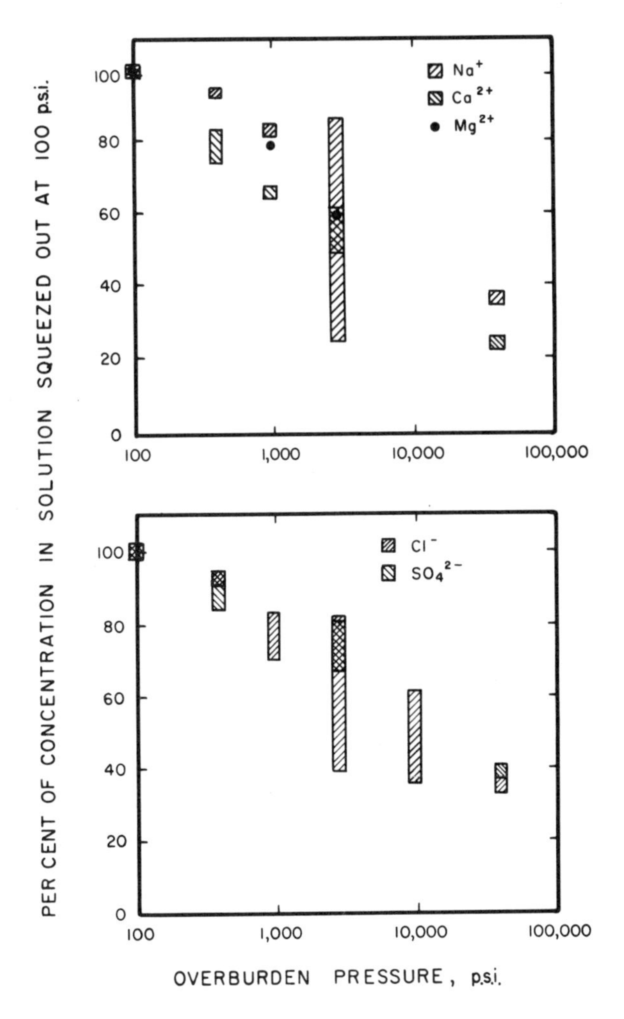

Fig.123. Content of various cations and anions expelled at different overburden pressures from a sea-water saturated montmorillonite clay (A.P.I. no.25, Upton, Wyo.). (After Rieke et al., 1964.)

shifted to higher pressures (Kryukov and Zhuchkova, 1963). According to Chilingarian and Rieke (1968), the chemistry of squeezed-out solutions begins to change appreciably when the remaining moisture content is about 20–25% for kaolinite and about 50–70% for montmorillonite clay.

Kazintsev (1968, p.186) observed a gradual decrease in chlorine concentration on squeezing a sample of Maykop clay (Eastern Pre-Caucasus) having an initial moisture content of 20–25%; the final moisture content after compaction constituted 8.83–10.88% (Fig.127).

Kazintsev also determined the effect of temperature (heating to 80°C) on concentra-

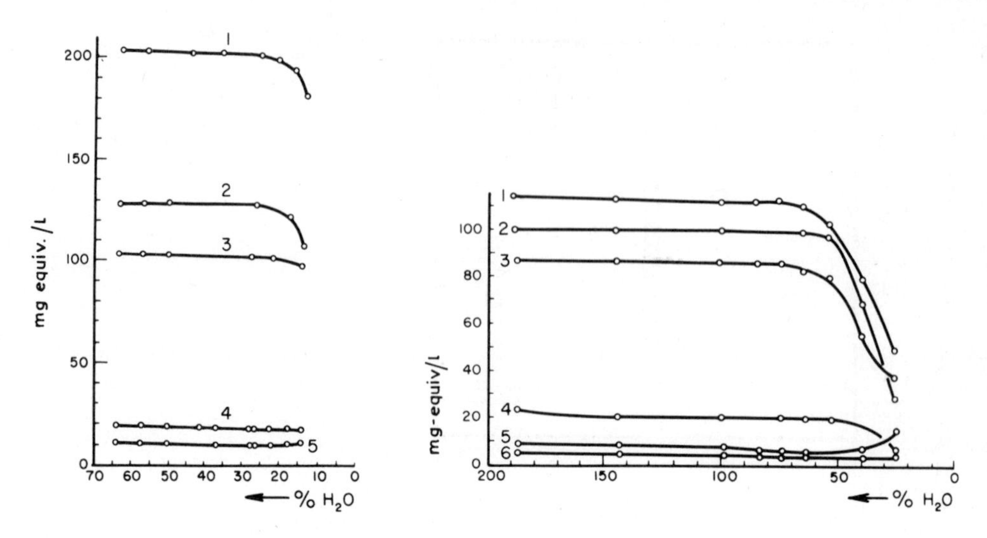

Fig.124. Changes in composition of solutions squeezed out of kaolinite clay. Concentration is plotted in mg-equiv./l on the ordinate. *1* = Na^+; *2* = SO_4^{2-}; *3* = Cl^-; *4* = Ca^{2+}; *5* = Mg^{2+}. (After Kryukov and Zhuchkova, 1963, p.97.)

Fig.125. Changes in composition of solutions squeezed out of bentonite. The concentrations in mg-equiv./l are plotted on the ordinate. *1* = $\kappa \cdot 10^4$, specific conductivity of solution; *2* = Na^+; *3* = Cl^-; *4* = SO_4^{2-}; *5* = Mg^{2+}; *6* = Ca^{2+}. (After Kryukov and Zhuchkova, 1963, p.38.)

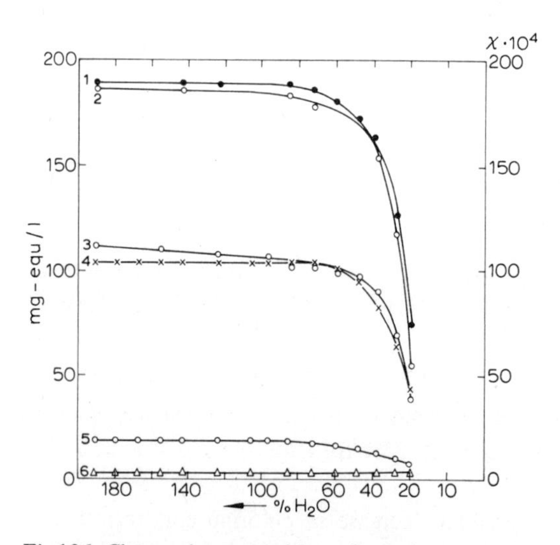

Fig.126. Changes in composition of solutions extruded out of askangel. Askangel is a colloidal variety of askanite clay found in the Georgian S.S.R. It is a residual product of weathering of andesite rocks in a humid, subtropical climate. The concentration is plotted on the ordinate in mg-equiv./l. *1* = Na^+; *2* = κ, specific conductivity of solution; *3* = Cl^-; *4* = SO_4^{2-}; *5* = Ca^{2+}; and *6* = Mg^{2+}. (After Kryukov and Zhuchkova, 1963, p.98.)

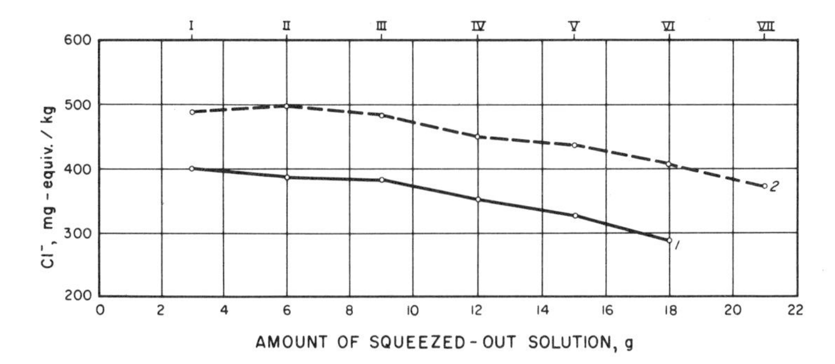

Fig.127. Variation in chlorine content in subsequent fractions (*I–VII*) of squeezed-out interstitial solutions of Maykop clay, eastern Pre-Caucasus. *1* = depth of 42 m, Divnoe area; *2* = depth of 158 m, Divnoe area. Cl^- content in mg-equiv./kg is plotted on the ordinate and the amount of extruded water in g on the abscissa. (After Kazintsev, 1968, fig.1, p.186.)

tion of various ions in squeezed-out solutions (Fig.128). The concentrations of Cl^- and Na^+ decrease with increasing pressure, whereas temperature does not seem to have any appreciable effect. The Mg^{2+} ion concentration increases (about 1½ times) with increasing pressure, but the absolute values are lower at high temperatures than at low. The concentration of K^+ decreases with pressure. The concentrations of K^+, Li^+, and I^+ are higher in solutions expelled at higher temperatures, whereas that of SO_4^{2-} is lower.

Krasintseva and Korunova (1968, p.191) studied the variation in the chemistry of solutions expelled from unlithified marine muds from Black Sea. At room temperature the results are presented in Fig.129. The chlorine concentration definitely decreases with increasing pressure, whereas concentrations of some components go through a maximum at pressures of 500–1,000 kg/cm^2. The Br^- and B contents increase with increasing

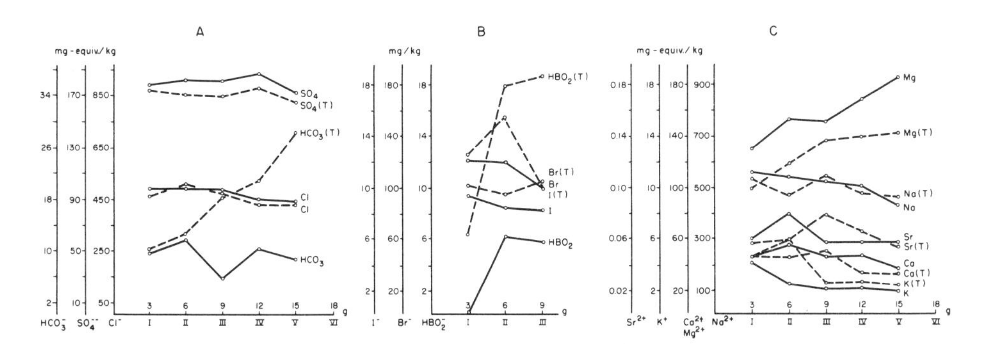

Fig.128. Changes in concentration of anions, cations and microcomponents with increasing compaction pressure in subsequent fractions (*I–VII*) of extruded interstitial solutions. Maykop clay, depth of 158 m, Divnoe area, eastern Pre-Caucasus. Solid curves = room temperatures; dashed curves = heated to 80°C. The amount of extruded solutions in g is plotted on the abscissa. (After Kazintsev, 1968, fig.2, p.188.)

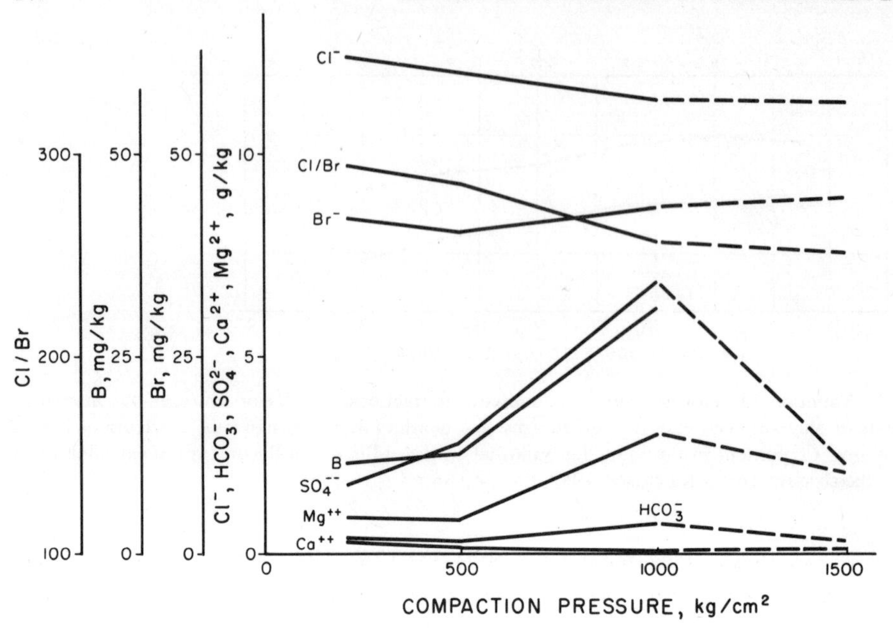

Fig.129. Relationship between concentration of various ions in interstitial solutions squeezed out of marine mud and compaction pressure at room temperature. (After Krasintseva and Korunova, 1968, fig.2, p.195.)

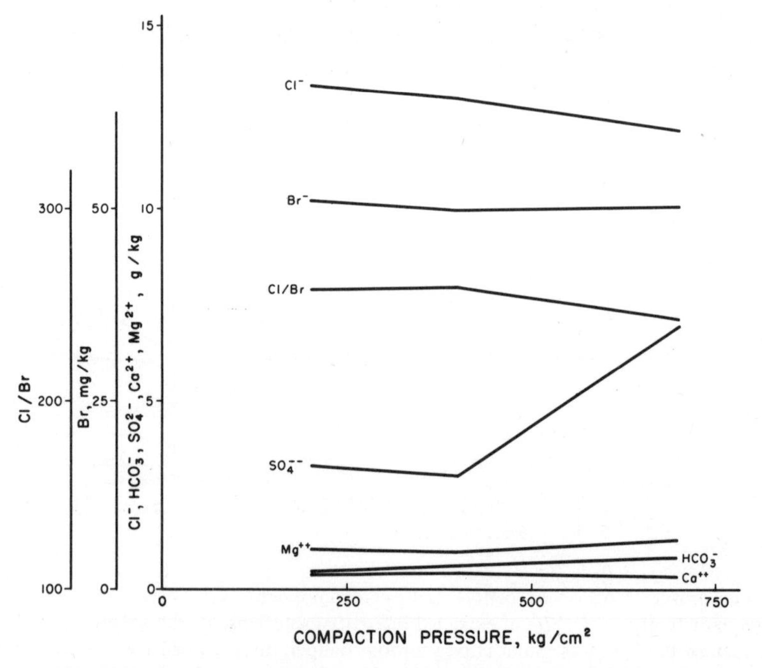

Fig.130. Variation in concentration of various ions in interstitial solutions squeezed out of marine mud with increasing pressure at 80°C. (After Krasintseva and Korunova, 1968, fig.3, p.196.)

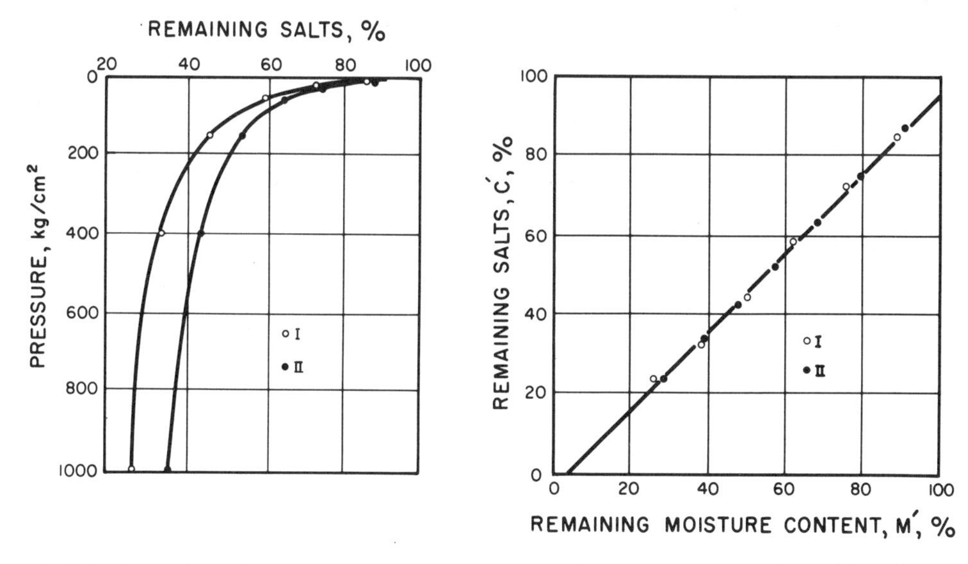

Fig.131. Geostatic leaching curve showing the relationship between percentage of remaining salts and compaction pressure, based on experimental data. *I* = experiment no.1, *II* = experiment no.2. Remaining salts, C', % = (content after compaction, mg-equiv./l) / (content prior to compaction, mg-equiv./l) × 100. (After Kotova and Pavlov, 1968, fig.1, p.59.)

Fig.132. Relationship between remaining moisture content M' (% of original) and amount of remaining salts, C' (% of original). *I* = experiment no.1, *II* = experiment no.2.
C', % = (content of salts after compaction, mg-equiv./l)/(content of salts prior to compaction, mg-equiv./l)×100. M', % = (volume of pore water after compaction)/(volume of pore water prior to compaction)×100. (After Kotova and Pavlov, 1968, fig.2, p.59.)

compaction pressure. In Fig.130, Krasintseva and Korunova presented relationship between concentration of various ions and compaction pressure at 80°C. Increasing temperature seems to decrease the amount of Mg^{2+} in extruded solutions.

On the basis of experimental and theoretical studies, Kotova and Pavlov (1968) showed that with increasing overburden pressure the amount of salts in clays decreases. They termed this leaching effect "geostatic leaching" (Figs.131 and 132). The percentage of remaining moisture content (% of initial) at 10 and 1,000 kg/cm² were 99% and 88%, respectively, in experiment no.1, and 88% and 39% in experiment no.2. The initial moisture content was 28.9% in experiment no.1 and 20.5% in experiment no.2.

Kotova and Pavlov (1968, p.61) presented the following equation:

$$C = C_0[1 - \exp(-\lambda/p^n)] \qquad (5\text{-}16)$$

where C is mineralization of pore water (mg-equiv./l; g/l, etc.) at pressure p; C_0 is initial mineralization of pore water at $p = 0$; λ and n are constants which depend on physico-chemical, geological, structural, and other variables. In their experiments, λ was equal to 4.0–4.45 and n to 0.087–0.116 (see Fig.133).

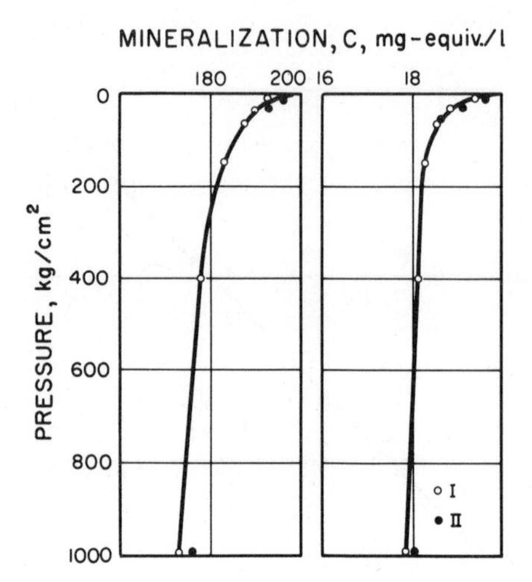

Fig.133. Variation in mineralization of pore waters remaining in formation after application of overburden pressure. *I* = theoretical curve, *II* = actual experimental data. (After Kotova and Pavlov, 1968, fig.3, p.61.)

McKelvey et al. (1957) showed that the electrolyte content of water expelled from exchange resins decreases with decreasing porosity. Based on data reported in the literature, Fertl and Timko (1970c) plotted the Na^+ content of expelled pore water during compaction versus the void ratio (Fig.134). This figure also shows that the electrolyte content of expelled water decreases with increasing overburden pressure.

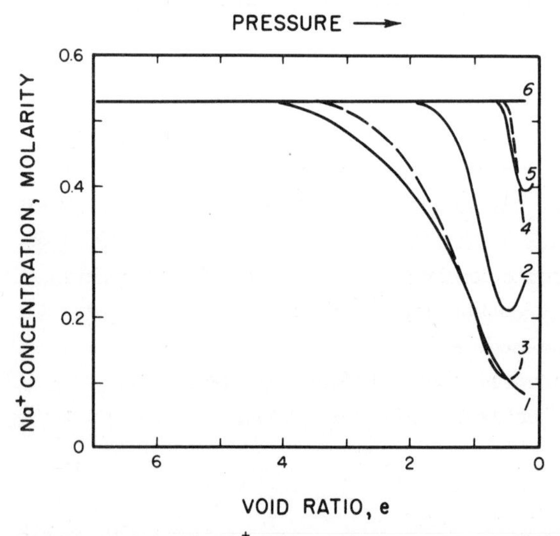

Fig.134. Variation in Na^+ content of pore water expelled during compaction with increasing pressure (decreasing void ratio). *1* = bentonite, theoretical; *2* = bentonite, experimental; *3* = montmorillonite, experimental; *4* = illite, theoretical; *5* = illite, experimental; *6* = kaolinite. (In: Fertl and Timko, 1970c, p.15, based on data after Warner, 1964, fig.6A, and Von Engelhardt and Gaida, 1963.)

According to Warner (1964), the double-layer theory predicts that the electrolyte content of expelled interstitial waters should decrease with increasing compaction pressure in the case of highly colloidal clays, when the interaction between the diffuse layers begins to occur. The average concentrations of negative and positive ions at any given point between the two clay plates can be estimated on using Boltzman's equations if the potential is known:

$$n_- = n \exp(-v_- e\psi/KT) \tag{5-17}$$

$$n_+ = n \exp(v_+ e\psi/KT) \tag{5-18}$$

where n = number of ions per cm^3 far from the plate; n_+ and n_- = number of positive and negative ions per cm^3 at any given point; e = elementary charge of an electron in e.s.u. (= $4.8 \cdot 10^{-10}$ e.s.u.); ψ = electric potential in volts; v = valency of ions; K = Boltzman's constant (= $1.38 \cdot 10^{-16}$ erg/degree); and T = absolute temperature in degrees Kelvin (°C + 273). Surface potential, ψ_o, can be calculated on using the following equation:

$$\psi_o = \frac{2KT}{ve} \sinh^{-1} \frac{\sigma}{\sqrt{2nDKT/\pi}} \tag{5-19}$$

where σ = surface charge density and D = dielectric constant of medium.

If X = 1/diffuse layer thickness, then:

$$X = \sqrt{\frac{8\pi n e^2 v^2}{DKT}} \tag{5-20}$$

The repulsive force, P_R, between two plates per cm^2 can be obtained on using the equation:

$$P_R = 2nKT(\cosh U - 1) \tag{5-21}$$

where $U = ve\psi_d/KT$ and ψ_d = electric potential at midpoint between two plates. The potential at the midpoint between two plates, ψ_d, can be calculated on using the equation presented by Verwey and Overbeek (1948, p.67).

The variation in electrolyte content of expelled waters depends on the surface potential and particle spacing. The salinity, C, of water (molarity) expelled at a certain time is approximately equal to:

$$C = C_o e^{-U} \tag{5-22}$$

where C_o is electrolyte concentration (molarity) in equilibrium solution.

Warner (1964) calculated the electrolyte concentrations of expelled water from bentonite and illite on the basis of double-layer theory, and also determined experimentally the electrolyte concentrations of water expelled from these two clays compacted in NaCl solutions having approximately the same Cl^- content as sea water. The theoretical and experimental results agreed reasonably well for bentonite down to a void ratio of 0.5 and very well for illite clay down to a void ratio of 0.25.

No change in salinity was observed by Warner (1964) for waters expelled from kaolinite. According to him, the packing of the particles in the latter case is such that the platy surfaces of most of the kaolinite particles probably did not ever approach each other at a spacing equivalent to their combined double-layer thickness (about 8 Å in 0.55 NaCl solution). Warner (1964) calculated that even at a pressure of 60,000 p.s.i., the platy surfaces of the kaolinite particles in 0.55 NaCl solution would still be separated by 90 Å. A second explanation offered by him is that the planar surface charge of the kaolinite would be very small, if exchange sites for kaolinite are located primarily on broken edges.

Thus, according to many investigators, the salinity of squeezed-out solutions progressively decreases with increasing overburden pressure. Consequently, the mineralization of interstitial solutions in shales is possibly less than that of waters in associated sandstones, because practically all of the interstitial fluids were expelled in many of these experiments. Future improved leaching techniques, coupled with material-balance calculations, probably will enable scientists to determine whether solid salts are left behind or not. The mineralization of solutions moving upward through a thick shale sequence as a result of compaction probably will progressively increase in salinity. It should be remembered, however, that if water from a sandstone bed moves through a shale layer into another sandstone bed, the water in the latter bed may be less mineralized because of filtration through a charged-net membrane.

The reasons for the gradual decrease in the mineralization of squeezed-out solutions can also be explained as shown in Fig.135. According to Pol'ster et al. (1967), when the capillary shown in this figure is filled with water, the density of water next to the capillary walls is maximum, whereas along the centerline of the capillary the density is lowest and approaches a normal one if the radius of the capillary is large enough (Fig.135A). According to Martin (1960, p.32), the density of adsorbed water on sodium montmorillonite varies with the amount of water present. The highest densities of bound (adsorbed) water occur when the H_2O/clay weight ratio for Na-montmorillonite is less than 0.5 (Martin, 1962, p.32). The dissolving capacity of water is inversely proportional to density (Fig.135B). As a result of compaction, the radius of the capillary is decreased and the water is squeezed out (Fig.135C), with the least bound water close to the center of the capillary being removed first. The remaining adsorbed waters poor in electrolytes are squeezed out at the end.

The above explanation is a subject for debate, however, because some investigators assign lower values to the density of adsorbed water than to the free water (see Anderson and Low, 1958; Grim, 1968, p.246; Cebell and Chilingarian, 1972). This problem remains unsolved because of difficulties involved in measuring the density of adsorbed water.

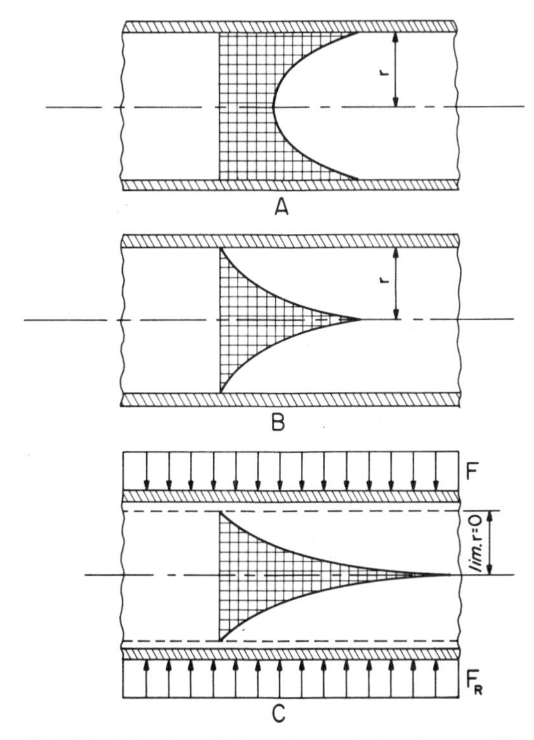

Fig.135. Behavior of electrolyte solution in a capillary having radius *r*. (After Pol'ster et al., 1967, fig.32, p.72.) A. Distribution of interstitial fluid density; maximum density occurs at the capillary walls. B. Distribution of dissolving capacity of the interstitial fluid; maximum dissolving capacity is at the center of capillary. C. Compaction of the capillary by a force *F*, illustrating that the part of the interstitial fluid squeezed out first is the one with the highest dissolving capacity (i.e., maximum salinity).

Some investigators, however, disagree with the above-described findings. For example, the findings of Manheim (1966), who used pressures ranging from 580 to 12,000 p.s.i., indicate that pressure does not appreciably affect the composition of extracted waters. Shishkina (1968) on studying interstitial solutions in marine muds from the Atlantic and Pacific Oceans and from the Black Sea did not observe any appreciable changes in the chemistry of squeezed-out solutions up to pressures of 1,260 kg/cm^2 in some samples and up to pressures of 3,000 kg/cm^2 in others. There was some increase in Ca^{2+} concentration at a pressure range of 675–1,080 kg/cm^2 and then there was a decrease at higher pressures. Shishkina stated that at very low pressures at which 80–85% of interstitial water is squeezed out, there are no changes in concentration. An important question which one should ask here, however, is: Are not we interested in the remaining 15–20% water?

Kryukov et al. (1962) observed more or less constant concentration of Cl^- up to a pressure of 594 kg/cm^2 for askangel, 1,420 kg/cm^2 for bentonite, and 3,850 kg/cm^2 for

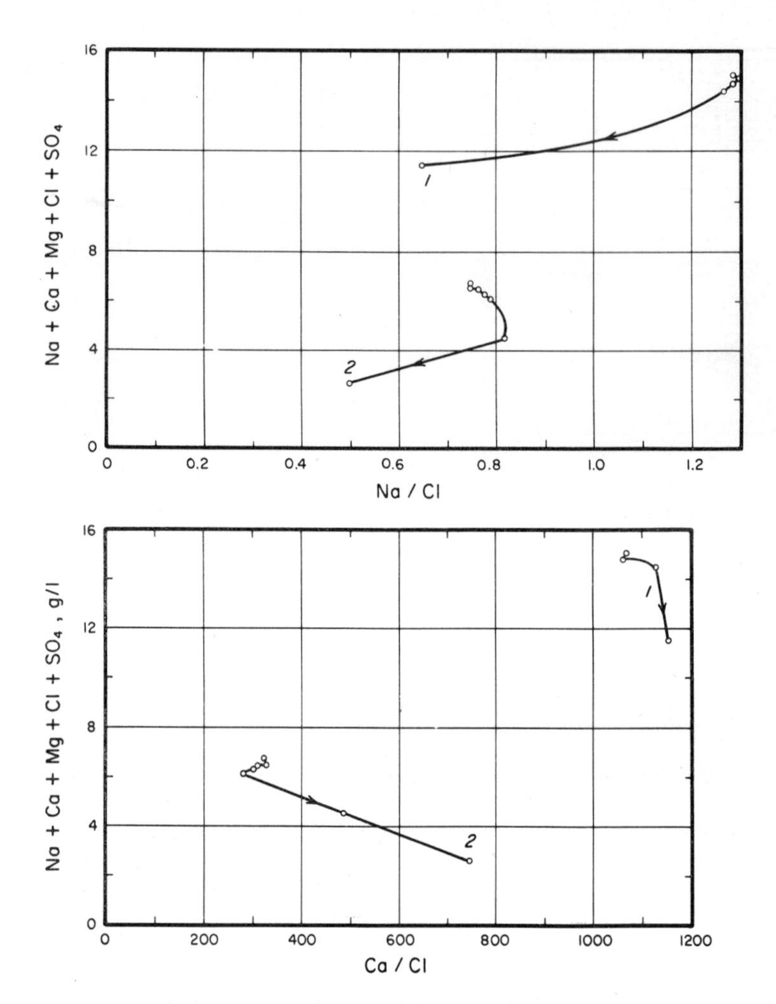

Fig.136. Variation in the Na/Cl and Ca/Cl weight ratios with total solids (g/l) content of expelled fluids and overburden pressure. *1* = kaolinite clay; *2* = montmorillonite clay. Arrows point in direction of increasing pressures; pressure was increased to 20,000 p.s.i. for kaolinite clay and to 25,000 p.s.i. for montmorillonite clay. (After Chilingar et al., 1969.)

kaolinite. For other ions changes start to occur at pressures of 600–700 kg/cm^2 and reach considerable proportions at pressures of 1,500–2,000 kg/cm^2.

It has been shown in Figs.117 and 118 that there is a tendency for the calcium ions to increase as salinity increases. According to White (1965, p.361), there is also a tendency for the calcium ions to increase relative to the sodium ions as salinity increases, because the Na^+ ions are more mobile than the Ca^{2+} ions. The Ca/Cl ratio, with a few exceptions, also increases with salinity (total solids). As noted by White (1965, p.350), fine-grained sediments are not equally permeable to all ions, and it seems reasonable to expect less

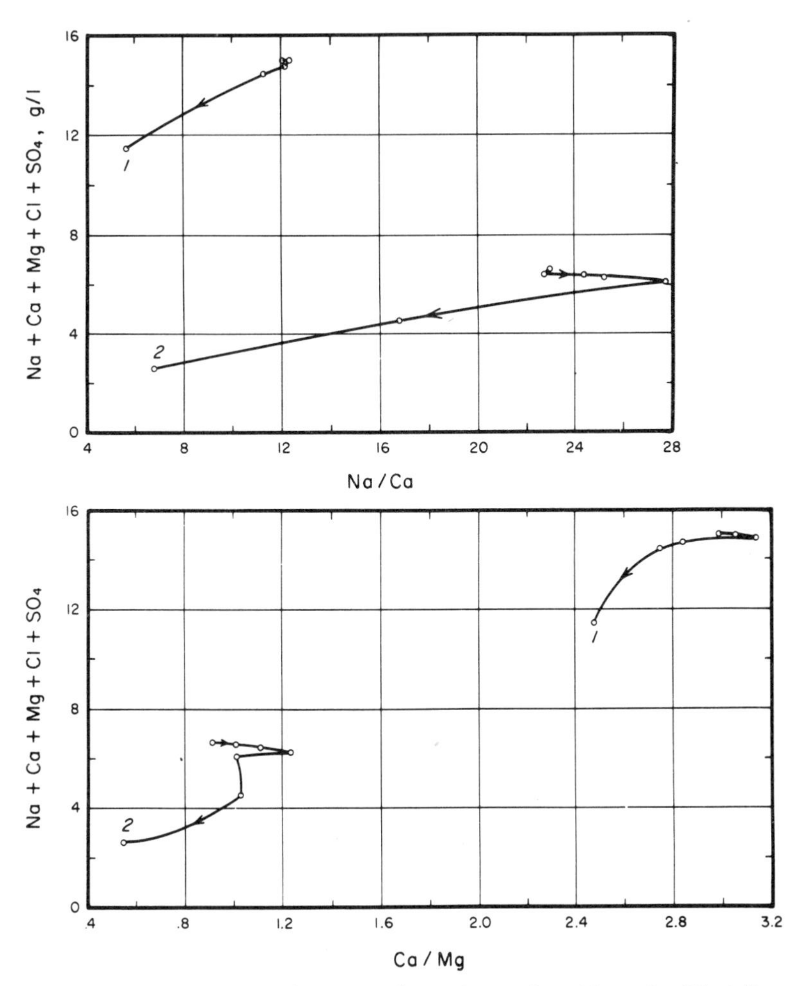

Fig.137. Variation in the Na/Ca and Ca/Mg weight ratios with total solids (g/l) content of squeezed-out fluids and overburden pressure. *1* = kaolinite clay and *2* = montmorillonite clay. Arrows point in direction of increasing pressures; pressure was raised to 20,000 p.s.i. for kaolinite clay and to 25,000 p.s.i. for montmorillonite clay. (After Chilingar et al., 1969.)

mobility for calcium (radius of hydrated ion = 9.6 Å) than for sodium (radius of hydrated ion = 5.6 Å). According to White (1965, p.351), there is strong evidence that molecular water is considerably more mobile than chloride ions.

According to Chilingar et al. (1969), the Na/Cl ratio generally decreases (for montmorillonite) with decreasing total solids and increasing compacting pressure (Fig.136). In the case of kaolinite, the Na/Cl ratio also decreases with decreasing total solids and increasing pressure. The Ca/Cl ratio generally increases with decreasing total solids and increasing pressure. For kaolinite clay, the Ca/Cl ratio increases with decreasing salinity

TABLE XXXII

Variation in composition of interstitial solution squeezed out of montmorillonite clay (API no.25). Composition of sea water used in saturating the sample is also given
(After Chilingar et al., 1969)

Ions	*Composition in p.p.m.*		
	sea water	*0–5,000 p.s.i.*	*0–10,000 p.s.i.*
Ca^{2+}	380	280	720
Mg^{2+}	650	17	320
Na^{+}	10,200	14,400	17,000
K^{+}	390	660	610
SO_4^{2-}	1,350	7,100	7,600
Cl^{-}	18,000	19,500	23,600
Total solids	30,970	41,957	49,850
Na/Cl	0.5667	0.7385	0.7203
Ca/Cl	0.0211	0.0144	0.0305
K/Cl	0.02167	0.03385	0.02585
Na/Ca	26.842	51.43	23.611
Ca/Mg	0.585	16.47	2.5

TABLE XXXIII

Variation in composition of interstitial solution squeezed out of montmorillonite clay (API no.25). The composition of sea water used in saturating the sample is given, but differs from that used in Table XXXII. Na/Cl, Ca/Cl, K/Cl, Sr/Cl, Na/Ca, and Ca/Mg ratios are presented
(After Chilingar et al., 1969)

Ions	*Composition in p.p.m.*			
	sea water	*0–1,000 p.s.i.*	*1,000–3,000 p.s.i.*	*3,000–10,000 p.s.i.*
Ca^{2+}	540	880	890	1,060
Mg^{2+}	1,580	915	995	1,535
Sr^{2+}	40	58	64	280
Na^{+}	10,500	16,200	17,400	20,700
K^{+}	390	295	300	364
SO_4^{2-}	4,520	5,140	9,250	11,200
Cl^{-}	20,000	22,500	23,000	28,500
Total solids	37,570	45,988	52,699	63,639
Na/Cl	0.525	0.720	0.731	0.726
Ca/Cl	0.025	0.0391	0.0374	0.0372
K/Cl	0.0195	0.0131	0.0126	0.0128
Sr/Cl	0.002	0.0025	0.00269	0.00952
Na/Ca	21.0	18.41	19.55	19.53
Ca/Mg	0.316	0.962	0.894	0.691

and increasing pressure. The Na/Ca ratio generally decreases with decreasing total solids and increasing pressure for both kaolinite and montmorillonite clays (Fig.137). The Ca/Mg ratio generally decreases with decreasing total solids and increasing pressure for both kaolinite and montmorillonite clays (Fig.137).

Chilingar et al. (1969) claim that when the samples were squeezed rapidly (pressure increased to a desired value immediately), the over-all salinity of the squeezed-out solutions was greater on using higher pressures (up to 10,000 p.s.i.). This possibly could occur in tectonically active areas. As shown in Table XXXII, the concentration of various ions (except K^+) in solutions expelled from montmorillonite clay saturated with sea water was greater on using 10,000 p.s.i. than at 5,000 p.s.i. It is possible, however, that because of rapid squeezing equilibrium was not reached and, if measured, the remaining moisture content would have corresponded to a lower compaction pressure.

Results presented in Table XXXIII also show that the concentration of various ions in the effluent increases with increasing overburden pressures. According to Chilingar et al. (1969), the results presented in Tables XXXI and XXXII suggest that the salinity of the effluent solutions first increases before starting to decrease. The pressures at which the salinity will start decreasing were possibly not reached in these experiments. Inasmuch as

TABLE XXXIV

Variation in the composition of the supernatant liquid and interstitial solution centrifuged out of montmorillonite clay (API no.25). The chlorinity ratios (Ca/Cl, K/Cl and Na/Cl) are presented along with the Na/Ca and Ca/Mg ratios

(After Rieke, 1970)

Ions	*Composition in p.p.m.*				
	sea water (S_t) (V_t=10 ml)	*supernatant liquid* (S_1) (V_1=2.95 ml)	*centrifuged liquid* (S_2) (V_2=2.9 ml)	*remaining liquid* (S_3) (V_3=4.15 ml)	*remaining liquid* (S_3) * (V_3=4.15 ml)
Ca^{2+}	480	444	462	518.2	431
Mg^{2+}	1,283	765	794	1,992	744
K^+	427	260	274	652.6	250
Na^+	10,554	13,949	14,813	5,164	13,345
SO_4^{2-}	2,172	4,380	4,471	**	4,292
Cl^-	19,574	20,355	21,823	16,202	19,329
Total solids	34,490	40,153	42,661	24,530 +SO_4^{2-} (4,471?)	38,391
Na/Cl	0.539	0.685	0.678	0.319	0.69
Ca/Cl	0.0245	0.0218	0.0212	0.0320	0.0223
K/Cl	0.0218	0.0128	0.0126	0.0403	0.0129
Na/Ca	21.9	31.4	32.0	9.966	30.96
Ca/Mg	0.374	0.58	0.581	0.26	0.579

* Remaining liquid composition was calculated using the supernatant liquid as the starting fluid.

** The results are not reported because the clay tested appears to have high SO_4^{2-} content.

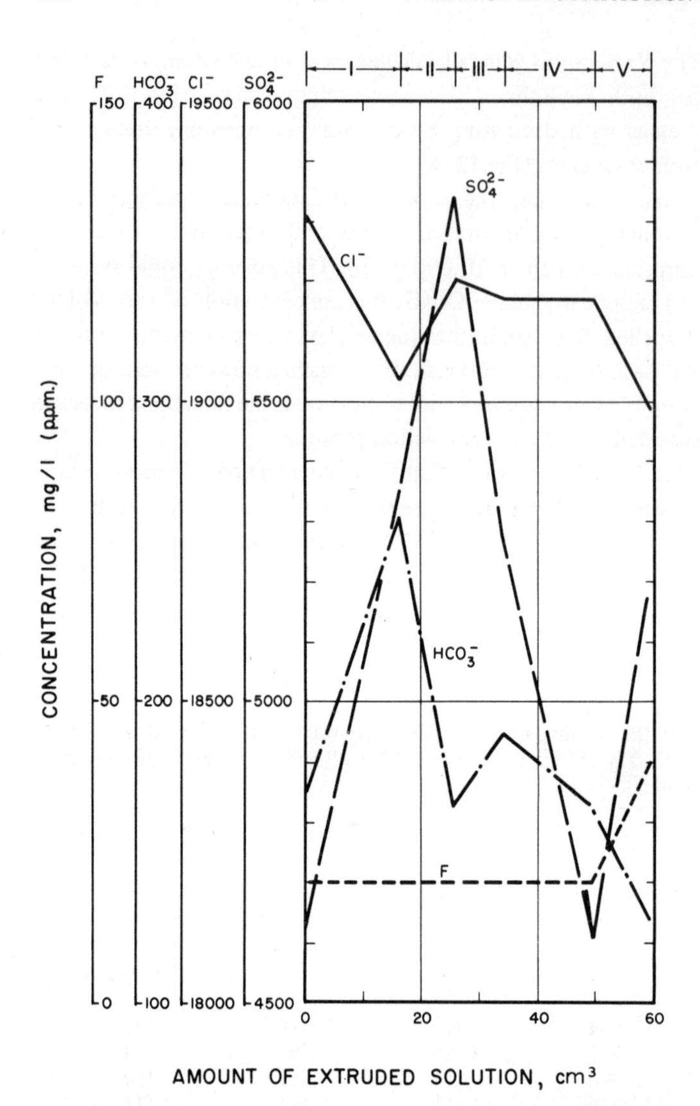

Fig.138. Variation in concentration of anions with increasing compaction pressure in subsequent fractions of extruded solutions from montmorillonite clay saturated in sea water. (After Sawabini et al., 1972.)

solutions were squeezed out rapidly, however, the rate of removal may be an important factor that has to be considered. On squeezing the saturated clays rapidly, the portion of fluid close to the water vent is squeezed out at lower pressures. At higher pressures the fluids deeper inside the clay body also have a chance to contribute, but only the more saline portion of the fluid.

To further investigate this problem, 10 ml of sea water were mixed with 3.1958 g of

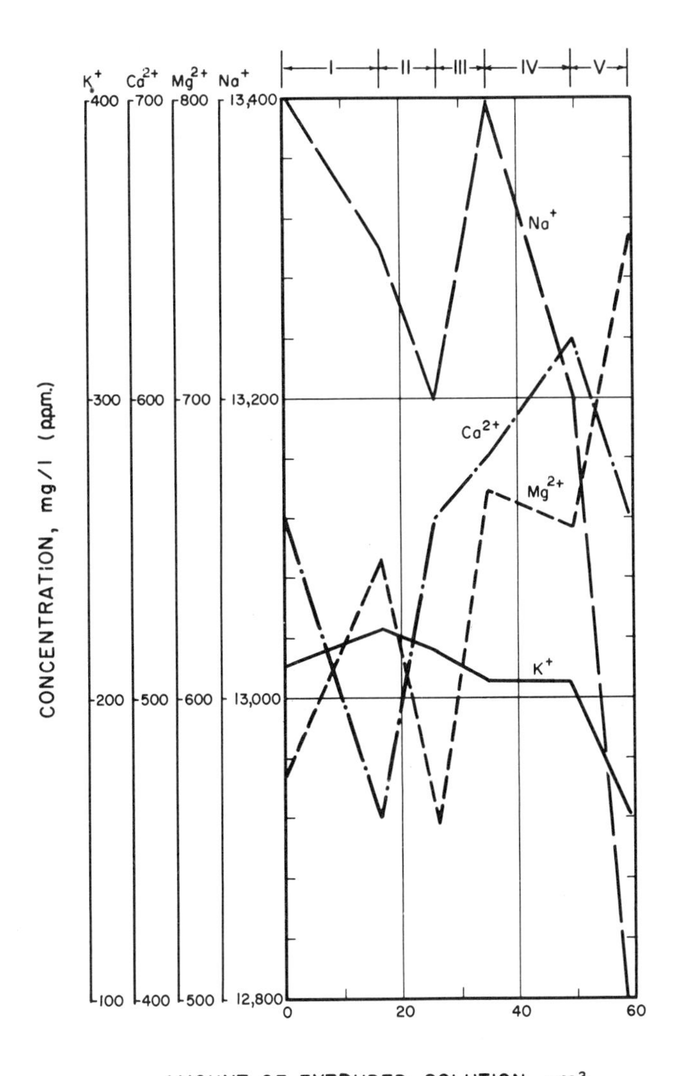

Fig.139. Variation in concentration of cations with increasing compaction pressure in subsequent fractions of extruded solutions from montmorillonite clay saturated in sea water. (After Sawabini et al., 1972.)

clay and allowed to hydrate for 4 days (Rieke, 1970). At the end of this period the supernatant fluid was decanted and analyzed. The remaining mud was centrifuged at 1,500 r.p.m. for 15 min (≈ 6 p.s.i.) and the expelled fluid was also analyzed. The results are presented in Table XXXIV. Assuming no ion adsorption, the fourth column in this table shows the composition of remaining fluid in the mud, calculated using a material balance equation ($S_t V_t = S_1 V_1 + S_2 V_2 + S_3 V_3$, where S is the salt content in p.p.m. and V

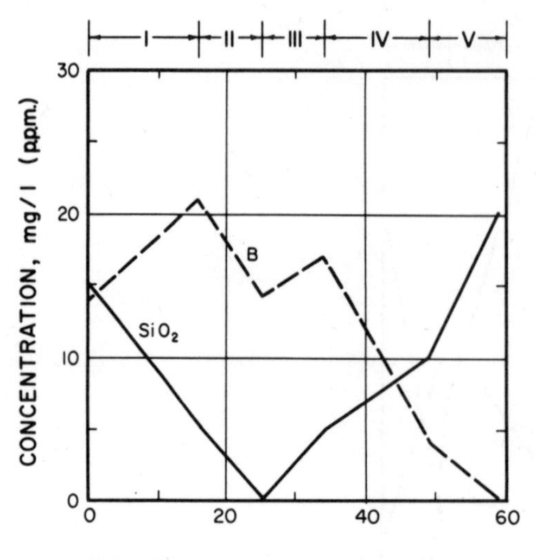

Fig.140. Variation in concentration of SiO_2 and B with increasing compaction pressure in subsequent fractions of extruded solutions from montmorillonite clay saturated in sea water. (After Sawabini et al., 1972.)

is the volume in ml). It was assumed that clay did not contribute any appreciable amount of dissolved solids. These results indicate that the total salinity of initially squeezed-out solutions, in this case, first increases and the remaining interstitial water has lower salinity. The last column in Table XXXIV was calculated using the following material balance equation: $S_1(V_2 + V_3) = S_2V_2 + S_3V_3$. In the latter case it was assumed that supernatant, equilibrated fluid has the same composition as the initial pore fluid. The results indicate that the pore solution remaining after centrifuging has lower salinity than the centrifuged liquid.

Nemykin (1968) found that the salinity of the water in centrifuged sandstone samples is higher than the salinity of the water that was used to saturate the samples. He explained this phenomenon by the presence of double electrical layer at the surface of the rock particles.

Sawabini et al. (1972) showed that the concentration of solutions expelled at the initial stages of compaction are slightly higher than that of interstitial solution initially present in montmorillonite clay saturated in sea water for one week (Figs.138–141). The sample was shaken vigorously twice a day, and it was assumed that the supernatant liquid has the same composition as the free interstitial water.

The findings of Sawabini et al. (1972) suggest that the concentration of extruded solution goes through a maximum (peak), or at least remains more or less constant, before starting to decrease with increasing overburden pressure. The remaining moisture content (% dry weight) of the montmorillonite sample at the end of the experiment was

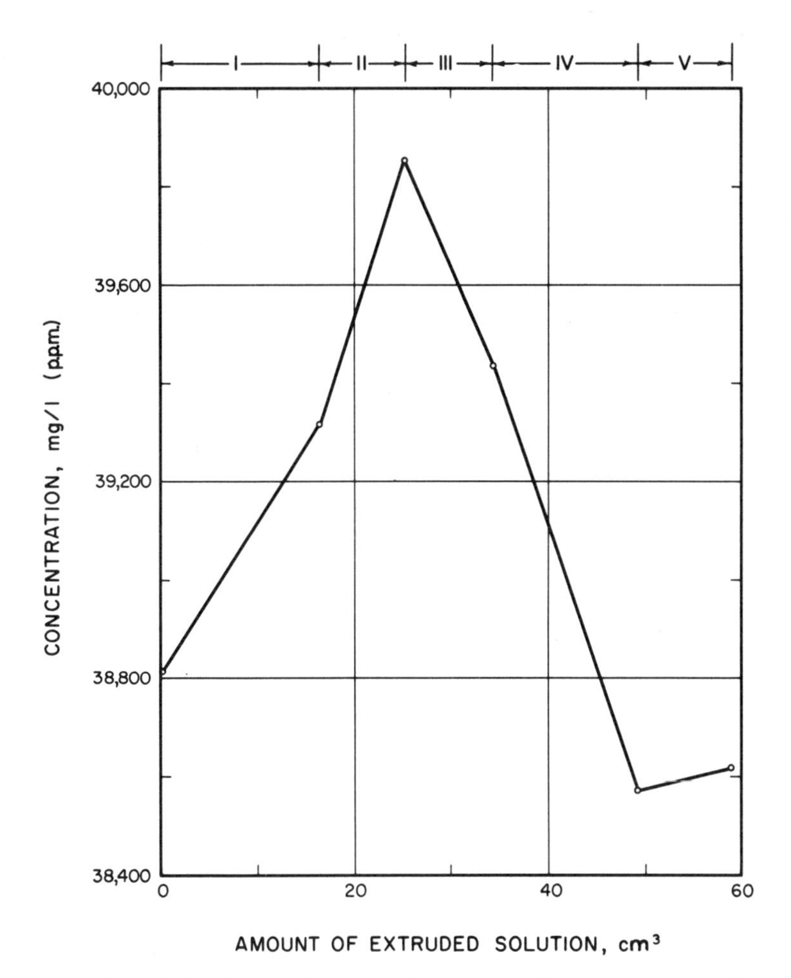

Fig.141. Variation in total dissolved solids content with increasing compaction pressure in subsequent fractions of extruded solutions from montmorillonite clay saturated in sea water. (After Sawabini et al., 1972.)

62%, which corresponds to a pressure of about 500 p.s.i. There is, however, a tendency for the Mg^{2+} cation concentration to increase with increasing overburden pressure as the fluid layers which are closer to the clay plates are squeezed out.

All these results show that great care should be exercised on studying the chemistry of interstitial solutions in marine muds. Possibly, one should prepare a continuous curve (pressure, or remaining-moisture content, versus chemistry of solutions squeezed out) for each different ion ranging from low pressures to very high pressures (≈200,000 p.s.i.). At these high pressures most of the interstitial fluid is removed. Leaching techniques should then be used to determine the amounts of remaining salts.

BRIEF REVIEW OF DIAGENETIC CHANGES OF PORE WATERS

Early diagenetic changes in marine sediments

It is important to know the extent to which the chemistry of interstitial fluids is affected by compaction as compared to other diagenetic processes. In the light of the foregoing discussions, it is also important to know to what extent the observed differences between ocean and interstitial waters in sediments are real and to what extent they are due to laboratory extraction techniques. The experimental results obtained by Chilingar et al. (1969) in the laboratory, as shown previously, indicate that interstitial water in compacted shales is probably fresher than that in associated sandstones, and that the salinity of water in undercompacted shales should be higher than in well-compacted ones. It is proposed that in the low-pressure extraction techniques employed by several investigators, ions may have been enriched relative to the original waters, and, therefore, the reported increase in salinity is merely a laboratory artifact. This suggestion is based on the observation discussed previously that during moderate compaction, clays give off fluids enriched in electrolytes relative to the original depositional water (e.g., Von Engelhardt, 1961; Von Engelhardt and Gaida, 1963; Rieke et al., 1964; and Table XXXIV). Siever et al. (1965) attributed the increase of the chloride ion to the effects of compacted sediments acting as semipermeable membranes: concentration of the salts in retained waters with removal of fresher waters.

Von Engelhardt and Gaida (1963) reported that for compaction pressures between 30 and 800 atm the concentration of electrolyte in the pore solution of montmorillonite clay diminishes with increasing compaction. They explained this reduction in electrolyte concentration as a result of the electrochemical properties of base, exchanging clays. At higher pressures ranging from 800 to 3,200 atm, however, there was an increase in salt concentration within the remaining pore solution. This increase was attributed to the trapping of small droplets of solution in the highly compressed montmorillonite clay pore system. The clay acted as a barrier to ion movement.

Many investigators have shown that even within the spans of short sediment cores various chemical changes can be measured (e.g., Emery and Rittenberg, 1952; Siever et al., 1965). Friedman et al. (1968) studied interstitial fluids in ocean sediments from the Atlantic ocean off Long Island, New York. They found that the Ca/Cl, K/Cl and Rb/Cl ratios were higher in the interstitial waters than in the overlying ocean water. The Mg/Cl and Li/Cl ratios were about the same, whereas the Sr/Cl ratio was higher in the overlying waters than in the interstitial water. All these ratios, except the Sr/Cl ratio, increased with depth in the sediments (Friedman et al., 1968, p.1315). The increase in the Ca/Cl ratio was attributed to the dissolution of small amounts of aragonite from shell material in the enclosing noncarbonate sediments. If true, it is surprising that the Sr/Cl ratio decreases as the Ca/Cl ratio increases if base exchange is not involved. Strontium's ionic radius, however, lies between those of calcium and potassium and Sr^{2+} might be expected to substitute in the clays for both Ca^{2+} and K^+. Increase in the K/Cl and Rb/Cl ratios may result from the dissolution of potassium-feldspar. Friedman et al. (1968) also noted that the chlorinity of interstitial waters increases regardless of whether or not the core samples were composed of sands or clays. It seems unlikely that loosely consolidated marine sands could act as semipermeable membranes.

Based on findings of Shishkina (1959), Strakhov (1960) recognized two hydrochemical modifications of pore fluids in Recent marine sediments which give rise to two types of waters: (1) sodium-bicarbonate and (2) calcium-chloride ($Cl^- > Na^+ + Mg^{2+}$).

(1) *Sodium-bicarbonate waters.* In the Okhotsk Sea, metamorphization of buried water results in (a) decreasing sulfate content (until its complete disappearance) and (b) increasing HCO_3^- concentration. There is a decrease in concentration of Ca^{2+} and Mg^{2+} cations, and some increase in Na^+ and K^+ contents, as compared to those of bottom water. This is probably due to formation of diagenetic beidellite from micas, and of calcite. In addition, there is accumulation of ammonium resulting from decomposition of organic matter.

(2) *Calcium-chloride waters.* In the Black Sea, the reduction in sulfate-ion content in the interstitial waters is not accompanied by an increase in alkalinity ($HCO_3^- + CO_3^{2-}$). Although sulfates are reduced, the water is not enriched in carbonates because of their precipitation. The Na^+, K^+ and Mg^{2+} contents decrease, as compared to those in sea water, whereas concentration of Ca^{2+} and NH_4^+ ions markedly increases with depth. The decrease in Mg^{2+} content is due to formation of authigenic iron–magnesium carbonates. In deeper horizons, decrease in Na^+ content is almost exactly compensated by increase in Ca^{2+} concentration owing to base exchange. Ion filtration and compaction should also be taken into account in explaining these variations (see Degens and Chilingar, 1967).

The intensity and nature of metamorphization of waters are directly related to amounts of buried organic matter. Although during early diagenesis the relative proportions of various ions, as compared to those of sea water, change considerably, the total mineralization does not change appreciably.

TABLE XXXV
Content (mg-equiv./l) of various ions in solutions squeezed out from Black Sea sediments at 5,700 p.s.i. Core no.13, 1,301 cm long, sea depth of 2,122 m
(After Shishkina, 1959, p.37)

Depth (cm)	Cl^-	SO_4^{2-}	Na^+	Ca^{2+}	Mg^{2+}	K^+
Black Sea [1]	271	27.2	232	12.3	53	5.3
0– 18	349	33.4	302	9.2	66	7.5
139–180	337	26.2	292	14.7	61	4.8
400–430	278	2.1	216	25.7	40	1.5
576–610	237	3.8	175	31.8	n.d.	n.d.
763–780	193	3.0	131	39.7	27	trace

[1] Average values, after Alekin (1953, p.269). Composition of sea water changes with depth; for example, Cl^- content is 10.27‰ at the surface and 12.64‰ at a depth of 2,000 m; n.d. = no determination.

Some results obtained by Shishkina (1959) on studying the interstitial waters of recent sediments of the Black Sea are presented in Table XXXV. In this particular case, she found a general decrease in concentration of various ions with depth, with the exception of the Ca^{2+} cation. Possibly, there was an additional source for the Ca^{2+} ions at greater depth. One should also not lose sight of the possibility that the chemistry of Black Sea water was changing with time.

Late diagenetic changes in sediments

The gradual increase in the concentration of the interstitial solutions during diagenesis has been shown by Philipp (1961) in the Eldingen Oil Field, east of Celle, Germany. The compositions of water from wells producing from different structural depths, and also of the water below the oil–water boundary, are presented in Table XXXVI. Well no.9 was drilled on the crest of the structure, where poorly-cemented sandstone of an early diagenetic stage has a porosity of 32%. Wells no.6 and no.11 follow downdip in the direction toward the oil–water boundary, where the sandstones exhibit advanced stages of diagenesis. Some migration of oil occurred during the early stages of diagenesis and trapped the interstitial water, and further migration occurred during the later stages of diagenesis trapping the more saline water. In Table XXXVI one can observe a gradual increase in the concentration of salts, accompanied by a relative increase in Ca^{2+} and relative decrease in Mg^{2+} cation content.

Rittenhouse (1967) showed that the relationship between bromine and total-solids content can possibly be used to determine the origin of some oil-field brines, especially those which owe their high salinity to evaporation or to solution of salt. There is a possibility that the bromine content increases during diagenesis. For example, Tageeva

TABLE XXXVI
Chemical composition of Eldingen oil-field waters near Celle, northern Germany
(After Philipp, 1961)

	Well no.9	*Well no.6*	*Well no.11*	*Well no.49*
Depth to the top of sandstone (m)	1,379	1,379	1,388	>1,388
Water type	interstitial	interstitial	interstitial	free
Dissolved solids (mg/l)	11.8	23.5	28.5	169
(mmole/l)	210	380	460	2,716
Content of cations in % (mmole/l)				
Na^+	88.5	94.3	93.7	92.5
Ca^{2+}	4.2	4.5	5.0	6.0
Mg^{2+}	7.0	0.9	1.0	1.4
K^+	0.3	0.3	0.3	0.1
SO_4^{2-}/Cl^- molar ratio	0.0000	0.00308	0.00341	0.000112

(1958) found that interstitial water in Recent Caspian Sea sediments from a depth of 15–40 m contains about 65% more bromine (relative to chloride) than does Caspian Sea water. The effect of compaction on the chemistry of squeezed-out solutions also should not be overlooked as a possible explanation.

Some chemical changes involving silicates

Siever et al. (1965) determined the silicate concentration in interstitial water in several sediment cores. Anikouchine (1967, p.508) plotted Siever et al.'s data (Fig.142) showing that SiO_2 concentration increases with depth in the upper few hundred centimeters of marine sediments. Anikouchine (1967) presented a mathematical description of the distribution of a dissolved chemical species in interstitial fluid of clayey marine sediments:

$$\frac{\partial c}{\partial t} = D\,\frac{\partial^2 c}{\partial h^2} - v_h \frac{\partial c}{\partial h} + k_1(c - c_f) \tag{5-17}$$

where D = constant coefficient of diffusivity, h = height above the fixed basement, v_h = velocity of expelled interstitial water, $\partial c/\partial h$ = concentration gradient, k_1 = first-order reaction-rate constant, c = concentration at h, and c_f = final or saturation concentration. The second-order term describes Fickian diffusion, advection is described by the product of v_h and the concentration gradient, and the reaction term is the product of the first-order reaction-rate constant k_1 and the extent-of-completion parameter $(c-c_f)$. The above equation was solved to obtain a steady state distribution of dissolved silicate across the sediment–water interface, and the agreement of Anikouchine's theoretical predictions and Siever's empirical data for SiO_2 becomes essentially constant below a depth of

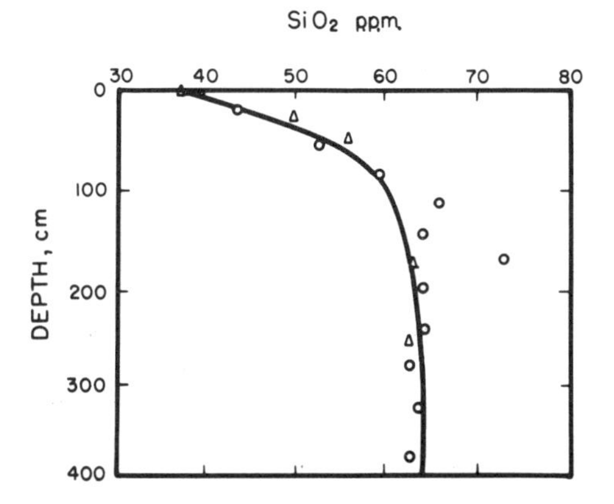

Fig.142. Theoretical distribution of silicate (△ core no.5, Atlantis 258, Cape Cod) in interstitial water fitted to Siever et al. (1965) data (○ core L-139, Gulf of California). (After Anikouchine, 1967, fig.2, p.508.) Not all of Siever et al. data on SiO_2 from the other cores match as well Anikouchine's theoretical curve.

250 cm. The effect of compaction on SiO_2 concentration in interstitial fluids has not been extensively studied (see Fig.140).

Silicate waters, which are rare, can be stable only at very low salt concentrations (up to 100 mg/l). With their movement into deeper horizons, they are converted to sodium-bicarbonate water (Filatov, 1961):

$$Na_2SiO_3 + CO_2 + H_2O = Na_2CO_3 + SiO_2{\cdot}H_2O$$

$$NaAlO_2 + H_2CO_3 + H_2O = NaHCO_3 + Al(OH)_3$$

With further migration, calcium-bicarbonate type waters can form:

$$CaSiO_3 + 2H_2CO_3 = SiO_2{\cdot}H_2O + Ca(HCO_3)_2$$

Some electrolytes decompose silicates. For example, $MgCl_2$ decomposes calcium and iron silicates.

Decomposition of feldspar and formation of kaolinite and of quartz possibly can take place through the action of interstitial solution as shown in the following formula (Von Engelhardt, 1967, p.504):

$$2KAlSi_3O_8 + 16H_2O \rightarrow 2K^+ + 2Al^{3+} + 8OH^- + 6H_4SiO_4$$

$$\rightarrow Al_2(OH)_4 + Si_2O_5 + 4SiO_2 + 2K^+ + 2OH^- + 13H_2O$$

It is assumed, in this case, that Si dissolves in the form of orthosilicic acid, which does not dissociate easily.

Some variations in Ca/Mg ratio of interstitial waters

Gurevich (1960) showed that the Ca/Mg ratio of interstitial waters decreases with decreasing age of rocks. In explaining this phenomenon, Gurevich subscribed to Marignac's reaction [$2CaCO_3 + MgCl_2 \rightleftharpoons CaMg(CO_3)_2 + CaCl_2$], to the long duration of reaction, and to repeated water exchange, which is unlikely. Whether the above reaction could occur in nature or not, should be checked experimentally. The analogous reaction of Kurnakov (in: Posokhov, 1966, pp.10, 102, etc.) is as follows:

$$2Ca(HCO_3)_2 + MgCl_2 \rightleftharpoons CaMg(CO_3)_2 + CaCl_2 + 2CO_2 + H_2O$$

Valyashko (1962), however, showed that this reaction is so slow that $Ca(HCO_3)_2$ has time to decompose to $CaCO_3 + H_2O + CO_2$ (also see Chilingar and Bissell, 1963). Valyashko (1962, p.55) obtained individual, rhombohedral crystals of dolomite (identified by crystallo-optical analysis) in the laboratory at atmospheric conditions (low CO_2 pressure). Dolomite admixture in calcite constituted only up to 1–2%. Whether or not at high CO_2 pressure this reaction would give rise to extensive dolomite precipitation is not known.

Valyashko (1962, p.47) presented the main reaction which occurs between calcium-bicarbonate and sulfate solutions:

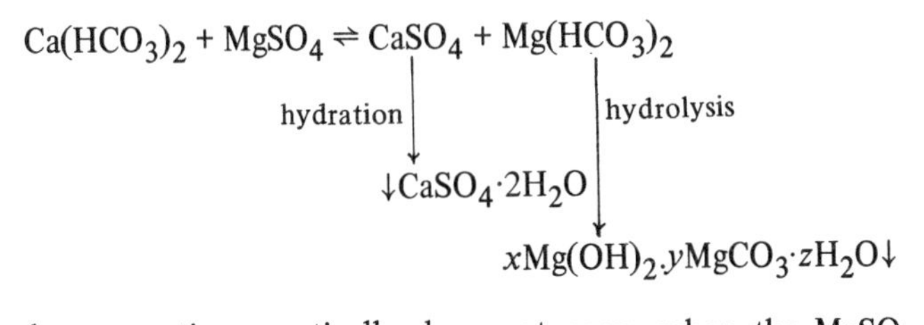

The above reaction practically does not occur when the $MgSO_4$ concentration in the solution is small; instead, there is decomposition of calcium bicarbonate with formation of calcite as follows: $Ca(HCO_3)_2 \rightarrow CaCO_3 + H_2O + CO_2$. As the $MgSO_4$ concentration increases, first gypsum starts to form, followed by basic carbonates of magnesium. At this time, $CaCO_3$ practically disappears from the bottom phase. In addition, Haidinger's reaction also occurs, but very slowly: $2Ca(HCO_3)_2 + MgSO_4 \rightarrow CaMg(CO_3)_2 + CaSO_4 + 2H_2O + 2CO_2$. Possibly, this reaction could account for certain extensive dolomite formation, but at higher CO_2 pressures than those used by Valyashko (1962, p.55) (atmospheric conditions), who obtained only individual, rhombohedral crystals of dolomite. At high pCO_2 this reaction will probably proceed to the right, whereas at low pCO_2 (present-day atmosphere) it could proceed to the left (dedolomitization). This statement is correct as long as pCO_2 does not exceed the equilibrium pCO_2 for a given value of dissolved $MgSO_4$. Relative concentrations of $CaSO_4$ and $MgSO_4$ are obviously very important. Haidinger's reaction is cited by many scientists to explain dolomite formation.

One should also not lose sight of the possible variation of the Ca/Mg ratio of sea water with age (Chilingar, 1956), which in part is suggested by the increase in the Ca/Mg ratio of carbonate rocks with decreasing age of rocks. According to data presented by Ronov (1964, p.719), however, the percentages of Mg^{2+}, Ca^{2+}, Na^+, and K^+ cations in Azoic Sea water were 24, 29, 30, and 17%, respectively (sum of all cations = 100%), as compared to 10.7, 3.2, 83.1 and 3% in the present-day ocean water. Further research may resolve this problem.

Influence of microorganisms on the chemical composition of underground waters

The influence of microorganisms on the chemical composition of underground waters should not be overlooked, because bacteria are present at great depth; for example, a depth of 4,000 m according to Kuznetsov et al. (1962). Bacteria can withstand very high pressures ($\approx$ 3,000–4,000 atm) and temperatures up to 75–80°C.

According to Mekhtieva (1962), highly mineralized calcium-chloride waters (≈800 mg-equiv./l) are not favorable for bacterial growth. These waters have (1) very low contents, or are devoid, of sulfates; (2) high Br concentration; (3) (Cl–Na)/Mg coefficient >3; and (4) low P content. On the other hand, the less mineralized waters, enriched in sulfates, H_2S, and CO_2 and having coefficient (Cl–Na)/Mg < 3, are favorable media for bacterial growth.

As a result of sulfate reduction, there is an accumulation of bicarbonates and precipitation of Ca and Mg. The sulfate reduction can be presented in an equation form as follows:

$$C_6H_{12}O_6 + 3Na_2SO_4 \rightarrow 3CO_2 + 3Na_2CO_3 + 3H_2S + 3H_2O + Q$$

If SO_4^{2-} is precipitated in the sediments as anhydrite or gypsum, there should be a 1:1 loss of Ca^{2+} and SO_4^{2-} ions from solutions (Graf et al., 1966, p.21). If SO_4^{2-} is bacterially reduced to H_2S with an accompanying generation of HCO_3^- from decomposition of organic matter, then it may be assumed that the resulting solution will differ only in containing additional HCO_3^-, equivalent to the SO_4^{2-} that was reduced. Thode and Monster (1965) reached the conclusion that bacterial SO_4^{2-} reduction is the major cause of sulfur-isotope fractionation in nature. Emery and Rittenberg (1952) found that in marine sediments off the southern California coast the SO_4^{2-} content of interstitial water decreases rapidly with depth, reaching zero at a depth of 7 ft. below the water–sediment interface in one case. Much of the H_2S formed is believed to diffuse to the sediment surface. Bacterial SO_4^{2-} reduction is limited by the amount of organic matter present. Chebotarev (1955) showed that sulfate-rich subsurface waters occur at shallower depths than bicarbonate-rich ones and that the dominantly chloride waters occur considerably deeper than the other two. The effect of compaction on the SO_4^{2-} content of interstitial fluids should be thoroughly investigated (see Figs.124–126, 128–130, 138).

Formation waters adjacent to hydrocarbon reservoirs have more abundant and variable microflora than do the waters of the same horizon at some distance from the oil–water contact. Numerous authors discussed in detail the sulfate-reducing, denitrifying, etc. bacteria. For more detailed treatment of the subject, the reader is referred to the books by Posokhov (1966, pp.218–232) and Davis (1967).

HYDROCHEMICAL FACIES AND VERTICAL VARIATION IN CHEMISTRY OF INTERSTITIAL FLUIDS

The hydrochemical facies changes of underground fluids may be defined as the lateral variation in the chemical composition of these fluids in a given stratigraphic unit. Although the variation in hydrochemical facies has been investigated by many authors (e.g., Back, 1960, 1966; Graf et al., 1965; and Seaber, 1965) in ground-water aquifers, it has been neglected in petroliferous horizons. The chemical composition of brines from oil wells having approximately equal depth and penetrating the same formations, sometimes can vary greatly within the same oil field. On the other hand, fluid analyses of wells many miles apart sometimes bear a close chemical resemblance. This seems to depend directly on the structural history and transmissibility (the product of the permeability times the thickness of the formation) of the formation. The interpretation of regional variation of the hydrochemical facies on the basis of isopleth maps of ion distribution may lead to ambiguous results. This could possibly be due to the non-uniform distribution of the samples and the interval being scanned, as well as to the following important factors: (1) age of the fluid, (2) diagenetic history of the formation, (3) degree of compaction, (4) hydraulic dilution and concentration, (5) migration of fluids, (6) temperature, and (7) geochemistry of associated strata. Data on the lateral variation in composition of oil-field brines have been collected by various investigators: Rall and Wright, 1953; Wright et al., 1957; Hawkins et al., 1963a, 1963b; Graf et al., 1966; and Mast et al., 1968.

It is well documented that the chemical nature of ground water in a given aquifer progressively changes as the water moves away from the place of infiltration (Back, 1960, 1966; Seaber, 1965). Differences in chemical concentrations in a ground-water aquifer can be related to: (1) mineralogy of the sediments, (2) nature of the original interstitial fluid, (3) grain size of sediment, (4) organic matter present in the sediments, and (5) amounts and types of soluble salts.

Subsurface information on brine chemistry is based normally on a number of tests from different localities. Samples from each well require separate sampling and analysis, and the areal distribution of brine composition cannot be accurately determined unless a sufficient number of samples has been obtained.

The composition of interstitial waters that are associated with petroleum can vary dramatically because of the numerous possible reactions. According to Dickey (1966), the salinity of waters associated with petroleum generally increases with depth. There are so many exceptions, however, that one cannot generalize (Chilingarian and Rieke, 1968). According to Collins (1970), the Ca/Na ratio usually increases with the depth and age of the associated rocks, whereas the Mg/Na ratio decreases.

Until recently, little study has been undertaken on the vertical variation in chemical composition of fluids in the oil fields. If changes in the fluid chemistry of rocks is somewhat regular and systematic, the vertical and lateral distribution of the chemical constituents may be represented by some suitable mathematical model (see Anikouchine, 1967). At the present time little can be said about the geochemical significance of the chemical frequency distributions, which often tend to be multi-modal for a given formation on the craton (Fig.143). Relationship between total solids and calcium content for various brines in the same formation is shown in Fig.118.

Mekhtiev (1956) showed that in the Azerbayjan S.S.R. oil fields (U.S.S.R.) mineralization of waters is decreasing with stratigraphic depth and calcium-chloride waters are gradually replaced by bicarbonate waters (Tables XXXVII–XXXIX).

As shown in Table XXXVII, the most highly mineralized waters in Bibieybat deposits

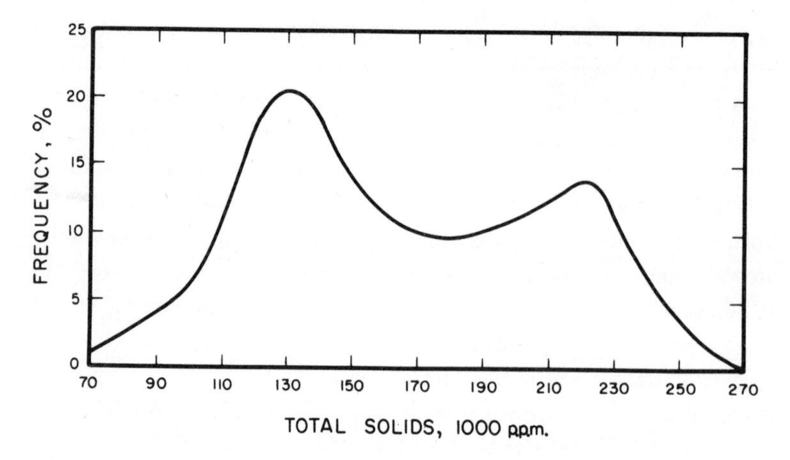

Fig.143. Bimodal frequency distribution of total dissolved solids in the Bartlesville Sandstone of Oklahoma (Pennsylvanian). Strong multimodal distributions are noticed in such sandstones on the craton. (Data after Wright et al., 1957.)

TABLE XXXVII
Hydrochemical characteristics of oil-field waters of Bibieybat (Azerbayjan S.S.R.)
(After Mekhtiev, 1956, p.292)

Horizon	*Salinity of water* (°Bé)	*Total minerlization* (mg-equiv./l)	Cl^- (mg-equiv./l)	SO_4^{2-}	HCO_3^-	*Characteristic coefficients (Palmer)*	
						S_2	A
I	10.2	334	180	0.6	0.8	17	–
IV	10	367	199	0.7	1.2	22	–
VII	11.3	375	197	0.7	1.1	18	–
X	11.3	364	182	0.3	1.0	15.10	–
XI	10.2	365	164	0.6	0.7	14.50	–
XII	9.3	330	157	0.4	1.0	9.70	–
XIII	7.5	249	133	0.3	2.5	1.10	–
XIV ser.	6.9	248	120	0.1	2.8	2	–
XVI ser.	4.2	140	52	0.3	8.8	–	19.5
NKG	3.9	130	52	0.3	1.6	–	21.4
NKP_1	3.5	130	36	0.3	3.8	–	23.6
I-KS	2.5	67	25	0.8	6.3	–	40.9
III-KS	2.2	60	17	0.4	0.3	–	41.3
V-KS	1.9	52	16	0.4	0.9	–	44
PK_1	1.8	46	12	0.3	6.9	–	44
PK_3	1.6	45	12	0.4	5.8	–	36.5

TABLE XXXVIII
Hydrochemical characteristics of formation waters of the petroliferous Balakhano–Sabunchino–Ramaninskiy region
(After Mekhtiev, 1956, p.293)

Horizon	*Salinity of water* (°Bé)	*Total mineralization* (mg-equiv./l)	*Cl⁻* (mg-equiv./l)	HCO_3^- (mg-equiv./l)	*Characteristic coefficients*	
					S_2	*A*
Surakhan Fm.	3.51	103	49	2	18.54	0
I hor.	15	480	238	–	90.00	0
II hor.	12.5–14.2	425	211	1	14.58	0
III hor.	11.5–13.4	427	212	7	15.12	0
IV hor.	10.8	337	166	2	13.80	0
IV a,b	6.0–10.7	268	130	3	5.58	0
IV c	8.0	249	121	3	4.10	0
IV e	5.7–7.0	213	101	5	0	1.04
V	4.7–5.9	169	77	7	0	6.14
VI	4.3–5.5	169	76	8	0	7.46
VII	4.5	135	57	8	0	11.54
VIII	4.5	137	60	8	0	11.40
IX	4.8	138	60	8	0	11.70
NKP	4.5–5.95	147	65	8	0	8.98
I-KS_1	4.9	93	35	11	0	22.88
II-KS_4	–	51	13	11	0	47.04
II-KS_5	1.8	51	16	9	0	37.12
PK	2.0	51	15	8	0	36.38

occur in bed VII (375 mg-equiv./l). Hard waters are replaced by alkaline waters at the boundary of Formation XIV. Sulfate ions are present in negligible amounts, whereas the HCO_3^- content increases with stratigraphic depth from 1 to 14–16 mg-equiv./l in the KS Formation.

In the Balakhano–Sabunchino–Ramaninskiy area, the transition from hard to alkaline waters occurs in Horizon IV (Table XXXVIII). Mineralization decreases with depth from 480 mg-equiv./l (Horizon I) to 50 mg-equiv./l (PK Formation). The content of bicarbonate ions gradually increases with depth reaching 10–16 mg-equiv./l in the KS Formation.

Decrease in mineralization with depth in the Kala region is clearly shown in Table XXXIX. Mekhtiev (1956) explained this decrease in salinity, with depth, to infiltration of atmospheric and sea waters into deeper reservoir rocks.

The sulfate content in these waters is usually less than 0.5% equivalent. The absence of the SO_4^{2-} anion points to its destruction, which leaves H_2S behind. The latter either: (1) remains in the free state, (2) is oxidized giving rise to free sulfur, or (3) forms sulfides, mainly those of iron.

Kissin (1964) found that there is a decrease in the mineralization of solutions squeezed out of Maykop shales (Pre-Caucasus, U.S.S.R.) with increasing pressure (Table XL). The

TABLE XXXIX
Hydrochemical characteristics of waters from petroliferous formations of the Kala region (Azerbayjan S.S.R.)
(After Mekhtiev, 1956, p.294)

Horizon	*Salinity of water* (°Bé)	*Total mineralization* (mg-equiv./l)	Cl^- (mg-equiv./l)	SO_4^{2-}	HCO_3^-	*Characteristic coefficients* S_2	*A*
C	15.5–14.7	509–513	299–262	0.1–1.2	0.3–0.6	21.2–21.36	–
D	15.3–16.8	559–518	281–268	0.1–1.3	0.1–0.7	20.9–22.60	–
II	13.9–14.7	518–437	242–212	0.1–0.9	0.1–0.7	12.5–282.4	–
III	13.9–12.6	470–439	235–219	0.1–0.3	0.1–0.5	18.5–208.2	–
IV	12.7–11.7	422–393	208–203	0.1–0.4	0.2–0.6	17.1–191.0	–
IV_S	12.4– 9.6	412–382	206–192	0.1–0.2	0.1–0.2	15.7–17.5	–
V	10.7– 8.8	349–297	175–155	0.1–0.1	0.6–1.3	10.7–13.2	–
VI	9.5– 8.2	386–282	192–141	0 –0.6	0.6–1.5	10.4– 12.3	–
NKG_1	5.6– 5.5	180–177	88.6–86.6	0 –0.1	1.5–1.8	0	0 – 3.3
NKP	3.7– 4.2	181–129	87.1–68.3	0	3.3–5.4	0	0.76– 7.5
KS_{1-2}	3.5– 3.2	106– 96	45.1–39.8	0	4.3–4.6	0	12.3 –15.4
KS	2.7– 2	75– 58	59 –16.3	0.1–0.6	4.3–5.5	0	22.3 –35.5
KS_{11-12}	2.8– 1.9	69– 48	19.1–14.7	0.1–0.2	6.4–13.7	0	26 –43.4
PK_1	2.3– 1.9	82– 42	25 –14	0.1–0.1	5.5–3.9	0	31 –49.3

clay samples were squeezed at room temperature and high pressures (6,000–8,000 kg/cm^2). The interstitial solutions in clays are of sodium-sulfate type and contain higher amounts of sulfate ion than do formation waters.

A definite relationship exists between the clay density, degree of mineralization and water pressure in reservoir rocks of the Maykop Formation (Fig.145). More deeply buried waters are less mineralized and are of sodium-bicarbonate type (Table XLI).

On the other hand, in terrigenous chalky deposits, with highly compacted clays, there is a gradual increase in mineralization of water with burial depth and increasing distance from recharge areas (Fig.144).

TABLE XL
Composition of pore solutions squeezed out from clays of Maykop age (Takhta-Kugul'tinskiy Deposit, Borehole 76)
(After I.G. Kissin, 1964, in: Pol'ster et al., 1967, p.71)

Depth of sample (m)	*Content* (mg-equiv./kg) CO_2	SO_4^{2-}	HCO_3^-	Ca^{2+}	Mg^{2+}	Na^++K^+	Σ (mg-equiv. /kg)	*Pressure* (kg/cm^2)
440	594.6	613.0	9.5	38.8	372.5	806.7	2,436	580–1,430
440	549.1	593.3	–	29.3	401.7	711.4	2,285	1,800–6,800
431	588.0	198.3	2.9	49.9	123.2	616.2	1,578	290–5,800
433	609.5	176.3	2.4	41.1	139.4	607.7	1,576	290–5,800

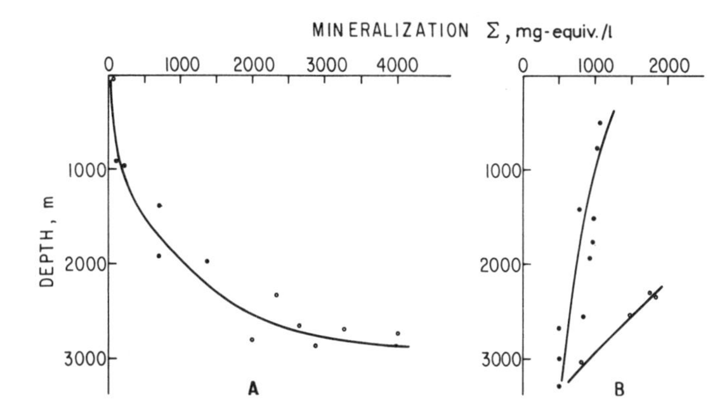

Fig.144. Relationship between mineralization of water and depth of burial. A. Chalky terrigenous deposits. B. Maykop deposits. Solid circles = areas adjacent to Caucasus slope; open circles = platform slope of Pre-Caucasus. (After Pol'ster et al., 1967, fig.34, p.74.)

Jones (1967, 1968, 1969) stated that the salinity of water in sandstone aquifers ranges from less than 1,000 p.p.m. to 300,000 p.p.m. with the salinity diminishing progressively below depths of 8,000–12,000 ft. in the overpressured formations of the Gulf Coast geosyncline. There is a decrease in the salinity of interstitial fluids with depth below the pressure seal in these geopressured reservoirs.

TABLE XLI

Changes in mineralization of waters of Lower Maykop deposits in eastern Pre-Caucasus area (After Pol'ster et al., 1967, p.73)

Region	*Distance from area of open deposits of rocks* (km)	*Area*	*Bore-hole no.*	*Depth of sampling* (m)	*Mineralization* (mg-equiv./l)
Caucasus Monocline	6– 7	Datykh	3	–	850
		Kirovskaya	26	477– 485	1,036
		Kirovskaya	2	747– 758	966
Frontal Mountains	15– 20	Karabulak	12	1,940–1,922	979
		Achaluki	4	1,728–1,736	936
		Achaluki	63	1,500–1,511	969
		Khayan-Kort	2	3,301–3,305	556
Between Rivers Terek and Sulak	50–100	Kraynovka	1	2,524–2,530	803
		Kraynovka	2	2,809–2,810	606
		Kraynovka	2	2,873–2,868	482
Near-Kumsk	125–130	Ozek-Suat	6	2,300–2,332	1,679
		Achikulak	4	2,532–2,497	1,849

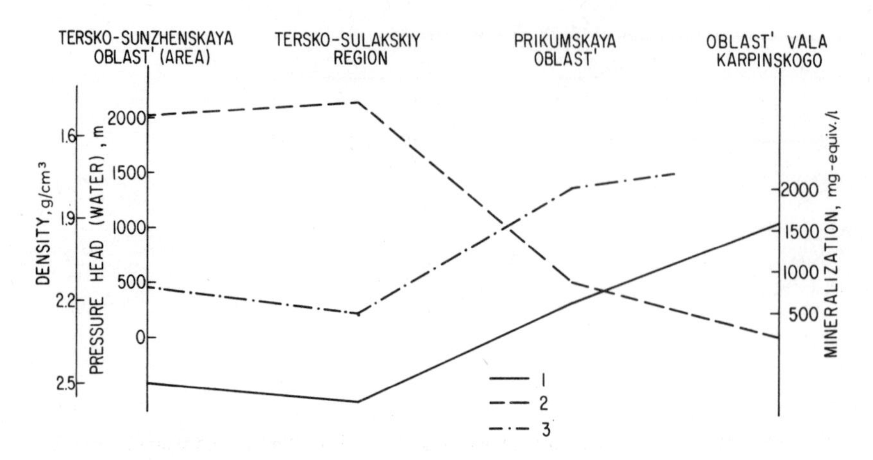

Fig.145. Relationship between density of Paleogene clays and water pressure in reservoir rocks. 1 = density (g/cm^3) of clays; 2 = water pressure in the reservoir rocks; and 3 = mineralization of waters (mg-equiv./l) in reservoir rocks. (After Pol'ster et al., 1967, fig.33, p.74.)

SALINITY DISTRIBUTION IN SANDSTONES AND ASSOCIATED SHALES

In 1947, De Sitter reported that the salinity of formation waters in sandstones varied from that of fresh water to ten times the salinity of sea water. The distribution of the salinity of interstitial waters present in young geosynclinal sediments along the Gulf Coast has been well-documented by Timm and Maricelli (1953), Myers (1963) and Fowler (1968).

Timm and Maricelli (1953, p.394) stated that high salinities up to 4½ times that of normal sea water characterize the interstitial solutions in Miocene–Pliocene sediments, where the relative quantity of undercompacted shale is small. In Eocene–Oligocene sediments, where the relative quantity of shale is large and the degree of compaction is high, interstitial solutions have salinities as low as one-half that of normal sea water. Fig.146 illustrates their concept that the formation waters in down-dip, interfingering, marine sandstone members, which have proportionately less volume than the associated massive shales, have lower salinities than that of sea water. More massive sands up-dip have salinities greater than that of sea water. Salinity was determined by using the following techniques: electrical resistivity, complete chemical analysis, and titration (see Gullikson et al., 1961a). Calculations showed that all water samples, of which complete mineral analyses were made, are secondary saline according to the Palmer's system of water analyses interpretation.

Myers (1963) studied the chemical properties of formation waters, down to a depth of 12,400 ft., in four producing oil wells in Matagorda County, Texas. The salinities of interstitial waters ranging from 5,000 p.p.m. to 12,500 p.p.m. are found below 10,000 ft. in each of the four wells, as compared to salinities of ≈ 70,000 p.p.m. above this depth.

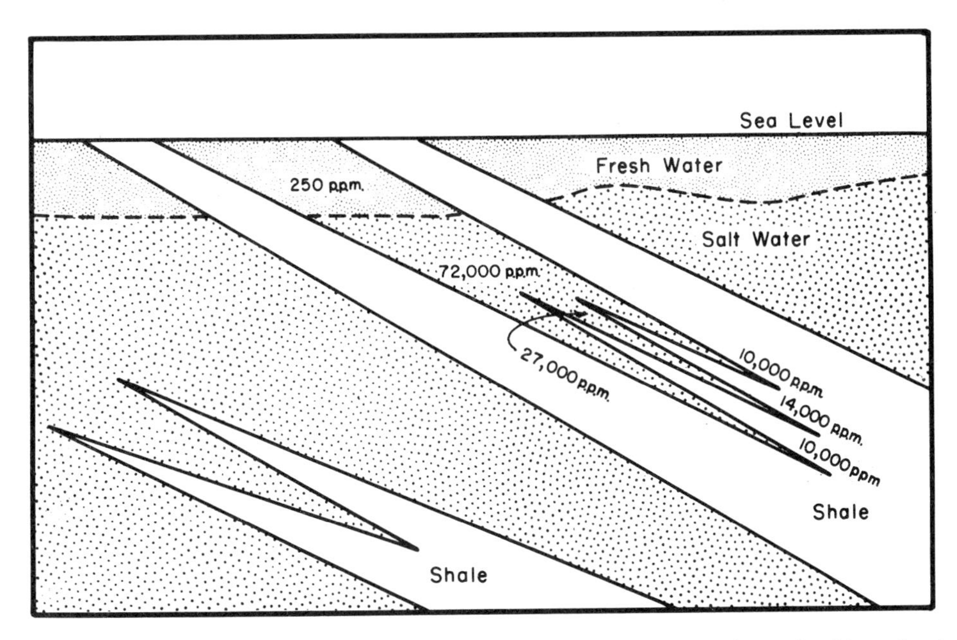

Fig.146. Idealized typical cross section of some sands and shales in southwest Louisiana showing generalized salinity relationships. (After Timm and Maricelli, 1953, p.396, 397, and 408. Courtesy of American Association of Petroleum Geologists.)

He commented that in this deeper section, the proportion of massive shale is large and the sands are near their downdip limits (become thinner). These results were in close accord with those of Timm and Maricelli (1953).

Some investigators including Hottman and Johnson (1965) observed that the sands with abnormally high pore-water pressures as associated with undercompacted shales having very high porosity. In an excellent paper, Dickey et al. (1968) observed that faults, which transect oil reservoirs, form pressure discontinuities and act as seals for zones of high fluid pressure for long periods of time. The high porosity of shales in such zones is reflected by the high values of conductivity. The depth marking the beginning of the abnormally high fluid pressures in the sandstones coincides with abnormal increase in conductivity of associated undercompacted shales (Williams et al., 1965; Wallace, 1965). Yet, calculations by the writers indicate that possibly high porosity of shales alone could not account for this abnormal increase in conductivity; the salinity of interstitial waters also appears to be important.

Although the process of clay compaction is continuous until the complete (?) lithification of clays, the volume of squeezed-out fluids into reservoir rocks gradually decreases. Consequently, there is a gradual decrease in excessive pore-water pressure in highly permeable horizons in deep parts of the basin until, finally, it becomes equal to the hydrostatic pressure. After that, theoretically the movement of waters occurs in the opposite

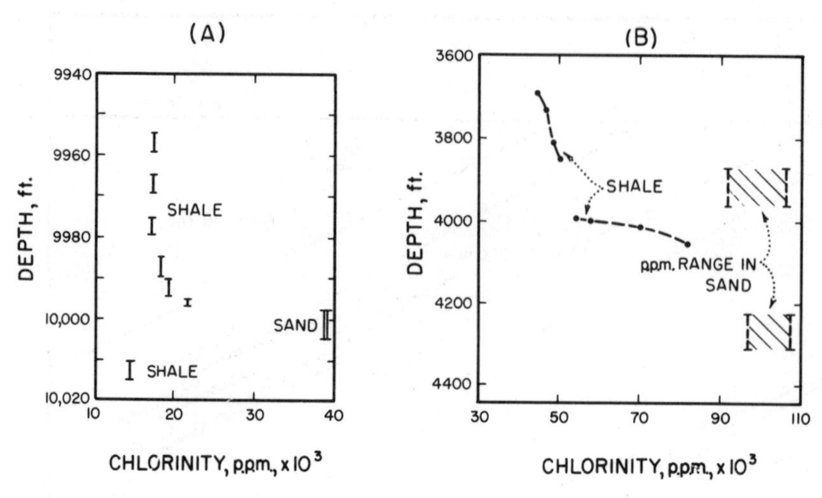

Fig.147. Chloride concentrations in shales and sands. (In: Fertl and Timko, 1970, fig.4, p.15, based on data by Hedberg, 1967.)

direction, from the periphery of the basin (recharge areas). The entire hydrodynamic system strives to attain equilibrium, which is controlled by areal distribution and size of recharge and discharge areas.

The findings of Fowler (1968) for the Chocolate Bayou Field, Brazoria County, Texas, seem to suggest that the salinity of water in undercompacted shales is higher than in well-compacted ones. He discovered a definite correlation between the high salinity of interstitial fluids and abnormally high pressures. This is possibly owing to the fact that undercompacted shales did not have a chance to contribute their fresher water to associated sandstones. In addition, he studied the variation in salinity of produced water with time. The typical pattern is one of decreasing salinity with time, and the freshest water is found in sands receiving most of this water from associated shales. This is in agreement with the experimental results of several investigators, as discussed earlier, which indicate that salinity of waters in shales is possibly less than that in associated sands.

The pore-water salinities in shales and in associated sandstones are compared in Fig.147. The examples include field case studies from the Middle East and Texas. Some data from an offshore well in Louisiana have been given by Fertl and Timko (1970c). In all cases, the salinities of interstitial fluids in shales are considerably lower than those in associated sandstones.

SQUEEZING OF OIL AND BITUMENS OUT OF MUDS AND SHALES

The possible effect of interstitial fluids on the formation of hydrocarbon deposits is of considerable importance and is discussed briefly here. Among petroleum engineers and geologists it is generally agreed that hydrocarbons are transported mainly by water. There is considerable disagreement, however, as to the state in which hydrocarbons migrate. Pol'ster et al. (1967) published an excellent book on the migration of pore fluids.

Compaction as a possible cause of primary migration of oil

Linetskiy (1956, in: Dvali, 1959, p.210), on the basis of results obtained by Lomtadze, concluded that removal of free water occurs at depths of 200–400 m, and that removal of bound water starts at the following pressures:

	Pressure (kg/cm^2)	Moisture content (%)
Bentonite	500	32.5
Kaolin	580	8
Cambrian clay	300	10
Clear quartz sand	540	15.5

These results, however, do not agree with those obtained by Chilingar and Knight (1960) and cited previously.

As shown in Fig.30, the moisture content at an overburden pressure of 1,000 p.s.i., for example, is 50% for montmorillonite clay (hydrated in distilled water) as compared to 16% for kaolinite clay (hydrated in distilled water). This suggests that compaction possibly plays an important role in primary migration of oil from source rocks (Weaver, 1967); namely, oil is pushed out by the available water (oriented?). Possibly, many oil shales containing non-hydratable clays remain as such because of lack of water to do the job.

On increasing pressure to 1,000 kg/cm^2 (14,223 p.s.i.), Lomtadze (in: Eremenko, 1960, p.537) found that 57% of the oil content is squeezed out of Cambrian clay with an original oil and water saturation of 14% each. In another series of tests, Cambrian clay contained 7% oil and 21% of water. In this case, at a pressure of 1,000 kg/cm^2 only 47% of the original oil was squeezed out. Cambrian clay, bentonite, and kaolin having initial oil saturations of 29.2, 32.2, and 52.4%, respectively, were subjected to a pressure of 1,000 kg/cm^2 by Lomtadze (in: Porfir'ev and Brod, 1959, p.29). Upon compaction, the remaining oil saturations were 9.1, 10.2, and 15.3, respectively.

The following experiment was undertaken by Chilingar (1961) using kaolinite clay which contained 20% of oil and 20% of water by weight in one series of tests and 10% oil and 30% water in another series. The remaining oil saturation after one month of compaction at 14,000 p.s.i. was 6.1% in the first instance and 4.9% in the second. Thus, it appears that with larger initial concentrations of oil, a greater percentage of oil can be squeezed out.

Lomtadze (1956, in: Porfir'ev and Brod, 1959) also squeezed samples of clays with various bitumen contents (determined by chloroform extraction). The results, which are

TABLE XLII
Effect of compaction to 100 kg/cm^2 (1,422 p.s.i.) on the bitumen content of some clays and shales (chloroform extraction was employed to remove the hydrocarbons from the samples)
(After Lomtadze, 1956, in: Porfir'ev and Brod, 1959, p.30)

Sediment and location	*Bitumen content (% of dry weight of sediment)*	
	before compaction	*after compaction*
Lacustrine–marine mud from Saki area, U.S.S.R.	0.27	0.28
"Ioldian" clay, Karelia, U.S.S.R.		
sample 1	0.017	0.099
sample 2	0.021	0.077
Maykop clay, Krasnodar area, U.S.S.R.		
sample 1	0.05	0.05
sample 2	0.13	0.18
Domanican shales, Ukhta, U.S.S.R.	2.88	2.49

presented in Table XLII, indicate that some bitumens are squeezed out when their initial concentration is high; whereas at low initial concentrations of bitumens, compaction only makes them more available for extraction.

The degree of hydrocarbon solubility in water is different for various hydrocarbons (Pol'ster et al., 1967). Generally an increase in their molecular weight decreases the solubility of hydrocarbons in aqueous solutions. On migrating through muds or shales as a result of compaction, subterranean fluids may increase considerably in salinity. The solubility of hydrocarbons is increased by the presence of higher quantities of electrolytes; however, brines saturated with certain inorganic salts will eventually release their hydrocarbons. In the case of an emulsion transfer, salts act as demulsifying agents. In both circumstances, an organic or oil phase will develop as soon as the solubility product of NaCl is exceeded (Degens et al., 1962).

Temperature gradient as a driving force for migration of petroleum

According to Watts (1963), the temperature gradient in sediments may act as a driving force for migration of petroleum into reservoir rocks. The temperature gradient could be effective due to its influence on both the adsorption and diffusion processes involved in petroleum migration. Organic molecules, which are abundant in a wide variety of argillaceous sediments, appear to adsorb on fine particles of the sediments. It seems that far more organic compounds are contained in sedimentary rocks in a dispersed form than in a concentrated state in reservoir rocks (Meinschein, 1961; Hunt, 1961; Vassoevich, in: Chilingar, 1964). Baker (1959, 1960) proposed an accumulation process whereby hydrocarbons are taken into aqueous solution in micellar form from sedimentary rocks and migrate in solution to reservoir rocks. Migration may take place in solution either by the diffusion of single molecules or by the diffusion of micelles in a colloidal solution (Baker, 1960). The rate of any such diffusion will be affected by the temperature gradient owing to thermal diffusion (Watts, 1963, p.926). Inasmuch as the temperature of the earth's crust shows a general decrease towards the surface, this temperature gradient could provide a suitable driving force for the migration of petroleum into a reservoir. Watts (1963) suggested that the effect of surfactants (surface-active agents), which are commonly present and assist micelle formation, and the ionic strength of the aqueous phase on the adsorption and diffusion should be investigated.

Chemistry of oil-field water dissolved in crude oil

One possible relationship between the chemistry of oil-field brines and associated crude oil may be due to the fact that there is high solubility of water in liquid hydrocarbons at high temperatures and pressures, and that the effect of temperature (and water solubility) on the selectivity for water over salt is notable. The extract phase contains less salt as the temperature and, consequently, the solubility of water are increased (Barton and Fenske, 1970, p.20). For example above 600°F, the extracted desalted water which was in solution in hydrocarbons contained 120 p.p.m. (by weight) or less of salt, well below the drinking-water standards of 250 mg/l of NaCl and 500 mg/l of total dissolved solids (Barton and Fenske, 1970, p.20). For C_6 hydrocarbons, *n*-hexane and benzene, the solubilities of water are 7.4–26% by weight (28–60 mole % water) at 432° and 515°F, respectively. Inasmuch as increasing pressure lowers the solubility of water in hydrocarbons, the low-salinity water will separate out from the hydrocarbon phase if exposed to high overburden pressures. Lowering of temperature will have the same effect; for example, when fluids migrate upwards in a sedimentary column. According to Hess et al. (1967), water solubility is more sensitive to temperature than to pressure changes.

Relationship of trace elements in crude oil and associated shales

Bonham (1956) reported that nickel and vanadium concentrations in Pennsylvanian crude oils from the Seminole Basin in Oklahoma reflected the paleogeography of the basin. Highest concentrations of these trace metals occurred near the ancient shorelines; the concentration decreased progressively basinward.

Twenty-five samples of crude oil from the Chesterian (Upper Mississippian) Waltersburg Formation and 11 core samples of Waltersburg shales in the Illinois Basin were analyzed for trace concentrations of nickel and vanadium (Mast et al., 1968). Sediment deposition in the basin during Chesterian time was rhythmic, with sandstone and shale units separated by units of limestone and shale. Samples of shales were spectrochemically analyzed to establish a relationship between the nickel and vanadium trace concentrations (Mast et al., 1968, p.14). No regular variation occurs laterally in the concentrations of nickel and vanadium in the Waltersburg shales. The vanadium concentration is almost twice the nickel concentration in shale, whereas the vanadium concentration is approximately one-third the nickel concentration in the oil. The isocons (lines of equal concentration) of the trace quantities of nickel and vanadium in the crudes are perpendicular to the depositional strike of the basin. The evidence is inconclusive as to whether or not the nickel and vanadium contents of the crudes are related to those of shales. In future studies, the composition of associated oil-field brines should also be considered.

REFERENCES

Adamson, L.G., 1966. *Application of Electrokinetic Phenomena in Dewatering, Consolidation and Stabilization of Soils and Weak Rocks in Civil and Petroleum Engineering, and Augmenting Reservoir Energy During Petroleum Production.* Thesis, Univ. Southern Calif., 406 pp.

Adamson, L.G., Chilingar, G.V., Beeson, C.M. and Armstrong, R.A., 1966. Electrokinetic dewatering, consolidation and stabilization of soils. *Eng. Geol.,* 1(4): 291–304.

Alekin, O.A., 1953. *Osnovy Gidrokhimii* (Principles of Hydrochemistry). Gidrometeoizdat, Leningrad, 296 pp.

Amba, S.A., 1963. *Use of Direct Electrical Current for Increasing the Flow Rate of Reservoir Fluids During Petroleum Recovery.* Thesis, Univ. Southern Calif., 255 pp.

Anderson, D.M. and Low, P.F., 1958. The density of water adsorbed by lithium-, sodium-, and potassium-bentonite. *Soil Sci. Soc. Am. Proc.,* 22(2): 99–103.

Anikouchine, W.A., 1967. Dissolved chemical substances in compacting marine sediments. *Geophys. Res.,* 72(2): 505–509.

Athy, L.F., 1930. Compaction and oil migration. *Bull. Am. Assoc. Pet. Geologists,* 14(1): 25–35.

Back, W., 1960. Origin of hydrochemical facies of ground water in the Atlantic Coastal Plains. *Int. Geol. Congr., Copenh., Rept. 21st Sess., Norden,* Part 1 : 87–95.

Back, W., 1966. Hydrochemical facies and ground-water flow patterns in northern part of Atlantic Coastal Plain. *U.S., Geol. Surv., Profess. Pap.,* 498-A: 42 pp.

Baker, E.G., 1959. Origin and migration of oil. *Science,* 129: 871–874.

Baker, E.G., 1960. A hypothesis concerning the accumulation of sediment hydrocarbons to form crude oil. *Geochim. Cosmochim. Acta,* 19: 309–317.

Barton, P. and Fenske, M.R., 1970. Hydrocarbon extraction of saline waters. *Ind. Eng. Chem. Process Des. Dev.,* 9(1): 18–25.

Bentor, Y.K., 1961. Some geochemical aspects of the Dead Sea and the question of its age. *Geochim. Cosmochim. Acta,* 25:239–260.

Berry, F.A.F., 1968. Membrane filtration or hydrodynamics of brine fluids in Paleozoic rocks of Kansas and Nebraska. *Chem. Geol.,* 4(1–2): 295–301.

Bischoff, J.L. and Ku, Teh-Lung, 1970. Pore fluids of recent sediments. 1. Oxidizing sediments of 20°N Continental Rise to Mid-Atlantic Ridge. *J. Sediment. Petrol.,* 40(3): 960–972.

Bogomolov, G.V. et al. (Editors), 1968. *Pore Solutions and Methods of Their Study* (Symposium). Izd. Nauka i Tekhnika, Minsk, 231 pp.

Bonham, L.C., 1956. Geochemical investigation of crude oils. *Bull. Am. Assoc. Pet. Geologists,* 40(5): 897–908.

Borovitskiy, V.P., 1969. Effect of natural electric potentials on migration of moisture and contained components in active layer. In: O.P. Bulygin et al. (Editors), *Questions of Hydrogeology and Hydrogeochemistry* (Symposium). Akad. Nauk. S.S.S.R. (Siberian Branch), (4): 180–186.

Buneeva, A.N., Kryukov, P.A. and Rengarten, E.V., 1947. Experiment in squeezing out of solutions from sedimentary rocks. *Dokl. Akad. Nauk. S.S.S.R.,* 57(7): 707–709.

Burst Jr., J.F., 1959. Diagenesis of Gulf Coast clayey sediments and its possible relation to petroleum migration. *Bull. Am. Assoc. Pet. Geologists,* 53(1): 73–93.

Cebell, W.A. and Chilingarian, G.V., 1972. Some data on compressibility and density anomalies in halloysite, hectorite and illite clays. *Bull. Am. Assoc. Pet. Geologists,* 56(4): 796–802.

Chave, K.E., 1960. Evidence on history of sea water from chemistry of deeper subsurface waters of ancient basins. *Bull. Am. Assoc. Pet. Geologists,* 44(3): 357–370.

Chebotarev, I.I., 1955. Metamorphism of natural waters in the crust of weathering, 1–3. *Geochim. Cosmochim. Acta,* 8: 22–48; 137–170.

Chilingar, G.V., 1956. Relationship between Ca/Mg ratio and geologic age. *Bull. Am. Assoc. Pet. Geologists,* 40: 2256–2266.

Chilingar, G.V., 1961. Notes on compaction. *J. Alberta Soc. Pet. Geologists,* 9: 158–161.

Chilingar, G.V., 1964. Established facts concerning origin of oil as viewed by N.B. Vassoevich. *Compass,* 41(4): 254–256.

Chilingar, G.V. and Bissell, H.J., 1963. Formation of dolomite in sulfate-chloride solutions. *J. Sediment. Petrol.,* 33(3): 801–803.

Chilingar, G.V. and Degens, E.T., 1964. Notes on chemistry of oil-field waters. *Bol. Assoc. Mexicana Geol. Petrols.,* 15(7–8): 177–193.

Chilingar, G.V. and Knight, L., 1960. Relationship between pressure and moisture content of kaolinite, illite and montmorillonite clays. *Bull. Am. Assoc. Pet. Geologists,* 44(1): 101–106.

Chilingar, G.V., Rieke III, H.H., Sawabini, S.T. and Ershagi, I., 1969. Chemistry of interstitial solutions in shales versus that in associated sandstones. *Soc. Pet. Engrs., Am. Inst. Min. Metall. Engrs., 44th Ann. Fall Meet., Denver, Colo.,* Paper no.2527: 8 pp.

Chilingarian, G.V. and Rieke III, H.H., 1968. Data on consolidation of fine-grained sediments. *J. Sediment. Petrol.,* 38(3): 811–816.

Coleman, J.M., 1967. Consolidation and early diagenesis of clays. *Ann. Symp. Abnorm. Subsurface Pressures, 1st,* Louisiana State Univ., Baton Rouge, La.

Collins, A.G., 1970. Geochemistry of some petroleum-associated waters from Louisiana. *U.S., Bur. Mines, Rep. Invest.,* 7326: 1–31.

Davis, J.B., 1967. *Petroleum Microbiology.* Elsevier, Amsterdam, 618 pp.

Davis, L.E., 1955. Electrochemical properties of clays. In: J.A. Park and M.D. Turner (Editors), *Clays and Clay Technology, Proc. Natl. Conf. Clays Clay Technol., 1st, State of Calif., Div. Mines Bull.,* 169: 47–53.

Degens, E.T. and Chilingar, G.V., 1967. Diagenesis of subsurface waters. In: G. Larsen and G. V. Chilingar (Editors), *Diagenesis in Sediments.* Elsevier, Amsterdam, pp.477–502.

Degens, E.T. and Epstein, S., 1962. Relationship between $^{18}O/^{16}O$ ratios in coexisting carbonates, cherts, and diatomites. *Bull. Am. Assoc. Pet. Geologists,* 46(4): 534–542.

Degens, E.T., Chilingar, G.V. and Pierce, W.D., 1962. Sobre el origen del petroleo dentro de concreciones de carbonato de calcio de edad Miocenica, formadas en aguas dulces. *Bol. Asoc. Mexicana Geol. Petrols.,* 14: 275–292.

Degens, E.T., Hunt, J.M., Reuter, J.H. and Reed, W.E., 1964. Data on the distribution of amino-acids and oxygen isotopes in petroleum brine waters of various geologic ages. *Sedimentology,* 3: 199–225.

DeSitter, L.U., 1947. Diagenesis of oil-field brines. *Bull. Am. Assoc. Pet. Geologists,* 31(11): 2030–2040.

Dickey, P.A., 1966. Patterns of chemical composition in deep subsurface waters. *Bull. Am. Assoc. Pet. Geologists,* 50(11): 2472–2477.

Dickey, P.A., Shriram, C.R. and Paine, W.R., 1968. Abnormal pressures in deep wells of southwestern Louisiana. *Science,* 160(3828): 609–615.

Durov, S.A., 1948. Classification of natural waters and graphical presentation of their composition. *Dokl. Akad. Nauk S.S.S.R.,* 59(1): 87–90.

Dvali, M.F., 1959. Possible factors and processes of primary oil migration. *Tr. VNIGRI,* 1959(132): 204–241.

Emery, K.O. and Rittenberg, S.C., 1952. Early diagenesis of California Basin sediments in relation to origin of oil. *Bull. Am. Assoc. Pet. Geologists,* 36(5) : 735–806.

Eremenko, N.A. (Editor), 1960. *Geology of Petroleum. I. Principles of Geology of Petroleum.* Gostoptekhizdat, Moscow, 592 pp.

Eremenko, N.A. and Neruchev, S.G., 1968. Primary migration during process of burial and lithogenesis of sediments. *Geol. Nefti i Gaza*, 1968(9): 5–8.

Fertl, W.H. and Timko, D.J., 1970a. Occurrence and significance of abnormal pressure formations. *Oil Gas J.*, 68(1): 97–108.

Fertl, W.H. and Timko, D.J., 1970b. How abnormal pressure detection techniques are applied. *Oil Gas J.*, 68(2): 62–71.

Fertl, W.H. and Timko, D.J., 1970c. Association of salinity variations and geopressures in soft and hard rock. In: *Soc. Prof. Well. Log. Analysts, 11th Ann. Logging Symp., May 3–6,* pp.1–24.

Filatov, K.V., 1961. Silicate waters and their position in horizontal hydrochemical zonation. *Dokl. Akad. Nauk S.S.S.R.*, 138(3): 663–666.

Fowler Jr., W.A., 1968. Pressure, hydrocarbon accumulation and salinities, Chocolate Bayou Field, Brazoria County, Texas. *Soc. Pet. Engrs., Am. Inst. Min. Metall. Engrs.*, Preprint, SPE 2226: 9 pp.

Friedman, G.M., Fabriacand, B.P., Imbimbo, E.S., Brey, M.E. and Sanders, J.E., 1968. Chemical changes in interstitial waters from continental shelf sediments. *J. Sediment. Petrol.*, 38(4): 1313–1319.

Glasstone, S., 1946. *Textbook of Physical Chemistry.* Van Nostrand, New York, N.Y., 1320 pp.

Gorrel, H.A., 1958. Classification of formation waters based on sodium chloride content. *Bull. Am. Assoc. Pet. Geologists,* 42(10): 2513.

Guoy, G., 1910. Sur la constitution de la charge électrique à la surface d'un électrolyte. *Ann. Phys. (Paris),* Sér.4, 9: 457–468.

Graf, D.L., Friedman, T. and Meents, W.F., 1965. The origin of saline formation waters. II. Isotopic fractionation by shale micropore systems. *Illinois State Geol. Surv., Circ.,* 393: 32 pp.

Graf, D.L., Meents, W.F., Friedman, T. and Shimp, N.F., 1966. The origin of saline formation waters. III. Calcium chloride waters. *Illinois State Geol. Surv., Circ.,* 397: 60 pp.

Grim, R.E., 1968. *Clay Mineralogy.* McGraw-Hill, New York, N.Y., 596 pp.

Gullikson, D.M., Caraway, W.H. and Gates, G.L., 1961a. Chemical analysis and electrical resistivity of selected California oil-field waters. *U.S., Bur. Mines, Rep. Invest.,* 5736: 21 pp.

Gullikson, D.M., Caraway, W.H. and Gates, G.L., 1961b. Applying modern instrumental techniques to oilfield water analysis. *U.S. Bur. Mines, Rep. Invest.,* 5737: 45 pp.

Gurevich, V.T., 1960. About metamorphization of underground waters in the process of catagenesis. *Geol. i Geochim.,* 3: 259–263.

Hanshaw, B.B., 1962. *Membrane Properties of Compacted Clays.* Thesis, Harvard Univ., Cambridge, Mass., 113 pp.

Hanshaw, B.B. and Zen, E-An, 1965. Osmotic equilibrium and overthrust faulting. *Geol. Soc. Am. Bull.,* 76(12): 1379–1386.

Hawkins, M.E., Dietzman, W.D. and Seward, J.M., 1963a. Analysis of brines from oil-productive formations in South Arkansas and North Louisiana. *U.S., Bur. Mines, Rep. Invest.,* 6282: 28 pp.

Hawkins, M.E., Jones, C.W. and Pearson, C.A., 1963b. Analysis of brines from oil-productive formations in Mississippi and Alabama. *U.S., Bur. Mines, Rep. Invest.,* 6167: 22 pp.

Hawkins, M.E., Dietzman, W.D. and Pearson, C.A., 1964. Chemical analysis and electrical resistivities of oilfield brines from field in East Texas. *U.S., Bur. Mines, Rep. Invest.,* 6422: 20 pp.

Hedberg, W.H., 1967. *Pore-Water Chlorinities of Subsurface Shales.* Ph.D. Diss., Univ. Wisconsin, Madison, Wisc., 121 pp.

Hess, H.V., Guptill Jr., F.E. and Carter, N.D., 1967. *U.S. Patent,* 3,325,400.

Ho, C. and Coleman, J.M., 1969. Consolidation and cementation of recent sediments in the Atchafalaya Basin. *Geol. Soc. Am. Bull.,* 80(2): 183–191.

Hottman, C.E. and Johnson, R.K., 1965. Estimation of formation pressures from log-derived shale properties. *J. Pet. Tech.,* 16(6): 717–722.

Hunt, J.M., 1961. Distribution of hydrocarbons in sedimentary rocks. *Geochim. Cosmochim. Acta,* 22: 37–49.

Jones, P.H., 1967. The hydrodynamics of aquifer systems in the Gulf Basin. *Ann. Symp. Abnorm. Subsurface Pressures, 1st*, Louisiana State Univ., Baton Rouge, La., pp. 91–201.
Jones, P.H., 1968. Hydrodynamics of geopressure in northern Gulf of Mexico Basin. *Soc. Pet. Engrs., Am. Inst. Min. Metall. Engrs.*, Preprint, SPE 2207: 10 pp.
Jones, P.H., 1969. Hydrodynamics of geopressure in the northern Gulf of Mexico Basin. *J. Pet. Tech.*, 21: 803–810.
Kartsev, A.A., Vagin, S.B. and Baskov, E.A., 1969. *Paleohydrogeology.* Nedra, Moscow, 151 pp.
Kazintsev, E.A., 1968. Pore solutions of Maykop Formation of Eastern Pre-Caucasus and methods of squeezing of pore waters at high temperatures. In: G.V. Bogomolov et al. (Editors), *Pore Solutions and Methods of Their Study* (A Symposium). Izd. Nauka i Tekhnika, Minsk, pp.178–190.
Kissin, I.G., 1964. *Eastern Pre-Caucasus Artesian Basin.* Nauka, Moscow.
Komarova, N.A. and Knyazeva, N.V., 1968. Extrusion of soil solutions by centrifuge method. In: G.V. Bogomolov et al. (Editors), *Pore Solutions and Methods of Their Study* (A Symposium). Izd. Nauki i Tekhnika, Minsk, pp.205–214.
Konzewitsch, N., 1967. *Estudio de las Clasificaciones Propuestas para Aguas Naturales Segun su Composicion Quimica.* Secretaria de Estado de Energia y Mineria, Buenos Aires, Republica Argentina, 108 pp.
Kotova, I.S. and Pavlov, A.N., 1968. About leaching of impervious rocks on increasing geostatic pressure. In: G.V. Bogomolov et al. (Editors), *Pore Solutions and Methods of Their Study* (A Symposium). Izd. Nauka i Tekhnika, Minsk, pp.55–68.
Krasintseva, V.V. and Korunova, V.V., 1968. Influence of pressure and temperature on composition of extruded solutions during mud compaction. In: G.V. Bogomolov et al. (Editors), *Pore Solutions and Methods of Their Study* (A Symposium). Izd. Nauka i Tekhnika, Minsk, pp.191–204.
Krejci-Graf, K., Hecht, F. and Palser, W., 1957. Über Ölfeldwasser des Wiener Beckens. *Geol. Jahrb.*, 74: 161–209.
Kruyt, H.R. (Editor), 1949. *Reversible Systems. Colloid Science. II.* Elsevier, Amsterdam, 733 pp.
Kryukov, P.A., 1961. Toward procedures of squeezing out of solutions from sedimentary rocks. *Gidrokhim. Inst. Novocherk., Gidrokhim. Mater.*, 33: 191–197.
Kryukov, P.A. and Komarova, N.A., 1954. Concerning squeezing out of water from clays at very high pressures. *Dokl. Akad. Nauk S.S.S.R.*, 99(4): 617–619.
Kryukov, P.A. and Zhuchkova, A.A., 1963. Physical-chemical phenomena associated with driving out of solutions from rocks. In: *Present Day Concepts of Bound Water in Rocks.* Izd. Akad. Nauk S.S.S.R., Laboratory of Hydrogeological Problems of F.P. Savarenskiy, Moscow, pp.95–105.
Kryukov, P.A., Zhuchkova, A.A. and Rengarten, E.V., 1962. Change in the composition of solutions pressed from clays and ion exchange resins. *Dokl. Akad. Nauk. S.S.S.R.*, 144(4): 1363–1365.
Kuznetsov, S.I., Ivanov, M.V. and Lyalikova, N.N., 1962. *Introduction to Geologic Microbiology.* Izd. Akad. Nauk S.S.S.R., Moscow.
Lang, W.J., 1967. The influence of pressure on the electrical resistivity of clay-water systems. *Clays Clay Miner., Proc. Natl. Conf. Clays Clay Miner., 14(1965)*: 11 pp.
Larsen, T.J., 1967. Purification of subsurface waters by reverse omosis. *J. Am. Water Works Assoc.*, 59(12): 1527–1548.
Larsen, G. and Chilingar, G.V. (Editors), 1967. *Diagenesis in Sediments.* Elsevier, Amsterdam, 524 pp.
Lindgren, W., 1933. *Mineral Deposits.* McGraw-Hill, New York, N.Y., 4th ed., 930 pp.
Lomonosov, I.S. and Pinneker, E.V., 1969. Experiment in systematizing mineral waters using as an example Siberian Platform. *Izd. Akad. Nauk S.S.S.R., Ser. Geol.*, 1969(12): 112–118.
Long, G., Neglia, S. and Rubino, E., 1970. Pore fluids in shales and its geochemical significance, pp.191–217. In: G.D. Hobson and G.C. Speers (Editors), *Advances in Organic Geochemistry.* Pergamon, New York, N.Y., 577 pp.
Magara, K., 1968. Compaction and migration of fluids in Miocene mudstone, Nagaoka Plain, Japan. *Bull. Am. Assoc. Pet. Geologists*, 52(12): 2466–2501.
Mangelsdorf Jr., P.C., Manheim, F.T. and Gieskes, J.M.T.M., 1970. Role of gravity, temperature gradients, and ion-exchange media in formation of fossil brines. *Bull. Am. Assoc. Pet. Geologists*, 54(4): 617–626.

Manheim, F.T., 1966. A hydraulic squeezer for obtaining interstitial water from consolidated and unconsolidated sediments. *U.S., Geol. Surv., Profess. Pap.*, 550–C: 256–261.

Manheim, F.T. and Sayles, F.L., 1970. Brines and interstitial brackish water in drill cores from the deep Gulf of Mexico. *Science,* 170: 57–61.

Martin, R.T., 1960. Water vapor sorption on kaolinite: Entropy of adsorption. *Clays Clay Miner., Proc. Natl. Conf. Clays Clay Miner., 8(1959),* pp.102–114.

Martin, R.T., 1962. Adsorbed water on clay: A review. *Clays Clay Miner., Proc. Natl. Conf. Clays Clay Miner., 9(1960),* pp.28–70.

Mast, R.F., Shimp, N.F. and Witherspoon, P.A., 1968. Geochemical trends in Chesterian (Upper Mississippian) Waltersburg crudes of the Illinois Basin. *Illinois State Geol. Surv., Circ.*, 421: 27 pp.

McKelvey, J.G. and Milne, I.H., 1962. Flow of salt solutions through compacted clay. *Clays Clay Miner., Proc. Natl. Conf. Clays Clay Miner., 11(1962),* pp.248–259.

McKelvey, J.G., Spiegler, K.S. and Wyllie, M.R.J., 1957. Salt filtering by ion-exchange grains and membranes. *J. Phys. Chem.*, 61: 174–178.

Meade, R.H., 1963. Factors influencing the pore volume of fine-grained sediments under low-to-moderate overburden loads. *Sedimentology,* 2: 235–242.

Meinschein, W.G., 1961. Significance of hydrocarbons in sediments and petroleum. *Geochim. Cosmochim. Acta,* 22: 58–64.

Mekhtiev, Sh.F., 1956. *Questions on Origin of Oil and Formation of Petroleum Deposits of Azerbayjan.* Izd. Akad. Nauk Azerb. S.S.R., Baku, 320 pp.

Mekhtieva, V.L., 1962. Distribution of micro-organisms in formation waters of Kuybyshev along the Volga and adjacent regions. *Geokhimiya,* 1962(8): 707–719.

Myers, R.L., 1963. *Dynamic Phenomena of Sediment Compaction in Matagorda County, Texas.* Thesis, Univ. Houston, Houston, Texas, 62 pp.

Mysels, K.J., 1959. *Introduction to Colloid Chemistry.* Interscience, New York, N.Y., 475 pp.

Nemykin, A.K., 1968. Effect of surface conductivity on the electrical resistivity of rocks. *Razved. Geofiz.*, 1968(26): 101–109.

Nightingale, E.R., 1959. Phenomenological theory of ion solvation. Effective radii of hydrated ions. *J. Phys. Chem.*, 63(9): 1381–1387.

Overton, H.L. and Zanier, A.M., 1970. Hydratable shales and the salinity high enigma. *Soc. Pet. Engrs., Am. Inst. Min. Metall. Engrs., 45th Ann. Fall Meet., Houston, Tex., Oct. 4–7,* Preprint SPE 2989: 19 pp.

Philipp, W., 1961. Struktur und Lagerstattengeschichte des Erdölfeldes Eldingen. *Z. Dtsch. Geol. Ges.*, 112: 414–482.

Pol'ster, L.A., Viskovskiy, Yu.A., Guseva, A.N., Parnov, E.I. and Plaskova, A.G., 1967. *Physicochemical Characteristics and Hydrogeological Factors of Migration of Natural Solutions (In Relation to Study of Oil- and Gas-Bearing Basins).* Nedra, Leningrad, 172 pp.

Porfir'ev, V.B. and Brod, I.O. (Editors), 1959. *Problem of Oil Migration and Formation of Oil and Gas Deposits.* Gostoptekhizdat, Akad. Nauk S.S.S.R., Moscow, 423 pp.

Posokhov, E.V., 1966. *Formation of Chemical Composition of Underground Waters (basic factors).* Gidrometeorologischeskoe Izd., Leningrad, 258 pp.

Powers, M.O., 1967. Fluid-release mechanisms in compacting marine mudrocks and their importance in oil exploration. *Bull. Am. Assoc. Pet. Geologists,* 51(7): 1240–1254.

Rainwater, F.H. and White, W.F., 1958. The solusphere – its inferences and study. *Geochim. Cosmochim. Acta,* 14: 244–249.

Rall, C.G. and Wright, J., 1953. Analysis of formation brines in Kansas. *U.S., Bur. Mines, Rep. Invest.*, 4974: 40 pp.

Rieke, III, H.H., 1970. *Compaction of Argillaceous Sediments (20–500,000 p.s.i.).* Thesis, Univ. Southern Calif., Los Angeles, Calif., 682 pp.

Rieke III, H.H., Chilingar, G.V. and Robertson Jr., J.O., 1964. High-pressure (up to 500,000 p.s.i.) compaction studies on various clays. *Int. Geol. Congr., 22nd, New Delhi,* 15: 22–38.

Rieke III, H.H., Chilingar, G.V. and Adamson, L.G., 1966. Notes on application of electrokinetic phenomena in soil stabilization. *Proc. Int. Clay Conf.*, Jerusalem, Israel, 1: 381–389.

Rittenhouse, G., 1967. Bromine in oil-field waters and its use in determining possibilities of origin of these waters. *Bull. Am. Assoc. Pet. Geologists,* 51(12): 2430–2440.

Ronov, A.B., 1964. General tendencies in evolution of composition of earth crust, ocean and atmosphere. *Geokhimiya,* 1964(8): 715–743.

Russell, W.L., 1933. Subsurface concentration of chloride brines. *Bull. Am. Assoc. Pet. Geologists,* 17(10): 1213–1228.

Sawabini, C.T., Chilingarian, G.V. and Rieke, H.H., 1972. Effect of compaction on chemistry of solutions expelled from montmorillonite clay saturated in sea water. In press.

Seaber, P.R., 1965. Variations in chemical characteristics of water in Englishtown Formation, New Jersey. *U.S., Geol. Surv., Profess. Papers,* 498-B: 35 pp.

Serruya, C., Picard, L. and Chilingarian, G.V., 1967. Possible role of electrical currents and potentials during diagenesis (electrodiagenesis). *J. Sediment. Petrol.,* 37(2): 695–698.

Shishkina, O.V., 1959. Metamorphization of chemical composition of mud waters in the Black Sea. In: N.M. Strakhov (Editor), *Toward Knowledge of Diagenesis of Sediments.* Izd. Akad. Nauk S.S.S.R., Moscow, pp.29–50.

Shishkina, O.V., 1968. Methods of investigating marine and ocean mud waters. In: G.V. Bogomolov et al. (Editors), *Pore Solutions and Methods of Their Study (A Symposium).* Izd. Nauka i Tekhnika, Minsk, pp.167–176.

Siever, R., Beck, K.C. and Berner, R.A., 1965. Composition of interstitial waters of modern sediments. *J. Geol.,* 73(1): 39–73.

Stankevich, E.F., 1959. About underground waters of calcium-chloride type with high $CaSO_4$ content in Ural–Volga area. *Dokl. Akad. Nauk S.S.S.R.,* 124(4).

Steinfink, H. and Gebhart, J.E., 1962. Compression apparatus for powder X-ray diffractometry. *Rev. Sci. Instrum.,* 35(5): 542–544.

Stern, O., 1924. Zur Theorie der elektrolytischen Doppelschicht. *Z. Elektrochem. und Angew. Phys. Chem.,* 30: 508–516.

Stiff Jr., H.A., 1951. The interpretation of chemical water analysis by means of patterns. *J. Pet. Tech.,* 3(10): 15–16.

Strakhov, N.M., 1960. *Principles of Theory of Lithogenesis.* 1. Izd. Akad. Nauk S.S.S.R., Moscow, 212 pp.

Tageeva, H.V., 1958. The geochemistry of the clay sediments of the Caspian Sea. *Proc. Acad. Sci. U.S.S.R., Geochem. Sect.,* 1958(121): 1056–1060.

Teorell, T., 1935. An attempt to formulate a quantitative theory of membrane permeability. *Proc. Soc. Exp. Biol. Med.,* 33: 282–285.

Thode, H.G. and Monster, J., 1965. Sulfur-isotope geochemistry of petroleum, evaporites, and ancient seas. In: A. Young and J.E. Galley (Editors), *Fluids in Subsurface Environments.* Am. Assoc. Pet. Geologists, Tulsa, Okla., pp.367–377.

Timm, B.C. and Maricelli, J.J., 1953. Formation waters in southwest Louisiana. *Bull. Am. Assoc. Pet. Geologists,* 37(2): 394–409.

Tolstikhin, N.I., 1932. Toward question of graphical representation of analysis of waters. In: V. Golubyatnikov (Editor), *Sampling of Mineral Deposits,* Gosgeolizdat, Moscow–Leningrad, pp.1–8.

Tolstikhin, N.I. et al., 1954. Hydrogeology. In: N.B. Vassoevich (Editor), *Companion of Field Petroleum Geologists, II.* Gostoptekhizdat, Leningrad, 2nd ed., pp.101–145.

Valyashko, M.G., 1962. *Geochemical Regularities in the Formation of Potassium Salt Deposits.* Izd. Moskov. Univ., A.P. Vinogradov (Editor), 397 pp.

Valyashko, M.G. and Polivanova, A.I., 1968. Highly mineralized waters in the system of natural waters, their genesis, peculiarities and distribution. In: *Genesis of Mineral and Thermal Waters. Int. Geol. Congr., 23rd, Prague, Rep. Sov. Geologists,* Akad. Nauk S.S.S.R., 2: 113–115.

Van Everdingen, R.O., 1968. Mobility of main ion species in reverse osmosis and the modification of subsurface brines. *Can. J. Earth Sci.,* 5(8): 1253–1260.

Van Olphen, H., 1963. *An Introduction to Clay Colloid Chemistry.* Wiley, New York, N.Y., 301 pp.

Vassoevich, N.B. (Editor), 1954. *Companion of Field Petroleum Geologists,* II. Gostoptekhizdat, Leningrad, 2nd ed., 564 pp.

Verwey, E.J.W. and Overbeek, J.Th.G., 1948. *Theory of the Stability of Lyophobic Colloids – The Interaction of Soil Particles Having an Electric Double Layer.* Elsevier, Amsterdam, 199 pp.

Von Engelhardt, W., 1961. Zum Chemismus der Porenlösung der Sedimente. *Bull. Geol. Inst. Univ. Upsala,* 40: 189–204.

Von Engelhardt, W., 1967. Interstitial solutions and diagenesis in sediments. In: G. Larsen and G.V. Chilingar (Editors), *Diagenesis in Sediments.* Elsevier, Amsterdam, pp.503–521.

Von Engelhardt, W. and Gaida, K.H., 1963. Concentration changes of pore solutions during the compaction of clay sediments. *J. Sediment. Petrol.,* 33(4): 919–930.

Wallace, W.E., 1965. Application of electric log measured pressure to drilling problems and a new simplified chart for well site pressure computation. *The Log Analyst,* 6: 4–10.

Warner, D.L., 1964. *An Analysis of the Influence of Physical-Chemical Factors Upon the Consolidation of Fine-Grained Clastic Sediments.* Thesis, Univ. California, Berkeley, Calif., 136 pp.

Watts, H., 1963. The possible role of adsorption and diffusion in the accumulation of crude petroleum deposits; a hypothesis. *Geochim. Cosmochim. Acta,* 27: 925–928.

Weaver, C.E., 1967. The significance of clay minerals in sediments. In: B. Nagy and U. Colombo (Editors), *Fundamental Aspects of Petroleum Geochemistry.* Elsevier, Amsterdam, pp.37–75.

White, D.E., 1965. Saline waters of sedimentary rocks. In: A. Young and J.E. Galley (Editors), *Fluids in Subsurface Environments. Am. Assoc. Pet. Geologists, Mem.,* 4: 342–366.

Williams, D.G., Brown, W.O. and Wood, J.J., 1965. Cutting drilling costs in high-pressure areas. *Oil Gas J.,* 63(41): 145–152.

Wright, J., Pearson, C., Kurt, E.T. and Watkins, J.W., 1957. Analysis of brines from oil-productive formations in Oklahoma. *U.S., Bur. Mines, Rep. Invest.,* 5326: 71 pp.

Wyllie, M.R.J., 1955. Role of clay in well-log interpretation. In: J.A. Pask and M.D. Turner (Editors), *Clays and Clay Technology. Calif. Div. Mines Bull.,* 169: 282–305.

Chapter 6

SUBSIDENCE

INTRODUCTION

Compaction of argillaceous sediments, natural or artificially induced, is of considerable importance in the understanding of geologic subsidence. Physical changes occurring in the sediments during subsidence of a depositional basin are directly related to the evolution of stress within the basin. Information on the thickness of sediments, facies changes, fluid displacement phenomena, distribution of unconformities, structural geometry, stress–strain relations, and abnormally high fluid pressure zones is needed for a satisfactory analysis of subsidence. Rock mechanics principles provide the concepts which are essential to deducing significant structural interpretations concerning stress from these data (Currie, 1967, p.50).

Some quantitative methods for measuring the magnitude and direction of stress in rocks were discussed in Chapter 3. The methods which are employed in determining the stress distribution in an active, sinking sedimentary basin can be grouped into three general categories: (1) direct measurement of stress (hydraulic fracturing and strain relief approaches), (2) inferred stress distribution from structural geometry (study of fracture patterns, fault displacements, and settlement features), and (3) inferred stress distribution from structural processes (application of rock mechanics principles) (Currie, 1967). Whereas a measure of stress distribution in rocks during one interval of time may assist one to anticipate structural features, such as a system of fractures that contributes to productive secondary porosity and permeability, the future history of these fractures cannot be predicted (Currie, 1967). Permeability and porosity can be changed by subsequent structural events, such as folding and faulting, or by physicochemical events, such as recrystallization, solution and cementation. Thus, the capacity of the sediments to contain and conduct fluids may be changed. The hydrodynamic conditions which exist during subsidence may control the physical and chemical mechanisms involved in fluid migration from shales. Present knowledge on compaction indicates that relative displacements of considerable magnitude can result from compaction of shales.

Sedimentologists should be aware of the following two possible situations: (1) a subsiding basin with compacting–subsiding sediments, which is a very complex situation; and (2) a stable basin with compacting–subsiding sediments. It appears that the entire relationship between the origin of depositional basins and subsidence of the crust and the mantle, along with the sediments, has not been completely understood by geologists. Not even in the early and geometrically simple stages of basin development (initial subsidence with accompanying deposition and compaction of sediments) will a uniform stress distribution exist throughout the sedimentary sequence. During the structural development of a basin, sediments within a portion of the basin will be deformed by orogenic activity in a mobile belt (an excellent example of this is the Ventura Basin in California). The origin of depositional basins and their subsidence are briefly discussed in the present chapter. Some of the limitations, restrictions, and background about the role subsidence plays in the origin of geosynclines are also mentioned.

Inasmuch as a better understanding of man-induced subsidence may help explain some of the processes involved in geologic subsidence, this subject is also covered in the present chapter. Although much has been learned about subsidence, a great deal more remains to be understood. As pointed out by Marsden and Davis (1967, p.100), the problem requires not only expanded field measurements, but also investigations by means of controlled, small-scale experiments. The latter are perhaps even more important. As a result, answers may be obtained faster by telescoping the time scale. Accurate quantitative information about the variables involved in land subsidence is indeed badly needed.

ORIGIN OF SEDIMENTARY BASINS AND GEOSYNCLINES

As pointed out by Scheidegger and O'Keefe (1967, p.6275), a fundamental question which needs to be answered is: What is the mechanical cause for the formation of geosynclines? In the older geologic literature, it was commonly assumed that the formation of a subsiding depositional basin was caused by the deposition of large amounts of sediments. This sedimentary load, while compacting, would depress the earth's crust so as to create a trough into which more sediments would be deposited. This process would finally produce the huge thicknesses of sediments found in basins such as the Gulf Coast Tertiary basin or the Cordilleran and Appalachian geosynclines (both Paleozoic) in U.S.A.

Bucher (1933) and Holmes (1944) have shown that this model is unsound in its present form if the concept of isostasy is to hold true. Holmes (1944, p.380) postulated that the weight of any column of material from the surface of the earth to a point somewhere in the mantle must be the same as that of any other column. Scheidegger and O'Keefe (1967, p.6275) showed that if this was true then no deep geosynclines could be formed in this manner by shallow-water deposition.

Hsu (1958) explored the above premise by hypothesizing that depositional basins might be areas of thinner crust. He showed that if the earth's crust was thinned by some mechanical effect other than deposition itself, it would be possible to accomodate the required depths of sediments. In all these cases, the deflection of the bottom boundary of the crust (usually identified with the Mohorovičić discontinuity, i.e., a seismic discontinuity, located approximately 35 km below the continents, separating the crust from the

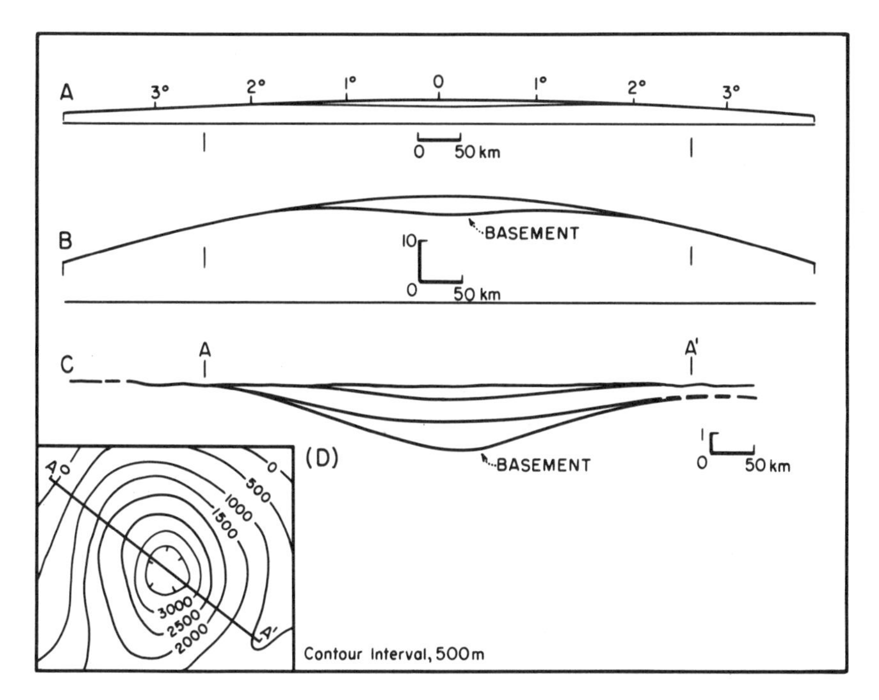

Fig. 148. Cross section of the Michigan Basin. A. True-scale cross section. B. Exaggerated curvature and thickness. C. No curvature, exaggerated thickness. D. Diagram of basement surface. (After Dallmus, 1958, fig.2, p.886; simplified by Currie, 1967, fig.4, p.45. Courtesy of the American Association of Petroleum Geologists.)

mantle) is always arched upward in geosynclinal regions, which is contrary to seismic evidence (Scheidegger and O'Keefe, 1967, p.6276).

Dallmus (1958, p.888) pointed out that the size, shape, and dynamic condition of the earth automatically impose definite limits on the vertical displacements which may take place in and on the crust of the earth. He stated that the shape of the earth sets the pattern for the shape of any local departures from the spherical shape caused by local disturbance in the stability of the crust. The cross-sectional shape of a depositional basin evolves from an arc of the earth's surface and during subsidence, rocks of the basin floor must experience compression until the original surface becomes coincident with the chord of the initial arc (Fig.148). Subsidence of a portion of a spherical surface means shortening, because the arc is longer than the corresponding chord. As successive layers of sediment are warped downward, they too will undergo general compression (Currie, 1967). Because of the pressure differential between the compressional central area of a dynamic basin and the tensional rim, there should be a continuous and diminishing expulsion of interstitial fluids from the central portion towards the rim until compaction ceases (Dallmus, 1958, p.883). As long as the basin is subsiding, the expulsion of interstitial fluids from the fine-grained clastics takes place in an up-dip direction into the coarse-grained clastic sediments, preferentially parallel to the bedding. If most of the sediments are argillaceous, then there is little or no continuous permeability in an up-dip direction in the basin. Consequently, hydrocarbons are trapped at random throughout such a basin (Dallmus, 1958, p.883).

Dallmus (1958, p.884) classified depositional basins as being dynamic or sedimentary. A dynamic basin is created where any portion of the earth's crust is actively sinking as a unit with respect to the center of the earth. A primary dynamic basin, by definition, is formed by flattening of the original surface to a curvature less convex than the curvature of the earth. Its areal limits are, therefore, defined by the shape of the deformed profile upon which the sediments accumulate (Dallmus, 1958, p.891). A sedimentary basin is defined by Dallmus (1958, p.884) as an existing topographic depression receiving sediments; and its size and shape are controlled only by the existing topography. A sedimentary basin may comprise several dynamic basins or parts of such basins, whereas a dynamic basin may be divided into separate sedimentary basins by pre-existing topography at the time of subsidence (see also Bissell, 1970, pp.285–287). Sediments laid down in a dynamic basin are subjected to two types of stresses: (1) tangential dynamic stresses caused by vertical movements imposed on a spherical shell, and (2) small vertical stresses imposed by the static load. These two types of stresses are independent of each other.

Secondary dynamic basins as described by Dallmus (1958, p.883) are graben and half-graben formed by normal faulting on top of actively rising large regional uplifts.

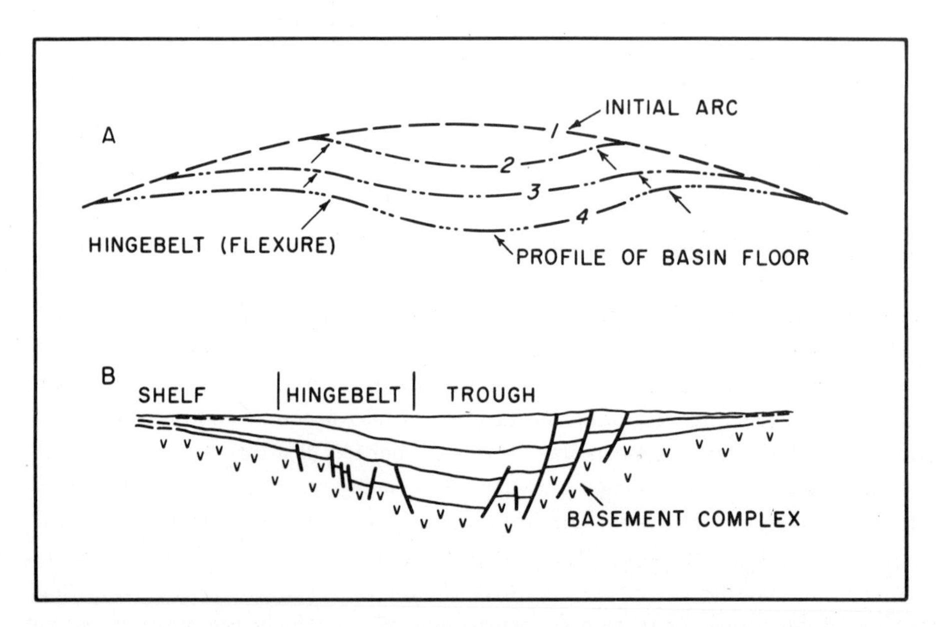

Fig.149. Generalized development of shelf, hingebelt and trough area during growth of a basin structure. A. Possible stages in basin subsidence. B. Common form of basin structure (after Weeks, 1952). (After Currie, 1967, fig.5, p.45. Courtesy of World Petroleum Congress.)

During the growth of basins, such features are in tension normal to their long axis and in compression parallel to the long axis. The size and shape of secondary dynamic basins are related in the first place to the size and shape of the uplift upon which they occur. Currie (1967, p.46) noted that observed variations in thickness and character of basinal sediments suggest that rates of subsidence and consequent departure from the initial arc are by no means uniform throughout a basin. Areas of slow departure become depositional shelves, areas of rapid departure become basin troughs, and transitional areas comprise hingebelts or flexure zones (Fig.149).

Throughout the course of basin development, bending of strata constitutes a common process of rock deformation (Currie, 1967, p.46). Flexing of sedimentary strata may result from processes other than basin subsidence. Currie (1967, p.47) pointed out that bending will occur in strata overlying an area in which differential compaction of sediments is in progress (Fig.150).

The means by which sedimentation can cause a depression is shown by Scheidegger and O'Keefe (1967, p.6277). They used a differential compaction mechanism to illustrate

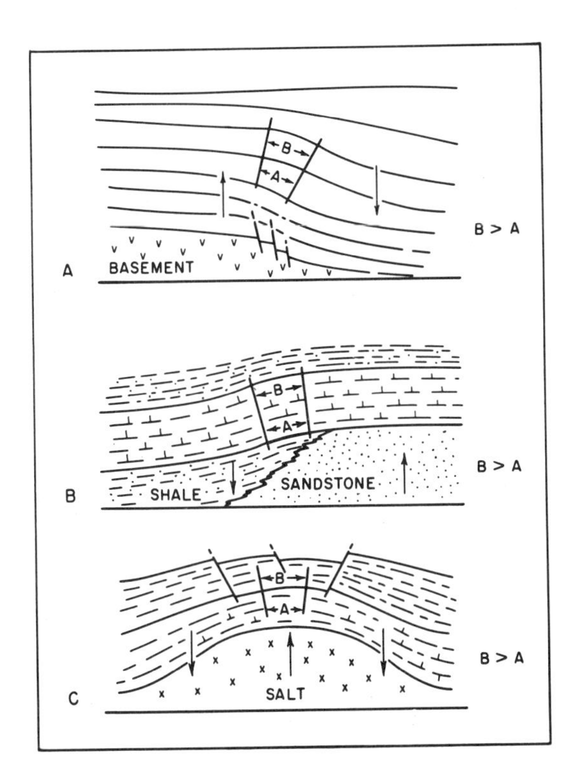

Fig.150. Relative displacements associated with bending of strata above a flexure zone. A. Illustrates relative displacements that arise from bending of strata about a flexure. B. Bending occurring in strata overlying an area in which differential compaction of sediments is taking place. C. Bending associated with domal structures. (After Currie, 1967, fig.6, p.45. Courtesy of World Petroleum Congress.)

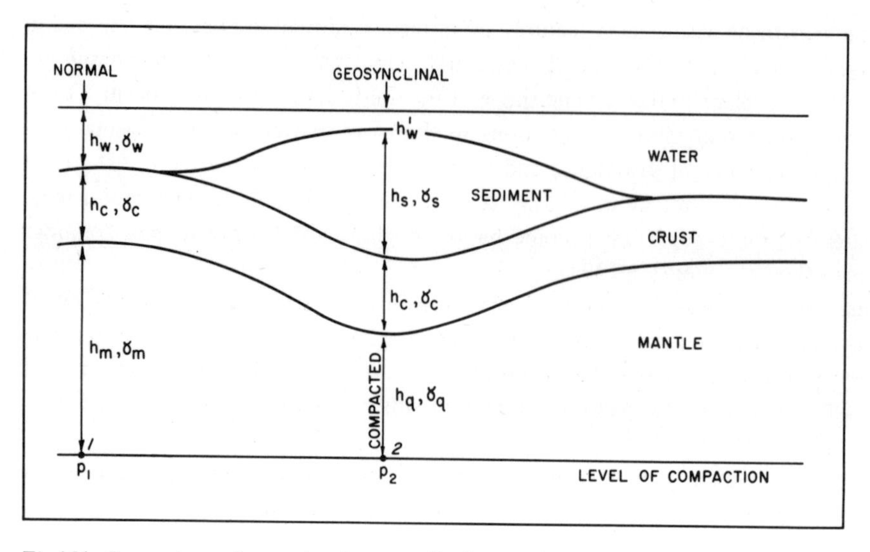

Fig.151. Comparison of normal and geosynclinal upper layers of the earth, according to Scheidegger and O'Keefe's differentiation–compaction mechanism. (Modified after Scheidegger and O'Keefe, 1967, fig.2, p.6276. Courtesy of *Journal of Geophysical Research*.)

that the old idea of a depression in the crust resulting from the deposition of sediments in shallow water need not be invalid because of the principle of isostasy. Their model consisted of compensating the sedimentary load on the crust by the migration of a low melting-point fraction in the mantle. This fraction may have nearly the same density as the sediments. Scheidegger and O'Keefe (1967, p.6276) stated that this process would most likely take place in the mantle, because the Mohorovičić discontinuity appears to be depressed beneath such depositional basins. The isostatic level of compensation of the overburden load is assumed to be at a level beneath the Mohorovičić discontinuity (Fig.151). Scheidegger and O'Keefe (1967) assumed that the specific weight of the compacted mantle beneath the geosyncline (γ_q) is greater than the normal specific weight (γ_m) because of losing the lighter fraction through migration.

As shown in Fig.151, in order to satisfy isostatic principles, the pressure at point *1* must be equal to the pressure at point 2 ($p_1 = p_2$). Thus:

$$p_1 = h_w\gamma_w + h_c\gamma_c + h_m\gamma_m = p_2 = h'_w\gamma_w + h_s\gamma_s + h_c\gamma_c + h_q\gamma_q \tag{6-1}$$

where γ_w, γ_s, and γ_c are the specific weights of water, sediment, and crust, respectively. Inasmuch as:

$$h_w + h_c + h_m = h'_w + h_s + h_c + h_q \tag{6-2}$$

then:

$$h_m = h'_w + h_s + h_q - h_w \tag{6-3}$$

Multiplying through by γ_m, eq.6-3 becomes:

$$h_m\gamma_m = h'_w\gamma_m + h_s\gamma_m + h_q\gamma_m - h_w\gamma_m \tag{6-4}$$

Combining eq.6-4 and eq.6-1 and solving for h_q yield:

$$h_q = h_s \frac{\gamma_m - \gamma_s}{\gamma_q - \gamma_m} - (h_w - h'_w)\left(\frac{\gamma_m - \gamma_w}{\gamma_q - \gamma_m}\right) \tag{6-5}$$

If h'_w is equal to zero then:

$$h_q = h_s \frac{\gamma_m - \gamma_s}{\gamma_q - \gamma_m} - h_w \frac{\gamma_m - \gamma_w}{\gamma_q - \gamma_m} \tag{6-6}$$

As shown in Fig.152, the overburden pressure versus depth curve beneath a geosyncline differs from that for the normal crust, which is due to different density distributions. If the geosynclinal density distribution enhances the differentiation–compaction mechanism, then the process would go on as long as there is a sediment supply on top (Scheidegger and O'Keefe, 1967, p.6277). Fig.152 shows that the total mass is less over

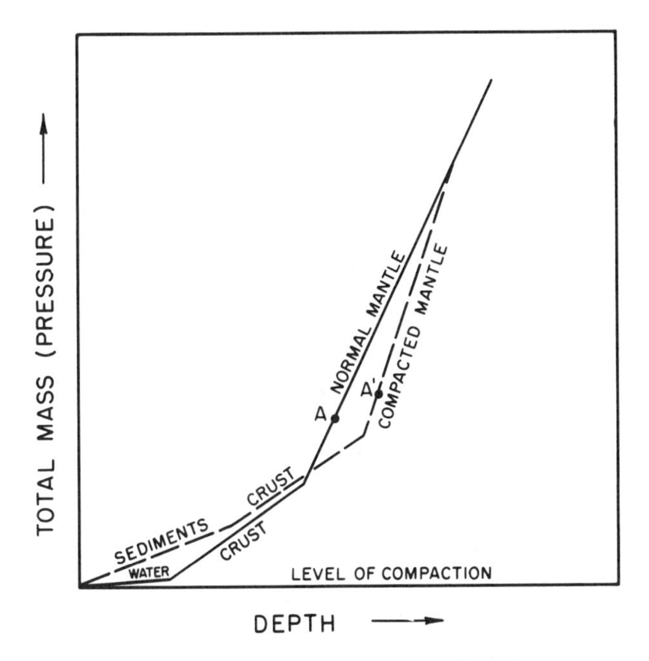

Fig.152. Total mass (pressure) in normal (solid line) and geosynclinal (dashed line) region (h'_w = 0). $A \rightarrow A'$ are corresponding points before and after the geosyncline formation; pressure is higher at A' than at A. (After Scheidegger and O'Keefe, 1967, fig.3, p.6277. Courtesy of *Journal of Geophysical Research.*)

the region of the compacted–differentiated mantle than in the normal mantle. Scheidegger and O'Keefe (1967) stated that inasmuch as corresponding points are deeper in the "compacted" crust than in the normal crust, the pressure at such corresponding points is higher in the geosyncline than in the normal crust. These authors did not investigate the problem of where the "liquid fraction" in the mantle has migrated.

HYDROGEOLOGICAL CYCLE

Kartsev et al. (1969, p.22) described a simple hydrogeological cycle (Fig.153), which starts with tectonic depression and transgression, followed by a period of subsequent uplift and regression, and terminates prior to the initiation of a new depression and regression. The first stage of a hydrogeological cycle (sedimentation stage) terminates when a sedimentation basin, upon ceasing to subside, is uplifted and denudation of the water-bearing horizons occurs (Fig.153A).

During the second stage (infiltration stage) there is infiltration of atmospheric waters (epigenetic waters), which gradually displace and replace original connate waters (syngenetic waters) (Fig.153B). This stage ends as a result of new tectonic depression of the basin, with accumulation of younger sediments; as a result, the infiltration of atmospheric waters terminates. At this time, a new hydrogeological cycle is initiated (Fig.153C).

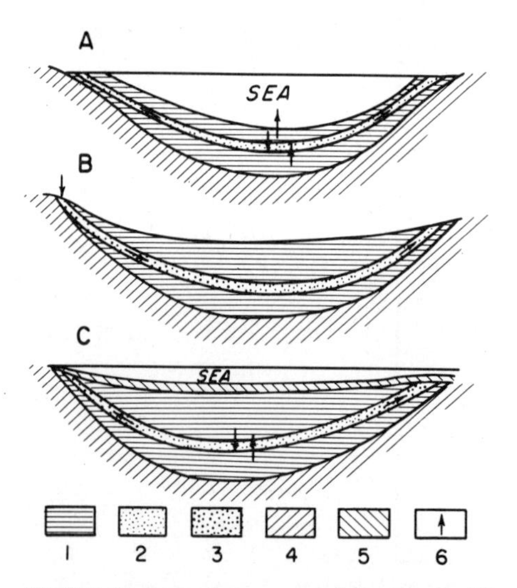

Fig.153. Hydrogeologic cycle of Kartsev. A. First sedimentation stage. B. Infiltration stage. C. Second (subsequent) sedimentation stage. *1* = muds and clays formed during sedimentation stage A and containing syngenetic waters; *2* = coarse-grained rocks containing syngenetic waters; *3* = reservoir rocks containing epigenetic waters (infiltrated atmospheric waters); *4* = bottom of depositional basin; *5* = muds and clays formed during stage C; and *6* = direction of water movement. (In: Kartsev et al., 1969, fig.6, p.24.)

It is important to note that during the second hydrogeological cycle, some remaining syngenetic waters of older shales and mudstones may be squeezed out into older coarse-grained rocks, thus replacing the epigenetic waters. Subsequent infiltration of atmospheric waters into both older and younger coarse-grained rocks further complicates the situation, and obscures the effect of compaction on the chemistry of interstitial fluids in coarse-grained rocks.

During the sedimentation stage most of the water movement occurs from argillaceous sediments into sands, whereas during the infiltration stage the major movement is from sands into shales. In both cases, there is a secondary movement of fluids in the opposite direction.

SUBSIDENCE AS A RESULT OF FLUID WITHDRAWAL

Present-day land subsidence is usually caused by the removal of fluids (water and/or oil). A comprehensive treatment of the subject was presented in two volumes in 1969 at the International Symposium on Land Subsidence in Tokyo, Japan, sponsored by the I.A.S.H. (International Association of Scientific Hydrology), International Society for Soil Mechanics and Foundation Engineering, and UNESCO. The main lithological and structural characteristics of the subsiding areas include the following:

(1) Sediments are unconsolidated and lack appreciable cementation.

(2) Sediment section is thick.

(3) Porosity of the sands is high: 20–40%.

(4) Sands are interbedded with clays, fine silts and/or siltstones, and shales.

(5) Fluid production is voluminous.

(6) Standing fluid levels in the wells exhibit large drops.

(7) In the case of water-producing areas, aquifers cover large areas and are shallow and flat-lying.

(8) Subsidence rate is cyclic, controlled by seasonal fluid-level fluctuation.

(9) Age of sediments is Pliocene or younger in the case of water-producing horizons and Miocene or younger in the case of oil-producing areas.

(10) Producing formations are located at shallow depth: 300–1,000 m.

(11) Overburden is composed of structurally weak sediments.

(12) In oil-producing areas, the reservoir beds have flat or gentle dips at the structure crest.

(13) Tension-type faulting, often with a graben central block, are present.

Horizontal surface movement is common to most present-day subsiding areas. The subsiding surface area is placed in tension peripherally and in compression at the center. These stresses cause horizontal movement with all peripheral points vectoring towards the subsiding center. The degree of horizontal movement is a function of the depths and thicknesses of the compacting horizons and the magnitude of the subsidence.

According to Allen et al. (1971), subsidence due to withdrawal of fluids occurs when (a) reservoir fluid pressures are lowered, (b) reservoir rocks are compactable (usually uncemented) and/or are unable effectively to resist deformation upon the transfer of load from the fluid phase to the grain-to-grain contacts, and (c) overburden lacks internal self-support and can easily deform downward.

When the hydrostatic head is lowered, the overburden support is decreased and grain-to-grain load increases. As a result, sands and silts compact by grain rearrangement and crushing, whereas plastic flow occurs in argillaceous sediments. Water from clays and shales moves into associated sands and, consequently, there is a decrease in volume of fine-grained sediments. The relative contribution of sands and of clays to compaction varies with depth and with the geologic history. According to Allen et al. (1971), at very shallow depths clays and silts are usually the major compacting materials, whereas at greater depths (300–1,000 m) sands constitute the major compacting material.

Susceptibility of the formation to subsidence is dependent upon many factors, such as the degree of compaction due to previous depth of burial during geologic time, types of clays, shape and size of sand grains, and relative proportions of interbedded clays and sands.

Load transfer occurs as fluid level is lowered. The two concepts used in calculating the overburden load are as follows:

(1) The effective stress acts in a dynamic situation, with downward seepage of fluid through the overburden (Lofgren and Klausing, 1969).

(2) The static load represents the effective weight of the overburden material.

The latter concept is the easiest to use, because the former approach requires knowledge of the magnitude of the volumetric rate of fluid flow and permeability of the sediments.

As pointed out by Allen et al. (1971), the concept of overburden load and load transfer is extremely important, because upon fluid removal subsidence would not occur unless there is a load transfer. The concept of a static overburden load (geostatic pressure) has been widely accepted (see eq.3-3 and eq.3-4). The maximum amount of load transfer possible at a particular depth is equal to the fluid pressure (hydrostatic pressure) at that point.

The manner in which the load change could occur upon production of fluids is illustrated in Fig.154. Initially, the geostatic pressure gradient (0.91 p.s.i./ft.) is equal to the sum of the intergranular pressure gradient (0.48 p.s.i./ft.) and hydrostatic pressure gradient (0.43 p.s.i./ft.). Assuming no residual fluid in the pores, the buoyant effect of the water is lost and the intergranular load is increased as the fluid level is lowered from *A* to *B*, for example. Geostatic load decreases (curve *3* shifts to curve*3b*, Fig. 154) as water is removed. The intergranular and geostatic loads are equal if the pores are dry (curve *2b*, Fig. 154). Compaction can occur if the intergranular load is increased.

In the case of a confined aquifer, which has a relatively impermeable cover (cap rock), as the fluid level is lowered from *A* to *C*, the intergranular load gradually increases until it becomes equal to the geostatic load (curve *2* shifts to curve *2c*, Fig.154). If pore spaces

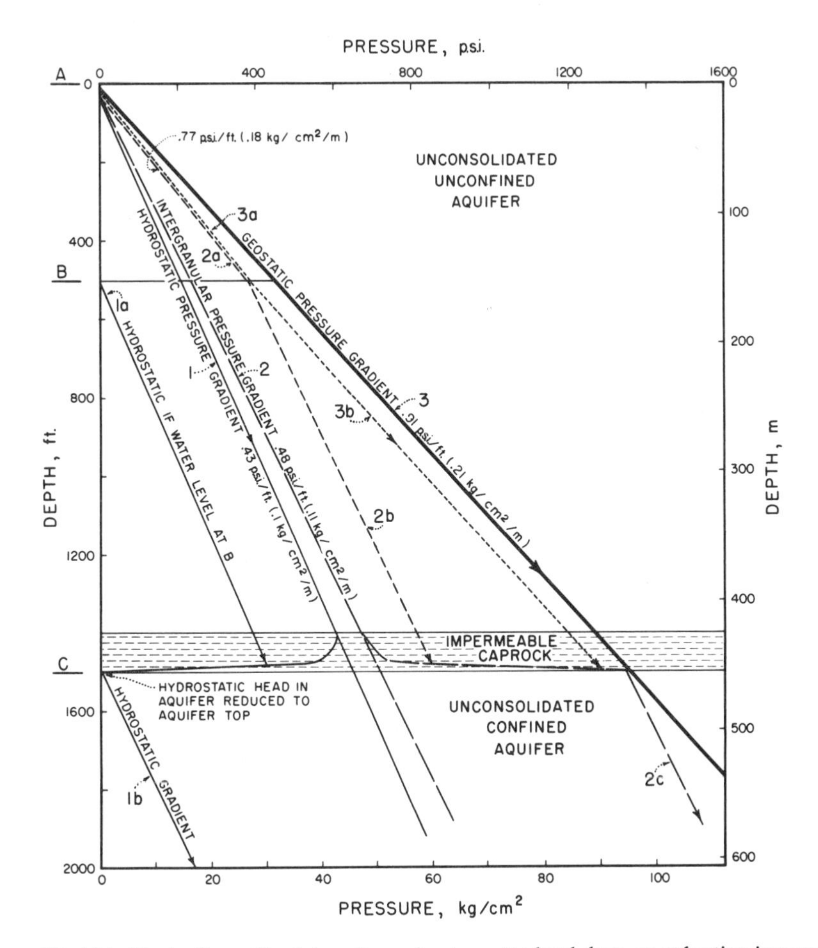

Fig.154. Illustration of load transfer owing to water-level drop or reduction in pore-water pressure in unconfined and confined aquifers. Geostatic, hydrostatic, and intergranular pressure gradients are plotted assuming that solids and water have specific gravities of 2.7 and 1.0, respectively, and that porosity is equal to 35%. (Modified after Allen et al., 1971, fig.4, p.285. Courtesy of *Enciclopedia della Scienza e della Tecnica, Mondadori.*)

still contain some residual water, the intergranular load and geostatic load are not equal below the level *C*.

Upon the reduction in pore-water pressure and consequent load transfer in the aquifers, pressure gradients are set up across the interfaces of interbedded siltstones, shales, and clays. As a result, water movement occurs from these fine-grained beds into coarse-grained aquifers. The volumetric rate of flow depends on the permeability of clays and silts, pore-water pressure drop, length of the drainage paths, viscosity of water, and cross-sectional area of flow. According to Allen et al. (1971), in shallow, unconsolidated sediments consisting of interbedded clays, silts, and sands, which have void ratios of

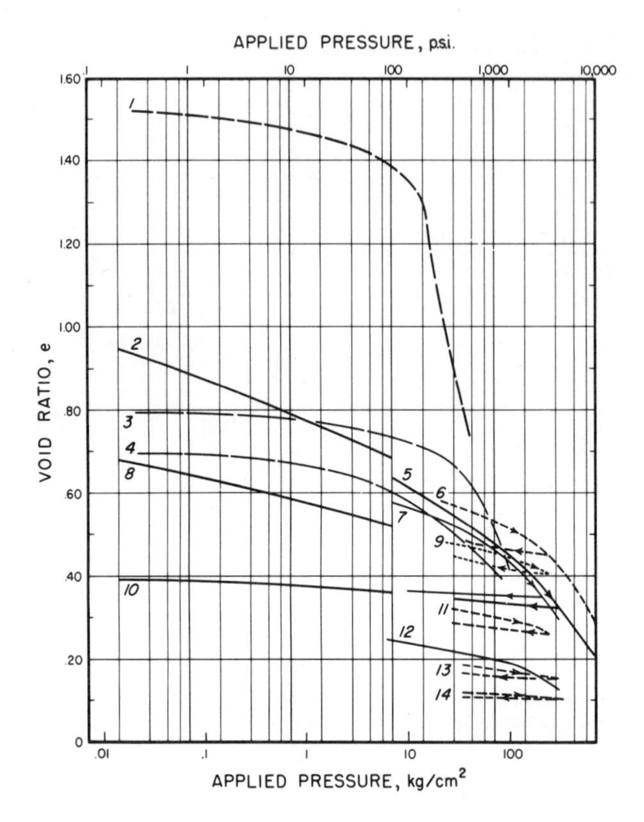

Fig.155. Relationship between void ratio and applied pressure for sand, silt and clay cores obtained at different depths from various areas. *1* = Corcoran Clay (depth of 425 ft.); *2* = very loose sand; *3* = Corcoran Clay (depth of 735 ft.); *4* = silt (depth of 1,345 ft.); *5* = average Wilmington (California) sands (depth of 2,000–4,000 ft.); *6* = average Wilmington (California) siltstones (depth of 2,000–2,900 ft.); *7* = sand from Maracaibo, Venezuela (depth of 3,100 ft.); *8* = intermediate compacted sand; *9* = average Wilmington (California) siltstone (depth of 3,000 ft.); *10* = very compacted sand; *11* = average Wilmington (California) siltstone (depth of 3,100–3,500 ft.); *12* = clay from Maracaibo, Venezuela (depth of 3,104 ft.); *13*, *14* = average Wilmington (California) siltstones (depth of 3,600–6,000 ft.). (After Allen and Mayuga, 1969.)

about 0.6 or greater, clays and silts are the major compactible materials upon dewatering. On the other hand, at depths of 300 m or greater and/or where the void ratios are below 0.6, sands constitute the major compacting material.

In Fig.155, void ratio is plotted versus applied pressure for sand, silt, and clay cores. At void ratios of 0.6 or greater and pressures of about 30 kg/cm^2, sands are as compactable as clays, or even more compactable. The clays having high void ratios are very compactable at high pressures. Void ratio-versus-pressure data obtained by Roberts (1969) shows that in the 1,000–20,000 p.s.i. pressure range, certain sands may be at least as compressible as the typical clays, if not more compressible (Fig.156; also see Fig.95).

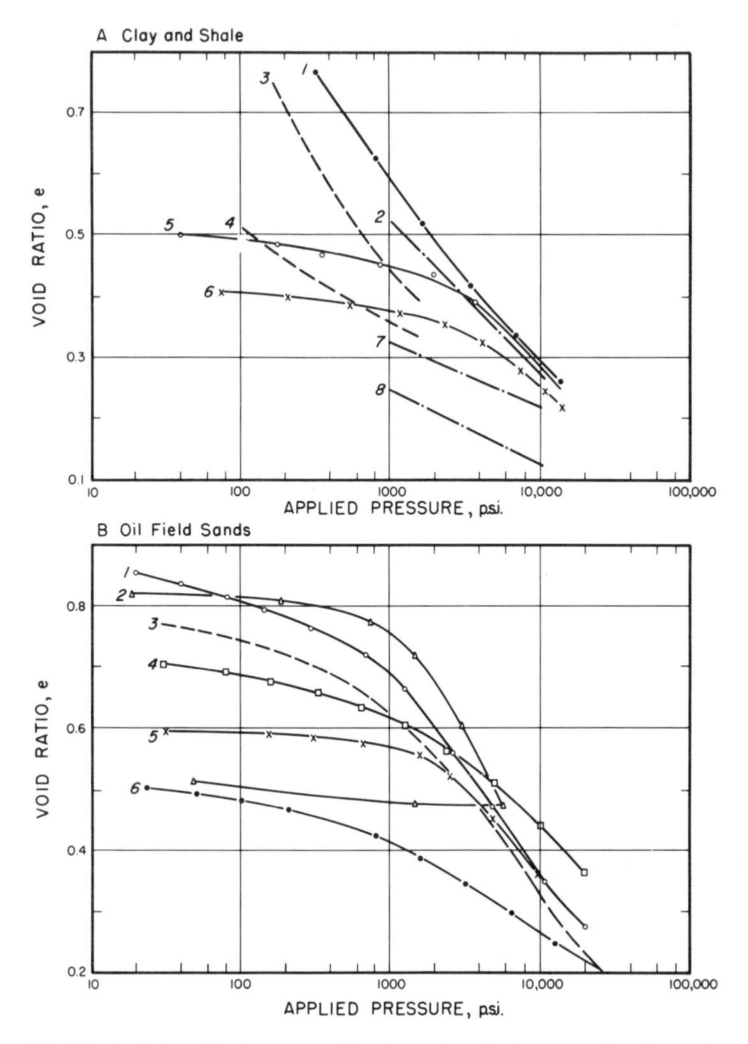

Fig.156. Relationship between void ratio and applied pressure for clay and shale (A) and for sands (B).

A: *1* = undisturbed Boston Blue Clay; *2, 7, 8* = clay cores from Venezuela obtained from depths of 2,486 to 4,769 ft. below ground surface; *3, 4* = Skempton's (1953) compression curves for plastic clays with liquid limits of 80% and 30%, respectively (*P.I.* of 50 and 12, respectively); *5* = undisturbed shale (C11); and *6* = undisturbed shale (C5).

B: *1* = remolded (25.10); *2, 6* = remolded (M.I.T., Geol. Dept.); *3* = undisturbed (1.1); *4* = undisturbed (25.1); *5* = remolded (25.13); and *7* = undisturbed (14.1).

(After Roberts, 1969, fig.1, p.369, and fig.4, p.372.)

At a depth of 3,000 ft., Boston Blue Clay could undergo about 6% compression (Fig. 156), i.e., for an initial stratum thickness of 100 ft. a total settlement of approximately 6 ft. may occur. Compression of the oil sand, which was disturbed and repacked into an initially loose condition could result in a settlement only 15% less than that of the Blue Clay (Roberts, 1969, p.375).

At a depth of 5,000 ft., the Blue Clay could undergo 5.5–6% compression. At this depth, various sands could undergo 1–7.5% compression (Fig.156 and Roberts, 1969, p.375).

At a depth of 8,000 ft. the Blue Clay could undergo about 5% compression, whereas various sands could undergo compressions varying from about 2 to 10%. At this depth, an 20–40 mesh Ottawa Sand is about twice as compressible as the Blue Clay (Roberts, 1969, p.375). Compressibilities of clays and shales are discussed in detail in Chapter 4.

Near-surface subsidence

Subsidence owing to the shrinkage of near-surface, highly carbonaceous material, such as peat bogs and swamp deposits, has been documented in various parts of the world (Allen et al., 1971). It is caused by the loss of the buoyant support of the water and the consequent change in skeletal loading. Oxidation and chemical changes, which take place after dehydration, also contribute to the shrinkage.

Another type of near-surface subsidence, called *hydrocompaction,* is caused by the addition of water to certain water-deficient, surficial deposits such as loess and mud flows. According to Lofgren (1969), this type of subsidence is caused by the loss of intergranular bonding in low-density deposits. At the present time, it commonly occurs in areas having light rainfall and rapid surface run-off. Rapid compaction occurs when deposits of this type are wetted by water (i.e., floods) and proceeds from the surface downward.

MATHEMATICAL ANALYSES OF SUBSIDENCE

Among the recent mathematical analyses of subsidence, those presented by Fukuo (1969), Ter-Martirosyan and Ferronsky (1969), and Sandhu and Wilson (1969) are noteworthy.

Fukuo (1969) derived the dynamic theory for the deformation of a granular solid saturated with a liquid, on assuming that the liquid is a Newtonian viscous fluid and that the solid grain skeleton is a linear viscoelastic solid. Three fundamental equations were presented by him, i.e., the equations of motion of liquid and skeleton and the equation of continuity between the particles and liquid. On assuming that solid grains and liquid are incompressible and the deformation of soil is a quasi-static process, these are fundamental equations of three-dimensional consolidation and of flow of confined ground water in a viscoelastic aquifer. Fukuo also showed a theoretical example of rheological deformation of an infinite-confined aquifer with uniform thickness as a result of pumping water at a constant rate. It is interesting to note here that Fukuo postulated a nonelastic deformation of soil as opposed to Terzaghi (1925ab, 1926, 1936), Biot (1941) and Mikasa (1963) who assumed elastic isotropy of stress–strain relations for the soil skeleton. Mikasa's theory of soft layer consolidation is quite useful, but he analyzed only a one-dimensional problem under constant load with a quasi-static method as was done by Terzaghi.

Ter-Martirosyan and Ferronsky (1969) presented a mathematical analysis of the one-dimensional and plane axi-symmetrical problem of bed consolidation as a result of pumping fluids (oil, water, and gas). The deforming elastic-creeping porous medium in their analysis is assumed to contain compressible fluids. These authors concluded that one can forecast the earth settlement on the basis of theoretical solutions.

Sandhu and Wilson (1969) applied the finite-element method to the problem of land subsidence. They viewed the settlement of land mass as an immediate or time dependent surface deformation caused by direct application of surface loads or by the loss of support associated with the withdrawal of pore fluids. Their method allows treatment of practically all cases involving static or quasi-static subsidence and permits consideration of complex geometrical configurations and arbitrary boundary conditions. Inasmuch as this method is applicable to two- or three-dimensional deformation, it takes into account horizontal as well as vertical movements. Sandhu and Wilson considered non-homogeneity, anisotropy, viscoelasticity and creep, temperature effects, residual stresses, and plastic behavior.

Okumura (1969) analyzed land subsidence in the Niigata area, Japan, by means of consolidation theory. Pumping the underground water from the deep sandy layers for extracting methane gas caused a time-dependent loading on the clay layers. An analytical solution of consolidation under the load increasing linearly with time was obtained in terms of the vertical strain. Okumura also developed a numerical method of analysis for the load decreasing linearly and then becoming constant, considering the difference in values of soil constants for consolidation and for rebound. These two methods were combined for analyzing the subsidence and the calculated results compared favorably with the observed settlement.

Jacquin–Poulet computer model

Jacquin and Poulet (1970) proposed a simple computer model which characterizes the structure, history of water circulation, and development of abnormally high geopressures of a depositional basin undergoing subsidence. Their model sedimentary basin had the following geometric and compositional characteristics: (1) it was shaped like an inverted cone with a circular base having a diameter of 600 km (Fig.157); (2) depth in the center of the basin varied with time from 0 to 3,000 m at the terminal stage; (3) total duration of the basin's life was $150 \cdot 10^6$ years; (4) sedimentary deposits were composed of successive sand and clay strata; and (5) as a result of burial, water was gradually expelled from the clay.

Jacquin and Poulet (1970) assumed that a single sand layer having a constant thickness and porosity was deposited initially on the impermeable basement. The sand layer was then covered by a thick clay layer, which represented the source rock for water and

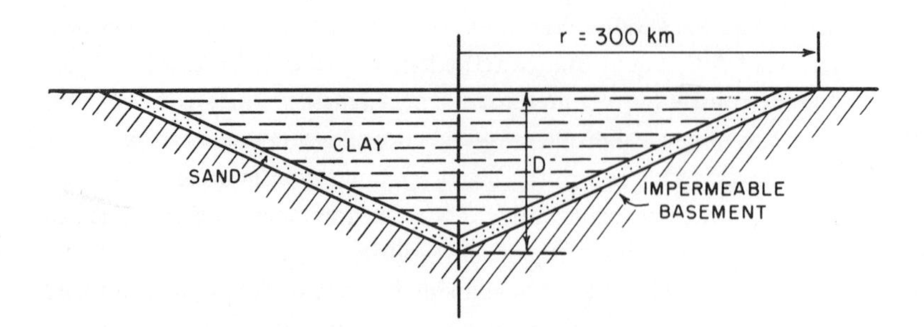

Fig.157. Schematic diagram of a depositional basin. (After Jacquin and Poulet, 1970, fig.1. Courtesy of the Society of Petroleum Engineers of AIME.)

hydrocarbons and was assumed to be highly compressible and to possess low permeability. The sand layer, on the other hand, was assumed to be a permeable, not very compressible, homogeneous conduit in which secondary migration of hydrocarbons took place. At a given point, the compressible clay was assumed to have at all times a porosity, which depends only on the depth at this point. The hyperbolic relationship between porosity and depth can be expressed as:

$$\phi = \frac{1}{LD + A} \tag{6-7}$$

where ϕ is fractional porosity, D is depth of the point under consideration, and L and A are constants. Boundary values are: $\phi = 0.9$ for $D = 0$ and $\phi = 0.2$ for $D = 1{,}000$ m. Compression of the clay leads to the expulsion of the fluid filling up the clay pores. This fluid consists of a water phase possibly containing small amounts of hydrocarbons in a form that Jacquin and Poulet (1970, p.2) did not specify.

It was further assumed that the clay was deposited at the center of the basin at a constant rate of sedimentation (expressed in mass per time unit) and with constant properties such as porosity, grain size and morphology. The permeability of the clay was assumed by Jacquin and Poulet (1970) to depend only on the porosity according to an empirical relation such as:

$$k = \lambda \cdot \phi^5 \tag{6-8}$$

where k is permeability, ϕ is porosity and λ is a constant. In their calculations, Jacquin and Poulet (1970, p.5) assumed a value of 10^{-5} for constant λ (expressed in 1/darcy).

Jacquin and Poulet (1970, p.2) assumed that the fluid circulating in the basin is a simple viscous fluid in a monophasic state, or at least it makes up a phase saturation of which at all points is very close to one, so that the corresponding relative permeability can be assumed to be equal to one. Fluid flow is horizontal in sands and is vertical in clays.

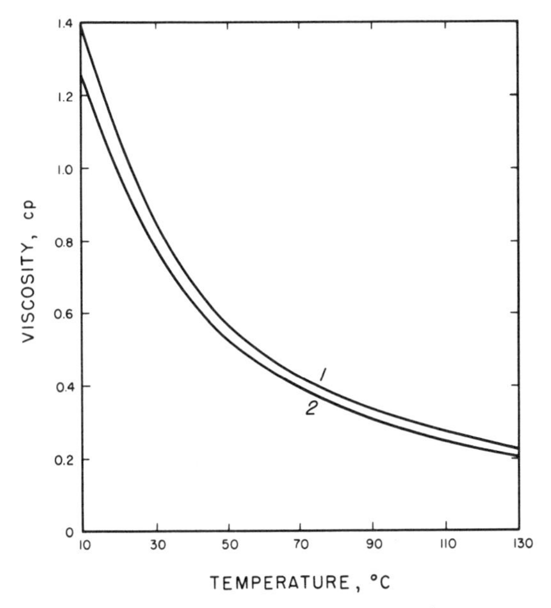

Fig.158. Relationship between viscosity and temperature for (*1*) 60,000-p.p.m. brine and (*2*) pure water. (After Jacquin and Poulet, 1970, fig.2.)

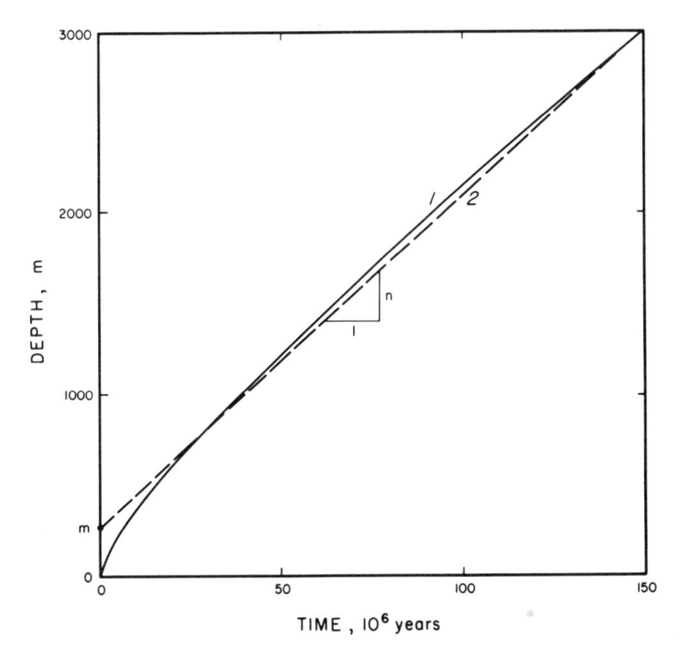

Fig.159. Evolution of basin center depth (D) with time (t). *1* = calculated evolution; *2* = linear approximation ($D = m + nt$). (After Jacquin and Poulet, 1970, fig.3.)

Fluid viscosity at any given point in the model depends only on the temperature at that point (Fig.158). Jacquin and Poulet (1970) assumed a geothermal gradient of 3°C/100 m and a surface temperature of 20°C. Viscosity changes of the fluid owing to the overburden pressure and salinity were considered to be negligible by these authors.

Depth–time relationship. Fig.159 illustrates the evolution of depth with geologic time at the center of the basin. The slope of the curve in Fig.159 gives the rate of burial and is at a maximum at the initial moment of sedimentation. The burial rate becomes essentially constant once the depth reaches 1,000 m, which corresponds to time $t = 40 \cdot 10^6$ years. The linear equation describing the depth–time relationship is expressed as:

$$D = m + nt \tag{6-9}$$

where D = depth at the center of the basin in meters, t = time in years, n = slope of the line or the rate of burial, and m = intercept on y-axis. Thus, D = 1,000 m when $t = 40 \cdot 10^6$ years and D = 3,000 m when $t = 150 \cdot 10^6$ years.

According to Jacquin and Poulet (1970, p.3), a depth of 1,000 m corresponds to a degree of burial beyond which the organic material of the sediments can be considered capable of becoming transformed into hydrocarbons.

Flow of water in clay. During compaction of clay, water is expelled from the pore space of the clay and is forced vertically upward toward the sea bottom and vertically downward toward the clay–sand contact in the model. At every moment, there is a surface Σ within the clay mass where the rate of flow of water is nil. This surface Σ separates the upward flowing zone from the zone in which the flow is downward (Fig.160). The surface Σ is determined by assuming that the pressure drops Δp_1 and Δp_2 for flow along the same vertical line in each of these two zones are equal (Jacquin and Poulet, 1970, p.3). This approach is based on Terzaghi's one-dimensional consolidation theory (Taylor, 1948, pp.225–238).

Fig.160 shows the relationship between the local depth of the sand–clay contact, h, undergoing downward movement, and the local thickness of clay present, z, under the surface Σ. According to Jacquin and Poulet (1970, p.3), as soon as h reaches a value of 300 m, the relationship between z and h can be considered to be linear:

$$z = m' + n'h \tag{6-10}$$

where m' = the intercept on y-axis and n' = slope of the line.

At any given time during the evolution of the model basin, the flow of fluid in the clay can be calculated at an arbitrary point M. Point M is located a distance, r, from the vertical axis through the center of the basin and at a depth y. Depending on the value of y, the flow of water at M will be either in the upward or downward direction. When

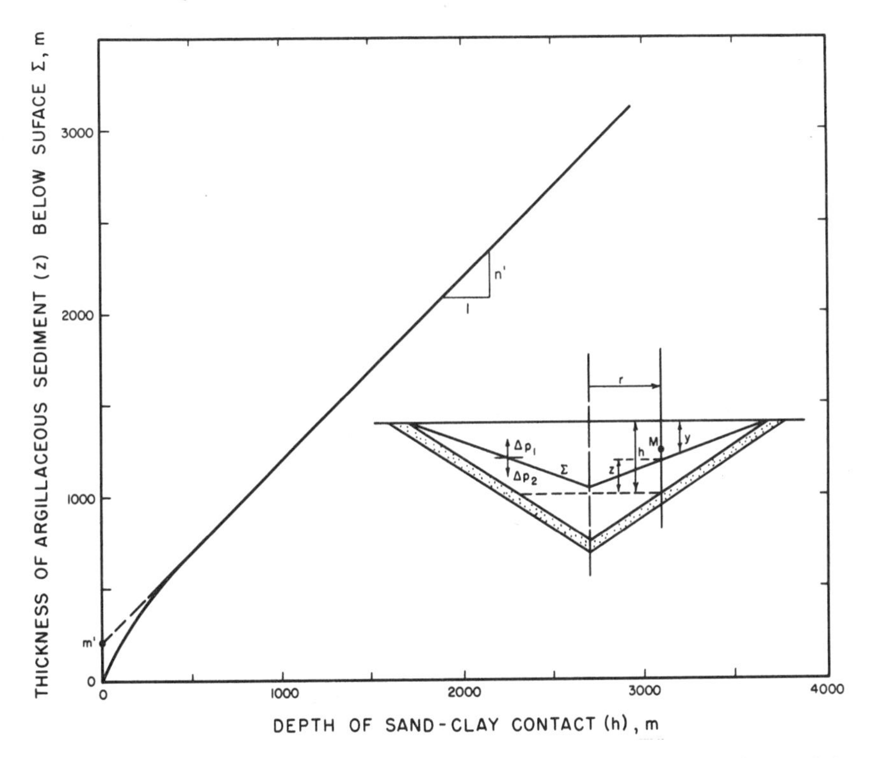

Fig.160. Relationship between local depth of sand-clay contact, h, and thickness of the argillaceous sediment, z, located under the surface Σ and feeding the sand bed. (After Jacquin and Poulet, 1970, fig.4.)

$(h-y) > z$, the flow is upward; whereas when $(h-y) < z$, the flow is downward. At $(h-y) = z$ there is no flow of fluid.

During the basin evolution, the value of $(h-y)$, which represents the thickness of argillaceous sediments deposited prior to the clay present at point M, will decrease as a result of progressive expulsion of water. The value of z, which depends on h according to eq.6-10 (see Fig.160), will increase with time. Depending on the position of M at the moment when $t = 150 \cdot 10^6$ years, there are three possible cases of fluid flow conditions that can occur in a subsiding basin (Jacquin and Poulet, 1970, p.4):

(1) Point M is located at the clay–sand contact. The clay present at M was deposited at the very beginning of the basin's existence; therefore, $y = h$ and $(h-y) = 0 < z$ when $t > 0$. Fluid flow at this point is downward into the sand layer (Fig.161A).

(2) Point M is located above the surface Σ, and $(h-y) > z$. Inasmuch as the value of $(h-y)$ can only decrease with time, whereas z continues to increase, the value of $(h-y)$ always remains greater than z. The clay at point M was deposited at a moment, t_0, corresponding to an advanced stage in the basin's evolution; thus, it has always been subject to upward drainage (Fig.161B).

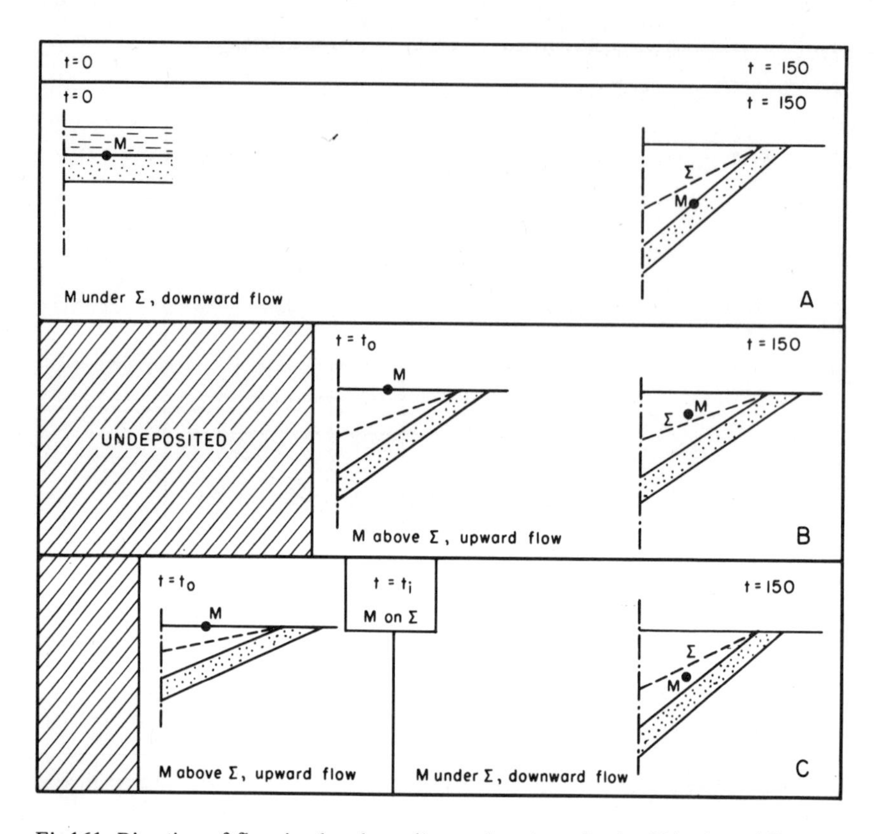

Fig.161. Direction of flow in clay depending on location of point M in three different cases. A. Point M is located at the clay-sand contact. B. Point M is located above the surface Σ. C. Point M is located above and below the surface Σ. (After C. Jacquin, personal communication, 1970.)

(3) Point M is located between the surface Σ and the clay–sand contact. For this case $(h-y)_{t=150} < (z)_{t=150}$. Downward flow of water existed at point M in the clay, at least toward the end of the basin's existence.

At the moment t_0 of deposition of the clay present at M, $(y)_{t_0} = 0$; therefore, $(h-y)_{t=t_0} = (h)_{t_0}$. Inasmuch as z is still less than h, then $(h-y)_{t=t_0} > (z)_{t_0}$. Immediately after its deposition, the clay present at M was subject to upward flow. An inversion of the flow direction occurred at some time t_i, when M was located on the Σ plane (Fig.161C).

Jacquin and Poulet (1970) calculated the cumulative volumes of water which have circulated through a 1-m^2 horizontal surface centered around point M, for different values of basin radius, r, and final depth, y, at point M. Volumes of water moving both in the upward and downward directions were plotted versus time in millions of years. These authors determined porosity at a known depth using eq.6-7 and then calculated permeability using eq.6-8 (empirical relation). After that, flow rates were calculated. In their computer model, Jacquin and Poulet used 1-m thick slices. They also calculated pressure drop above the Σ plane (Δp_1) and that below the Σ plane (Δp_2).

Flow through clay gives rise to high pore pressures compared to normal hydrostatic pressure. Along a given vertical line, this overpressure is nil at the clay–sand contact and at the top of the clay deposit; it is maximum at the point of intersection with the Σ plane. Jacquin and Poulet's model enables one to reconstruct the history of water circulation in clays, which is quite complex, and to estimate the distribution and order of magnitude of overpressures, which agree closely with standard observations made in actual sedimentary basins.

According to the study by Jacquin and Poulet (1970), the circulation of water through sand horizons occurs at very low speeds ranging from 1 to 10 mm/year, but the volumes of water involved are large: 10^{12}–10^{13} m^3 for a period between 40 and 150 million years.

Studies similar to that by Jacquin and Poulet may provide better understanding of oil formation (diagenesis), migration, and accumulation.

REFERENCES

Allen, D.R. and Mayuga, M.N., 1969. The mechanics of compaction and rebound, Wilmington Oil Field, Long Beach, California, U.S.A. In: *Land Subsidence. I.A.S.H.–Unesco, Publ.* no.89 AIHS, 2: 410–423.

Allen, D.R., Chilingar, G.V., Mayuga, M.N. and Sawabini, C.T., 1971. Studio e previsione della subsidenza. *Enciclopedia della Scienza e della Tecnica.* Arnoldo Mondadori Editore, pp.281–292.

Biot, M.A., 1941. General theory of three-dimensional consolidation. *J. Appl. Phys.,* 12(5): 155–164.

Bissell, H.J., 1970. Realms of Permian tectonism and sedimentation in western Utah and eastern Nevada. *Bull. Am. Assoc. Pet. Geologists,* 54: 285–312.

Bucher, W.H., 1933. *The Deformation of the Earth's Crust.* Princeton Univ. Press, Princeton, New Jersey (republished in 1964 by Hafner, New York, N.Y., 518 pp.).

Currie, J.B., 1967. Evolution of stress in rocks of a sedimentary basin. *Rock Mechanics in Oilfield Geology, Drilling and Production. Proc. World Petrol. Congr., Mexico City, Mexico.* Elsevier, Amsterdam, pp.41–51.

Dallmus, K.F., 1958. Mechanics of basin evolution and its relation to the habitat of oil in the basin. In: L.G. Weeks (Editor), *Habitat of Oil.* Am. Assoc. Pet. Geologists, Tulsa, Okla., pp.883–931.

Fukuo, Y., 1969. Visco-elastic theory of the deformation of confined aquifer. In: *Land Subsidence. I.A.S.H.–Unesco, Publ.* No.89 AIHS, 2:547–562.

Holmes, A., 1944. *Principles of Physical Geology,* 1st ed. Nelson, London. (2nd ed. 1965, 1288 pp.).

Hsu, K.J., 1958. Isostasy and a theory of geosynclines. *Am. J. Sci.,* 256: 305.

Jacquin, C. and Poulet, M.J., 1970. Study of the hydrodynamic pattern in a sedimentary basin subject to subsidence. *45th Ann. Fall Meet. Soc. Pet. Engrs., Am. Inst. Min. Metall. Engrs., Houston, Texas,* Presented Paper No. SPE 2988, 6 pp.

Kartsev, A.A., Vagin, S.B. and Baskov, E.A., 1969. *Paleohydrogeology.* Nedra, Moscow, 150 pp.

Lofgren, B.E., 1969. Field measurements of aquifer-system compaction, San Joaquin Valley, California, U.S.A. In: *Land Subsidence. I.A.S.H.–Unesco, Publ.* no.88 AIHS, 1: 272–284.

Lofgren, B.E. and Klausing, R.L., 1969. Land subsidence due to ground-water withdrawal, Tulare–Wasco area, California. *U.S., Geol. Surv., Profess. Pap.,* 437B: 103 pp.

Marsden, S.S. and Davis, S.N., 1967. Geological subsidence. *Sci. Am.,* 216(6): 93–100.

Mikasa, M., 1963. *Consolidation of Soft Clays, A New Consolidation Theory and its Application.* Res. Inst. Kajima Construction Co. Ltd., Publ. Office, Japan.

Okumura, T., 1969. Analysis of land subsidence in Niigata. In: *Land Subsidence. I.A.S.H.–Unesco, Publ.* no.88 AIHS, 1: 130–143.

Poland, J.F. and Davis, G.H., 1969, Land subsidence due to withdrawal of fluids. In: *Reviews in Engineering Geology*, II. Geol. Soc. Am., New York, N.Y., pp.187–269.

Roberts, J.E., 1969. Sand compression as a factor in oil-field subsidence. In: *Land Subsidence. I.A.S.H.–Unesco, Publ.* no.89 AIHS, 2: 368–376.

Sandhu, R.S. and Wilson, E.L., 1969. Finite element analysis of land subsidence. In: *Land Subsidence. I.A.S.H.–Unesco, Publ.* no. 89 AIHS, 2: 393–400.

Scheidegger, A.E. and O'Keefe, J.A., 1967. On the possibility of the origination of geosyncline by deposition. *J. Geophys. Res.*, 72(24): 6275–6278.

Taylor, D.W., 1948. *Fundamentals of Soil Mechanics.* Wiley, New York, N.Y., 700 pp.

Ter-Martirosyan, Z.G. and Ferronsky, V.I., 1969. Some problems of time–soil compaction in pumping liquid from a bed. In: *Land Subsidence. I.A.S.H.–Unesco, Publ.* no.88 AIHS, 1: 303–314.

Terzaghi, K., 1925a. *Erdbaumechanik auf Bodenphysikalischer Grundlage.* Deuticke, Leipzig, 399 pp.

Terzaghi, K., 1925b. Principles of soil mechanics, I: Phenomena of cohesion of clay. *Eng. News Rec.*, 95: 742–746.

Terzaghi, K., 1926. Simplified soil tests for subgrades and their physical significance. *Public Roads,* 7: 153–162.

Terzaghi, K., 1936. Simple tests determine hydrostatic uplift. *Eng. News Rec.,* 116 (June 18): 872–875.

Terzaghi, K. and Peck, R.B., 1948. *Soil Mechanics in Engineering Practice.* Wiley, New York, N.Y., 566 pp.

Weeks, L.G., 1952. Factors of sedimentary basin development that control oil occurrence. *Bull. Am. Assoc. Pet. Geologists,* 36: 2071–2124.

Chapter 7

ABNORMAL GEOPRESSURES

INTRODUCTION

In previous chapters emphasis was placed mainly on the argillaceous sedimentary rocks that maintain a normal hydrostatic fluid pressure within pore spaces during compaction. The purpose of this chapter is to provide sedimentologists with a better understanding of the origin of abnormal fluid pressures, their significance to the study of compaction, and where they occur in the subsurface. Prior to 1930, few well-established records of abnormally high fluid pressures were known (Rubey and Hubbert, 1959). Among the earliest clear-cut examples published were the high pressures in steeply folded Tertiary sediments in the Khaur Field of West Pakistan (Keep and Ward, 1934). In recent years many geologists and engineers have investigated these abnormal pressures; however, few studies have been published relating to the geology of this phenomenon. Hubbert and Rubey (1959), Thomeer and Bottema (1961), Wallace (1969), and Fertl and Timko (1970a,b,c) have established that abnormal formation pressures, exceeding the normal hydrostatic pressure gradient of the interstitial fluids, occur in many locations throughout the world. Prediction of these pressures is an important factor in the successful planning of exploration and drilling programs by oil companies.

By definition, normal formation pressure is generally equated with hydrostatic force. Normal formation pressure at a certain depth can be defined as a pressure that approximately equals the hydrostatic head of water of equal depth. Were this formation open to the atmosphere, then the water column extending from the formation to ground level should balance the formation pressure. Abnormal pressures in the subsurface can be expressed conveniently in terms of a hydrostatic pressure gradient or fluid pressure/overburden stress ratio (λ), which is the ratio of the observed fluid pressure in a reservoir or aquifer to the pressure due to the weight of overlying deposits, computed for the depth at which the pressure is measured. Petrostatic, geostatic, or lithostatic pressure gradient refer to the pressure exerted by the weight of the overlying rock (Hubbert and Rubey, 1959; Thomeer and Bottema, 1961).

By the mid-1930's, approximately 20 wells in the Texas–Louisiana Gulf Coast region had encountered abnormal pore pressures at depth of 8,000–12,000 ft. In 1938, Cannon and Craze published a study on reservoir pressures in the Texas–Louisiana Gulf Coast area. They found that a normal hydrostatic pressure gradient in the post-Cretaceous

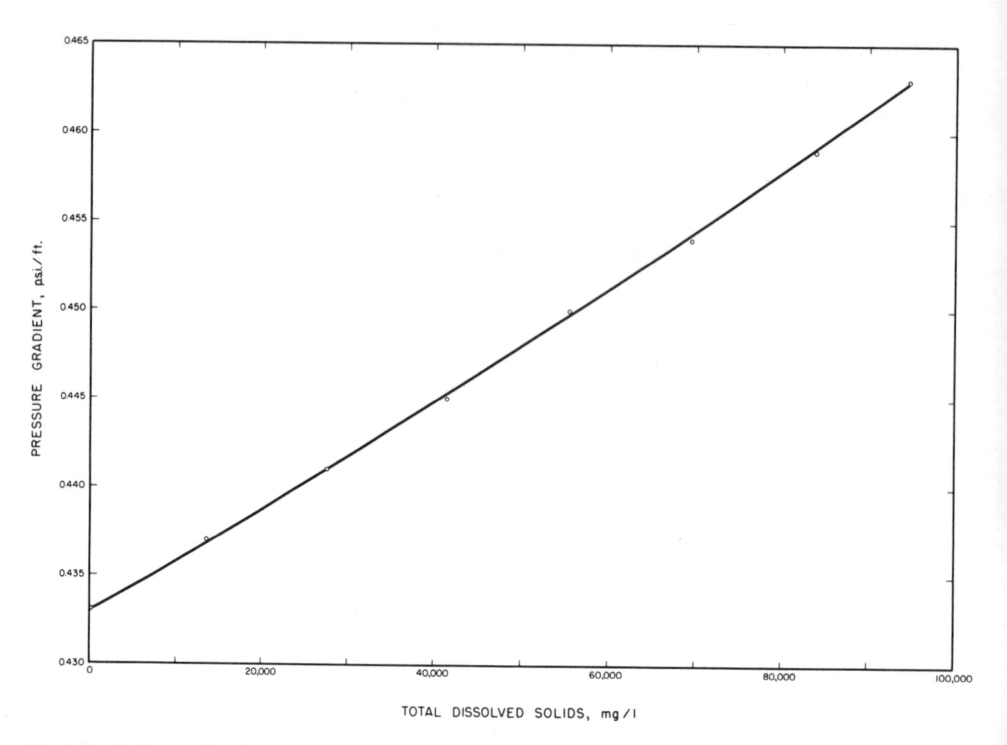

Fig.162. Relationship between total solids content in p.p.m. (mg/l) and fluid pressure gradient in p.s.i./ft. Compressibilities of water and gas in solution were not taken into consideration. Pressure gradient averages about 0.0043 p.s.i./ft. per increase of 0.01 g/cm^3 in density. (After Levorsen, table A-1, p.663. Courtesy of W.H. Freeman and Co.)

sediments of the Gulf of Mexico basin is about 0.465 p.s.i./ft. (0.106 bar per meter of depth). For the most part, these were original shut-in fluid pressures measured in the producing wells at oil–water contacts.

Dickinson (1953) and Fertl and Timko (1970a) defined abnormal geopressures as any pressure which exceeds the hydrostatic pressure of a column of brine containing about 80,000 p.p.m. (mg/l)[1] total solids (Fig.162). Jones (1969) mentioned that the term "geopressure" was first employed by Charles Stewart of the Shell Oil Co. to describe abnormally high subsurface fluid pressures. It should be pointed out that such a term is misleading, inasmuch as all subsurface pressures are geopressures. Terms such as abnormal geopressures, abnormal fluid pressures, or overpressures are more distinct and clear. Harper (1969) stated that all reservoir pressures are abnormal if they occur in the wrong place. From a drilling viewpoint any formation pressure gradient above 0.620 p.s.i./ft. is

[1] Most geologists and engineers use p.p.m. and mg/l interchangeably. This is technically incorrect; to change mg/l to p.p.m. one has to divide mg/l by the specific gravity of the brine.

considered abnormal because it commonly causes drilling problems (Harkins and Baugher, 1969). Most geologists consider the pressure gradients in excess of 0.465 p.s.i./ft. as being indicative of abnormally high fluid-pressure conditions in the Gulf Coast, whereas in other areas such as the Rocky Mountain region pressure gradients exceeding 0.433 p.s.i./ft. are considered abnormal (Finch, 1969). It should be noted that variations from the normal (expected) hydrostatic pressures occur commonly and the internal pore pressure may be either greater or smaller (Levorsen, 1958). In basins, such as the Williston Basin in North Dakota, Montana and southern Saskatchewan, Canada, where saline interstitial waters with up to 356,000 p.p.m. occur, the normal pressure gradient may be as high as about 0.512 p.s.i./ft. (Finch, 1969). According to Wallace (1969, p.971), shales surrounding abnormally pressured sandstone reservoirs and aquifers exhibit pore pressures of equal abnormality.

There is an upper limit to the hydrostatic pressure gradient and, as pointed out by Thomeer and Bottema (1961), hydrostatic pressures can never exceed the pressure exerted by the weight of the overlying rock column. The geostatic gradient is assumed to be equal to 1 p.s.i./ft., corresponding to an overburden density of 2.31 g/cm^3, which approximates the mean bulk density of water-filled sediments (Hubbert and Rubey, 1959, p.155). In reality, however, the geostatic gradient is variable because the density of the sediments and pore fluids can vary (see Chapter 2, Figs.12, 14, 15). For example, Keep and Ward (1934) reported a geostatic pressure gradient of 1.06 p.s.i./ft. for the Khaur Field in West Pakistan. Most of the abnormal hydrostatic pressure gradients reported from the Gulf Coast are equal to 0.8^+ p.s.i./ft., and there are a few known cases of 0.9^+ p.s.i./ft. (Wallace, 1969).

Hubbert and Rubey (1959) presented pressure data at the water–oil contact for eight producing fields in Iran (Table XLIII). The hydrostatic pressure gradient in these fields ranged from 0.71 to 0.98 p.s.i./ft. In the latter case, the geostatic load of 1.0 p.s.i./ft. is

TABLE XLIII

Hydrostatic pressure gradient at crest of oil reservoir and depth of overburden in eight Iranian oil fields (After Hubbert and Rubey, 1959, table 4, p.156. Courtesy of the Geological Society of America.)

Name of field	*Hydrostatic pressure gradient* (p.s.i./ft.)	*Depth of overburden* (ft.)
Masjid-i-Suleiman	0.88	640
Lali	0.76	3,900
Haft Gel	0.74	1,900
Naft Sefid	0.98	3,000
Agha Jari	0.71	4,800
Pazanun	0.68	5,700
Gach Saran	0.83	2,600
Naft-i-Shah	0.95	2,300

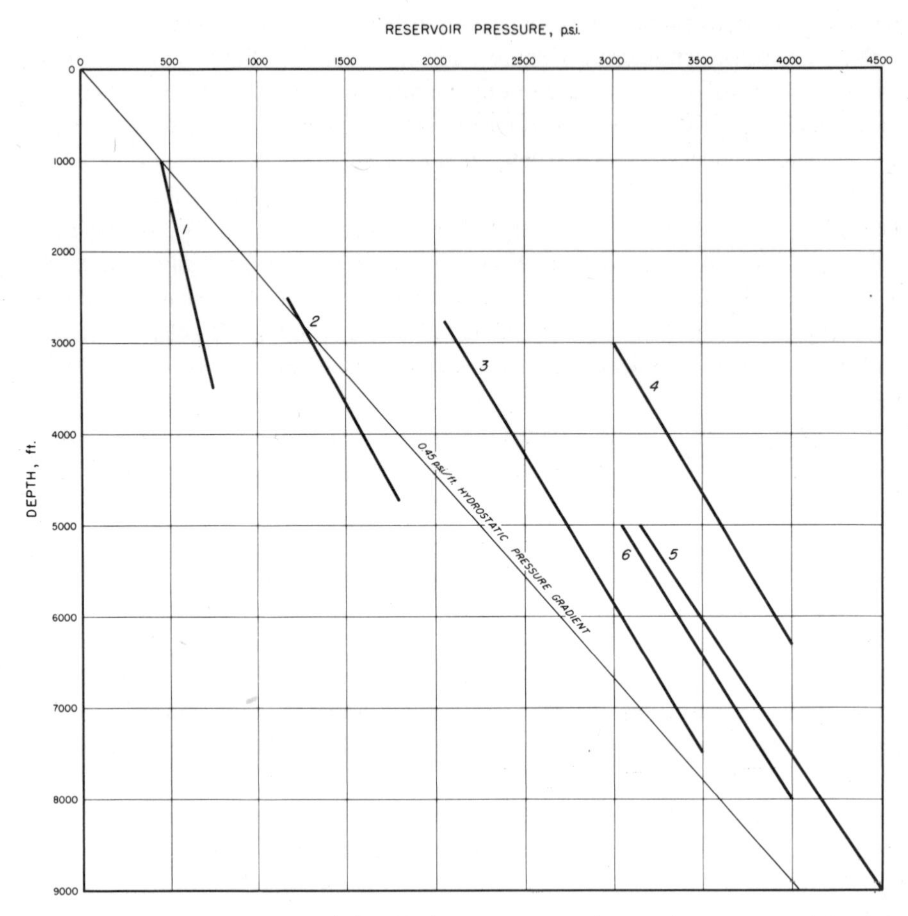

Fig.163. Diagram showing the relationship between measured reservoir pressures and depth in some Iranian oil fields. *1* = Masjid-i-Suleiman; *2* = Haft Gel; *3* = Gach Saran; *4* = Naft Sefid; *5* = Agha Jari; *6* = Lali. (Data from Lane, 1949, table 1, p.56.)

the sum of pore pressure (0.98 p.s.i./ft.) and grain-to-grain stress of 0.02 p.s.i./ft. (98% versus 2%). Fig.163 illustrates the fluid pressures in six Iranian oil fields where the pressure gradients lie below and above the normal hydrostatic gradient for the region (0.45 p.s.i./ft.). The hydrostatic pressure gradients encountered on drilling wells in the Agha Jari and Naft Sefid Fields of Iran (Hubbert and Rubey, 1959, p.155) are presented in Table XLIV. These pressure gradients occurred in the Lower Fars Formation, which is composed of marls, anhydrite, some limestones and considerable amounts of salt. In part, the abnormal pressures are due to the difference in pressure gradients between hydrocarbons and water at the crest of the structures.

TABLE XLIV
Variations in the hydrostatic pressure gradients of brines encountered during drilling at the Agha Jari and Naft Sefid Fields, Iran
(After Hubbert and Rubey, 1959, table 5, p.156. Courtesy of the Geological Society of America.)

Well	*Depth* (ft.)	λ = *fluid pressure/overburden stress ratio*
AJ 5	8354	0.85
AJ 10	4703	0.84
AJ 18	3010	0.90
AJ 22	6314	0.97
AJ 24	4681	0.83
AJ 25	7012	1.00
AJ 27	2445	0.86
AJ 36	4841	0.91
W 7	4870	1.00
W 11	4999	0.94
W 13	3147	0.97
W 14	4885	0.91
W 16	4538	0.87
W 21	5226	0.98
W 24	3712	0.91

ORIGIN OF ABNORMAL GEOPRESSURES

Considerable disagreement exists among earth scientists as to the mechanism responsible for creating abnormally high pressures (Levorsen, 1958; Hubbert and Rubey, 1959; Wallace, 1969; and Overton and Zanier, 1971). Several significant details concerning the abnormal geopressures in the Gulf Coast Basin have been determined from field research (Overton and Zanier, 1971). Abnormal pressures are primarily associated with (1) sediments in which the sand fraction constitutes $< 25\%$; (2) water salinities which decrease with depth; and (3) high temperatures ($> 200°$F).

The abnormal formation pressures in many sedimentary basins are normally attributed to one of the following causes: (1) continuous loading and incomplete gravitational compaction of sediments; (2) faulting; (3) phase changes in minerals during compaction; (4) salt and shale diapirism; (5) tectonic compression; (6) osmotic and diffusion pressures; and (7) geothermal temperature changes creating fluid-volume expansion or contraction. Some authors have suggested that other mechanisms might conceivably give rise to high pore pressures. These explanations are: (1) Existence of fossil pressures that corresponded to a previous greater depth of burial (Reed, 1946). (2) Invasion of water derived from magmatic intrusions (Platt, 1962). (3) Infiltration of gas, which is possibly the cause of some of the high pressures encountered during drilling offshore wells in Tertiary rocks of Indonesia (Aliyev, in: Tkhostov, 1963). Comins (in: Dickinson, 1951, p.17) believed that where the edge water in a reservoir had an abnormal pressure gradient and the formation had no communication with the outcrops, the high pressures could be attributed to vertically migrating gas from a deeper, higher-pressure source. (4) Solution and precipitation of minerals with a consequent change in pore volume (Levorsen, 1958). (5) Fractionation (and cracking) of heavy hydrocarbons to lighter fractions (Chaney, 1949). (6) Presence of an aquifer and an artesian head, which is indicative of recharge areas at high elevations. This is well known in hydrology and is thought to cause some of the high pore pressure found in Iranian limestones. Comins (in: Dickinson, 1951, p.17) pointed out that in Iran the occurrence and magnitude of oil seepages were related to the pressure gradient. Where λ was greater than 0.6, oil seepages were not present. (7) Occurrence of earthquakes (Levorsen, 1958). (8) Difference between the density of hydrocarbons (oil and gas) and that of pore water in a reservoir (Fig.164). According to Gretener (1969), the abnormal pressure can be calculated on using the following equation:

$$p_a = \Delta z(\rho_w - \rho_h)g \tag{7-1}$$

where Δz is the thickness (height) of the hydrocarbon zone above the oil–water contact;

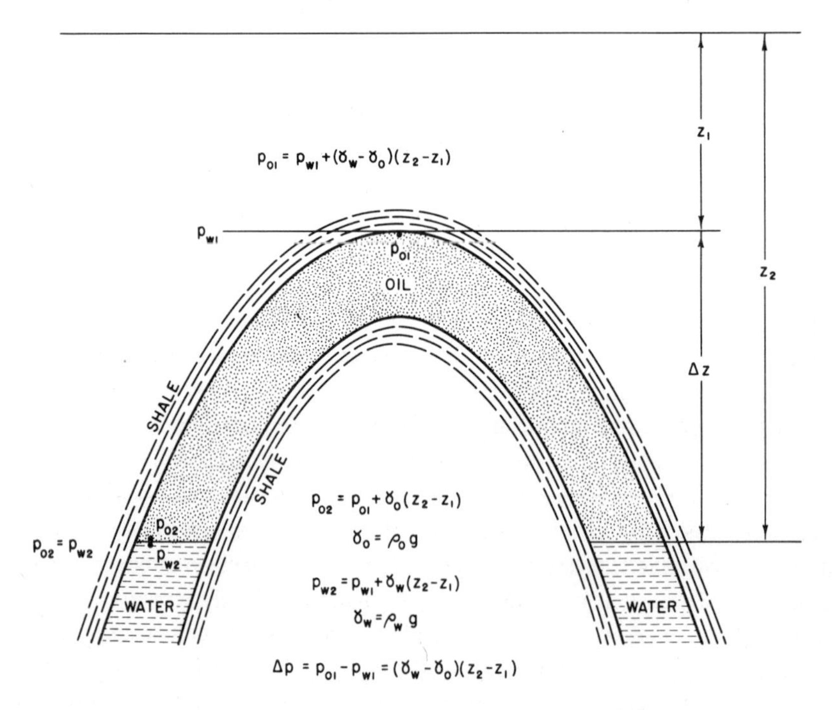

Fig.164. Cross-sectional view of an anticlinal reservoir sandwiched between two impervious shale bodies, showing abnormal pressures in hydrocarbon accumulation in hydrostatic water environment. (Modified after Hubbert and Rubey, 1959, p.150; and Gretener, 1969, p.267.)

ρ_w is the density of water; ρ_h is the density of hydrocarbons; and g is the acceleration of gravity. Gretener (1969, p.266) reported that Δz may attain a value of 5,000 ft. in some large Middle Eastern structures. Assuming a density contrast of 0.23 g/cm^3 between the oil and water, the maximum abnormal pressure produced may reach 500 p.s.i.; for an equal column of gas, it may be as high as 1,000 p.s.i.

Most overpressures resulting from the above causes are transient phenomena that will slowly dissipate with time, reestablishing the normal hydrostatic pressure gradient in the sediments.

Compaction as a cause of abnormal formation pressures

One of the most complete studies on the geologic aspects related to abnormal fluid pressures was made by Dickinson in 1951. He pointed out that in the Gulf Coast Tertiary province high pressures occur frequently in isolated Miocene and Pliocene porous sand beds found in thick shale sections, which are located below the main deltaic sand series. Location of the abnormal geopressures is controlled by the regional facies changes in the basin and appears to be independent of depth and geologic age of the formation. Three types of reservoir seals necessary to preserve abnormal pressures were described by Dickinson (1953). The reservoir types are (1) small reservoirs sealed by pinchouts; (2) large reservoirs sealed updip by faulting against thick shales and downdip by a regional facies changes; and (3) reservoirs formed by fault blocks. Excessive fluid pressures within sediments are, according to Dickinson (1953), dominated by two factors: (1) the compression of the fluids owing to compaction and (2) the resistance to fluid expulsion.

When the upper layer of a freshly-deposited argillaceous sediment is undergoing initial compaction at the bottom of the sea, the interstitial fluid is continuous with the overlying sea water and the pore pressure is essentially hydrostatic. The solid mineral grains comprising the sediment are in contact with each other; however, it is possible that some clay solids might not have direct contact owing to the presence of adsorbed water layers. Grain-to-grain contact pressure is equal to the total pressure due to the overburden load minus the pore pressure (buoyant force). Buoyancy effect, however, is minor in a thin shallow sediment (Thomeer and Bottema, 1961, p.1722). As the upper sediment layer is buried under subsequently deposited mud and sand layers, a gradual compaction takes place. If the sedimentation rate is slow, this sediment layer will then gradually adjust to the additional load imposed by the overlying sediments. As the mineral grains are pressed closer together, pore water is expelled. With time, fluid pressures will also adjust to some gradient if there is fluid intercommunication between adjacent sediment layers. Inasmuch as the argillaceous sediment has high porosity and is permeable in its initial state, the expelled fluids will flow in the direction of least resistance, usually into a porous sand layer. As long as the fluid can escape under normal loading conditions and porosity is intercommunicating, hydrostatic pressures will be encountered.

Rapid loading and continuous sedimentation. It is normally accepted that in areas like the Gulf Coast Basin, thick argillaceous sediments have to be deposited rapidly over considerable areas in order to prevent establishment of normal hydrostatic pressure equilibrium during compaction (Fertl and Timko, 1970a). Where rapid deposition of large quantities of clays and lesser amounts of sand occurs in a basin, the sand bodies will be completely surrounded by the shales. If the relative gravitational loading rate of sediments is high, permeability decreases rapidly and, as a result, the pore fluids cannot escape from the sands through the overlying argillaceous sediments. Inasmuch as the interstitial fluids help to support the increasing overburden load, further compaction of the formation is retarded or stopped. Thus, the formation becomes abnormally pressured because the fluids are subjected not only to hydrostatic forces, but also to the weight of newly-deposited sediment.

Rubey and Hubbert (1959, p.180) discussed rapid loading effects on saturated sediments. A sedimentary sequence subjected to rapid active loading undergoes a rapid buildup in pore pressure and sometimes possible leakage of the pressurized fluid. This buildup in pore pressure, p_a, is expressed as:

$$p_a = p - p_n \tag{7-2}$$

where p is the pore pressure and p_n is the normal hydrostatic pressure ($p_n = \rho_w g D$, where D is the depth).

If the overburden load on the argillaceous sediment increases by the amount ΔS and there is no pressure dissipation, the effective grain-to-grain stress σ' exerted on the sediment would remain unchanged. Thus, the increase in load would have to be equal to the increase in pore pressure of the system:

$$\Delta p_a = \Delta S \tag{7-3}$$

The rate at which the abnormal pressure would increase as the sedimentational load is being added continuously, can be expressed as:

$$\left(\frac{\mathrm{d}p_a}{\mathrm{d}t}\right)_1 = \frac{\mathrm{d}S}{\mathrm{d}t} \tag{7-4}$$

where $\mathrm{d}t$ is increment in time and $\mathrm{d}S/\mathrm{d}t$ is the rate of load addition. Rubey and Hubbert (1959, p.181) stated that this is the mechanism of pressure buildup in response to added sediments. Then, they considered the opposite process of pressure dissipation. As a first approximation of this complex process, Rubey and Hubbert (1959) assumed that the water-transmission properties of the clay do not change during pressure leakage. They deduced as a consequence of Darcy's law that the rate of discharge of the water at any instant would be proportional to the magnitude of the anomalous p_a. The rate of change of pressure due to fluid leakage is then equal to:

$$\left(\frac{\mathrm{d}p_a}{\mathrm{d}t}\right)_2 = \frac{-p_a}{T} \tag{7-5}$$

where T is the relaxation constant, a proportionality factor having the dimensions of time, which signifies the period of time it would take to dissipate completely the initial anomalous pressure, p_{ai}, at the initial rate of pressure decrease.

An expression can now be given describing the total rate of abnormal pressure increase in the presence of the above-described two processes:

$$\frac{\mathrm{d}p_a}{\mathrm{d}t} = \frac{\mathrm{d}S}{\mathrm{d}t} - \frac{p_a}{T} \tag{7-6}$$

On assuming that the rate of sedimentation is constant, i.e., the rate of sediment loading is constant,

$$\frac{\mathrm{d}S}{\mathrm{d}t} = K' \tag{7-7}$$

and on substituting eq.7-7 into eq.7-6 and rearranging the terms gives:

$$\frac{\mathrm{d}t}{T} = \frac{\mathrm{d}p_a}{K'T - p_a} \tag{7-8}$$

Integrating eq.7-8 from the initial conditions $t = 0, p_a = p_{ai}$ one obtains:

$$\frac{t}{T} = -\ln \frac{K'T - p_a}{K'T - p_{ai}} \tag{7-9}$$

or:

$$\mathrm{e}^{-t/T} = \frac{K'T - p_a}{K'T - p_{ai}} \tag{7-10}$$

Solving for p_a, gives:

$$p_a = K'T + (p_{ai} - K'T)\mathrm{e}^{-t/T} \tag{7-11}$$

Rubey and Hubbert (1959) explained that on assuming that the rate of loading K' and the relaxation time T are both constant, then t/T becomes large and the anomalous pressure, whether initially less than or greater than $K'T$, will approach $K'T$ as a limiting value.

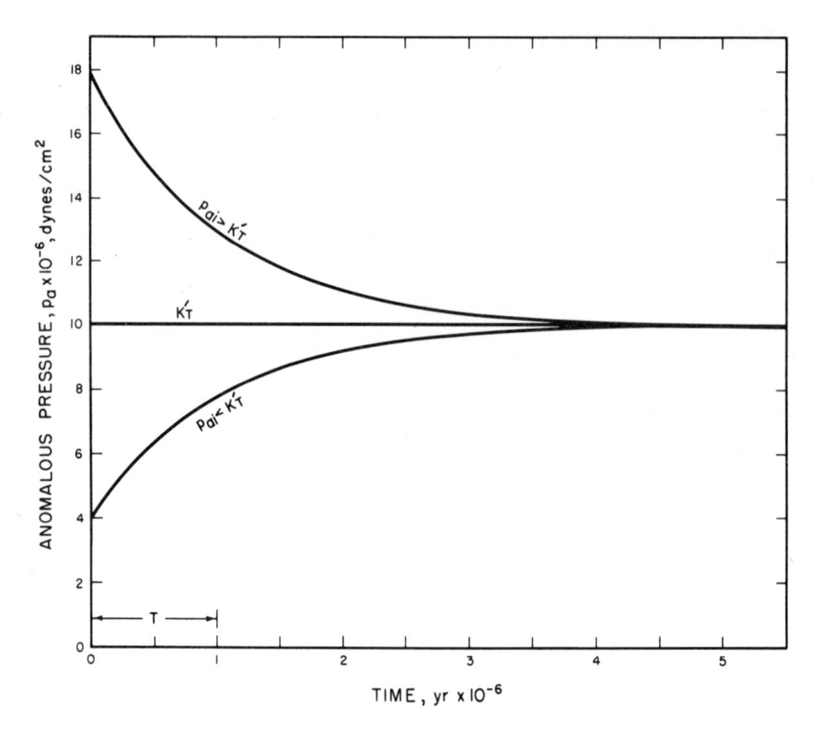

Fig.165. Relationship between anomalous pressure and time, showing tendency of pressure to approach the $K'T$ product with lapse of time (K' = rate of loading = 10 (dynes/cm^2)/year; and T = relaxation constant = 10^6 years. $p_a = K'T + (p_{ai} - K'T)e^{-t/T}$, where p_{ai} is the initial abnormal pressure and t is the time. (After Rubey and Hubbert, 1959, fig.5, p.182. Courtesy of the *Geological Society of America Bulletin.*)

Fig.165 illustrates the tendency of anomalous fluid pressures to approach the product of rate of sediment loading (K') multiplied by (T) with lapse of time. An approximation of the effects of loading rate on the development of abnormal fluid pressure can be made, using Fig.166.

When a new sediment load is deposited in a basin, the instantaneous effect is to increase the fluid pressure/overburden pressure ratio in the already buried sediments. From Fig.166 the amount of this instantaneous effect can be roughly estimated by constructing a line drawn horizontally to the right from any given original depth to a new depth at which this same point would lie after the thickness of new sediment has been added. If the pore fluids escape quickly in response to the loading, the compaction rate is rapid. On the other hand, if the pore-fluid escape is hindered by the low permeability, the compaction rate may be very slow (Rubey and Hubbert, 1959, p.182).

As pointed out by Rubey and Hubbert (1959, p.182), when the rate of sedimentation is slow with respect to the rate of escape of pore fluids from the sediments, then any excess fluid pressures created by the new load will be dissipated about as rapidly as

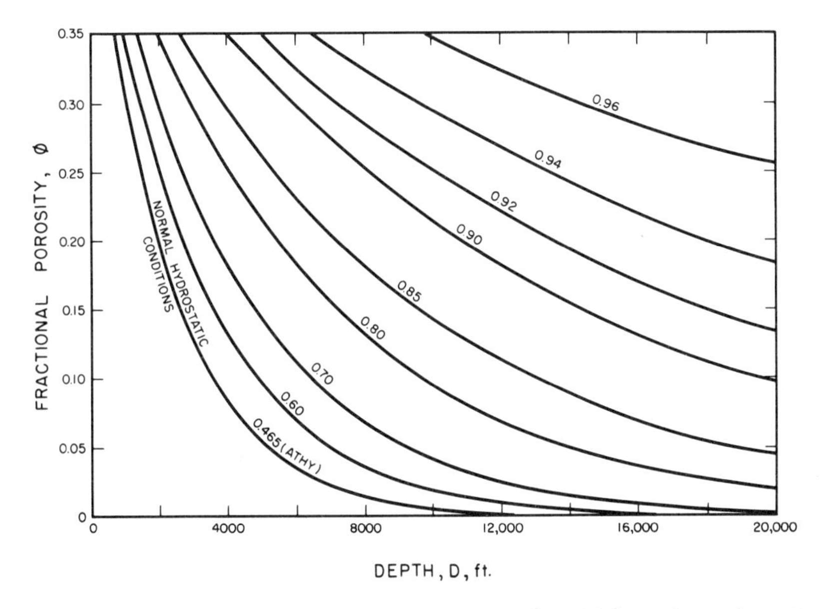

Fig.166. Relationship between porosity and depth of burial for various values of λ (fluid pressure/overburden pressure ratio) for an average shale or mudstone. Athy's curve (λ = 0.465) is assumed to represent "compaction equilibrium" condition. (After Rubey and Hubbert, 1959, fig.4, p.178. Courtesy of the *Geological Society of America Bulletin*.)

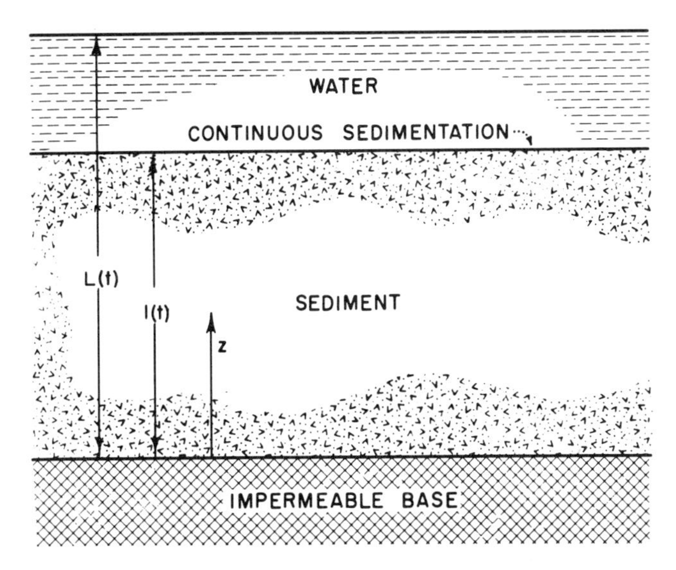

Fig.167. Schematic diagram of continuous sedimentation in water. (After Bredehoeft and Hanshaw, 1968, fig.4, p.1103. Courtesy of the *Geological Society of America Bulletin*.)

formed. Essentially hydrostatic pressures and compaction equilibrium will be maintained at all depths.

In contrast to a slow rate of loading, if the sedimentation rate is rapid, the possibility exists that the pore fluid can be dissipated rapidly enough to still maintain essentially hydrostatic pressure at depth, when permeable sands are abundant. This would not be the case if the majority of the buried sediments are very fine-grained clastics, especially at greater depths.

The progress of compaction in an argillaceous layer with time has been studied by several authors. Bredehoeft and Hanshaw (1968) presented a hydrodynamic compaction model based on a continuous sedimentation rate, which possibly describes the creation of high pressures in the Gulf Coast sediments. In most circumstances the maintenance and creation of an anomalous fluid pressure depends largely upon the low permeability and specific storage of the thick argillaceous rocks in the Gulf Coast deposits. Specific storage, S_s, is defined as the volume of water taken into storage or discharged per unit volume per unit change in head. Fig.167 is a schematic view of the continuous sedimentation model (in water), in which z is equal to zero at the impermeable boundary. The excess fluid head created within the sediments can be described by the following equation (Gibson, 1958, p.172):

$$\frac{\partial^2 h'}{\partial z^2} = \frac{S_s}{K} \frac{\partial h'}{\partial t} - \frac{\rho'}{\rho_w} \frac{\partial l}{\partial t} \qquad 0 \leqslant z \leqslant 1 \tag{7-12}$$

boundary conditions are:

$$l = 0 \qquad t = 0$$

$$\left. \frac{\partial h'}{\partial z} \right|_{z=0} = 0 \qquad t \geqslant 0$$

$$h'(l, t) = 0 \qquad t > 0$$

where h' is the excess fluid head $= p_a/\rho_w g$ (p_a is the transient pore pressure in excess over the normal hydrostatic pressure originally present); z is the vertical coordinate; S_s is the specific storage and is defined as the volume of water taken into storage or discharged per unit volume per unit change in head and is equal to $\rho_w g[(1/E_s) + (\phi/E_w)]$, where $\rho_w g$ = weight of unit volume of water, E_s = modulus of compression of the rock skeleton confined in situ ($= \sigma_z/\epsilon_z$ or ratio of vertical stress to vertical strain), ϕ = porosity, and E_w = bulk modulus of elasticity of water; K is hydraulic conductivity, which is equal to $(k\rho_w/\mu)g$, where k = permeability, ρ_w = density of fluid, g = acceleration of gravity, and μ = viscosity of fluid; t is time; $\rho' = \rho_b - \rho_w$, i.e., the difference between the bulk density,

ρ_b, of the sediments and the density of the pore fluid, ρ_w; and l is the thickness of the sediments.

Gibson (1958) pointed out that eq.7-12 is independent of the depth of water. Inasmuch as the depth of water may vary arbitrarily with time, $L(t)$, then excess head can be expressed as:

$$h' = h - L(t) \tag{7-13}$$

For continuous sedimentation in which the rate of sediment accumulation is equal to $\partial l/\partial t = \omega$, where ω is constant, the solution to eq.7-12, following Gibson (1958, p.175), is:

$$\frac{h'\rho_w}{l\rho'} = 1 - \left(\frac{\pi K t}{S_s l^2}\right)^{-1/2} \exp\left[-\left(\frac{z}{l}\right)^2 \frac{S_s l^2}{4Kt}\right]$$

$$\times \int_0^\infty \xi' \tanh\left(\frac{\xi' S_s l^2}{2Kt}\right) \cosh\left[\left(\frac{z}{l}\right)\frac{\xi' S_s l^2}{2Kt}\right] \exp\left(-\frac{\xi' S_s l^2}{4Kt}\right) \mathrm{d}\xi' \tag{7-14}$$

where $\xi' = z/l(t)$. Gibson (1958, p.171) stated that it is unlikely that closed solutions can be obtained for arbitrary rates of sediment deposition.

Bredehoeft and Hanshaw (1968) plotted the above solution in dimensionless form (Fig.168). By employing Fig.168, they considered whether or not continuous sedimentation could result in the high fluid pressures found in the Gulf Coast deposits. Assuming

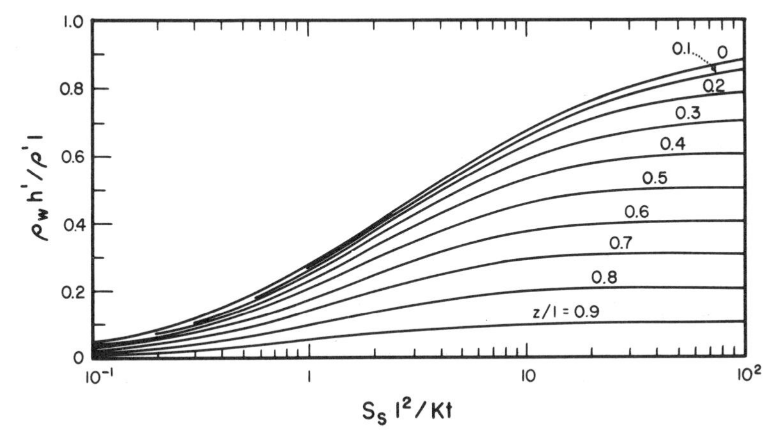

Fig.168. Dimensionless plot of the excess-head distribution in a layer of sediment that is increasing in thickness at a constant rate. (After Bredehoeft and Hanshaw, 1968, fig.5, p.1104. Courtesy of the *Geological Society of America Bulletin*.)

TABLE XLV

Values assumed by Bredehoeft and Hanshaw (1968, p.1105) in preparing pressure-versus-depth plots (Fig.169)

ω (rate of sedimentation)	= 500 m/10^6 years
t (period of sedimentation)	= $20 \cdot 10^6$ years
l (thickness of sediments = $t\omega$)	= 10,000 m
S_s (specific storage)	= $3 \cdot 10^{-5}$ cm^{-1}
ρ_b (average bulk density of sediments)	= 2.3 g/cm^3
ρ_w (density of pore water)	= 1 g/cm^3

reasonable values, as given in Table XLV, it was possible to calculate pressure versus depth using Fig.168. The results of this calculation are given in Fig.169 which indicates that negligible excess pore pressures can be created in sediments having a hydraulic conductivity of 10^{-6} cm/sec (1 md), or higher. As the hydraulic conductivity of the sediments decreases, however, significant high pressures are produced (Bredehoeft and Hanshaw, 1968, p.1105). Sediments with K values of 10^{-8} cm/sec (10^{-2} md), or lower, can be expected to have pressures approaching lithostatic conditions with the assumed rates of sedimentation.

Katz and Ibrahim (1971) presented a simplified mechanism for explaining shale com-

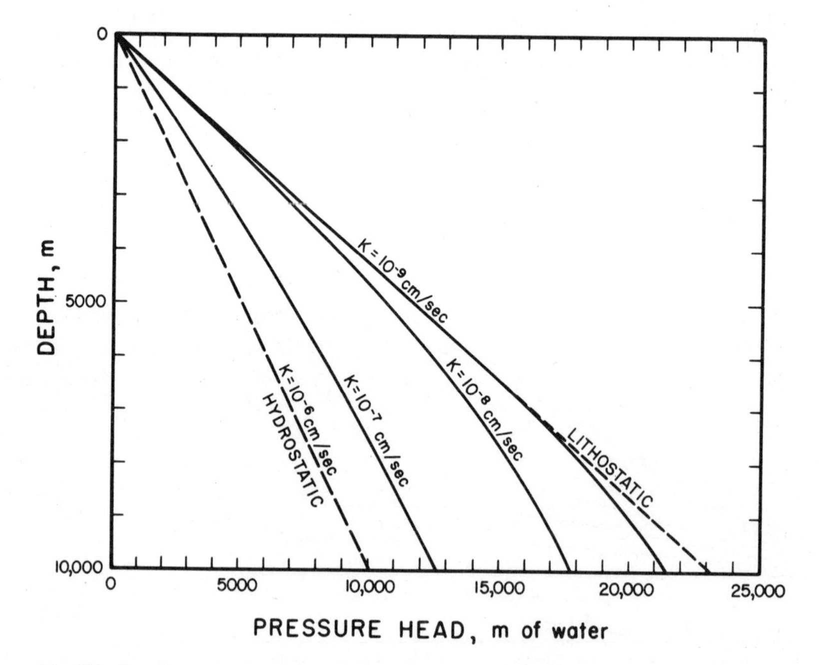

Fig.169. Depth versus pressure relation that approximates conditions in the Louisiana and Texas Gulf Coast basin receiving sediments at a constant rate. (After Bredehoeft and Hanshaw, 1968, fig.6, p.1104. Courtesy of the *Geological Society of America Bulletin*.)

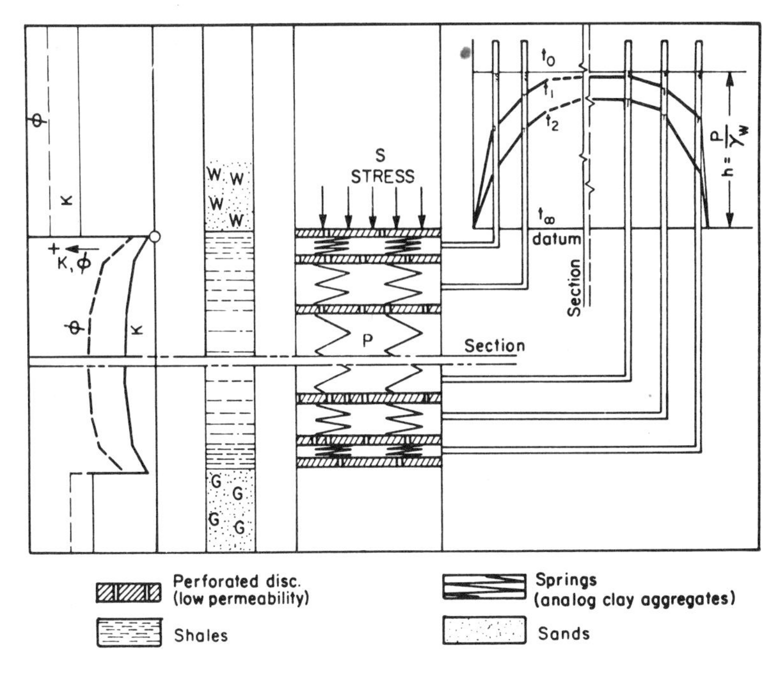

Fig.170. Schematic representation of shale compaction, porosity and permeability relationships, and creation of abnormally high pressures. k = permeability; ϕ = porosity; t = time; p = pore pressure; γ_W is specific weight of water; h = height to which fluid will rise in the tubes; W = water; G = gas. (After Katz and Ibrahim, 1971, fig.12. Courtesy of the Society of Petroleum Engineers of A.I.M.E.)

paction and the creation of abnormal fluid pressures. Their model is based on Terzaghi's conceptual piston and spring analogy (see Figs.49 and 51). Although Terzaghi's model has certain limitations, it can be modified to furnish solutions for correlating a rapid sedimentation rate with the entrapment of abnormally high fluid pressures like those found in Gulf Coast Tertiary shales. Fig.170 illustrates Katz and Ibrahim's model, which is based on the compaction of an argillaceous layer between two permeable sand layers. As in Terzaghi's model, the shale formation is represented by a series of springs and perforated disks. The perforated disks represent low-permeability shales, which restrict the escape of fluids; whereas the springs represent the deformable character of the clay matrix. In Katz and Ibrahim's model, sudden loading corresponds to a rapid sedimentation rate. The water occupying the space between the perforated disks represents the interstitial shale fluid. When stress is applied suddenly to the system, the water between the disks initially will support the entire load. After a brief period of time, the same fluid will be forced through the openings in the disks either in an upward or downward direction depending on the relative magnitudes of pressures inside and adjacent to the system. Fluid movement in this model is restricted to upward and downward directions. As the uppermost and lowermost disks of the model move slightly closer to the internal disks, the springs in-between will begin to carry part of the applied load (Fig.170). As a result, the fluid pressure between the external disks will decrease; but the closer these disks approach each other, the more difficult it will be for the pore water to escape from inside the system. Katz and Ibrahim (1971) mentioned that the gradual decrease of permeability from the center toward the top and bottom of the model could be represented either by a decrease of the number of perforations as a result of compaction or by an increase of the number of disks per unit length in the model. Fig.170 also shows the fluid-pressure distribution as a function of time. Higher fluid pressure is indicated to exist in the middle rather than in the upper or lower portions of the model. This means that it takes more time for the water in the middle of the model to escape than at the outer boundaries. The porosity and permeability distribution in this system is illustrated on the left-hand side of Fig.170.

Katz and Ibrahim's model shows agreement with most of the information gathered during drilling into abnormally pressured formations in the Gulf Coast. It illustrates the observed increase in porosity associated with undercompacted shales, extreme drop in permeability with confining lithostatic pressure, and entrapment of high interstitial fluid pressure in the shales.

Faulting as a cause of overpressured formations

Dickey et al. (1968) offered an alternate explanation of how high formation pressures originated in the Gulf Coast sediments. In southwestern Louisiana, according to Dickey et al., the pattern of high-pressure zones appears to be related to the peculiar patterns of faulting contemporaneous with sedimentation and compaction. This "growth fault" process prevents the expulsion of water from the pores of argillaceous sediments during compaction and diagenesis. The high pore pressures might have facilitated sliding and slumping of the sediments at the edge of the continental shelf. Growth faults have many of the characteristics associated with slump-type landslides, and Dickey et al. (1968) stated that they may indeed be due to the old slides that have ceased their activity and were buried by later sedimentation. The role of growth faults in the structural deformation of geopressured sediments is described by Ocamb (1961) and Thorsen (1963).

The Oligocene and Miocene sediments in southwestern Louisiana consist of three facies: (1) continental and deltaic facies consisting of massive sands; (2) neritic facies composed of alternating sands and shales; and (3) shale facies consisting of argillaceous sediments deposited on the outer shelf and slope. Shallow-water and continental sediments overlap the marine sequence deposited earlier in deeper water (Dickey et al., 1968). Abnormal pressures are first encountered in the neritic facies directly beneath the base of the more massive and continuous deltaic sands. Harkins and Baugher (1969) stated that in order for abnormal pressures to develop, the shales usually must be over 200 ft. in thickness. The intertonguing sand–shale facies forms down-slope from the deltaic facies and, therefore, as a prograding sequence, tends to rise stratigraphically in a

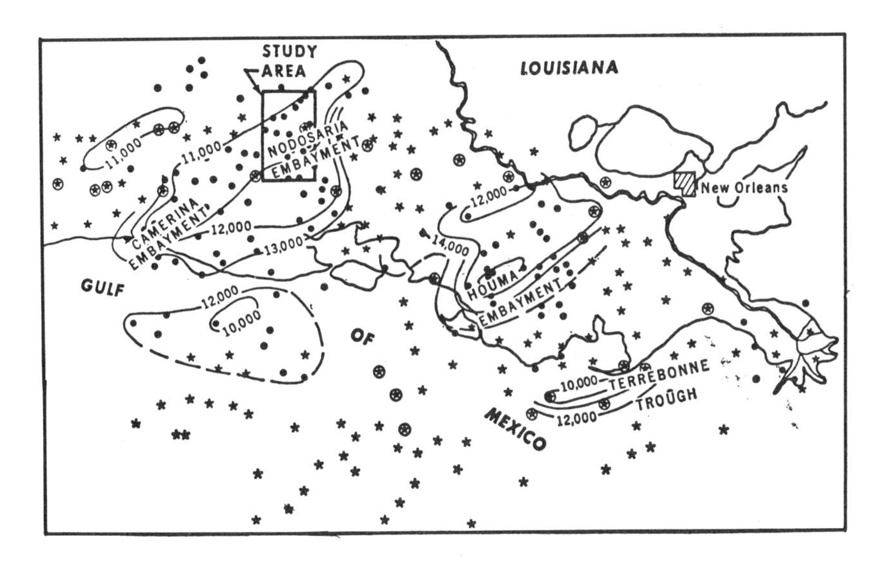

Fig.171. Map of southern Louisiana showing the location of fields with abnormally high reservoir pressures. Approximate depths to the top of the high-pressure zone are shown by highly generalized contours. (Depths in ft.) Stars = salt domes; encircled stars = salt domes with high pressures; full dots = fields without salt domes. (After Dickey et al., 1968, fig.7. Courtesy of *Science*, copyright 1968 by the American Association for the Advancement of Science.)

basinward direction (Harkins and Baugher, 1969). The stratigraphic units thicken seaward.

Geologic structure associated with this sedimentation pattern is dominated by growth faults lying roughly parallel to the coast between salt domes (Fig.171). "Embayments" are areas where salt domes are scarce and growth faulting has caused some stratigraphic units to be abnormally thick. Abnormal pressures are usually found at depths of 10,000–11,000 ft. Dickey et al. (1968) pointed out that the stratigraphic units are thicker on the downthrown side of the growth faults than they are on the upthrown side (Fig.172). Their explanation for this thickening of the sediments is that movement along the fault plane was continuous during sedimentation. The fault plane cut the sea floor while sediments were being swept over it, so that the downthrown block was covered with a thicker layer of sediment. As shown in Fig.172, grabens also commonly occur. Abnormal pressures are associated with this structure–facies relationship and rise stratigraphically in a basinward direction, being modified by growth faulting (Harkins and Baugher, 1969).

Dickey et al. (1968) proposed the following explanation of the relationship between abnormal fluid pressures and growth faulting. During compaction, pore fluids in the marine sediments migrate vertically upward towards the sea floor at a constant rate. As compaction progresses, the vertical permeability of the argillaceous sediments decreases rapidly, forcing the interstitial fluids to travel parallel to the bedding planes. If growth

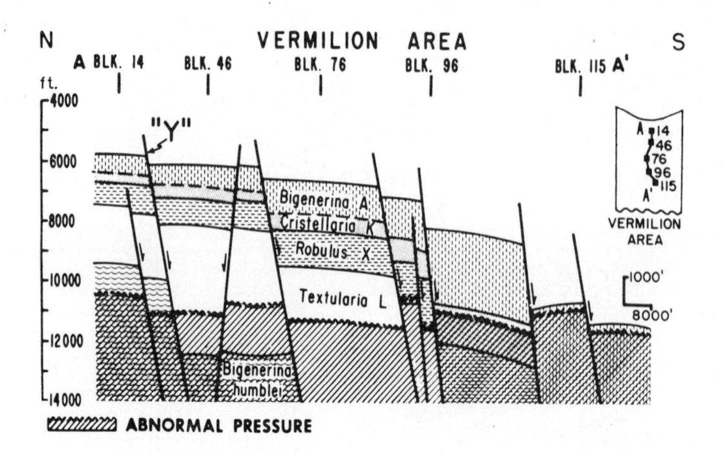

Fig.172. Schematic section illustrating stratigraphic rise of abnormal pressure across growth faults in the Vermilion area of southwestern Louisiana. (After Harkins and Baugher, 1969, fig.4, p.963. Courtesy of the Society of Petroleum Engineers of A.I.M.E.)

faulting occurs while abundant water is still present in the shales, the routes of updip fluid migration parallel to the bedding would be shut off by the fault plane. It should be noted that faults in unmetamorphosed rocks and semi-consolidated sediments seldom act as channels for fluid movement. Pressure buildup tests in producing oil and gas wells have shown that faults, which cut reservoirs, form pressure discontinuities and are excellent seals to fluid movement. As a result of cutoff because of faulting, the fluid has to sustain a heavier overburden load as sedimentation proceeds. Whenever the growth faulting occurred after most of the water had been expelled and the shales were already compacted, the abnormal pressures were observed to be much lower, or pressures were normal (Dickey et al., 1968).

Dickey et al. (1968) pointed out that inasmuch as growth faults have dip angles of less than 50°, wells frequently cross fault planes. It is possible for a well to encounter the abnormal-pressure zone and then, after crossing a fault plane, to enter a different fault block where pressures are normal.

Phase changes in minerals during compaction

Gypsum–anhydrite conversion. Hanshaw and Bredehoeft (1968) investigated a physico-chemical mechanism involving the phase change of gypsum to anhydrite with the release of structural water, which produces excessive pore pressures. The thermodynamics of the gypsum dehydration reaction indicates that conversion will probably occur at shallow depths and produce water at a constant rate (Fig.173). This reaction is similar to the dehydration of montmorillonite, inasmuch as the latter process also produces fluids at a constant rate under the proper dehydrating conditions.

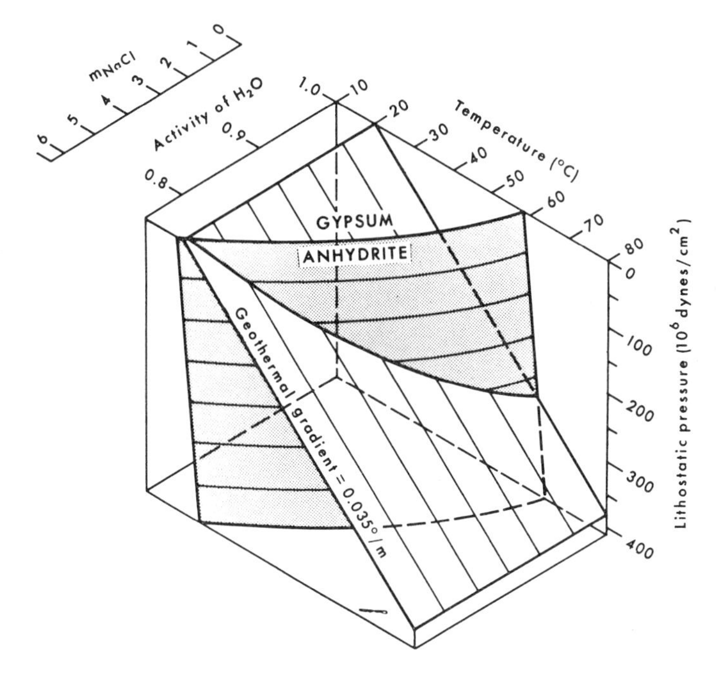

Fig.173. Pressure–temperature-activity of H_2O diagram for the ternary system $CaSO_4$–H_2O–*i*, where *i* is any third component which lowers the activity of H_2O. Contours on the stippled phase-boundary plane are 250-m depth contours. $p_W = p_S$. p_S = pressure on solid phase. (After Hanshaw and Bredehoeft, 1968, fig.7, p.1115. Courtesy of the *Geological Society of America Bulletin.*)

Heard and Rubey (1966) calculated that the excess water produced in a 50-ft. gypsum bed as a result of conversion to anhydrite is equivalent to a water layer 24.25 ft. in thickness. If the anhydrite layer behaves elastically, it will take up some of the load as the fluid is expelled into the surrounding sediments. The pore pressure within the anhydrite bed will not remain lithostatic, but will diminish with time (Hanshaw and Bredehoeft, 1968, p.1111).

In analyzing the effectiveness of any phase-change mechanism, the following variables have to be considered: (1) depth, (2) temperature, (3) rate of burial, (4) heat flow, (5) enthalpy requirements of the reaction, (6) permeability, (7) specific storage of the surrounding material, and (8) chemistry of the solution in contact with the solid phases.

According to both MacDonald (1953) and Stewart (1963), gypsum is seldom found below a depth of 600–900 m. These field data are consistent with the data presented in Fig.173 by Hanshaw and Bredehoeft (1968) on assuming a geothermal gradient of 0.035°C/m. In a distilled-water environment, the maximum depth at which gypsum could exist is 1,150 m. One is not likely to encounter such an environment in nature, however. For the purpose of discussion, Hanshaw and Bredehoeft (1968, p.1114) assumed that gypsum would be buried with a pore solution that was initially similar in composition to

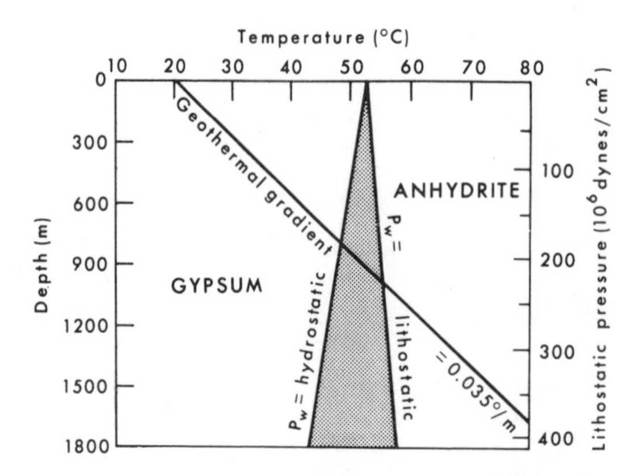

Fig.174. Pressure–temperature relationship between gypsum and anhydrite where activity of H_2O = 0.93 (≈ 2*M* NaCl). (After Hanshaw and Bredehoeft, 1968, fig.8, p.1116. Courtesy of the *Geological Society of America Bulletin.*)

that of sea water which has been evaporated to the point where gypsum precipitates. This would be equivalent to the activity of H_2O of 0.93 (about 2 *M* NaCl). Fig.174 depicts the pressure–temperature relationship for equilibrium between gypsum and anhydrite in a solution with an activity of H_2O equal to 0.93. Hanshaw and Bredehoeft (1968) stated that apparently there is enough geothermal heat available, at the rate of burial of sediments in the Gulf of Mexico, such that the rate of reaction of the system will not be buffered on the univariant phase-boundary curve (Fig.174). Pressure and temperature conditions within the gypsum layer follow the geothermal temperature-gradient curve for the region and when the bed reaches the temperature of the univariant phase boundary for the condition of p_w = hydrostatic, gypsum starts to dehydrate (p_w = pressure on fluid phase).

Clay-mineral dehydration and transformation. Another mechanism, which may originate abnormal fluid pressures during sedimentation and compaction, involves the release of interlayer water from clay minerals undergoing dehydration and/or conversion. Powers (1967, p.1249) postulated that fluid transfer could take place during the alteration of montmorillonite to illite.

Montmorillonitic clays adsorb more hydrogen-bonded oriented water than do the other non-expanding clays. During compaction, the last remaining water layers are probably desorbed by electrostatic-attraction forces associated with potassium fixation within the clay structure. Water from the interlayer position is transferred as free water to the pores of the shale. Powers (1967, p.1244) stated that this water may undergo a density decrease (from approximately 1.4 g/cm^3 to around 1 g/cm^3) with a corresponding increase in volume. At the moment of transfer, it is assumed that all the pore space is filled with fluid. The additional water entering the pores would increase the pore pressure in the shale. Powers (1967, p.1249) calculated that the volume increase would range from a probable low of 2.5% for sandy shales to as much as 20% for pure clayey shales. Fluid pressure buildup would depend on the formation rate of water, having normal density, and on the simultaneous rate of escape of this normal water from the shale. Low fluid escape rates, or no escape, will increase the pore pressure in the formation to an abnormal value. The effect of compressibility of the released water at formation temperatures must be taken into consideration. Powers (1967) suggested that abnormal pressures created by this mechanism could easily balance or exceed the overburden weight, thereby placing the overburden in a state of flotation similar to that proposed by Hubbert and Rubey in 1959. Weaver and Beck (1969, p.77) believed that although the release of interlayer water is an important factor in formation of high fluid pressures, it is not a necessary factor.

Fig.175 illustrates the distribution of water in the subsurface based on Powers (1959) and Burst (1969) models. Powers (1959) proposed a two-stage subsurface dehydration system for montmorillonitic sediments (Fig.175A). Burst (1969) reinterpreted Powers' data and developed a water-escape curve composed of three distinct dehydration stages (Fig.175B). Fig.175C describes the amount of water present in an argillaceous sediment composed of 80% clay, of which 80% is montmorillonite. As shown in Fig.175C, the total water content of a shale decreases continuously downward, thereby constantly restricting the possible avenues for fluid distribution (Burst, 1969, p.87). At a certain depth, as shown at the base of Fig.175C, the fluid channels become disconnected and permeability

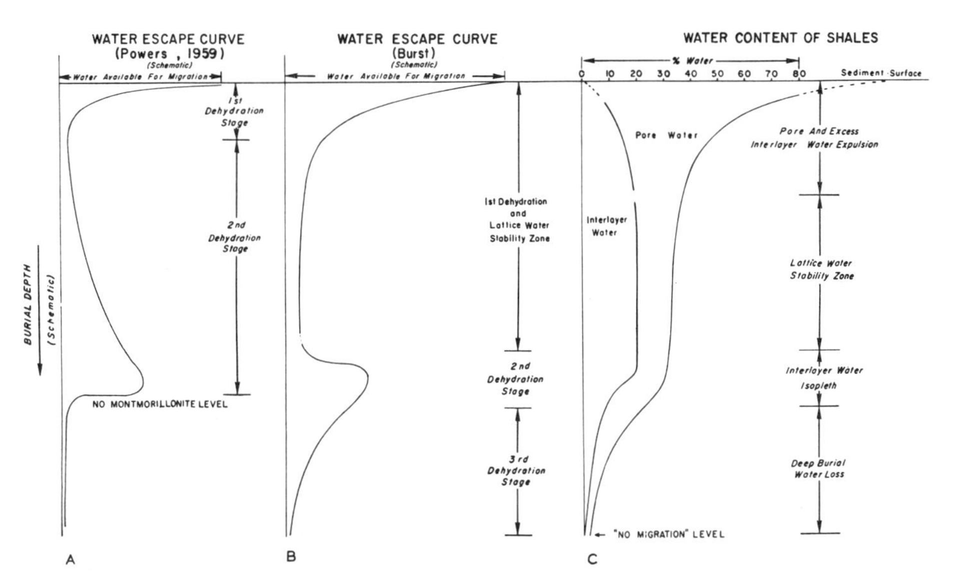

Fig.175. Subsurface water redistribution. A. Curve from Powers (1959) illustrating two-stage dehydration of montmorillonitic sediments. B. Curve from Powers adjusted and reinterpreted as three-stage system. C. Water content (%) at various stages of dehydration. (After Burst, 1969, fig.9, p.86. Courtesy of the *American Association of Petroleum Geologists Bulletin*.)

declines to the point where fluid migration is no longer possible. The second stage of dehydration shows a large increase in free fluid at depth. Burst (1969, p.87) believed that the amount of water in movement during the second stage is 10–15% of the compacted bulk volume and represents a significant fluid displacement capable of redistributing any mobile subsurface component.

Burst (1969) postulated that interlayer water discharge from montmorillonitic sediments during compaction was essentially a temperature-dependent phase change in which a silicate lattice hydrate (montmorillonite) begins to dehydrate at some critical temperature in the range of 200°–230°F, regardless of burial depth. He stated that temperature is a more effective dehydration agent than either depth of burial or age, and that the next-to-last (penultimate) layer of water is removed at a temperature of approximately 221°F. Weaver and Beck (1969) pointed out that many authors assumed that if montmorillonite clay can be dehydrated at 100°–110°C under one atmosphere of pressure, then it can be dehydrated at similar temperatures at depth.

Khitarov and Pugin (1966) claimed to have created experimentally a montmorillonite with two layers of water remaining at 150°C and 5,000 atm, and a montmorillonite having one water layer at 200°C and 10,000 atm. It appears that in their experiments the pore water was removed (open-system). In the presence of pore water (closed-system) one might expect that higher temperatures would be required for dehydration (Weaver and Beck, 1969). Clay-mineral studies in the New Zealand and Salton Sea, California, geothermal fields indicated that expanded montmorillonite layers maintain their swelling ability up to temperatures of 210°–230°C (Steiner, 1967; Muffler and White, 1969). Weaver and Beck (1969, p.78) stated that dehydration of montmorillonite below temperatures of 200°C probably is due to chemical changes.

Clay-mineral diagenesis. Many investigators have pointed out that clay-mineral diagenesis indicators might prove to be related to high-pressure zones as well as being important markers in locating petroliferous zones (Burst, 1969; Fertl and Timko, 1970a). The most commonly observed alteration of clay minerals with depth is transformation of montmorillonite (having swelling lattice) to metamontmorillonite (mixed layers) and, finally, to illite (with non-swelling lattice). Several grades (commonly four) of metamontmorillonites, which represent different dehydration states and mixing of montmorillonite and illite layers, are usually recognized. Dunoyer de Segonzac (1964) and Weaver and Beck (1969) have observed that expandable layers are converted to illitic and chloritic layers.

A considerable amount of evidence accumulated by various investigators shows that during deep burial and compaction there is a diagenetic conversion of montmorillonite to illite and, possibly, kaolinite is altered to chlorite. Burst (1959) noted a progressive modification in the structure of montmorillonite and its eventual disappearance with increasing burial depth in the Wilcox Formation (Gulf Coast, U.S.A.). He proposed that this change was a benefication of degraded and fragmental mineral lattices by gradual fixation of potassium and magnesium to form illite and chlorite, respectively. Disappearance of montmorillonite in Gulf Coast sediments was caused by the conversion to illite as a result of substitution of Mg^{2+} for Al^{3+} in the silicate structure, with the consequent fixation of interlayer potassium (Powers, 1959, 1967). Other workers have noted the lack of non-interlayered montmorillonite in deeply buried sediments (Powers, 1959; Weaver, 1960, 1961).

Van Moort (1971) investigated the gradual change in clay minerals with depth in Papua, New Guinea. A change from 60% montmorillonite and 40% illite (randomly interstratified clay minerals) at 3,500 ft. depth to 20% montmorillonite and 80% illite at 10,200 ft. was observed in a complete section of Mesozoic shales. From 10,800 ft. downward there is only 10–20% expandable material. Van Moort (1971, p.17) stated that the change in clay mineralogy is not related to systematic sedimentary facies changes or metamorphic processes. Burial depth, time, and the composition of interstitial brines are more likely to be factors that have determined the final clay mineral assemblages.

Füchtbauer and Goldschmidt (1963) and Dunoyer de Segonzac (1964) noted a decrease in the kaolinite/chlorite ratio with depth and suggested the destruction of kaolinite and the formation of chlorite. Dunoyer de Segonzac (1964) studied a 4,000-m section of Upper Cretaceous shales in the Cameroun. He concluded that with increasing burial depth, montmorillonite is converted to illite through an intermediate mixed-layer phase. Kaolinite decreases in abundance and eventually disappears, whereas chlorite appears and then increases in abundance. Kossovskaya (1960) also described the conversion of montmorillonite, by the way of mixed-layer illite–montmorillonite, to illite with increasing burial depth. Powers (1959, 1967) concluded that mixed layering of montmorillonite and illite is dependent on the porosity, chemistry, permeability and, probably, the temperature of the sediment from the time of its deposition and compaction to the time of its burial to a maximum depth. Burst (1969) reinterpreted his earlier mineralogical data and now believes that thermal dehydration could explain the observed changes in the mixed-layers; structural conversion involves only dehydration to one water layer (see Perry and Hower, 1970).

Virtually all of these earlier studies consisted only of X-ray diffraction determinations of the distribution of clay minerals as a function of depth. Perry and Hower (1970) indicated that a reinterpretation of the X-ray diffraction characteristics of mixed-layered clays leads to a somewhat different understanding of the mineralogical changes involved. They studied the mineralogy of shale samples, ranging from Pleistocene to Eocene in age, from five wells located in the Texas–Louisiana Gulf Coast. It was observed that the mixed illite–montmorillonite layering dominated the clay mineralogy and undergoes a monotonic decrease in expandability from about 80% to a lower limit of 20% of montmorillonite layers with increasing depth. Mica (discrete illite) and kaolinite phases are present throughout and their origin is believed to be detrital (Perry and Hower, 1970). The mica content decreases with depth, whereas the kaolinite shows no systematic variation. Detrital chlorite was present in appreciable amounts in one well only, from shallow-water facies.

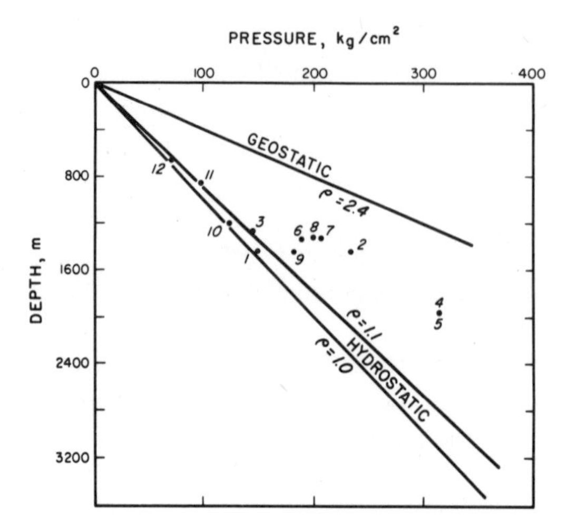

Fig.176. Relationship between the formation pressures and depths of reservoirs in Cambay Basin of India. *1–3* = Kalol; *4–8* = Navagam; *9* = Cambay; *10, 11* = Ankleshvar; and *12* = Kosamba. (After Eremenko and Neruchev, 1968, fig.2, p.8.)

According to N. Batachariya (in: Eremenko and Neruchev, 1968, p.7), the montmorillonite/kaolinite (M/K) and montmorillonite/chlorite (M/C) ratios vary with depth as follows: (1) at a depth interval of 1,200 to 1,500 m, M/K ratio varies from 1.0 to 0.4, and chlorites are absent; (2) at a depth interval of 1,500 to 1,650 m, M/K ratio changes from 0.4 to 0.1, whereas M/C ratio varies from 0.7 to 3.3 and higher; and (3) at depths below 1,650 m, both M/K and M/C ratios are equal to zero owing to the absence of montmorillonite.

The samples collected from an Eocene formation at depth intervals of 629–3,190 m, showed disappearance of montmorillonite at a depth below 1,412 m. Consequently, in this basin, having an elevated geothermal gradient, which reaches 6.5°C/100 m (about twice as high as normal), the transformation of hydratable clays (montmorillonites) terminated at considerably shallower depths (down to 1,500 m). It is interesting to note that abnormally high pressures also occur at a depth of 1,400 m (Fig.176).

Perry and Hower (1970) performed numerous whole-rock chemical analyses on the well samples, which showed no systematic variation in composition with depth except for a decrease in calcium and magnesium contents caused by the solution of carbonates. Potassium concentration increased progressively with depth in the clay-size fraction, indicating a redistribution of potassium within the rock. Perry and Hower (1970) attributed the source of the potassium to the breakdown of mica with depth. Apparently this diagenetic reaction is independent of stratigraphic boundaries and the geologic age of the sediments, even though in some places it gives that appearance, because, in general, the geologically older sediments have been buried deeper than the younger ones. Temperature seems to be more important than pressure in governing the reaction. The decrease in expandability with depth is more rapid in areas having higher geothermal gradients. Dunoyer de Segonzac (1964) also performed chemical analyses of unfractionated sediments, but unfortunately omitted the alkalis.

Chemical and mineralogical study of a series of sidewall samples from a well in the South Pass area of Louisiana and from the Mississippian Springer Shale in the Anadarko Basin of Oklahoma was made by Weaver and Beck (1969). The potassium, magnesium and iron were found to be less abundant both in absolute values and relative to aluminium content in the Mississippian and Tertiary shales than in the Precambrian and Paleozoic shales. Weaver and Beck (1969, p.66) noted that both the ancient shales and recent muds have appreciable amounts of dioctahedral aluminium-rich chlorite and that the younger muds will eventually have a higher proportion unless ions are added from outside the system.

To the writers' knowledge, the complete conversion of mixed-layer illite–montmorillonite to 10 Å illite has never been documented in a wellbore; however, it has been observed in the field where igneous dikes cut through montmorillonitic sediments. To date, there is no documented, conclusive experimental evidence that this conversion of clays is possible and that it is not a depositional phenomenon. Day et al. (1967) and Louden and Woods (1970) conducted experiments on the conversion of illite, montmorillonite, kaolinite, and various heterogeneous clay-mineral mixtures under elevated temperatures and pressures, and reported numerous changes which occurred in the clay minerals. Louden and Woods (1970) data indicated that during drilling, illite in formation shale samples downhole reconverted to montmorillonite. They also used very high temperatures (> 400°F) in the autoclave cell. These authors, however, did not present conclusive evidence in the form of X-ray diffraction patterns. The writers of this book have worked extensively with sodium-montmorillonite at elevated temperatures and pressures using different chemical systems, but were not able to confirm in the laboratory the transformations under subsurface conditions as reported by Burst (1959) or Perry and Hower (1970). Thus, all the present theories are based only on field evidence. The laboratory experiments are inconclusive, possibly owing to unrealistic formation conditions, improper scaling of the chemical environment and geologic time, and uncoordinated X-ray diffraction techniques and interpretation of the results before and after exposing the samples to high temperatures and pressures.

Bayliss and Levinson (1970) reported that the major clay minerals from shales in the Mackenzie River delta, Canada, consist mainly of disordered kaolin and illite in about equal proportions. These two clay minerals comprise about 95% of the total clay minerals. A small amount of a montmorillonite–chlorite randomly mixed layer, which is predominantly montmorillonite, occurs at the top and gradually changes to a pure chlorite at a depth of 7,000 ft.

Weaver and Beck (1969) cast doubt on the dehydration and clay conversion mechanism as being able to create abnormal fluid pressures in the subsurface. They stated that there is no obvious clay-mineral change coinciding with the top of the high-pressure zones. In addition, Weaver and Beck (1969, p.76) pointed out that there are sedimentary formations, such as the Springer in the Anadarko Basin (Oklahoma, U.S.A.) and the Lewis in the Washakie Basin (Montana, U.S.A.), which exhibit modified montmorillonite structure but do not have abnormally high pressures. One cannot exclude, however, the possibility of pressure leakage with geologic time.

An important question which remains to be answered is why the outcropping Wilcox sediments are montmorillonite-rich, whereas the buried Wilcox sediments are poor in montmorillonites. It appears that temperature and depth of burial are more important than time alone as factors in the diagenesis of these sediments. Possibly, future research work will determine the relative importance of pressure and temperature factors upon achieving proper scaling in the laboratory experiments.

Additional data on the diagenesis and transformation of clay minerals can be found in two recent books, by Seidov and Alizade (1970) and Sarkisyan and Kotel'nikov (1971).

Creation and maintenance of abnormal pressures. The rate of fluid production as a result of mineral dehydration and conversion should also be considered from a quantitative viewpoint. Hanshaw and Bredehoeft (1968, p.1117) considered a hydrologic model in which there is a constant flux of water from the compacting formation. The creation of excess pore pressure in the formation and its maintenance with time is a boundary value problem expressed as:

$$\frac{\partial^2 h'}{\partial z^2} = \frac{S_s}{K} \frac{\partial h'}{\partial t} \tag{7-15}$$

for the conditions of $0 \leqslant z \leqslant 1$

$$h'(z,0) = 0 \qquad \text{at } t = 0$$

$$h'(0,t) = 0 \qquad \text{at } t > 0$$

$$\left.\frac{\partial h'}{\partial z}\right|_{z=l} = -\frac{q_0}{K} \qquad \text{at } t > 0$$

where h' is the excess head; z is the vertical coordinate; S_s is specific storage; K is hydraulic conductivity; t is time; l is thickness of sediments; and q_o is flow into or out of the confining layer per unit area. The vertical dimension is defined at the source layer as $z = l$ (Hanshaw and Bredehoeft, 1968, p.1118).

The solution to this problem is taken by analogy from conduction of heat in solids (Carslaw and Jaeger, 1959, p.113) as:

$$\frac{h'K}{q_0 l} = \frac{z}{l} - \frac{8}{\pi^2} \sum_{n=0}^{\infty} \frac{(-1)^n}{(2n+1)^2} \exp[-(2n+1)^2 \pi^2 K t/(4 S_s l^2)] \sin\left[\frac{(2n+1)\pi z}{2l}\right] \tag{7-16}$$

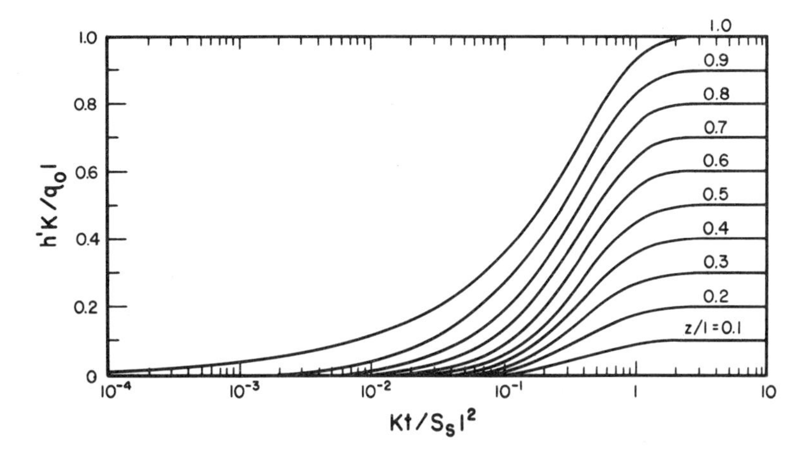

Fig.177. Dimensionless graph of the excess-head distribution for a finite layer with constant flux at one boundary. (After Hanshaw and Bredehoeft, 1968, fig.9, p.1118. Courtesy of the *Geological Society of America Bulletin*.)

Hanshaw and Bredehoeft (1968, p.1118) presented a graphical solution to eq.7-16 in Fig.177.

The above-discussed finite zone behaves as an infinite medium until the effect reaches the outer boundary. The problem is simplified for time before the change in head reaches the outer boundary by considering the head distribution in a semi-infinite medium. The problem as stated by Hanshaw and Bredehoeft (1968, p.1118) is:

$$\frac{\partial^2 h'}{\partial z^2} = \frac{S_s}{K}\frac{\partial h'}{\partial t} \qquad \text{where } 0 \leqslant z \leqslant \infty \tag{7-17}$$

$$h'(z,0) = 0 \qquad \text{at } t = 0$$

$$h'(\infty,0) = 0 \qquad \text{at } t > 0$$

$$\left.\frac{\partial h'}{\partial z}\right|_{z=0} = \frac{-q_0}{K} \qquad \text{at } t > 0$$

It is necessary to define $z = 0$ at the boundary of the media, which for the core of a semi-finite media is at the source layer.

Again by analogy the solution is obtained from Carslaw and Jaeger (1959, p.75):

$$\frac{Kh'}{q_0 z} = \frac{2\sqrt{Kt/S_s z^2}}{\sqrt{\pi}} \exp(-S_s z^2/4Kt) + \operatorname{erf}\left(\frac{1}{2\sqrt{Kt/S_s z^2}}\right) - 1 \tag{7-18}$$

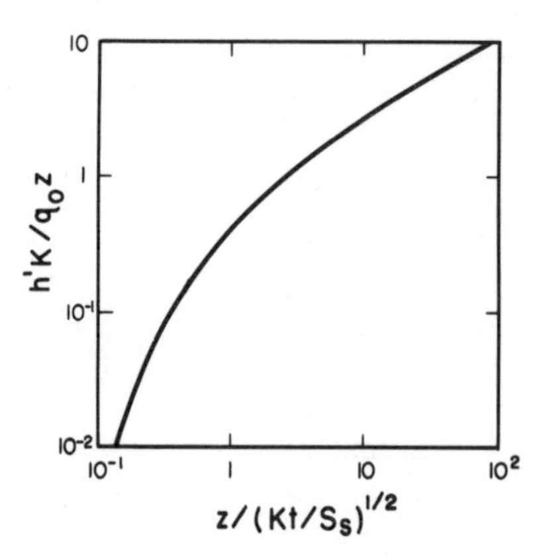

Fig.178. Dimensionless graph of the excess-head distribution in a semi-infinite medium with constant flux at the boundary. (After Hanshaw and Bredehoeft, 1968, fig.10, p.1119. Courtesy of the *Geological Society of America Bulletin*.)

In Fig.178 a solution to the above equation is presented. The head at the surface, $z = 0$, is given by Carslaw and Jaeger (1959, p.75) as:

$$\frac{h'K}{q_0} = 2\sqrt{Kt/\pi S_s} \tag{7-19}$$

Hanshaw and Bredehoeft (1968) reported that for small amounts of time, this solution (eq.7-19) gives the same results as eq.7-15 for the finite layer that is evaluated at the

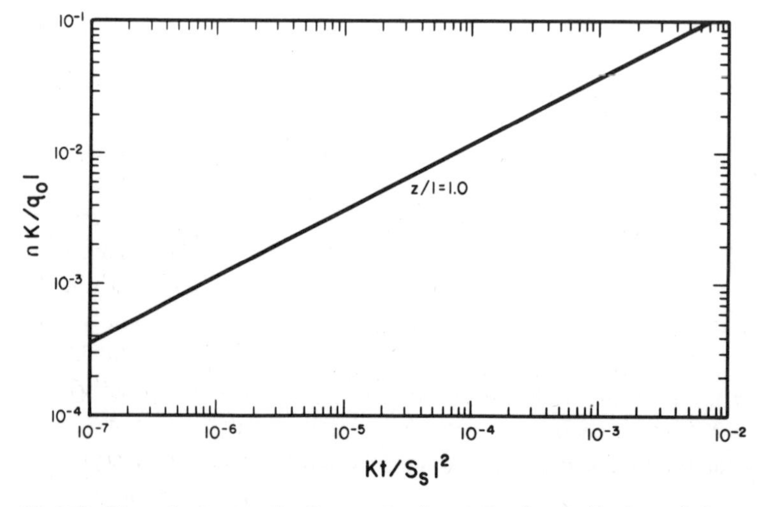

Fig.179. Dimensionless graph of excess-head variation for small values of time at a surface of constant flux (z/l = 1). The solution is the same for either the semi-infinite or the finite medium. (After Hanshaw and Bredehoeft, 1968, fig.11, p.1119. Courtesy of the *Geological Society of America Bulletin*.)

surface of constant flux ($z/l = 1$). Excess head at the surface of constant flux is presented in Fig.179 for very small amounts of time by solving eq.7-15. Hanshaw and Bredehoeft (1968, p.1119) used Figs.178 and 179 to compute the pressure for a source bed at 1,200 m depth, overlain and underlain by beds of low hydraulic conductivity. The conductivities of 10^{-12} cm/sec and 10^{-10} cm/sec were both used. They reported that the results obtained are extremely sensitive to hydraulic conductivities and flow rates.

Assuming a burial rate of 500 m/10^6 years, the phase transition period (time during which fluid will be produced) for a bed 15-m thick is 30,000 years; and upward or downward flow will be $3.85 \cdot 10^{-10}$ cm/sec. At a hydraulic conductivity of 10^{-12} cm/sec, Hanshaw and Bredehoeft (1968) calculated that it is possible for the fluid pressure to approach the lithostatic load. As the fluid pressure increases, it will decrease the rate of reaction and, therefore, decrease the fluid flux at low values of the hydraulic conductivity.If the hydraulic conductivity is increased to 10^{-10} cm/sec, there will be an insufficient quantity of fluid produced to create pressures much in excess of the hydrostatic pressure. The important variables are the burial rate, thickness of the gypsum bed, and hydraulic conductivity of the confining layer. The gypsum-dehydration mechanism in compacting sediments will produce high fluid pressures only if all the above variables are within certain definable limits.

This is also true for the montmorillonite-dehydration model. Hanshaw and Bredehoeft (1968, p.1117) assumed that if each cubic centimeter of sediment contains 2 g of montmorillonite, then dehydration of montmorillonite will produce 0.33 g/cm^3 of H_2O. Enthalpy data from Şudo et al. (1967) indicates that 178 cal./cm^3 is required to release the interlayer water. Employing the following assumed values of 10^{-6} cal./cm^2/sec for heat flow rate and $1.6 \cdot 10^{-9}$ cm/sec as a burial rate, $6.3 \cdot 10^{10}$ sec would be required to increase the depth by one cm. In that time, 630 cal./cm^2 are available from the usual flow of heat in the earth. There is more than enough heat available for the dehydration reaction to proceed (Hanshaw and Bredehoeft, 1968). Interlayer water would be released at the rate of $5.1 \cdot 10^{-10}$ cm/sec. In the phase transition of gypsum, the reaction went from a solid to a solid plus water; however, in the dehydration of montmorillonite the dense water expands. Not all of the water is moved from the reaction site. If one assumes that all interlayer water has a density of 1.4 g/cm^3 (Martin, 1962), expansion, upon release to water having normal density of 1 g/cm^3, will result in an increase in specific volume (reciprocal of density) of 28.5% (Hanshaw and Bredehoeft, 1968, p.1117). They calculated that the total flow (q_o) upward or downward will be equal to $0.285 \times 0.5 \times 5.1 \cdot 10^{-10}$ or $7.3 \cdot 10^{-11}$ cm/sec. In this case, a conductivity of 10^{-12} cm/sec and about 10^6 years are required to approach lithostatic pressure on the fluid. In an actively subsiding basin such as the Gulf Coast Basin, this mechanism could provide a significant increase in pore pressure if the amount of montmorillonite is high and the permeability is low.

Salt and shale diapirism

Harkins and Baugher (1969, p.962) discussed briefly the pressure anomalies associated with salt domes; special emphasis was placed on the relationship between high formation pressures and stratigraphy. Dickey et al. (1968) pointed out that the salt domes are scarce in the "embayments" in southern Louisiana where the abnormal pressures are most abundant (Fig.171). It is possible that in the embayment areas the shales did not compact well and still have abnormally low densities, less or slightly greater than the density of salt. Howard (1971, p.501) considered the effect of density contrast between salt and sediment on the direction of movement of the salt. Salt becomes ductile and flows in the manner of a viscous liquid more readily in an area having a steeper geothermal gradient. The low density and strength of salt, relative to other sediments, and the plastic behavior of salt would allow initiation of movement and development of salt structures (Johnson and Bredeson, 1971, p.207). When the density of the sediments overlying the salt exceeds the density of the salt, there is an upward movement of salt with a consequent formation of domes. Buoyancy tends to push the dome upward through the sediments and increases with increasing height of the dome. Buoyancy reaches a maximum at an elevation above the salt bed where the density of the sediment equals that of salt, and then diminishes at higher levels where sediments are less dense than salt.

If structural growth of the salt domes was contemporaneous with deposition and compaction of sediments, then the chance of abnormal pressure zones developing ahead of the dome would be negligible. Fig.180 illustrates a rapid rise in the pressure boundary reflecting earlier topography. Abnormal pressure associated with the sediment sheath could be misleading as shown by Fig.180. The term sheath refers to predominantly shale material which is out of place between the salt stock and the younger sedimentary rocks. Gilreath (1968, p.138) pointed out that highly pressured shale sheaths are associated with some salt domes. Sheaths originated either as a result of (1) the folding of the shale

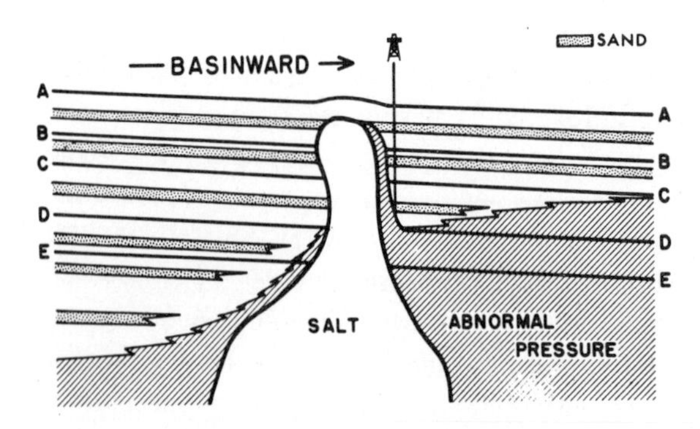

Fig.180. Schematic section through a piercement salt dome showing modification of abnormal pressure surface. (After Harkins and Baugher, 1969, p.964. Courtesy of the Society of Petroleum Engineers of A.I.M.E.)

upward by the rising salt dome, with younger sediments being deposited unconformably against the sheath, or (2) the faulting or folding of shale into its position between the salt and the flanking sediments (Johnson and Bredeson, 1971, p.208). Harkins and Baugher (1969) illustrated the abnormal pressures associated with the sheath in Fig.180. A well drilled into Formation *C* would encounter high pressures in Formation *D* sediments. The sheath sediments are out of place, having been dragged into their present position by the dome. Bed thinning on the flank of the dome due to the greater compaction or distortion by squeezing towards the center of the uplift does not appear to be significant. Johnson and Bredeson (1971, p.209) stated that bed thinning which occurred at the time of deposition, and as a result of local erosional unconformities after deposition, appears to be greater than that resulting from other causes.

Gilreath (1968, p.144) discussed briefly the origin of diapiric shale bodies along the Gulf Coast. Diapiric shale masses have the same characteristics as the overpressured shales, i.e., low electrical resistivity values on electric logs (0.5 ohm-m), low wet-bulk densities (2.1–2.3 g/cm^3), and high pressure gradients ($\approx$ 0.9 p.s.i./ft.). Their origin is believed to be similar to that of salt domes. Morgan et al. (1968, p.145) noted that rapid deposition at the mouth of the Mississippi River of thick localized sand masses, directly upon lighter, plastic clay, leads to instability. This instability is relieved by diapiric intrusion of the mud, with the resulting formation of mudlumps.

Tectonic compression

In contrast to the compaction by simple gravitational loading, a more drastic compression of the sediments can be produced by horizontal compressive stresses of tectonic origin (Hubbert and Rubey, 1959, p.153). The largest principal total stress σ_x is assumed to be horizontal, whereas the least total principal stress σ_z is vertical and equal to the weight of the overburden per unit area. The largest and least effective principal stresses can be expressed as:

$$\sigma_x' = p_{ex} = \sigma_x - p_p \tag{7-20}$$

$$\sigma_z' = p_{ez} = \sigma_z - p_p \tag{7-21}$$

where p_p is pore-water pressure.

As pointed out by Hubbert and Rubey (1959, p.153), if σ_z is fixed and σ_x' is increased more rapidly than the pore pressure can be dissipated by fluid leakage from the formation, then p_p will increase until it is at its maximum value of $p_p = \sigma_z$. Then the value of λ (the ratio of pore pressure to the effective stress) will be equal to one, whereby the superincumbent material could be moved tangentially with negligible frictional resistance. At the same time, σ_z' would be equal to zero, and σ_x' corresponding to failure of the rock would be reduced to the ordinary crushing strength of the sediments. Hubbert and Rubey

(1959, p.153) stated that an incompletely compacted water-filled clay or shale would respond to deformational stresses by an increase of the fluid pressure. This increase in pressure would in turn weaken the rock by reducing the values of the effective stresses necessary to cause the rock to fail.

The existence of conditions necessary for λ to approach or to be equal to unity, will depend upon the relative rates of two opposing processes: (1) the application rate of the lateral deformation stress, and (2) the pressure dissipation rate by leakage of the compressed fluid. Hubbert and Rubey (1959) stated that of the two processes for increasing the deformational stresses, i.e., (1) that of sedimentary loading in tectonically quiescent geosynclines and (2) that of the application of orogenic stresses, the latter appears to be more effective.

As discussed previously for sedimentary loading, the fluid pressure in the case of gravitational compaction should be able to approach $\lambda = 1$ only where the clay is so completely uncompacted that the effective overburden stress is very small in comparison to the total load.

In the case of lateral orogenic compression, where the overburden stress is the least stress, no such limitation exists. Thus, if the rate of increase of the applied stresses is sufficiently rapid as compared with the leakage rate, there is nothing except for stronger rocks to prevent p_p from becoming equal to σ_z and λ from becoming unity (Hubbert and Rubey, 1959, p.153).

Hubbert and Rubey (1959, p.154) discussed abnormal pressures occurring in tectonically active regions. In 1927, Anderson reported on a well being drilled on the Potwar plateau of the Punjab in West Pakistan that encountered abnormal water pressures during and after drilling. This well (Jhatla) was located in an area immediately south of overturned folds in the foothills of the Himalaya Mountains. Abnormal fluid pressures are also associated with steep folding and small thrust faults in the Khaur field in West Pakistan resulting in pressure gradients approaching 1 p.s.i./ft. (Keep and Ward, 1934; Pinfold, 1954). In 1937, Abraham implied that similar pressure conditions had been encountered in Burma. As reported by Baldwin (1944), abnormal fluid pressures exist in the Tupungato Field in the Andean foothills of Argentina. The high pressures noted in Trinidad by Reed (1946) and Suter (1954) are thought to be caused by the continuing tectonic activity in that region. Pressure gradients up to 0.8 p.s.i./ft. in sediments at depths less than 6,000 ft. are reported from Trinidad.

Watts (1948, p.191) reported that the D-7 zone (depth of 9,200 ft.) in the Ventura Avenue Field in California had an initial reservoir pressure nearly equal to the overburden pressure. The field, which has had a long history of overpressures, is at the apex of the Ventura Anticline. The normal hydrostatic gradient in this area was reported by Watts as being 0.44 p.s.i./ft., whereas in the high-pressure D-7 zone $\lambda = 0.9$. Inasmuch as the area is severely faulted and folded, Watts (1948, p.194) suggested that faults acted as seals and that the high pressures arose from the weight of the overburden. The possibility of the effect of lateral tectonic stresses in this tectonically active area should not be disregarded, however. Finch (1969) stated that abnormal pressures in the Antelope Field, North Dakota, originated as a result of lateral tectonic compression.

Rubey and Hubbert (1959, p.184) theorized that a stress of 1 kbar (14,500 p.s.i.) directed horizontally instead of vertically (overburden load at a depth of about 14,500 ft.) would have much the same effect on (1) the compaction and permeability of argillaceous sediments, (2) the expulsion of fluids, and (3) the development of abnormal fluid pressures. Rubey (1930, p.35) found that in the Upper Cretaceous rocks of northern Wyoming, mudstones were deformed by lateral compression. The more steeply dipping mudstones, in general, have a lower porosity, which indicates that as the beds were tilted and folded, the pore spaces between the clay particles were decreased with consequent fluid expulsion (Rubey, 1930, p.36). Much of the material presented earlier on gravitational compaction probably also applies to the development of high fluid pressures in argillaceous sediments as a result of lateral compression in orogenic belts.

Osmotic and diffusion pressures

Inasmuch as shale beds may act as semi-permeable membranes so that the dissolved electrolytes are held back, whereas H_2O passes through freely, Hanshaw and Zen (1965) suggested that osmosis might be a plausible mechanism for the creation of abnormally high fluid pressures in sediments. They suggested that if the dissolved components in a body of circulating ground water were filtered out by clay zones (membranes), then under equilibrium conditions on the influx side of the membrane the pressure should be anomalously high, whereas salinity will be high on the side from which the flow occurs. On establishment of osmotic equilibrium, the higher salt concentration is balanced by higher pressure, so that the chemical potential of H_2O is equal across the membrane. Osmotically-induced pressure would be most observable in formations with low fluid transmissibility or in formations which are surrounded by rocks having low transmissibility. Rocks with high permeability and good lateral continuity could dissipate osmotically-induced pressure rapidly. The possible role of argillaceous rocks as membranes has been suggested by many researchers, (e.g., Schlumberger et al., 1933; DeSitter, 1947; Wyllie, 1955). The effectiveness of argillaceous rocks as a semipermeable membrane has been demonstrated by Young and Low (1965).

Water flow under an osmotic gradient appears to be identical to water flow under an equivalent pressure gradient. Hanshaw and Zen (1965) assumed two saline solution densities of 1.07 g/cm^3 (= 0.465 p.s.i./ft. gradient) and 1.02 g/cm^3 (= 0.444 p.s.i./ft. gradient) in order to compute theoretical pore pressures generated by osmosis. At a temperature of 50°C, these solutions are capable of generating pressures of 280 bars and 380 bar, respectively, against a hypothetical reference saturated NaCl solution. These values are within the range of observed anomalous pressures. Hanshaw and Zen (1965, p.1383) prepared a graph relating the calculated osmotic pressure difference across an ideal membrane as a function of NaCl concentration and temperature in the binary system NaCl–H_2O.

Although the presence of evaporite beds or salt domes is not necessary for establishing anomalous pressure, they would help because a high NaCl concentration in the sediments could produce high osmotic pressures. Osmotic pressures generated across shale barriers in a sedimentary sequence depend upon the osmotic efficiency of the clay beds as well as upon the water salinity content. Hanshaw and Zen (1965) pointed out that high salt concentration membranes tend to lose efficiency owing to ionic transport across the membrane. A series of such membranes probably would be more effective than a single one. Jones (1969, p.806) proposed that in known overpressured reservoirs, stepwise increments of osmotic pressure with depth through a series of interbedded sands and clays could, as by a multistage pump, produce any of the formation pressures observed to date in the Tertiary sediments of the Gulf Coast Basin. Also the osmotic mechanism does not require deep burial of the sediments in order to be effective. Conceivably, osmotically-derived fluid pressures could equal or exceed that owing to the weight of the overburden and thus cause reservoir rupture and diapirism, especially where heating had reduced the load-bearing strength of the shales by increasing the free pore water content through mineral dehydration (Jones, 1969).

The authors believe that conventional osmosis probably was not responsible for overpressured zones occurring in the Gulf Coast Basin, because (1) the sedimentary sequence here contains massive thick shales, which are associated with high pressures, and (2) the high fluid pressures are associated with the fresher water rather than with the more saline fluids.

Overton and Zanier (1971) presented a model for the creation of overpressures based on diffusion pressure due to "reverse osmosis". Reverse osmosis occurs across a membrane whenever pressure is applied on the more saline side, causing the water to migrate toward the medium having lesser salt concentration (fresher water). Overton and Zanier (1971, p.107) pointed out that, strictly speaking, reverse osmosis was probably not effective in causing overpressures in the Gulf Coast Tertiary sediments, because the total gravitational load (driving force) is higher in the less saline deeper-water zones. Thus, diffusion pressure is a better term for this phenomenon.

The sequence of pressure buildup in a vertical direction (without faulting) using a diffusion pressure model can be summarized as follows (Overton and Zanier, 1971):

(1) Large amounts of shales are deposited in the marine basin.

(2) Compaction squeezes out the water of hydration from the shales. The rate of water expulsion is determined by the diffusion rate of Na^+ ions from the matrix.

(3) Diffusion of NaCl is time dependent and causes a maximum salinity level to be somewhat more shallow than the 200°F geotemperature level.

(4) Below the 225°F geotemperature level, the shales undergo chemical reactions such as ionic complexing and breakdown of organic compounds to yield CH_4 gas.

(5) Methane moves outward, causing banking of the Na^+ ion at shallower depths.

(6) At equilibrium, the methane expansion is minimized by the tendency of ions and water of hydration to move downward.

(7) The emerging methane gas represents a heat source, whereas the shale represents a temperature stabilizer due to the low thermal conductivity.

(8) Because of this heat source at fairly constant temperature, shallower shale masses, which are undergoing chemical reaction, exhibit larger temperature gradients.

(9) A geochemical caprock is created at the top of the salinity reversal (or maximum) because of the following factors which affect calcium solubility: (a) reduction in the temperature, (b) reduction in pressure of the expanding gas, and (c) increased water salinity. A large pressure drop causes increased deposition of calcium compounds in the caprock.

Significant differences in Overton and Zanier's (1971) vertical and horizontal diffusion pressure models (profiles) are due to tectonic stresses rather than overburden pressure. The sequence along the horizontal profile is as follows:

(1) Water salinity is maximized at points of active stress and failure, such as expanding diapirs and faults.

(2) Horizontal diffusion of salt through sands is more rapid than through shales, and ions pile up at the fault surfaces.

(3) Lateral gas expansion also causes salt banking at fault planes.

(4) The horizontal salinity gradients cause the amount of gas and the solubility of calcium compounds to decrease.

(5) Lateral dropout of calcium carbonate and pyrite at the fault surface could possibly form a pressure seal.

Overton (personal communication, 1970) stated that at depths greater than 8,000 ft., gas has pressures greater than that due to the hydrostatic head. This pressure usually is not transient, and is probably held by an electrochemical effect. Gas produced at shallow depths of 6,000 ft. or less has an original pressure of approximately hydrostatic. The average salt water density determines the gradient and the gas pressure. Overton observed that salinity contrasts across geopressure caprocks remain constant upon establishment of equilibrium. Contrasts of 10:1 (150,000 p.p.m. versus 10,000 p.p.m. NaCl) are commonly observed across the pressure caprock.

Below the caprock, organic material undergoes chemical changes with resulting release of methane gas. The diffusion of gas upward through the transition zone is accompanied by increasing amounts of heavier hydrocarbons. Hence, the entire phenomenon of the presence of gas and high pressures, in Overton's opinion, is geochemical in origin.

Geothermal temperature changes

Jones (1969, p.804) pointed out that abrupt changes in temperatures over short depth ranges are hydrologically critical to the geopressured regime, because the movement of water is the most important factor in sustaining terrestrial heat flow in sedimentary basins. Conventional maps of geothermal conditions, however, tend to obscure, rather than to identify, abrupt changes in temperature. An increase in the geothermal temperature, as the compacting sediments are subsiding in the basin, causes the pore fluids (gas, oil and water) to expand more than the enclosing rocks. Such an expansion would create abnormal fluid pressures in the sediments. There are three modes of heat transport through fluid-saturated sediments: (1) convective flow of interstitial fluids, (2) conduction through mineral grains and interstitial fluids, and (3) radiation. Jones (1969, p.807) listed several factors that have a direct bearing on the heat flux in sediments:

(1) Thermal conductivity and composition of the (a) mineral grains that form the rock matrix and (b) interstitial fluids.

(2) Specific heat of the pore fluids and solids.

(3) Porosity and pore distribution in the shales and sands.

(4) Density, viscosity and thermal expansion of the pore fluids.

(5) Thermal expansion of solids.

(6) Absolute temperature.

Lewis and Rose (1970) and Jones (1969) observed that in the Gulf Coast region the overpressured zones have abnormal temperature gradients. No relationship between the average geothermal gradient and pressure/depth ratio (geostatic ratio) in the Gulf Coast Tertiary sediments was found by Jones (1969, p.804) after studying 175 south Louisiana overpressured reservoirs above a depth of 11,000 ft. Nevertheless, the occurrence of abnormal pressures is commonly associated with a sharp increase in the geothermal gradient in the sealing clay member of the reservoir (C.E. Hottman, personal communication, 1966; in: Jones, 1969, p.804). According to Lewis and Rose (1970), the abnormally pressured shale zones constitute thermal barriers, because they are undercompacted and have high porosity in respect to the adjoining sediments. Reduction in the upward flow of water in these zones greatly reduces the rate of upward flow of heat and, consequently, the overpressured zones become heat storage areas. In addition, the insulating effect of water is three times greater than that of the shale matrix. The larger the amounts of fluid stored in the overpressured shales, the greater is the insulating value of the zone. Whenever there is an insulating layer in the earth's crust there can be a buildup of heat beneath this layer. Thus, the geothermal gradient is steepest in the portion of the beds above a permeable reservoir. Jones (1969, p.805) reported gradients as high as 6°F/100 ft. in such settings.

The steepness of the geothermal gradient varies inversely with the thickness of unconsolidated sediments in the structural basins (Jones, 1969, p.807). Geothermal gradients are large in the undercompacted shales overlying the reservoir sands and are very much reduced in the aquifers. The thermal conductivity of sediments varies inversely with the geothermal gradient, if the geothermal flux is uniform over broad areas. Langseth (1965) stated that the thermal conductivity of clay varies inversely with its water content, and Zierfuss and Van der Vliet (1956) discovered that the thermal conductivity of sand increases with porosity owing to the occurrence of convective heat transport in the wider

TABLE XLVI

Abnormally high-pressure aquifers in southern Louisiana and adjacent areas of the continental shelf (partial extract of data in exhibit H, southern Louisiana rate case AR 61–2, gas supply section, U.S. Federal Power Commission)

(After Jones, 1969, table 1, p.804. Courtesy of the Society of Petroleum Engineers of A.I.M.E.)

Depth (ft.)	*Field name*	*Temp.* (°F)	*Pressure* (p.s.i.)	*Geostatic ratio*
1,355– 1,430	Bay Marchand	91	680	0.502
2,674– ?	Southeast Pass	106	1,390	0.520
5,405– 5,433	Ship Shoal	140	3,132	0.579
6,268– 6,311	West Cameron	163	3,205	0.511
7,080– 7,090	West Cameron	173	3,712	0.524
7,483– 7,507	West Cameron	179	3,993	0.533
7,996– 8,013	West Cameron	187	4,370	0.546
8,400– 8,413	Vermilion Bay	208	4,580	0.545
8,700– 8,831	Vermilion Bay	211	4,680	0.538
9,012– 9,047	Eugene Island	182	4,715	0.523
9,033– 9,061	South Pelto	219	4,938	0.547
9,401– 9,422	Church Point	201	6,417	0.683
9,464– 9,533	South Pelto	225	5,416	0.572
9,824– 9,877	East Cameron	213	6,025	0.613
9,879– 9,906	East Cameron	213	6,001	0.607
10,005–10,047	Iowa	246	7,169	0.717
10,025–10,039	Jefferson Island	194	5,379	0.537
10,410–10,418	High Island	209	7,802	0.749
10,500–10,517	English Bayou	233	8,154	0.777
10,585–10,630	West Cameron	223	6,325	0.598
10,790–10,816	Raceland	208	6,792	0.629
10,800–10,906	Mud Lake	231	5,724	0.530
10,950–10,974	Church Point	243	7,686	0.702
11,200–11,389	Mud Lake	246	6,272	0.560
11,330–11,356	Rayne	217	6,900	0.609
11,650–11,679	Chalkley	233	9,345	0.802
11,933–11,943	Erath	231	6,602	0.553
11,950–11,995	Bayou Penchant	230	9,031	0.756
12,200–12,246	Lake Arthur	262	10,100	0.828
12,295–12,328	Thornwell	303	11,800	0.960
12,450–12,493	Belle Isle	231	6,690	0.537
12,550–12,562	Grand Isle	263	8,745	0.697
12,693–12,757	Vermilion	235	8,276	0.652
12,900–12,935	Bastian Bay	224	8,859	0.687
12,942–12,949	Bayou Chevruil	246	11,067	0.855
13,200–13,228	Lake Chicot	232	11,522	0.873
13,265–13,275	West Cameron	280	11,664	0.879
13,617–13,640	Thibodaux	237	10,418	0.765
13,700–13,735	Thornwell	272	12,282	0.896
13,708–13,761	West Delta	239	10,782	0.787

TABLE XLVI (continued)

Depth (ft.)	*Field name*	*Temp.* (°F)	*Pressure* (p.s.i.)	*Geostatic ratio*
13,753– ?	Caillou Island	270	7,113	0.517
13,937–13,950	Rousseau	241	10,635	0.763
14,145–14,178	Ship Shoal	261	7,109	0.503
14,150–14,225	Houma	253	10,790	0.763
14,300–14,341	Lake Sand	263	10,975	0.767
14,344–14,376	Garden City	259	12,096	0.843
14,594–14,606	Garden City	263	12,295	0.842
14,602–14,628	Constance Bayou	278	7,340	0.503
14,600–14,650	Lapeyrouse	264	9,075	0.622
14,700–14,731	Lapeyrouse	266	10,020	0.682
14,900–14,940	Lake Washington	266	10,180	0.683
15,050–15,084	Garden City	260	14,210	0.944
15,150–15,160	Deep Lake	332	9,390	0.620
15,249–15,289	Lake Pagie	268	10,819	0.709
15,318–15,375	Thornwell	315	11,376	0.743
15,336–15,407	Lake Arthur	329	13,933	0.909
15,580–15,595	Leleux	277	13,570	0.871
15,600–15,800	Deep Lake	366	9,885	0.634
15,871–15,880	Lake Sand	296	12,505	0.788
16,000–16,018	Lacassine	275	14,625	0.914
16,450–16,495	Hollywood	280	14,540	0.884
16,570–16,585	Weeks Island	266	9,495	0.573
17,300–17,340	Belle Isle	316	11,420	0.660
17,395–17,429	Lake Sand	318	13,477	0.775

pores. As pointed out by Bogomolov (1967), water carries out the major role in the redistribution and subtraction of heat in the geothermal field of the earth's sediments.

Jones (1969) stated that convective and conductive heat flow are important in the low-temperature range above depths of 10,000 ft. in the northern Gulf of Mexico Basin. Water temperatures in this area are greater than 250°F at depths ranging from 10,000 to 14,000 ft. (Jones, 1967). Table XLVI presents some reservoir temperatures for southern Louisiana. Lewis and Rose (1970) showed a range in average geothermal gradients from 1.6°F to 2.2°F/100 ft. for the Texas Gulf Coast.

Perhaps the most obvious feature of the geothermal-gradient maps of the northern Gulf Coast Basin is its conformity with the structural map. Elongate areas, beneath which the geothermal gradient is lowest, overlie the axis of the depositional basin (Gulf Coast geosyncline). Sediments which overlie the deepest portion of the Gulf Coast geosyncline would appear, then, to possess the highest thermal conductivity. Jones (1969, p.807) stated that if they do not, then they must form a thermal sink and are now storing heat energy received from below; their temperature must inevitably rise. The endothermic diagenetic processes occurring in these argillaceous sediments, such as the dehydration of montmorillonite, require the addition of heat and, thus, reduce the amount of heat flux to the overlying sediments.

Jones (1969, p.808) concluded that, by checking the upward flow of water from the saturated sediments beneath the shales, the sealing clay beds have caused a reduction of the geothermal flux above and overheating of the undercompacted sediments below.

ANATOMY OF AN ABNORMAL FLUID PRESSURE ZONE

Abnormal pressure zones can be described and studied by applying various formation-evaluation techniques during or after drilling a well into such a zone. Electric and nuclear well-logging devices, laboratory analysis of cores, side-wall samples and cuttings, drilling data, and reservoir transient-pressure buildup methods are commonly employed to investigate the overpressured formations.

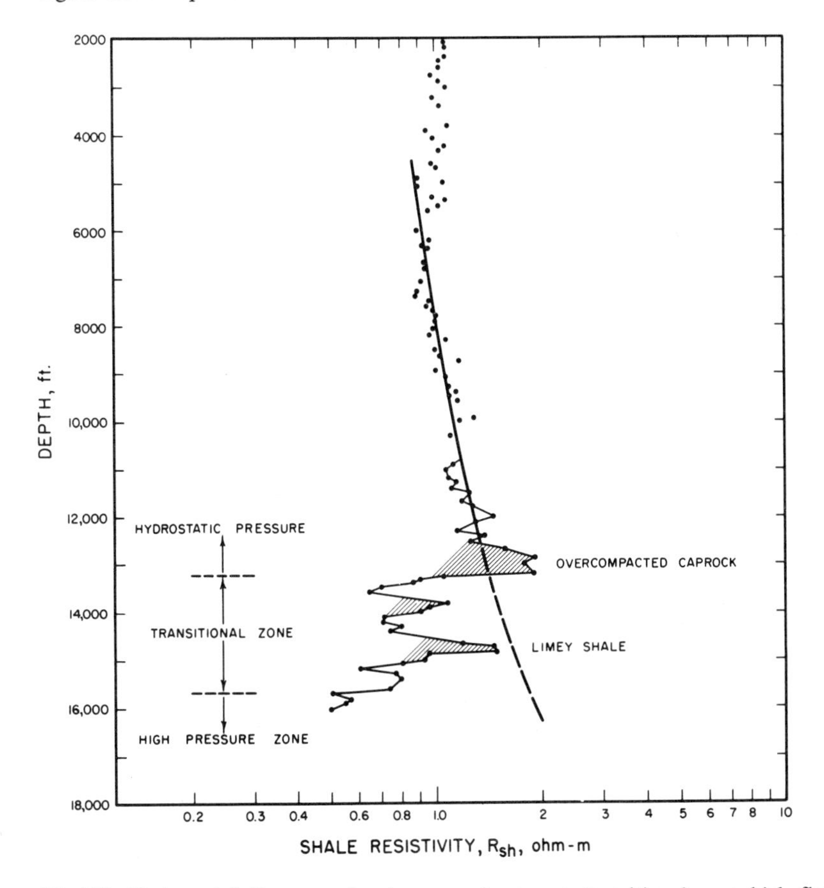

Fig.181. Shale resistivity curve showing normal pressure, transitional zone, high fluid-pressure zone, and calcareous caprocks in a Gulf Coast well. (After Fertl and Timko, 1970c, fig.6A. Courtesy of the Society of Petroleum Engineers of A.I.M.E.)

A considerable amount of field evidence indicates that most high-pressure intervals can be divided into three zones: caprock, transition pressure zone, and abnormal pressure zone. Fig.181 illustrates shale resistivities as determined by the short normal log. The resistivities, plotted versus depth, show a departure from the normal trend on entering the transition zone. A major departure of resistivities from the normal trend occurs in the high-pressure zone. Location of caprocks is clearly shown on the resistivity logs. The transition zone may be over 1,000 ft. in thickness or very thin (abrupt transition). Increases of almost 5,000 p.s.i. in pore pressures across shale sections only 300–400 ft. thick have been measured in the Gulf Coast wells. Fertl and Timko (1970c) noted that the presence of faults may or may not mark the pressure discontinuities.

Calcareous shale caprocks, which are definite permeability barriers, are commonly encountered at the boundary between normally compacted sediments and the overpressured environments. Caprocks can vary in thickness from a few inches to several hundred

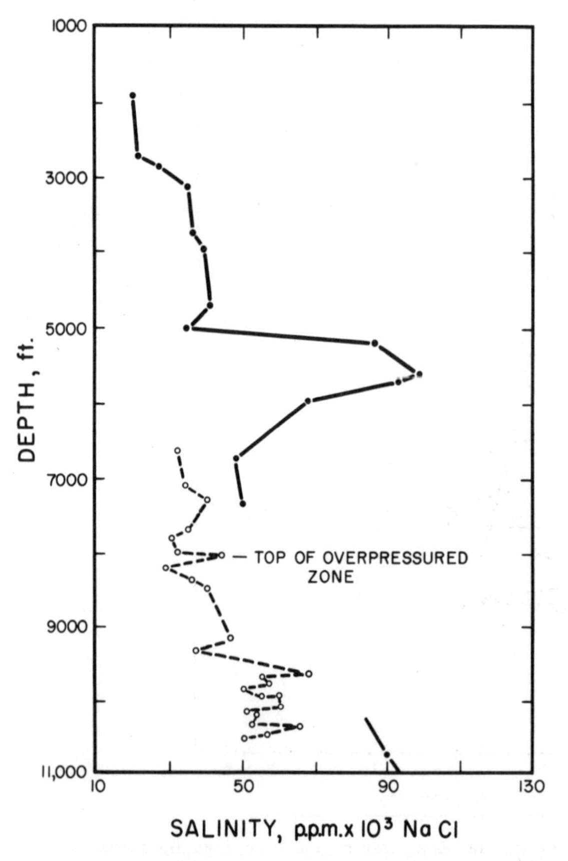

Fig.182. SP log-derived salinity trend in offshore Louisiana well. (After Fertl and Timko, 1970a, fig.4, p.106. Courtesy of *The Oil and Gas Journal.*) Solid line: data from logs; dashed line: data from side-wall core samples.

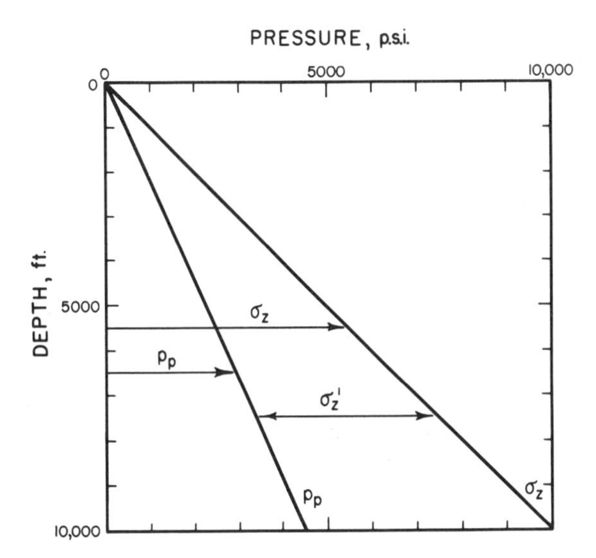

Fig.183. Total overburden stress (σ_z), pore-fluid pressure (p_w, σ_w, or p_p), and effective overburden pressure (σ'_z) versus depth (D), assuming that ρ_b = 2.3 g/cm^3 and ρ_w = 1.04 g/cm^3. (After Gretener, 1969, p.258. Courtesy of the *Bulletin of Canadian Petroleum Geologists.*)

feet. Occasionally, as shown in Fig.181, a series of such sealing barriers may be encountered. Fertl and Timko (1970c) stated that it is unlikely that the cemented caprock was deposited or formed prior to the compaction of sediments.

According to Overton and Timko (1969), salinity measurements from the spontaneous potential (SP) logs have shown that the salinities of the interstitial pore fluids in marine sands above overpressured zones are low. Fig.182 shows freshening of water in sands at a

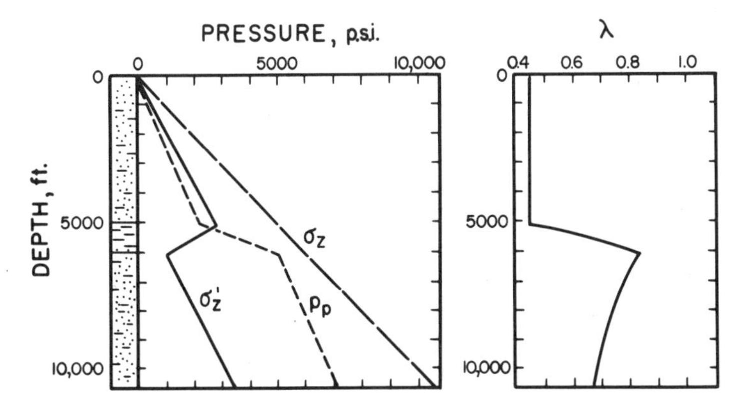

Fig.184. Schematic logs of the pore-fluid pressure (p_p), effective stress (σ'_z) and the total overburden stress (σ_z) in a section containing a leaky seal, assuming that ρ_b = 2.3 g/cm^3 and ρ_w = 1.04 g/cm^3. (After Gretener, 1969, fig.12, p.270. Courtesy of the *Bulletin of Canadian Petroleum Geologists.*)

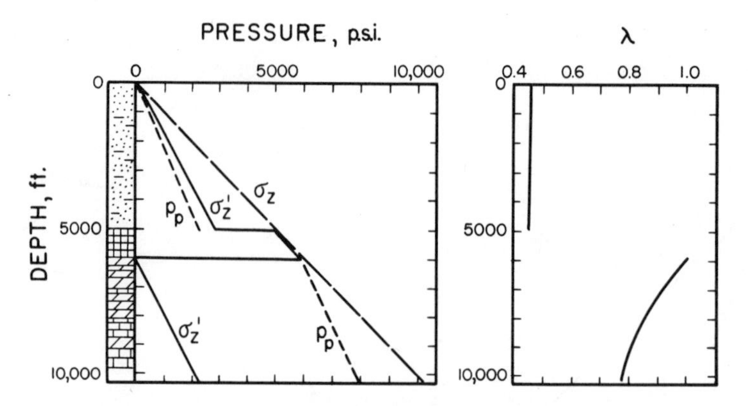

Fig.185. Schematic logs of the pore-fluid pressure (p_p), effective stress (σ'_z) and the total overburden stress (σ_z) in a section containing a perfect seal, such as an evaporite horizon (ρ_b = 2.3 g/cm^3 and ρ_w = 1.04 g/cm^3). (After Gretener, 1969, fig.13, p.271. Courtesy of the *Bulletin of Canadian Petroleum Geologists.*)

depth of 6,000 ft., which is above the top of an overpressured zone that starts at a depth of around 8,000 ft. (Fertl and Timko, 1970a, p.107). Possibly, any water that filters through a caprock into the overlying sands becomes fresher.

As described in Chapter 3, the pore-fluid pressure is strongly linked to the concept of effective stress. The relationship between the depth of burial and the total overburden stress and pore pressure is commonly presented in a graphical form as illustrated by Fig.183. Gretener (1969, p.270) presented a schematic example of the abnormal pressure relations that could occur in a thick shale sequence having low permeability (Fig.184). The condition of a highly impermeable sequence providing a nearly "perfect" seal, such as a salt bed, was depicted by Gretener (1969, p.271) in Fig.185. Within the salt bed, which is completely impervious and contains no interstitial water, the pore pressure is not defined and has no physical significance. Below the salt seal, the pore pressure/overburden stress ratio (λ) may acquire a value of unity, while the effective overburden stress approaches zero (Gretener, 1969). Conditions similar to this may exist in Iran as described by Mostofi and Gansser (1957).

Calculation of abnormal pressures from well logs

Magara (1968b, p.2) demonstrated that abnormal fluid pressures in the subsurface can be estimated from certain well logs, such as the sonic log or porosity logs, by using the following expression:

$$p_a = D_1 \rho_w g + (D_2 - D_1)\bar{\rho}_{bw} g \tag{7-21}$$

where p_a is the abnormal pressure, D_1 and D_2 ($D_2 > D_1$) are burial depths which are

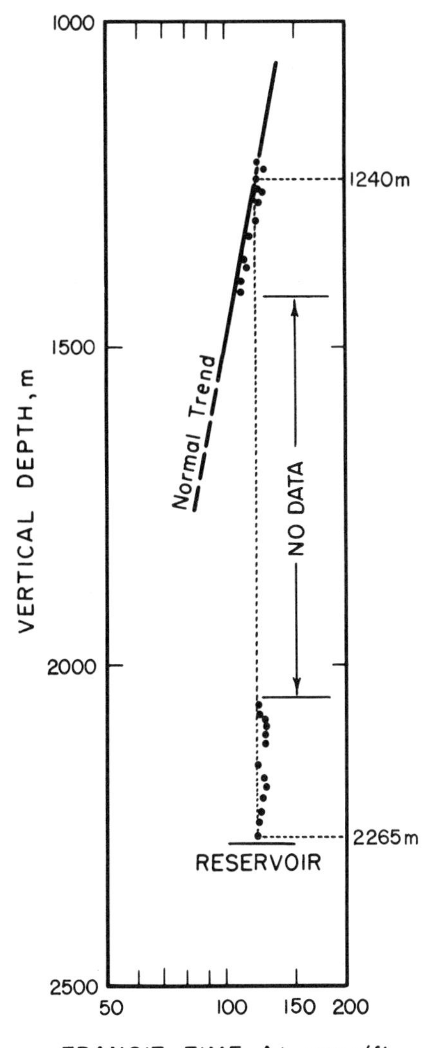

Fig.186. Δt (sonic travel time) versus depth plot for mudstone in Fujikawa SK-2D well, Japan. (After Magara, 1968c, fig.3, p.3. Courtesy of the *Bulletin of American Association of Petroleum Geologists.*)

determined from a porosity or travel time versus depth plots (the abnormally high pressure, p_a, at depth D_2 is equivalent to the normal hydrostatic pressure at D_1 plus the pressure due to the overburden load between depths D_1 and D_2; Δt is equal for depths D_1 and D_2), g is acceleration of gravity, ρ_w is the density of the interstitial fluid, and $\bar{\rho}_{bw}$ is the mean value of the water-saturated bulk density of the sediments.

Fig.186 is used here as an example for calculating an abnormal fluid pressure in a mudstone (Fujikawa SK-2D well, Nagaoka Plain, Japan). The transit (travel) time, Δt, from the sonic log is plotted versus depth. Inasmuch as Δt is approximately proportional to porosity in a pure shale interval, it can be used to estimate the fluid pressures (Magara, 1968b, p.3). Matsuzawa (1961, 1962) assumed reasonable values of $\bar{\rho}_{bw}$ and ρ_w for Japanese Tertiary rocks to be 2.25 g/cm^3 and 1.02 g/cm^3, respectively. The transit time, Δt, in the mudstone just above the abnormally high-pressure reservoir (D_2 = 2,265 m) is equal to the Δt of the mudstone with normal hydrostatic fluid pressure at a depth of 1,240 m (D_1 = 1,240 m). Magara (1968b) estimated the abnormal fluid pressure at the base of the mudstone (depth D_2) to be:

$$p_a = 1{,}240 \times 1.02 + (2{,}265 - 1{,}240) \times 2.25 = 358 \text{ kg/cm}^2$$

The initial bottom-hole pressure actually measured in the well was 362.97 kg/cm^2.

Fertl and Timko (1970a,b,c) also presented methods for detection and evaluation of abnormal pressures.

REFERENCES

Abraham, W.E.V., 1937. Geological aspects of deep drilling problems. *J. Inst. Pet. Technologists,* 23: 378–387.

Anderson, R.V.V., 1927. Tertiary stratigraphy and orogeny of the northern Punjab. *Bull. Geol. Soc. Am.,* 38: 665–720.

Baldwin, H.L., 1944. Tupungato oil field, Mendoza, Argentina. *Bull. Am. Assoc. Pet. Geologists,* 28: 1455–1484.

Bayliss, P. and Levinson, A.A., 1970. Clay mineralogy and boron determinations of the shales from the Reindeer well, Mackenzie River Delta, N.W.T., Canada. *Bull. Can. Pet. Geol.,* 18(1): 80–83.

Bogomolov, Y.G., 1967. Geotemperature regime. *Bull. Int. Assoc. Sci. Hydrol.,* 4: 86–91.

Bredehoeft, J.D. and Hanshaw, B.B., 1968. On the maintenance of anomalous fluid pressures: I. Thick sedimentary sequences. *Geol. Soc. Am. Bull.,* 79(9): 1097–1106.

Burst, J.F., 1959. Postdiagenetic clay mineral environmental relationships in the Gulf Coast Eocene. *Clays Clay Miner., Proc. Natl. Conf. Clays Clay Miner., 6(1957),* pp.327–341.

Burst, J.F., 1969. Diagenesis of Gulf Coast clayey sediments and its possible relation to petroleum migration. *Bull. Am. Assoc. Pet. Geologists,* 53(1): 73–93.

Cannon, G.E. and Craze, R.C., 1938. Excessive pressures and pressure variations with depth of petroleum reservoirs in the Gulf Coast region of Texas and Louisiana. *Am. Inst. Min. Metall. Engrs. Trans.,* 127: 31–37.

Carslaw, H.S. and Jaeger, J.C., 1959. *Conduction of Heat in Solids,* 2nd ed. Oxford Univ. Press, London, 510 pp.

Chaney, P.E., 1949. Abnormal pressures and lost circulation. In: *Drilling and Production Practice.* Am. Petrol. Inst., New York, N.Y., pp.145–148.

Day, J.J., McGlothlin, B.B. and Huitt, J.L., 1967. Laboratory study of rock softening and means of prevention during steam or hot water injection. *J. Pet. Tech.,* 19(5): 703–711.

DeSitter, L.U., 1947. Diagenesis of oil-field brines. *Bull. Am. Assoc. Pet. Geologists,* 31(11): 2030–2040.

Dickey, P.A., Shiram, C.R. and Paine, W.R., 1968. Abnormal pressures in deep wells of southwestern Louisiana. *Science,* 160: 609–615.

Dickinson, G., 1951. Geological aspects of abnormal reservoir pressures in the Gulf Coast region of Louisiana, U.S.A. *Proc. World Petrol. Congr.*, 3rd, Sect. 1, pp.1–17.

Dickinson, G., 1953. Reservoir pressures in Gulf Coast Louisiana. *Bull. Am. Assoc. Pet. Geologists*, 37(2): 410–432.

Dunoyer de Segonzac, G., 1964. Les Argiles du Cretace Supérieur dans le Bassin de Douala (Cameroun): Problèmes de diagenèse. *Bull. Serv. Carte Géol. Alsace–Lorraine*, 17(4): 287–310.

Eremenko, N.A. and Neruchev, S.G., 1968. Primary migration during process of burial and lithogenesis of sediments. *Geol. Nefti i Gaza*, 1968(9): 5–8.

Fertl, W.H. and Timko, D.J., 1970a. Occurrence and significance of abnormal-pressure formations. *Oil Gas J.*, Jan. 5: 97–108.

Fertl, W.H. and Timko, D.J., 1970b. How abnormal-pressure-detection techniques are applied. *Oil Gas J.*, Jan. 12: 62–71.

Fertl, W.H. and Timko, D.J., 1970c. Occurrence of cemented roof rock and geopressure caprock and its implication in petroleum geology and geohydrology. *Ann. Fall. Meet. 45th, Soc. Pet. Engrs., Houston, Texas*, SPE Paper No. 3085, 5 pp.

Finch, W.C., 1969. Abnormal pressure in the Antelope field, North Dakota. *J. Pet. Tech.*, 21(7): 821–826.

Füchtbauer, H. and Goldschmidt, H., 1963. Beobachtungen zur Tonmineral Diagenese. *Proc. Int. Conf. Clays, 1st, Stockholm*, pp.99–111.

Gibson, R.E., 1958. The progress of consolidation in a clay layer increasing in thickness with time. *Geotechnica*, 8: 171–182.

Gilreath, J.A., 1968. Electric-log characteristics of diapiric shale. In: J. Braunstein and G.D. O'Brien (Editors), *Diapirism and Diapirs. Am. Assoc. Pet. Geologists, Mem.*, 8: 137–144.

Gretener, P.E., 1969. Fluid pressure in porous media – its importance in geology; A review. *Bull. Can. Pet. Geologists*, 17(3): 255–295.

Hanshaw, B.B. and Bredehoeft, J.D., 1968. On the maintenance of anomalous fluid pressures: II. Source layer at depth. *Geol. Soc. Am. Bull.*, 79(9): 1107–1122.

Hanshaw, B.B. and Zen, E-an, 1965. Osmotic equilibrium and overthrust faulting. *Geol. Soc. Am. Bull.*, 76(12): 1379–1386.

Harkins, K.L. and Baugher III, J.W., 1969. Geological significance of abnormal formation pressures. *J. Pet. Tech.*, 21(8): 961–966.

Harper, D., 1969. New findings from overpressure detection curves in tectonically stressed beds. *Ann. Calif. Regional Meeting, 40th, Soc. Pet. Engrs.*, Paper no.2781, 7 pp.

Heard, H.C. and Rubey, W.W., 1966. Tectonic implications of gypsum dehydration. *Geol. Soc. Am. Bull.*, 77(7): 741–760.

Howard, D.J.C., 1971. Computer simulation models of salt domes. *Bull. Am. Assoc. Pet. Geologists*, 55(3): 495–513.

Hubbert, M.K. and Rubey, W.W., 1959. Role of fluid pressure in mechanics of overthrust faulting. I. Mechanics of fluid-filled porous solids and its application to overthrust faulting. *Bull. Geol. Soc. Am.*, 70(2): 167–206.

Jacquin, C., 1965. Interactions entre l'argile et les fluides – Écoulements à travers les argiles compactes. Étude bibliographique. *Rev. Inst. Franc. Pét. Ann. Combust.*, 20(10): 1475–1501.

Johnson, H.A. and Bredeson, D.H., 1971. Structural development of some shallow salt domes in Louisiana Miocene productive belt. *Bull. Am. Assoc. Pet. Geologists*, 55(2): 204–226.

Jones, P.H., 1967. Hydrology of Neogene deposits in the northern Gulf of Mexico basin. *Proc. Ann. Symp. Abnormal Subsurface Fluid Pressures, 1st, LSU, Baton Rouge, La.*, pp.91–201.

Jones, P.H., 1969. Hydrodynamics of geopressure in the northern Gulf of Mexico basin. *J. Pet. Tech.*, 21(7): 803–810.

Katz, D.L. and Ibrahim, M.A., 1971. Threshold displacement pressure considerations for caprocks of abnormal pressure reservoirs. *Conf. Drilling Rock Mech., 5th, Soc. Pet. Engrs.*, No.3222.

Keep, C.E. and Ward, H.L., 1934. Drilling against high rock pressures with particular reference to operations conducted in the Khaur field, Punjab. *J. Inst. Pet. Technologists*, 20: 990–1013.

Khitarov, N.I. and Pugin, V.A., 1966. Behavior of montmorillonites under elevated temperatures and pressures. *Geochem. Int.*, 3(4): 621–626.

Kossovskaya, A.G., 1960. Über die spezifischen epigenetischen Umwandlungen terrigener Gesteine in Tafelland und Geosynklinal-Gebieten. *Dokl. Akad. Nauk S.S.S.R.*, 130(1): 176–179.

Lane, A.C., 1922. Weight of sedimentary rocks per unit volume. *Bull. Geol. Soc. Am.*, 33: 353–370.

Lane, O.B.E., 1949. Drilling practices in Iran. *Oil Gas J.*, 48(13): 56–61.

Langseth, M.G., 1965. Techniques of measuring heat flow through the ocean floor. In: W.H.K. Lee (Editor), *Terrestrial Heat Flow. Geophys. Monograph*, 8: pp.58–77.

Levorsen, A.T., 1958. *Geology of Petroleum*. Freeman, San Francisco, Calif., 703 pp.

Lewis, C.R. and Rose, S.C., 1970. A theory relating high temperatures and overpressures. *J. Pet. Tech.*, 22(1): 11–16.

Louden, L.R. and Woods, E.W., 1970. Is shale mineralization a cause of formation damage? *World Oil*, 170(2): 55–58.

MacDonald, G.J.F., 1953. Anhydrite–gypsum equilibrium relationships. *Am. J. Sci.*, 251: 884–898.

Magara, K., 1968a. Considerations of upward and downward migrations of fluids. *J. Jap. Assoc. Pet. Tech.*, Part I, 33(4): 10–17; Part II, 33(4): 40–47 (in Japanese).

Magara, K., 1968b. Subsurface fluid pressure profile, Nagaoka Plain, Japan. *Bull. Jap. Pet. Inst.*, 10: 1–7.

Magara, K. 1968c. Compaction of fluids in Miocene mudstone, Nagaoka Plain, Japan. *Bull. Am. Assoc. Pet. Geologists*, 52(12): 2466–2501.

Martin, R.T., 1962. Adsorbed water on clay: A review. *Clays Clay Miner. Proc. Natl. Conf. Clays Clay Miner., 9th (1960)*, pp.28–70.

Matsuzawa, A., 1961. On the relationship between the density of sedimentary rocks and the subsurface geology. *Geophys. Explor.*, 14: 195 (in Japanese).

Matsuzawa, A., 1962. On the relationship between the density of sedimentary rocks and the subsurface geology. *Geophys. Explor.*, 15: 1 (in Japanese).

Morgan, J.P., Coleman, J.M. and Gagliano, S.M., 1968. Mudlumps: diapiric structures in Mississippi delta structures. In: J. Braunstein and G.D. O'Brien (Editors), *Diapirism and Diapirs. Am. Assoc. Pet. Geologists, Mem.* 8: 145–161.

Mostofi, B. and Gansser, A., 1957. The story behind the 5 Alborz. *Oil Gas J.*, 55(3): 78–84.

Moulenes, B., 1964. Origine des pressions anormales dans les gisements de pétrole. Étude bibliographique. *Rev. Inst. Franc. Pét. Ann. Combust.*, 19(2): 196–212.

Muffler, L.J. and White, D.E., 1969. Active metamorphism of Upper Cenozoic sediments in the Salton Sea geothermal field and the Salton Trough, southeastern California. *Bull. Geol. Soc. Am.*, 80(1): 157–182.

Ocamb, R.D., 1961. Growth faults of south Louisiana. *Trans. Gulf Coast Assoc. Geol. Soc.*, 11: 139–175.

Overton, H.L. and Timko, D.J., 1969. The salinity principles – a tectonic stress indicator in marine sands. *The Well Log Analyst*, (3): 34–43.

Overton, H.L. and Zanier, A.M., 1971. An osmotic model for gas and overpressured formations. *Conf. Drilling Rock Mech., 5th*, SPE Paper No.3221.

Perry, E. and Hower, J., 1970. Burial diagenesis in Gulf Coast pelitic sediments. *Clays Clay Miner.*, 18: 165–177.

Pinfold, E.S., 1954. Oil production from Upper Tertiary fresh-water deposits of West Pakistan. *Bull. Am. Assoc. Pet. Geologists*, 38: 1653–1660.

Platt, L.B., 1962. Fluid pressure in thrust faulting, a corollary. *Am. J. Sci.*, 260(2): 107–114.

Powers, M.C., 1959. Adjustments of clays to chemical change and the concept of the equivalence level. *Clays Clay Miner.*, 2: 309–326.

Powers, M.C., 1967. Fluid-release mechanisms in compacting marine mud rocks and their importance in oil exploration. *Bull. Am. Assoc. Pet. Geologists*, 51(7): 1240–1254.

Reed, P., 1946. Trinidad leaseholds applies advanced methods in drilling and production. *Oil Gas J.*, Oct. 5: 44–46.

Rubey, W.W., 1930. Lithologic studies of fine-grained upper Cretaceous sedimentary rocks of the Black Hills region. *U.S. Geol. Surv., Profess. Pap.*, 165-A, 54 pp.

Rubey, W.W. and Hubbert, M.K., 1959. Role of fluid pressure in mechanics of overthrust faulting. II. Overthrust belt in geosynclinal area of western Wyoming in light of fluid-pressure hypothesis. *Bull. Geol. Soc. Am.*, 70(2): 167–206.

Sarkisyan, S.G. and Kotel'nikov, D.D., 1971. *Clay Minerals and Problems of Oil–Gas Geology.* Nedra, Moscow, 183 pp.

Schlumberger, C., Schlumberger, M. and Leonardon, E., 1933. Electrical coring; a method of determining bottom-hole data by electrical measurements. *Am. Inst. Min. Metall. Engrs., Trans.*, 110: 237–272.

Seidov, A.G. and Alizade, Kh.A., 1970. *Mineralogy and Conditions of Formation of Bentonite Clays of Azerbayjan.* Izd. Elm, Akad. Nauk Azerbayjan. S.S.R., Baku, 190 pp.

Steiner, A., 1967. Clay minerals in hydrothermally altered rocks at Wairakei, New Zealand. *Clay Miner. Conf., 16th,* Abstr., pp.31–33.

Stewart, F.H., 1963. Marine evaporites. In: M. Fleischer (Editor), *Data of Geochemistry,* 6th ed. *U.S. Geol. Surv., Profess. Pap.*, 440-Y, 52 pp.

Sudo, T., Shimoda, S., Nishigaki, S. and Aoki, M., 1967. Energy changes in dehydration processes of clay minerals. *Clay Miner. Bull.*, 7: 33–42.

Suter, H.H., 1954. The general and economic geology of Trinidad, British West Indies. *Colon. Geol. Miner. Resour. Gt. Br.*, 2(3,4), 3(1): 134 pp.

Thomeer, J.H.M.A. and Bottema, J.A., 1961. Increasing occurrence of abnormally high reservoir pressures in boreholes, and drilling problems resulting therefrom. *Bull. Am. Assoc. Pet. Geologists,* 45(10): 1721–1730.

Thorsen, C.E., 1963. Age of growth faulting in southeast Louisiana. *Trans. Gulf Coast Assoc. Geol. Soc.*, 13: 103–110.

Tkhostov, B.A., 1963. *Initial Rock Pressures in Oil and Gas Deposits.* Pergamon, New York, N.Y., 118 pp.

Tkhostov, B.A., Vezirova, A.D., Vendel'shteyn, B.Yu. and Dobrynin, V.M., 1970. M.F. Mirchink (Editor), *Petroleum in Fractured Reservoirs.* Nedra, Leningrad, 220 pp.

Van Moort, J.C., 1971. A comparative study of the diagenetic alteration of clay minerals in Mesozoic shales from Papua, New Guinea, and in Tertiary shales from Louisiana, USA. *Clays Clay Miner.*, 19(1): 1–20.

Wallace, W.E., 1969. Water production from abnormally pressured gas reservoirs in south Louisiana. *J. Pet. Tech.*, 21(8): 969–982.

Watts, E.V., 1948. Some aspects of high pressures in the D-7 zone of the Ventura Avenue field. *Am. Inst. Min. Metall. Engrs.*, 174: 191–200.

Weaver, C.E., 1960. Possible uses of clay minerals in search for oil. *Bull. Am. Assoc. Pet. Geologists,* 44: 1505–1518.

Weaver, C.E., 1961. Clay mineralogy of the late Cretaceous rocks of the Washakie Basin. *Wyo. Geol. Assoc. Guideb., 16th Field Conf.*, pp.148–154.

Weaver, C.E. and Beck, K.C., 1969. *Changes in the Clay-Water System with Depth, Temperature and Time.* OWRR Proj. No.A-008-GA, Sch. Ceramic Eng., Georgia Inst. Tech., Atlanta, Ga., WRC-0769, 95 pp.

Wooding, R.A., 1959. Steady-state free thermal convection of liquid in a saturated permeable medium. *J. Fluid Mech.*, 2: 273–285.

Wyllie, M.R.J., 1955. Role of clay in well-log interpretation. In: J.A. Pask and M.D. Turner (Editors), *Clays and Clay Technology. Calif. Div. Mines Bull.*, 169: 282–305.

Young, A. and Low, P.F., 1965. Osmosis in argillaceous rocks. *Bull. Am. Assoc. Pet. Geologists,* 47(7): 1004–1008.

Zierfuss, H. and van der Vliet, G., 1956. Laboratory measurements of heat conductivity of sedimentary rocks. *Bull. Am. Assoc. Pet. Geologists,* 40(10): 2475–2488.

Chapter 8

EQUIPMENT AND TECHNIQUES USED IN COMPACTION STUDIES

DEVELOPMENT OF COMPACTION EQUIPMENT

Compaction studies prior to 1930

When a pressure of over 25 kb ($\approx$ 362,000 p.s.i.) is reached, most materials in the pressure vessel are solids at room temperature. The development of equipment to reach and hold such static high pressures was launched in the middle of the 19th Century by two French scientists, Louis Cailletet and E.H. Amagat. Other scientists, however, paved the way for Cailletet and Amagat. Cailletet (1872) was the first to be noticed by the scientific and engineering community as introducing a long range of pressures (up to 700 atm) through the use and development of the dead-weight or free-piston gauge, whereas Amagat's important work on the compression of gases brought about the development of methods of pressure measurements. The first extensive use of Cailletet's complex free-piston "primary" gauge ("manomètre à pistons libres") was by Amagat (1893), whose system involved the use of a second piston to measure the force expelling a smaller diameter high-pressure piston (Fig.187). This smaller piston was coaxially mounted above the larger piston. Thrust applied to the small piston was transmitted to the larger piston which was connected with an open mercury column several meters high. The mercury column automatically attains the height required to balance the applied high pressure. When the system is in equilibrium, the pressure acting under the large diameter piston is less than that acting on the first piston and their area ratio expresses this relationship. Leaks in the fluid line were partially eliminated by using a very viscous molasses as the pressure-transmitting fluid. The conduit for injection of oil was located much below the mercury level so that it would not interfere with the bronze pump. Owing to the leaks in the fluid transmitting lines in this system, Amagat measured with much difficulty the compressibilities of fluids under pressures up to 3,000 kg/cm^2 (42,669 p.s.i.). Piston friction was minimized by giving both pistons a rotary motion immediately before a pressure reading was recorded.

The first published high-pressure research with water was performed by John Canton (1762) in England, who worked with the compressibility of water and other fluids. That a fluid like water is compressible is not only of great scientific interest, but is important in reservoir engineering and in hydrocarbon migration studies. Tait (1889) investigated the

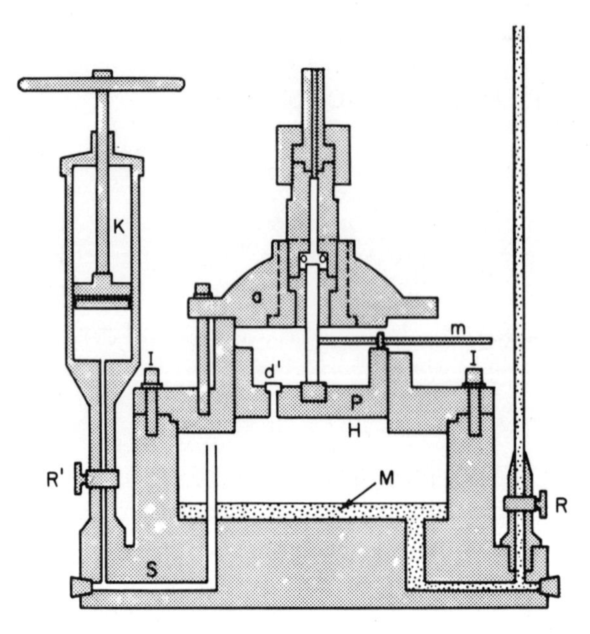

Fig.187. Amagat's double free-piston gauge. P = large piston; H = oil; M = mercury; R and R' are stopcocks; K = bronze pump; S = oil conduit; I = bolts; d' = valve for escape of air; m = lever arm; a = bronze head; oo' = sample cavity. (After Amagat, 1893, p.71.)

pressure effects on sea water during the Challenger expedition. The first geologically oriented high-pressure and high-temperature experiments were performed for the United States Geological Survey by Carl Barus (1889–1892). He developed in the laboratory pressures as great as 2,000 kb/cm^2 (28,446 p.s.i.) and temperatures up to 400°C in order to study the compressibility and thermodynamic properties of liquids. By using an elastic deformation gauge (a helical Bourdon spring), Barus was able to measure accurately the pressures developed by his complex screw-type compressor (Barus, 1892c, p.17). The pressure gauge was calibrated against the electrical resistance of mercury under identical conditions of pressure and temperature. This measurement was accurate to considerably better than 1%, but its accuracy was limited by hysteresis in the steel of the Bourdon spring (Bridgman, 1958, p.68).

The first genuine geological and geochemical investigations were initiated, at the turn of the century, by the scientists associated with the Geophysical Laboratory of Carnegie Institute, Washington, D.C. Experimental observation of rock flowage in the laboratory under high confining pressures at room temperatures has, for the most part, been neglected by geologists in North America. Frank D. Adams initiated experimental study into the deformation of rocks encased in thick-walled, close-fitting steel tubes under high confining pressures (Adams and Nicholson, 1901). Their research proved that rocks under a confining pressure are no longer brittle but exhibit plastic deformation, which occurs in

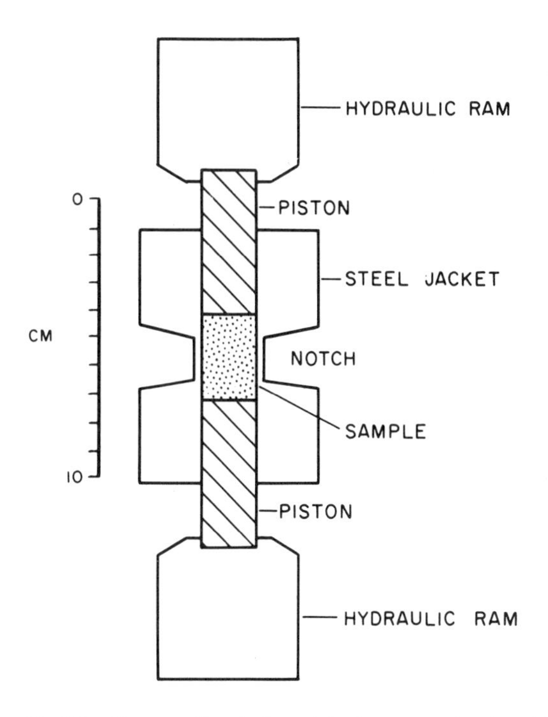

Fig.188. Cross-sectional view of testing apparatus in which the "strength" of the rocks was measured "qualitatively" under high confining pressures (Adams and Coker, 1910, fig.2, p.471). The confining pressure depended upon the amount of bulging of the nickel steel jacket when the rock sample was being compressed by the "Novo" pistons. At the start of the deformation, the confining pressure was approximately equal to zero.

nature. Fig.188 illustrates the bomb in which the various rock specimens were jacketed and deformed by simple uniaxial compression. In order that the conditions of differential pressure may be satisfactorily developed in the specimen, it was necessary to notch the nickel steel jacket around the central portion of the rock core (Fig.188). The pistons were machined from chromium tungsten steel (called "Novo") which had an ultimate compressive strength of 411,800 p.s.i.

Adams' jacketing idea was based upon Professor Friedrich Kick's work (Adams and Coker, 1910). Kick (1892, p.920) devised an encasing method using a copper, brass or iron jacket to secure the lateral pressure. This kept the unidirectional pressure concentrated on the sample (hydrostatic conditions) without bursting the bomb. Kick was led into the study of plastic deformation of brittle materials under confining pressure because of the observations of Albert Heim, who noted that in the formation of mountains, sedimentary rocks apparently were thrust up in the ductile state under high tectonic pressures. Kick successfully showed through laboratory experiments that brittle materials such as rocks and some metals can be made ductile under a hydrostatic confining pressure. Ammonites which show various stages of distortion in tectonically stressed and deformed sedimentary formations, were cited by Kick as excellent field examples of a brittle body having undergone plastic deformation.

Professor Percy W. Bridgman (1909), the Father of American high-pressure research modified the design and decreased the dimensions of Amagat's "manomètre à pistons libres" so that it became a reliable high-pressure apparatus. During the period from 1900 to 1923, Bridgman investigated the thermodynamic, phase, and electrical properties of mercury, water and selected solids up to 12 kb (174,050 p.s.i.). Bridgman formalized the first pressure scale by relating the true hydrostatic pressure condition of the "free" piston apparatus to the change in resistance of a manganin-wire (an alloy of copper, manganese and nickel) exposed to the same pressure. This work was based on prior research by Lisell (1903), who noted the resistance changes in metallic wires under pressure. Calibration of the manganin-wire gauge is accomplished by recording its resistance at a pressure at which mercury becomes solid (7,640 kg/cm^2 – previously used figure; 7,723 kg/cm^2 – new value at 0°C) and at the lowest solid–solid phase transition of bismuth (25,420 kg/cm^2 or 25,705 kg/cm^2 at 25°C and at 30°C, respectively). The manganin-wire gauge is well-suited for high-pressure measurements and probably would not be soon replaced.

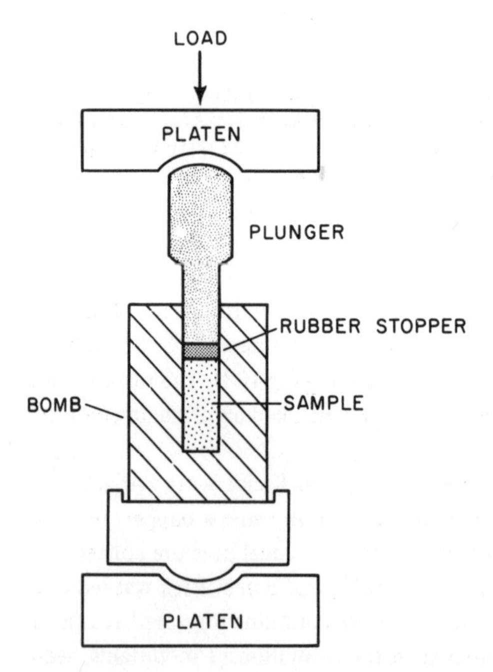

Fig.189. Schematic diagram of a simple high-pressure bomb used by Johnson and Adams (1912). The cross-hatched area represents the parts drawn to scale; all other parts are represented diagrammatically. The bomb centers itself between the platens of the press as pressure is applied to the plunger. Oil is used to decrease the friction between the plunger and the bomb's wall. The rubber stopper prevents any oil from slipping into the sample chamber and contaminating the sample.

Von Karman (1911) was one of the first investigators who constructed a high-pressure triaxial cell. Most notable of the early investigators at the Carnegie Institute of Washington, D.C., was Leason H. Adams, who employed a simple plunger in studying the effects of hydrostatic pressure on the density of homogeneous metallic and nonmetallic crystals (Fig.189). Adams et al. (1919) described the equipment used and the procedure employed to determine volume changes in solids, including quartz and calcite. The sample to be compressed was placed in kerosene inside a cylindrical borehole of a thick-walled steel cylinder. The pressure vessel was closed at the bottom except for a drainage plug and the resistance gauge connection, and fitted at its top with a movable piston. The piston was forced downward by means of a hydraulic press which transmits a uniform hydrostatic pressure through the kerosene to the sample. Measurement of the pressure inside the cell was accomplished by the change in electrical resistance of a manganin wire exposed to the internal pressure. Bridgman (1918) studied the stress–strain relationships in crystalline hollow rock cylinders, and obtained workable but only approximate results.

Adams and Williamson (1923) were the first investigators in the United States to conduct a comprehensive study on the bulk compressibilities of various minerals and rocks. In one set of experiments they coated the rock cores with tin foil to prevent access of the pressurizing fluid to the pores, and in another set they exposed the rock cores to the fluid. They found that at low pressures the compressibilities of the jacketed rocks were much greater than the compressibilities of the unjacketed rocks. At higher pressures, the bulk compressibilities of the jacketed samples decreased and approached asymptotically the compressibilities of the unjacketed specimens. In 1924, Bridgman measured the effect of pressure on the thermal conductivity of rocks.

Development of equipment after 1930

The major developments in high-pressure vessels occurred after 1930. Bridgman's (1931, 1937) second pressure scale was formalized through the use of high-pressure, flat opposed-anvil equipment, which gives a quasi-hydrostatic pressure measurement of the sharp electrical-resistance jumps in solids under pressure. The Bridgman "squeezer" was free of internal friction and used the idea of amplifying the low pressure of the hydraulic fluid by concentrating the total force on a small area on the face of the anvil (Fig.190). Pressures for the fixed points as determined by the anvils were higher by 30% than the pressures determined through the use of a piston-cylinder apparatus. This is attributed to a combination of a radial pressure gradient in the solid used to transmit the pressure to the material under study, and that the effective load-bearing area is less than the area of the anvil face (p is not equal to W/A as assumed by early investigators, where A = cross-sectional area of anvil face and W = load). When the pressure gradient was considered, the discrepancy between the calculated transition pressures (up to 90 kb) for the anvil and piston-cylinder devices disappeared.

In the middle thirties, D.T. Griggs performed a series of experiments to investigate the

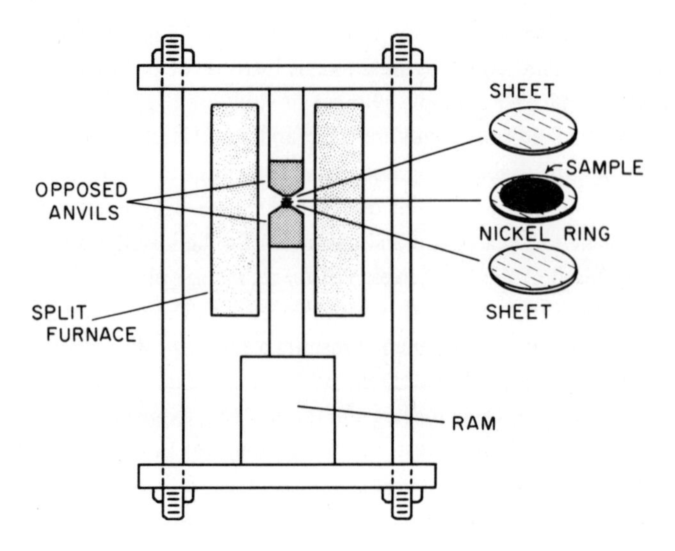

Fig.190. Bridgman's (1937) opposed-anvil apparatus which is shown diagrammatically with an enlarged view of the sample assembly. Force produced by the hydraulic ram is concentrated on the small faces of the anvils. The sample assembly is composed of a nickel sample ring (0.01-inch thick) sandwiched in between two platinum–rhodium sheets (0.001-inch thick). Pressures in excess of 100 kb may be created with hard carbide anvils.

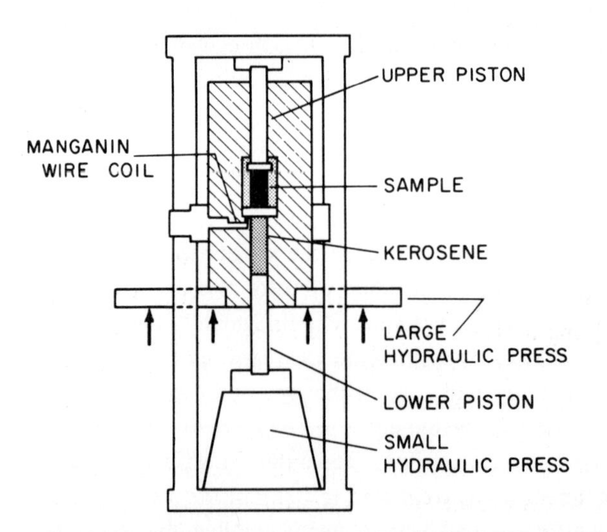

Fig.191. Schematic diagram of D.T. Griggs' (1936, fig.2, p.545) hydrostatic high-pressure apparatus. A large hydraulic press moves the pressure bomb upward, thereby forcing the upper piston down on the sample. The sample is surrounded by kerosene which is compressed by a smaller hydraulic press located at the base of the apparatus. Ring packing was employed so that a solid upper piston could be used; sample deformation was measured directly by measuring the displacement of this piston. The hydrostatic pressure was measured by the change in resistance of a manganin-wire coil, whereas the sample deformation was measured by a micrometer dial gauge attached to the upper piston.

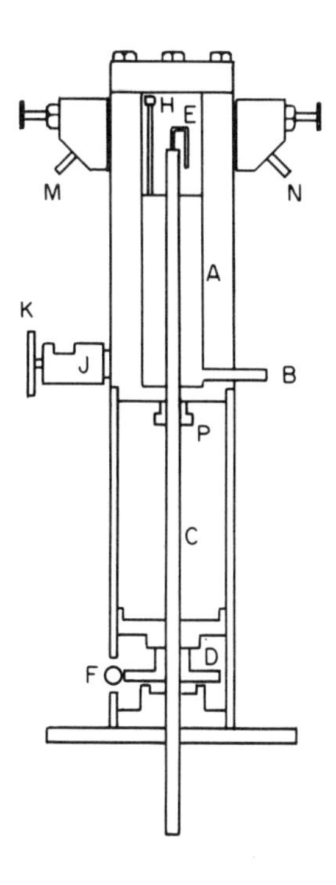

Fig.192. Variable volume P–V–T cell (Sage et al., 1934, fig.1, p.1218). *A* = steel pressure cell (inside diameter of 2 inches and inside height of 11 inches); *B* = valve through which mercury could be added or withdrawn from *A*; *C* = hollow rod which agitates the hydrocarbon mixture in the pressure cell *A*; *E* = electric contact on the rod *C* used to determine the volume of the space above the mercury surface; *F* and *D* = work gear and nut which drives rod *C*; *H* = agitating rod (rotated about 130 r.p.s.); *M* = liquid valve; *N* = gas valve for addition of gas to *A*.

physical processes which occur during rock deformation under high confining pressure (Fig.191). In designing the apparatus, Griggs took advantage of developments in high-pressure techniques by Bridgman, so that the range of pressure investigation was carried to four times the amount of confining pressure available in F.D. Adams' apparatus (Fig. 188). Griggs (1936) studied the relationship between creep and confining pressure (up to 11,000 atm or 161,656 p.s.i.) for the Solnhofen Limestone and a marble, and up to 12,000 atm (176, 350 p.s.i.) confining pressure for quartz.

Sage and Lacey during the interval from 1934 to 1947 designed and constructed various types of equilibrium cells which permitted the study of complex hydrocarbon mixtures at elevated temperatures and pressures. The first cell was a simple constant-volume bomb (Pomeroy et al., 1933). In 1934, Sage constructed the varying-volume P–V–T (pressure–volume–temperature) cell; the volume inside the pressure bomb was changed by injection of mercury through a valve located at the bottom of the cell (Fig.192). The maximum pressure and temperature reached in the first two P–V–T cells were 3,000 p.s.i. and 220°F, respectively. A new model of P–V–T apparatus differed from the earlier models in that the volumes of liquid and gas phases were calculated by knowing the height of the mercury column in the cell. The volume of the mercury was measured by using a potentiometer circuit (Sage et al., 1934). Their final model had a working temperature range from 0°F to 460°F at pressures up to 10,000 p.s.i. (Sage and Lacey, 1948).

Standing (1952) developed graphs relating the volumetric and phase behavior of oil reservoir hydrocarbon systems by measuring fluid properties in a P–V–T cell. Previously, the prediction of the volumetric behavior of naturally occurring hydrocarbon systems was accomplished by analytical means. The analytical approach, however, was not adequate to predict the thermodynamic behavior of complex multicomponent hydrocarbon systems which exist at high pressures and temperatures in petroleum reservoir rocks.

The first high-pressure apparatus used strictly in geochemical studies was a modified Bridgman 12-kb pressure cell designed by Yoder (1950). Yoder's pressure system was almost entirely identical in design to the apparatus constructed in 1910 and remodeled in 1933 by Bridgman. Yoder extended the earlier work performed on the high–low inversion of quartz by another Carnegie Institute scientist Gibson (1928). Argon, which is an inert gas, was used for pressure transmission because it has a lower compressibility than either helium, nitrogen, or hydrogen. In addition, argon does not readily cause embrittlement of the platinum furnace windings in the cell's bore.

The first serious attempt to create synthetic diamonds by Hannay in 1880 was not successful (Flint, 1968). During the 1930's and 1940's, Bridgman failed in his many attempts to convert graphite to diamond on using high pressures, even though his equipment operated in the diamond-stable pressure region. It was not until 1955 that diamond was synthesized at General Electric's research laboratory in Schenectady, New York. Allmanna Svenska Elektriska Aktieboget (ASEA), however, claims that diamonds were produced in 1953 at their laboratory in Sweden. At ASEA, Von Platen began his experimental work on the synthesis of diamonds in 1942. He designed the *cubic* anvil apparatus (Brit. Pat. 788,891 and 791,099), in which the pressure transmitted to the sample depends upon the shearing stresses in the pressure-transmitting media used (Liander and Lundblad, 1960) (see Fig.193). Fig.193 also illustrates the *belt* static-pressure apparatus developed at General Electric laboratories and employed in producing diamonds by direct transition from graphite (Hall, 1961). Two tapered pistons compress the confined graphite and catalyst metals in a short tapered cylinder at a high temperature.

The compressibility of liquid water had been measured by many scientists over the past several hundred years beginning with John Canton in 1762 and ending with the excellent recent work of George C. Kennedy at the University of California at Los Angeles. Fig.194 shows the high-pressure apparatus used by Kennedy in his early work on the pressure–volume–temperature relationship of distilled water. Heise Bourdon gauges were used to measure pressures up to 14 kb (203,060 p.s.i.), whereas the higher pressures ($>$ 14 kb) were determined by measuring the changes in resistance of a manganin coil surrounded by toluene in a copper syphon. The syphon was enclosed in a steel cartridge

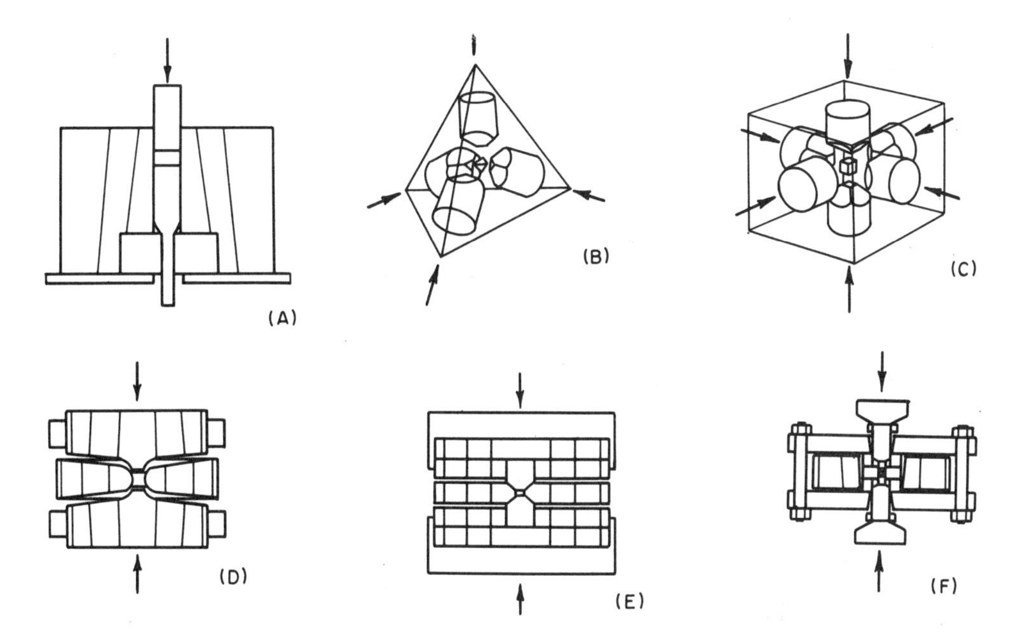

Fig.193. Schematic diagrams of various high-pressure systems used in laboratory investigation of materials under high confining pressures. A. Extruder. B. Tetrahedral anvil apparatus. C. Cubic anvil apparatus. D. Belt apparatus. E. Girdle apparatus. F. Stepped piston apparatus.

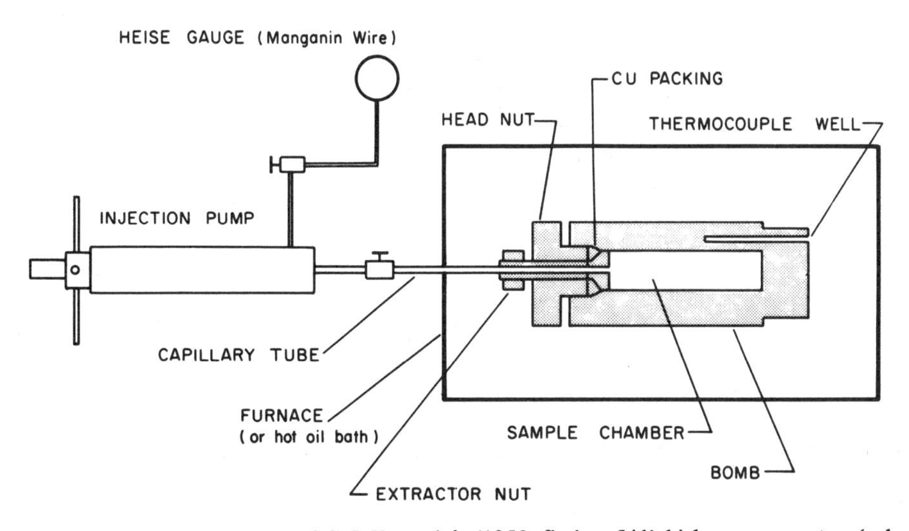

Fig.194. Schematic diagram of G.C. Kennedy's (1950, fig.1, p.541) high-pressure system (volumometer) used in P–V–T experiments on water.

which may be substituted for the Heise gauge. Temperatures in the cell were measured by determining the emf of a thermocouple (composed of 90% platinum and 10% rhodium), calibrated against the boiling points of water and sulphur and the melting points of antimony, sodium chloride, copper and gold. In a more recent study on the P–V–T relationships of sea water and water plus carbon dioxide, Newton and Kennedy (1965) and Kennedy and Holser (1966) utilized an improved apparatus and an entirely different operational procedure.

During the past few years, spectacular progress in high pressure and temperature research has led to important hypotheses about the types of minerals present and the behavior of rocks at varying depths in the earth's crust and interior. Stress–strain and deformation studies on various minerals and igneous and metamorphic rocks have been

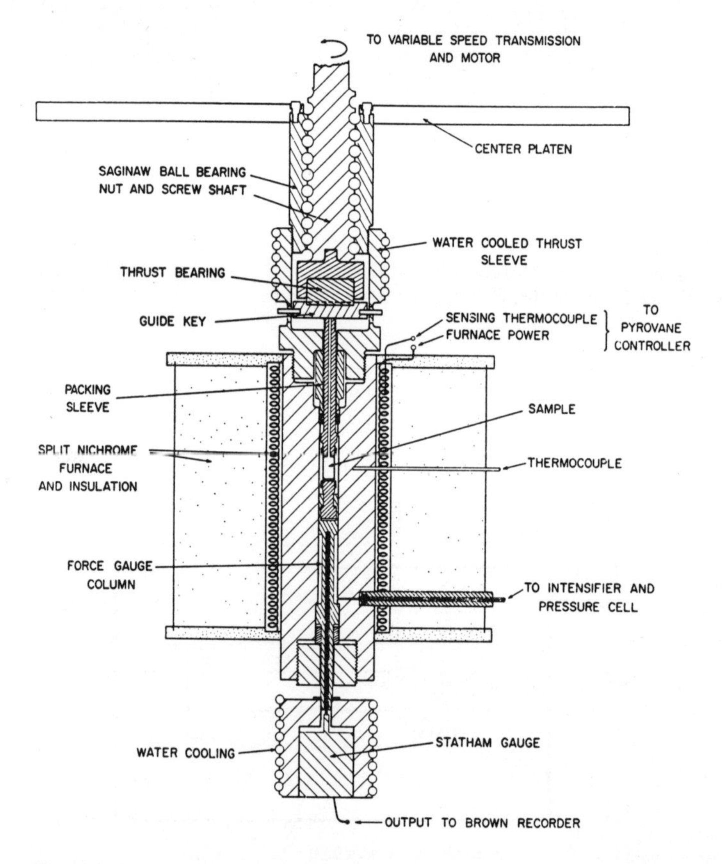

Fig.195. Schematic diagram of H.C. Heard's (1963, fig.2, p.166) 5-kb, 500°C high-pressure extension apparatus. The pressure cell features an external heating unit and a moving piston driven by a variable-speed transmission and motor unit which can apply a strain rate of 10^{-1} to 10^{-8} inch/inch sec^{-1}. The Statham force gauge (pressure transducer) output was recorded on an oscilloscope fitted with a polaroid recording camera or an X-Y recorder.

carried out under the direction of D.T. Griggs by Hugh C. Heard and Neville L. Carter at the University of California at Los Angeles, using single-piston and cubic-anvil apparatuses. The single-piston deformation apparatus (Fig.195) was developed for the investigation of the triaxial extension, at various constant strain rates, on jacketed cylinders of Yule Marble (Heard, 1963). Carter et al. (1964) described the use of the cubic-anvil apparatus, which was employed to study the plastic deformation and recrystallization of quartz. Heterogeneous strain was investigated using the single-piston device, presented in Fig.195, by deforming homogeneous, mechanically anisotropic bodies of calcite, marble and phyllite (Heard et al., 1965). In 1968, the maximum sustained pressure and temperature reached were 250 kb (3,626,000 p.s.i.) and 1000°C, respectively. Static pressures at 500 kb (7,252,000 p.s.i.) at 25°C, obtained on using the anvil-type apparatuses, have been reported. Pressures of 100 kb (1,450,380 p.s.i.) at temperatures up to 2,100°C also have been reported. A.E. Ringwood (Australian National University) developed a high-pressure apparatus (up to 250 kb) for studying the geochemical and density relationships of rocks and minerals in the earth's upper-mantle. Other scientists have obtained transient super-high pressures through the use of explosives or by impact of high-velocity projectiles (Deal, 1962). Such super-high pressures of 5,000 kb (72,520,000 p.s.i.) have been measured, although they only exist for a few millionths of a second.

CLASSIFICATION OF COMPACTION EQUIPMENT

Application of various kinds of equipment are presented in Table XLVII, whereas schematic diagrams of high-pressure apparatuses used in laboratory investigations of materials under high confining pressures are given in Fig.193. The reader is also referred to an excellent book by Tsiklis (1965) on the design and operational techniques of high and super-high pressure apparatuses. A general classification of compaction equipment is presented in Table XLVIII. It is hoped that classification with respect to type will be of assistance to researchers concerned. The apparatuses are classified on the basis of how the pressure is applied. The pressure-transmitting media in such devices may be either mechanical (ram or lever) or fluid – either liquid (such as oil or mercury) or gas (such as nitrogen or helium). The mechanical pressure system is subdivided on the basis of the relationship between the length of the piston and its diameter. An objective evaluation of the different material employed in the construction of high-pressure equipment is difficult to make because most authors did not include such data in their papers. Wherever possible, the material used in equipment construction should be presented along with the description of the device.

There are four basic requirements that should be met by the equipment used in investigating the chemical and mechanical properties of sediments or clays under high pressures, namely (1) application of a uniform known stress on the contained sample, (2) measurement of the resulting sample deformation, (3) channeling and collection of

TABLE XLVII
Application of compaction apparatuses [1] (see Fig.193)

Apparatus	*Approximate maximum pressure (p.s.i.)*	*Some proponents and investigators* [2]	*General application*
Consolidometer	2,000	Terzaghi, Casagrande	compression of clays and soils
Filter press (mech.)	300	Siever	chemistry of solutions squeezed out of marine sediments
Single-stage piston	20,000	Manheim	chemistry of solutions squeezed out of marine sediments
Extruder	500,000	Several	metal fabrication
Volumometer	200,000	Kennedy	$P-V-T-S^3$ relationships of water
Dual piston	500,000	Chilingar, Kryukov	clay compaction and fluid extraction
Duplex piston		Kryukov	clay compaction and fluid extraction
Opposed anvils	1,500,000	Bridgman	material science
Tetrahedral anvils	1,500,000	Hall, H.T.	mineral synthesis
Cubic anvils	1,500,000	Von Platen	diamond synthesis
Belt	1,500,000	Hall, H.T., Bundy	diamond synthesis
Girdle	1,500,000	Wilson	material science
Stepped piston	1,500,000	Giardini	diamond synthesis
P–V–T cell	15,000	Sage, Standing	P–V–T relationships of hydrocarbons
Filter press (gas)	1,000	Presley	chemistry of solutions squeezed out of marine sediments
Triaxial, hydrostatic	40,000	Hall, H.N., Fatt, Sawabini and Chilingar	petroleum engineering research

[1] See Comings (1956), Hall (1958), Wentrof (1962), and Tsiklis (1968) – selected bibliography.
[2] See references.
[3] Pressure–volume–temperature–salinity.

the expelled fluid, and (4) utilization of a sample of sufficient size so that representative information can be obtained. The main approach used by investigators has been one of innovation and modification of the two basic high-pressure apparatuses, namely, the simple plunger type and the consolidometer.

Currently there is considerable interest by marine geochemists in the design of a mechanical or gas "squeezer" which would extract the interstitial fluid from unconsolidated marine sediments. In the simple plunger apparatus a sample only undergoes deformation in response to the unidirectional applied pressure; a definite volume change occurs in the sample with increasing axial load on the sample. In the consolidometer, on the

TABLE XLVIII
General classification of apparatuses used in the investigation of geochemical and geodynamic problems (see Fig.193)

Pressure application	*Apparatus type*	
Mechanical pressure system		
Linear piston ($L \leqslant D$) [1]	"free" piston	
	consolidometer	(1) fixed ring
		(2) floating ring
	filter press (squeezer)	
Linear piston ($L > D$)	single piston	(1) single stage
		(2) plunger
		(3) extruder
		(4) volumometer
	dual piston	
	duplex piston	
Multiple piston ($L \geqslant D$)	anvil	(1) opposed (flat)
		(2) tetrahedral
		(3) cubic
		(4) dodecahedral
	belt	
	girdle	
	stepped piston	
Fluid pressure system		
Mercury	P–V–T cell	(1) visual cell
		(2) blind cell
Gas	filter press (squeezer)	
Kerosene	pressure cell	
Hydraulic fluid	triaxial, hydrostatic	

[1] L = length and D = diameter of the piston.

other hand, there are mass changes in addition to volume change, because fluid is expelled from the sample owing to the volumetric change within the sediment sample.

Consolidometers

There are two basic approaches in equipment design that are widely employed by engineers in the field of soil mechanics to achieve one-dimensional consolidation (compaction) of soils. Lambe (1951, p.76) classified the consolidometers either as a lever (wheel) mechanical system or as a mechanical jack system in which the method of axial loading consists of pressure being applied to the sample by a ram. Terzaghi (1925), who built the first consolidometer, which he called an oedometer, applied mechanical pressure to clays using a water-permeable sand filter. The maximum applied pressure was 285 p.s.i.

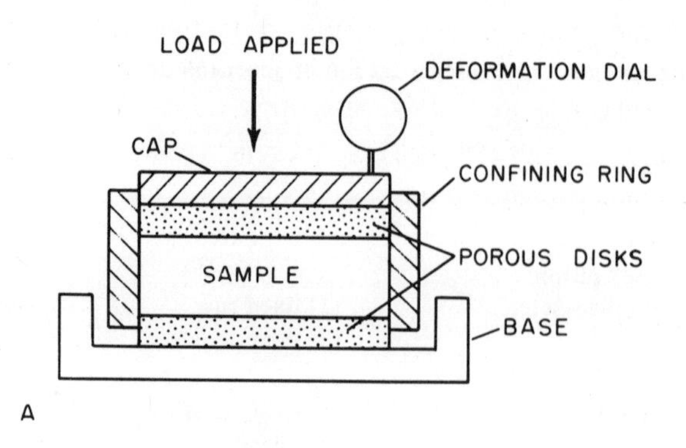

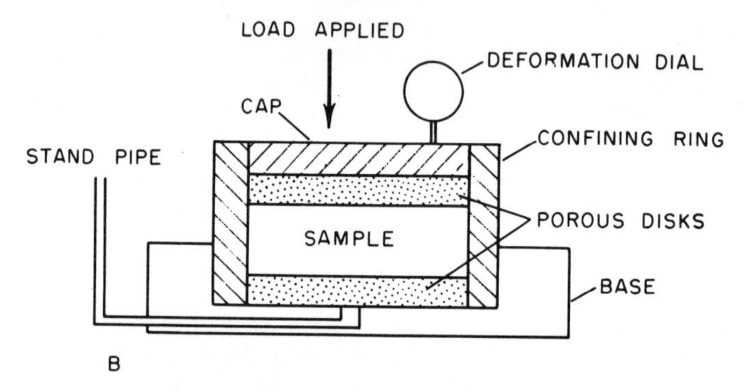

Fig.196. Schematic diagram of two kinds of ring containers used in one-dimensional consolidometers. A. Floating-ring container. B. Fixed-ring container.

He estimated that the capillary forces during air-drying of hydrated clays resulted in pressures of 2,400–4,800 p.s.i. In the early thirties, A. Casagrande designed a large-diameter consolidometer (in: Taylor, 1948, p.212).

The basic principles of the consolidometer are illustrated in Fig.196. In Lambe's classification, the lever system employs a lever arm upon which weights of known magnitude are hung, whereas the jack system uses a hydraulic ram or a platform scale. The consolidometer consists of a rigid metal cylinder with a closed base, containing a sample in the shape of a thick circular disk sandwiched between two porous stones. Two kinds of soil-sample ring containers (Fig.196) are used, the floating-ring container and the fixed-ring container.

Porous stones (generally ceramic, but sometimes natural stones) at the top and bottom of the sample allow free drainage of the interstitial fluid from the sample undergoing compaction. Their pores are not sufficiently large to allow any of the soil particles to pass through. In some consolidation tests one of the porous plates is replaced by a non-porous stone (natural rock) or plate in order to make the fluid drainage unidirectional. In the fixed-ring container, all the sample movement occurs downward relative to the container, whereas in the floating-ring container axial compression occurs toward the middle from both top and bottom of the sample. Friction between the container wall and the sample is lower in the floating-ring device than in the fixed-ring container. The main advantage of the fixed-ring device is that permeability tests can be performed on the sample. A consolidation test consists of instantaneous application of a constant load to the sample and the measurement of the resulting deformation with time. The general procedure is to subject the sample to small incremental loadings until the maximum load is reached. In the determination of the sediment or soil compressibility characteristics and the speed with which axial compression occurs, the time and the amount of sample deformation are recorded, the latter by use of an indicator dial gauge.

Extensive clay-compaction experiments have been performed by engineers and soil scientists in the low-pressure range (≤ 2,000 p.s.i.). The object of their experiments was to determine the physical properties of the soils and clays under stress. Water or other liquids are always used in these experiments for the purpose of retaining the clay's plasticity or in determining the permeability of the sample.

Macey (1940) used a high-pressure apparatus to determine the pressure–moisture relationship in hydrated clays and their conductivity (permeability) to interstitial fluids, which are squeezed out. His apparatus was essentially a one-dimensional consolidometer (Fig.197), and its design was based on the one used previously by Westman (1932). Westman's apparatus included a hydraulic press, a hollow steel cylinder designed for an internal hydraulic pressure of 22,000 p.s.i., and two equal-length permeable pistons

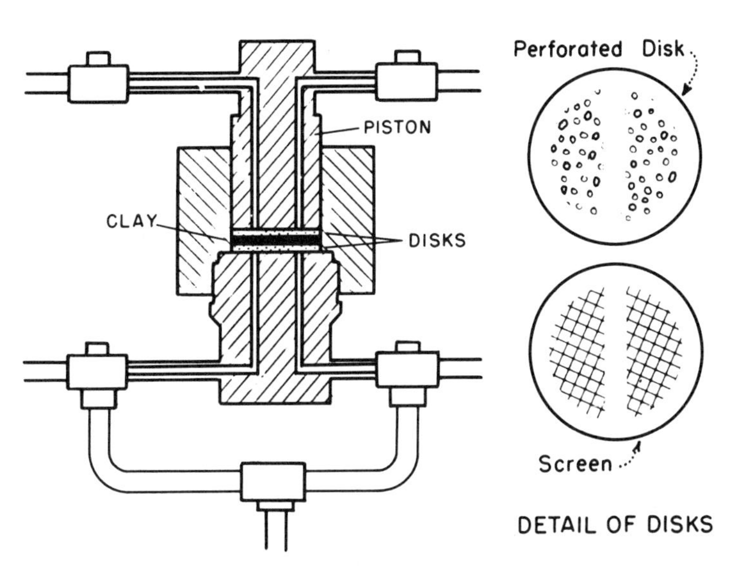

Fig.197. Details of the apparatus used by Macey (1940, fig.2, p.630) in determining the pressure–moisture-content relationship in clays and for the collection of the interstitial fluids squeezed out. Screen and perforated disk, with unperforated strip along the diameter, are also shown.

($L > D$). Each piston assembly consisted of a steel plunger, a perforated and grooved steel disk, a 40-mesh bronze screen, a filter-paper support (no.123 C.S. & S.) and one piece of no.50 Whatman filter paper. The 1¼-inch flat round steel disk contained 45 holes, 1/32 inch in diameter at 1/8-inch centers, separated by an unperforated strip ¼-inch wide along the diameter (shown in Fig.197). The grooved side of the disk was placed against the plunger so that each group of holes communicated only with one of the two conduits in the plunger. Macey's device consisted of a similar pressure cylinder that was closed at its bottom by a screw plug which took the place of Westman's lower piston. An accurately fitted piston applied an axial load from the top. Both the plug and piston contained doubly drilled conduits which led to an external system of lines through which water could be pushed downward through the clay specimen under pressure. The purpose of the hydraulic system was to remove any air in the apparatus at the start of the compaction experiment. As in Westman's apparatus, the clay sample was covered at its top and bottom with wet, finely-porous filter paper. Two drilled disks, with a series of channels (grooves) on one side, served to distribute the flow of water uniformly over the clay sample cross section. Mechanical pressure was applied by means of a lever arm and weights, in contrast to Westman's apparatus employing a hydraulic ram. Macey reported using a maximum pressure of only 617 p.s.i.

McKelvey and Milne (1962) studied the permeability of clay membranes to salt solutions under pressures up to 10,000 p.s.i. Fig.198 illustrates their confining steel cylinder which was fitted with a nylon liner. The liner was placed so that the input and output faces of the permeable pistons would be insulated electrically from each other to avoid short-circuiting the streaming potential (McKelvey and Milne, 1962, p.251). Bakelite insulating disks were inserted in the end plates to give additional electrical insulation. NaCl solutions of varying concentrations from a reservoir were pumped through a high-pressure steel

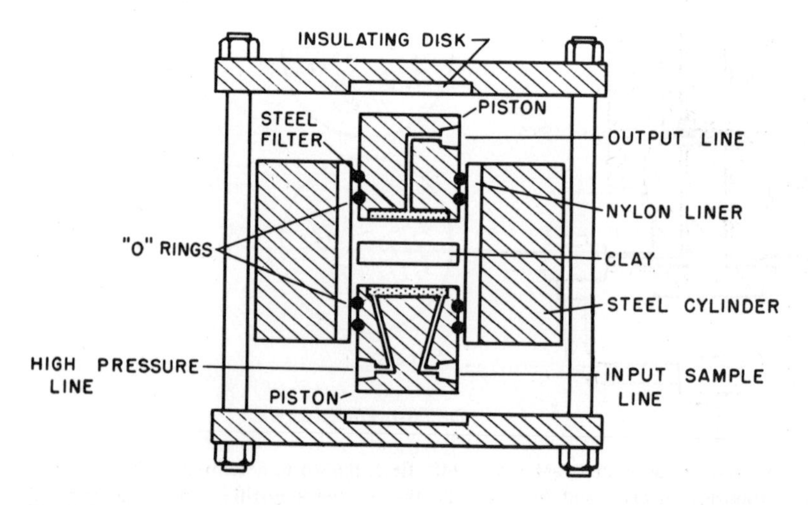

Fig.198. Schematic diagram of the McKelvey and Milne (1962, fig.2, p.251) flow cell.

tubing and forced through the clay membrane. The outflow from the sample was collected in a graduated pipette. An air-operated pump forced an oil column downward on top of the salt solution in the reservoir at pressures up to 10,000 p.s.i. There is some similarity between this apparatus and the one constructed by Westman (1932).

Von Engelhardt and Gaida (1963) described their compaction apparatus as consisting of a thick-walled hollow cylinder made from a nickel–iron–tungsten alloy resistant to corrosion by concentrated chloride solutions. As shown in Fig.199, their apparatus appears to be a consolidometer of the platform type in which the pressure is applied through the use of a hydraulic press. Load is applied to the lower steel piston *F* by the upper steel piston *G*, which is fitted on top of and around the lower piston. The clay is compressed by the lower piston, which is covered by a thin silver plate that acts as a packing against the cylinder wall. Another packing was made out of "Hydrofit" (*J*) and was positioned between the cylinder wall, lower piston, and the lower edge of the upper piston. A maximum pressure of approximately 3,200 atm (47,000 p.s.i.) was reached by using this device. The escaping interstitial fluid flowed downward through a nickel–iron–tungsten steel sieve plate *E*, which was covered with a membrane and a filter paper, and was collected for analysis. The movement of the piston and thickness of the com-

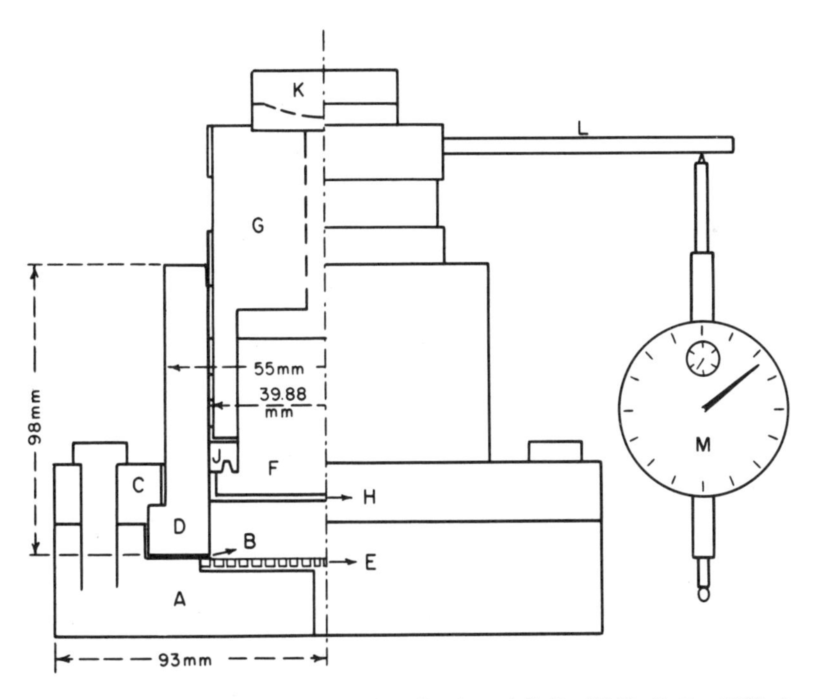

Fig.199. Schematic diagram of Von Engelhardt and Gaida (1963, fig.5, p.923) clay-compression apparatus. *A* = base; *B* = silver tightening ring; *C* = lock nut; *D* = side; *E* = sieve plate; *F* = lower steel piston; *G* = upper piston; *H* = conduit; *J* = "Hydrofit" packing; *K* = load block; *L* = lever arm; *M* = dial gauge.

pressed samples were measured with an accuracy of 2 microns, whereas pressure was measured with an accuracy of about 1 atm.

Dr. E.C. Robertson of the U.S. Geological Survey employed a consolidometer using a steel or carbide pressure cylinder in the study of carbonate sediment compaction. The maximum pressure used in these experiments was 14,500 p.s.i. on samples 1.2 cm in diameter. A large hydraulic press was used for high-pressure tests up to 290,000 p.s.i. (E.C. Robertson, personal communication, 1966).

Filter presses

In studying the chemistry of the interstitial fluids of hydrated clays and "soupy" marine sediments, the problem of extracting such fluids for chemical analysis was solved by various investigators by utilizing a consolidometer. Another method attempted was centrifuging samples at high speeds. It was shown, however, that centrifuging cannot achieve efficient separation of the fluid from the clay, because of low differential pressures. Various inexpensive filter presses for extracting interstitial water from argillaceous sediments and clays have been used by Shepard and Moore (1955), Siever (1962), Gann (1965), Hartman (1965), and Presley et al. (1967). Shepard and Moore (1955) and Siever (1962) used pressure squeezer designs based on a mechanical pressure application, whereas Hartman (1965), Gann (1965) and Presley et al. (1967) employed a gas pressure system to achieve the same results. These "squeezers" are satisfactory for obtaining fluid samples from clays at a given constant pressure, and their design derives its origin from the mud filter press that is used in the petroleum industry to determine the filtrate loss of drilling fluids. Most presses are constructed out of stainless steel, with or without chrome plating. Interstitial fluids extracted by these "squeezers" are satisfactory for analyses of alkali metals, alkaline earths and anions. These presses, however, should not be used for the determination of iron or certain other transitional elements in the extracted fluids, unless the stainless-steel sample cylinder and connecting parts are lined with an inert lining. Although equipment design criteria varied from investigator to investigator there are certain main points common to all: (1) the filter press must be reliable, (2) removal of water must be achieved in such a way that the composition is not changed, (3) the sediment sample must be large enough so that a sufficient amount of fluid can be obtained, (4) evaporation must not occur, and (5) the sample must not squeeze-by.

Shepard and Moore (1955, p.1537) mentioned the use of a filter press to obtain water samples from Recent sediments. Siever (1962) devised a modified type of drilling-fluid filter press in order to study the chemistry of interstitial fluids of modern marine sediments. Sediment was introduced into the cylinder and pressure was applied to an upper piston which forces the water out through a filter plate at the bottom of the press. The parts were constructed out of stainless steel with the exception of the piston and thrust bearing, which consist of bronze, and the neoprene O-ring and gaskets. Filter paper and a 72-mesh stainless-steel screen were employed to filter the clay. The pressure in Siever's apparatus was applied either by a mechanical manual screw device which forced the piston against the sediment (maximum pressure of 300 p.s.i.) or by a nitrogen gas-driven piston. An interesting modification was made on the filter press by surrounding the squeezer with a large cylinder containing an argon (inert) atmosphere. This was done so that the redox potential of the interstitial fluids could be determined; these are then compared with the in-situ measurements (Siever, 1962, p.330).

Gann (1965) used a commercial high-pressure gas-filter press to study the chemistry of fluids squeezed out of hydrated montmorillonite and illite clays. The Magcobar high-temperature (up to 500°F), high-pressure (up to 1,000 p.s.i.) filter press used by Gann is composed of a stainless-steel cell fitted with a yoke at its bottom and a cap at its top

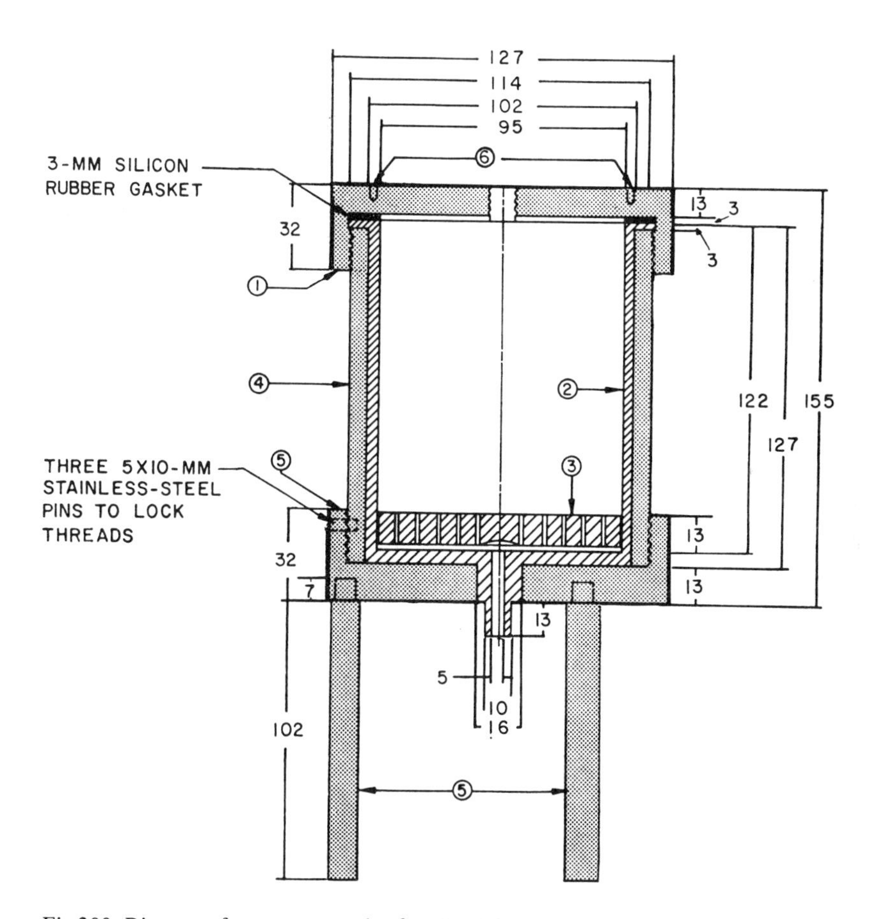

Fig.200. Diagram of a gas-pressured teflon-lined filter press. All dimensions are in mm. *1* = screw-type stainless-steel lid; *2* = inner jacket and outlet tube machined from a single piece of teflon rod; *3* = teflon filter plate with perforations; *4* = outer casing of stainless steel; *5* = stainless-steel base screwed permanently into the outer casing; *6* = two recesses in lid [tightening wrench consists of an aluminium bar (40 × 5 × 1.25 cm) with two pins]. (After Presley et al., 1967, fig.1, p.357.)

through which the gas enters the cell. The "soupy" clay is kept from running out through the yoke by filter paper placed on top of a fine-mesh screen. The filter screen is forced in position by screwing the yoke and pressure cell together. A neoprene O-ring fits between the screen and yoke to complete the seal. Nitrogen gas or air is used as the pressure medium.

Presley et al. (1967) developed a simple and efficient low-pressure (1,000 p.s.i.) gas-operated "squeezer" lined with teflon. The teflon lining insures that the fluids and sediment samples are not contaminated by iron and associated elements from the stainless steel cylinder (Fig.200). The filter press can accommodate about 300 g of sediment. The latter is placed on top of a hardened piece of filter paper, which covers the filter plate. A cap is wrench-tightened on the cylinder's top. Pressure is applied until occurrence of gas breakthrough (after 2–30 min according to Presley et al., 1967, p.357); at that time the pressure is reduced by opening a pressure-release valve. Nitrogen gas is introduced gradually at pressures that increase to a maximum of approximately 1,000 p.s.i. (70 kg/cm^2) over a period of about 10 sec. In most cases, the rate of interstitial-water expression ranges from 1 to 5 ml/min.

Linear piston apparatuses

Low-pressure squeezing, centrifuging, and leaching techniques generally prove inadequate when testing small clay samples. In the U.S.S.R. the high-pressure linear piston ($L > D$) devices have been used extensively for both consolidated rocks and unconsolidated sediments. The laboratory techniques involving such equipment developed generally independently both in the U.S.S.R. and in the United States. In 1947, Buneeva et al. squeezed interstitial fluid from Jurassic claystones containing 9% water by wet weight. Laboratory compaction and fluid extraction experiments using single-stage, dual-piston, or duplex-piston devices are usually performed at room temperature.

Kryukov and his co-workers in the U.S.S.R. are leaders in studying the compaction of sediments and the chemistry of the associated solutions squeezed out. They used several different types of compaction equipment. Kryukov's (1961) alcohol system is based on a miscible-liquid displacement process, which is very similar to the process used in reservoir engineering research. The types of alcohol that can be used in the experiment to displace the interstitial water are isopropyl alcohol (iPA), methyl alcohol (MA), normal amyl alcohol (nAA), normal butyl alcohol (nBA), secondary butyl alcohol (SBA) and tertiary butyl alcohol (TBA). It should be noted that each one of the above alcohols has varying degrees of solubility in the pore fluids depending upon their chemistry. If the sample contains both water and hydrocarbons, then the alcohol's solubility depends on the relative saturation of each phase in the sample. The principal concern is to maintain a "piston" type of displacement of the fluids by the alcohol.

With the aid of a hydraulic press, Kryukov (1961) achieved initial compression of the sample in the pressure cylinder, where a constant pressure was maintained during the

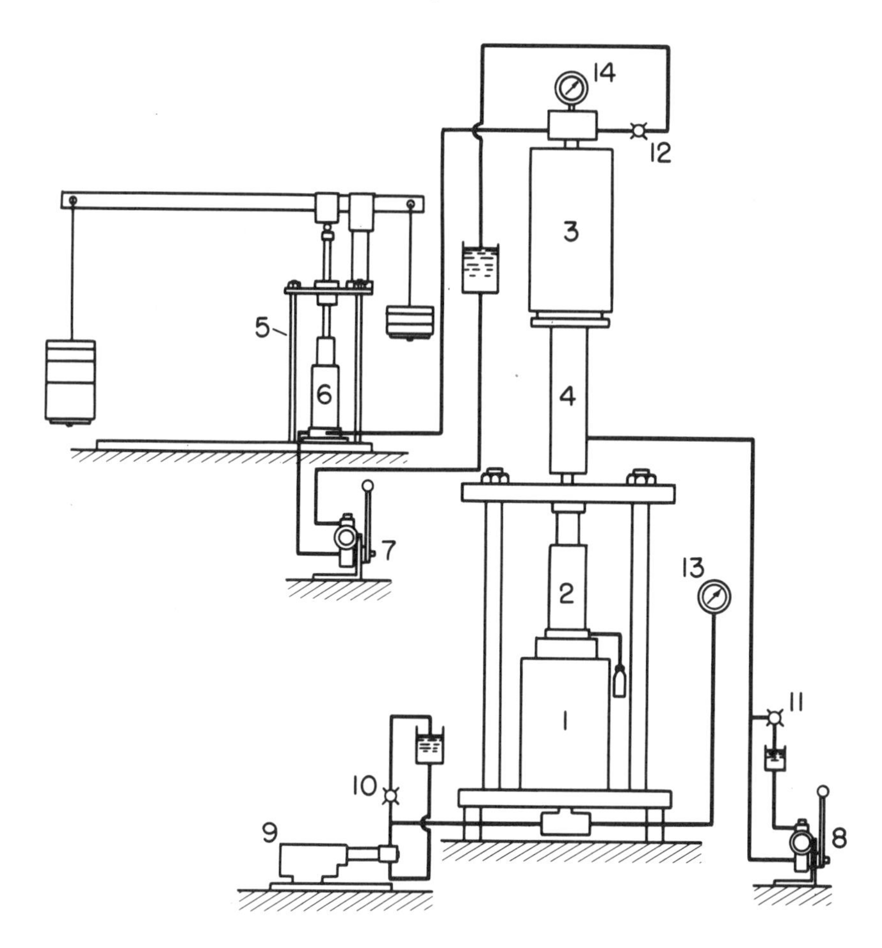

Fig.201. Schematic diagram of Kryukov's pressure system in which the interstitial solutions are replaced with alcohol. *1* = hydraulic press; *2* = sample chamber; *3* = low-pressure multiplier cylinder; *4* = high-pressure multiplier cylinder; *5* = lever press; *6* = auxiliary cylinder; *7, 8* = hand pump; *9* = pump (N-Zh-R – 1, manufactured in U.S.S.R.); *10, 11, 12* = valves; *13* = gauge (250 kg/cm^2); *14* = gauge (11 kg/cm^2). (After Kryukov, 1961, p.195.)

replacement process. The upper part of the apparatus is an intensifier consisting of a large-diameter cylinder using oil as the hydraulic fluid. Oil is supplied to the upper cylinder through a hand pump and the pressure is maintained constant with the aid of the regulating system shown in Fig.201. The regulating system consists of a lever arm exerting pressure by the use of weights on the oil in the auxiliary cylinder, which is connected in parallel with the upper large-diameter cylinder. Obviously, such a system could be automated. The alcohol is placed in the smaller-diameter multiplier cylinder and is forced into the sedimentary rock core (Fig.202A). The magnitude of pressure necessary to force the alcohol through the core must be determined experimentally. For dense sandstones

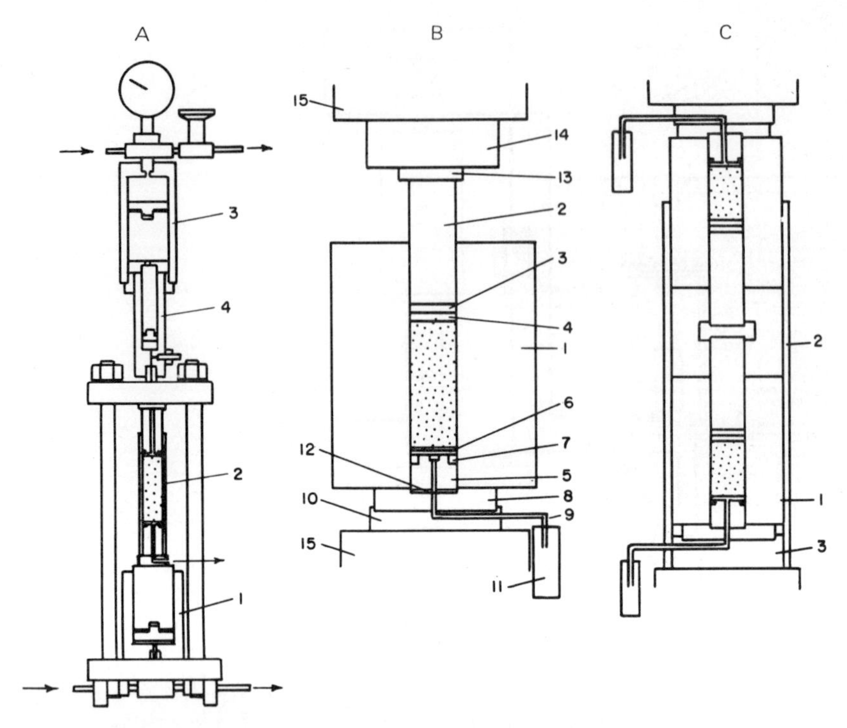

Fig.202. Various high-pressure apparatuses used by Kryukov and his associates.
A. Schematic diagram of apparatus for replacement of solutions. *1, 3* = cylinders of hydraulic presses; *2* = cylinder with sample; *4* = cylinder with alcohol. The lower and upper left inlets are for input of oil. Upper middle inlet is for introduction of alcohol. Lower middle outlet is attached to a receiver. Lower right outlet is connected to the manometer. (After Kryukov, 1961, p.194; 1964, p.14.)
B. Apparatus for squeezing out of solutions at pressures up to 10,000 kg/cm^2. *1* = cylinder; *2* = upper piston; *3* = ebonite cushion; *4* = rubber pad; *5* = lower piston; *6* = piston; *7* = rubber seal; *8* = bottom plate; *9* = steel capillary tube; *10* = support; *11* = receiver; *12* = vinyl chloride pad; *13* = soft steel plate; *14* = hardened steel cylinder; *15* = press landings. (After Kryukov, 1961, p.192.)
C. Schematic diagram of a duplex apparatus for squeezing out interstitial solutions. *1, 2* = guiding cases; *3* = base. (After Kryukov, 1961, p.193.)

having moisture contents of 5.7–10.2%, Kryukov (1961, 1964) used pressures up to 1,400 kg/cm^2 (20,000 p.s.i.) and forced the alcohol through at 420 kg/cm^2 (5,700 p.s.i.). The duration of displacement of the interstitial fluid was approximately 24 hours, whereas about three weeks were required in Kryukov's other system involving direct squeezing of the same amount of fluid (about 60%) using pressures up to 9,000 kg/cm^2.

Kryukov (1961, p.192) described a dual-piston apparatus which he used to squeeze fluids out of core samples weighing from 2 to 15 g at pressures up to 10,000 kg/cm^2 (142,200 p.s.i.) (Fig.202B). Kryukov's (1961, p.193) duplex apparatus is shown in Fig.202C. Although Kryukov was interested mainly in the chemical changes in the pore fluids with pressure, he has also experimented with the liquefaction of crystal hydrates at

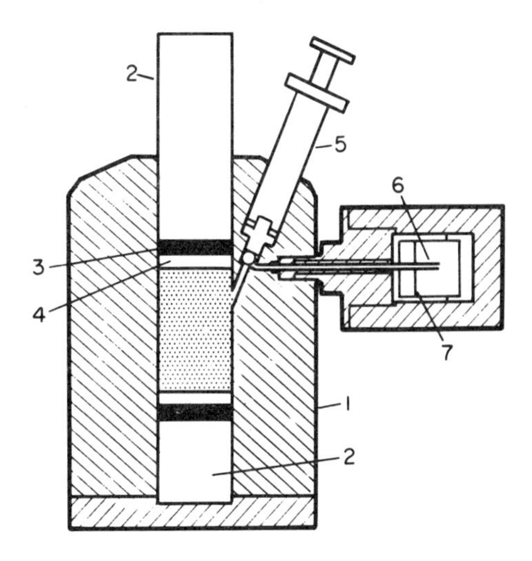

Fig.203. Schematic diagram of the high-pressure cell used by Kryukov to investigate the liquefaction of crystal hydrates at high pressures. *1* = Pressure cylinder (steel 2×13); *2* = upper and lower pistons (steel Sh×15); *3* = hard rubber washer (ebonite disk); *4* = rubber washer; *5* = valve; *6* = needle; *7* = collecting chamber. (After Kryukov, 1964, fig.12, p.32.)

high pressures and moderate temperatures. Fig.203 shows the pressure cell which he used in the latter investigations.

Manheim (1966) described a single-stage device, design of which was based on apparatus developed by P.A. Kryukov. The apparatus utilizes a commercial ram for the piston to which a machined base, with a filtering unit and fluid outlet, is fitted (Fig.204). All metal parts in the filter base are AISI-SAE 303 stainless steel, and the pressure cylinder and piston are made from chrome-plated steel. Teflon and rubber disks are installed just below the piston to prevent fluid by-passage upward around the piston when pressure is applied. Disposable plastic syringes are inserted directly into the base of the apparatus to receive the expelled fluid. The narrow effluent hole was bored to permit the fitting of the standard "Luer" taper of the syringe nose. The maximum load exerted on the sample in the small squeezer was about 22,000 p.s.i., using a 10-ton laboratory press. In the larger squeezer, however, a 20,000-lb. load will apply a pressure of only about 5,000 p.s.i. The filtering unit consists of a perforated plate and a stainless-steel screen. A porous (sintered) metal plate may be used to replace both screen and perforated plate.

Kazintsev (1968, pp.188–189) described a hermetically closed apparatus which enables heating the samples to temperatures above 100°C without evaporation of squeezed-out interstitial solutions (Fig.205). The receiver (syringe) is fastened with springs, which are previously regulated to expand on application of a definite load.

Van der Knaap and Van der Vlis (1967) developed a uniaxial compaction apparatus for use at high pressures (Fig.206). In this apparatus, the sample is enclosed in a cylinder,

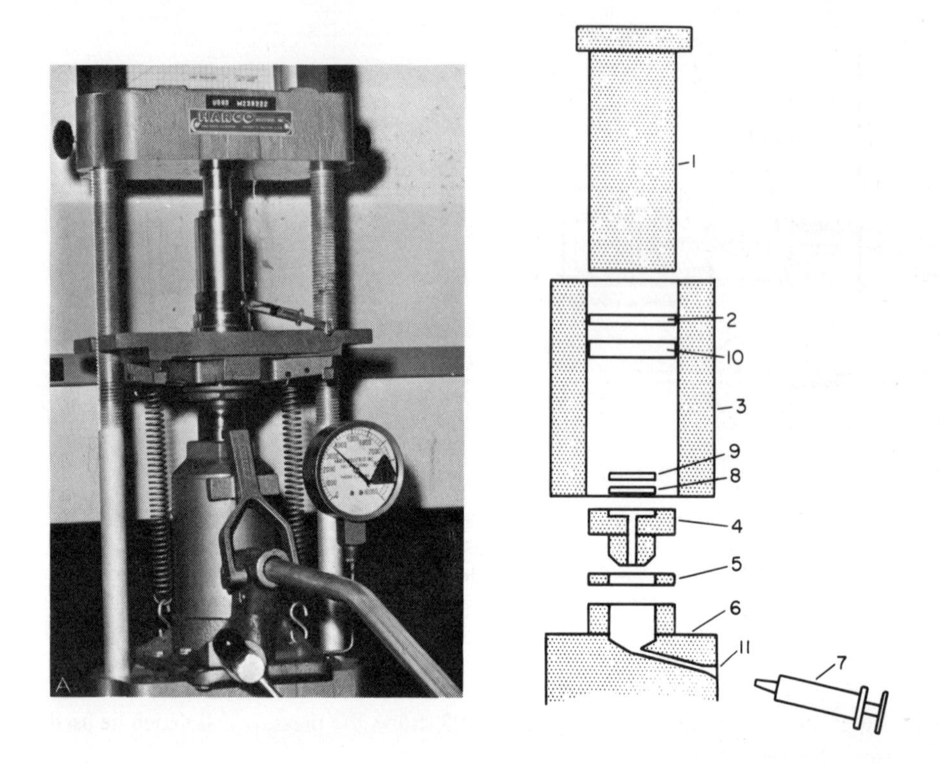

Fig.204. Photograph (A) and schematic diagram (B) showing components of a single-stage piston device of Manheim (1966). *1* = ram; *2* = teflon disk; *3* = cylinder; *4* = filter holder; *5* = rubber (Neoprene) washer, 3/16-inch thick; *6* = base; *7* = syringe; *8* = stainless steel wire-screen disk, 1/16-inch thick; *9* = perforated plate, 1/16-inch thick (filter paper support); *10* = rubber (Neoprene) disk; *11* = effluent passage reamed to fit nose of syringe. (Photo A, courtesy of K.T. Manheim, 1971.)

closed off at top and bottom by drainage disks made of porous steel. A vertical pressure is applied by means of a piston and fluids are drained via tubes. The walls of the cylinder are thick in order to keep the lateral expansion to a minimum; thus, a condition of no horizontal displacement is closely approached. The pore pressure is kept at atmospheric level while gradually raising the pressure of the piston. Variations in thickness of the sample are measured with dial gauges and differential transformers, which enable detection of length changes of fractions of a micron.

Chilingar et al. (1963) described a very high-pressure (up to 500,000 p.s.i.) compaction apparatus (Fig.207), which consists of three main parts: (1) The heavy steel frame built from AISI-SAE 1020, or 4130 steel. (2) A complete high-pressure plate consisting of three separate tapered cylinders: (a) hardened S-2, M-4, T-15 tool-steel pressure cylinder, (b) an inner supporting ring machined out of SAE E 4340, and (c) an outer ring built from AISI-SAE 1020 steel. These cylinders were pressed into one another at a 1.5%

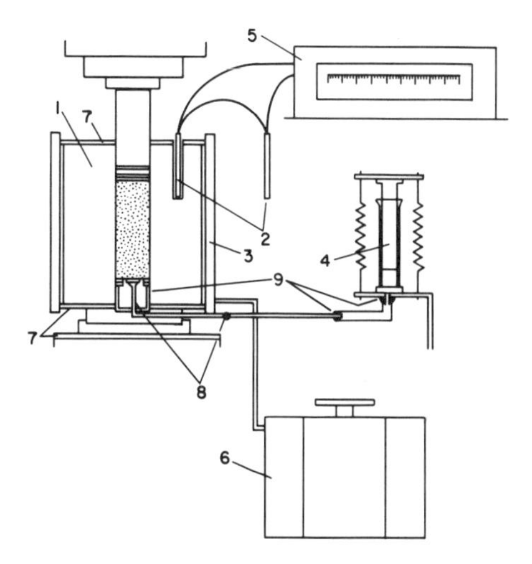

Fig.205. High-temperature compaction apparatus for squeezing out interstitial solutions. *1* = molding press; *2* = thermocouple; *3* = oven; *4* = receiver (syringe); *5* = temperature recorder; *6* = voltage regulator; *7* = asbestos; *8* = tin; *9* = epoxy resin. (After Kazintsev, 1968, fig.3, p.189.)

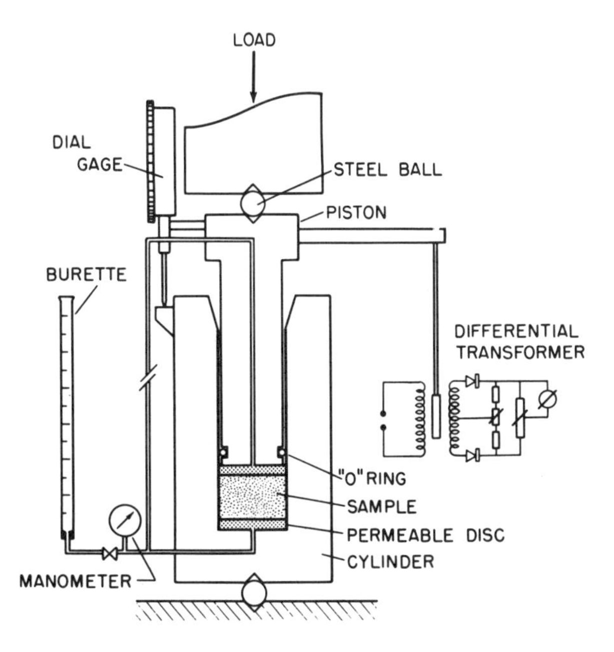

Fig.206. Compression-test apparatus of Van der Knaap and Van der Vlis (1967, fig.3, p.88).

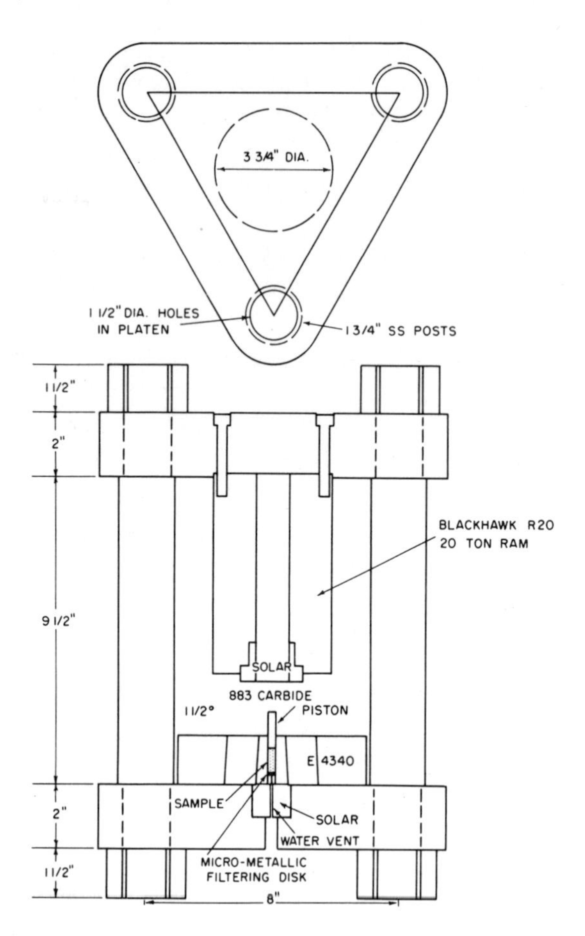

Fig.207. Schematic diagram of the 500,000-p.s.i. compaction unit. (After Chilingar et al., 1963a, fig.1, p.1043. Courtesy of Geological Society of America.)

interference fit. (3) A 20-ton "Blackhawk" RC-20 "Holloram" fastened to the upper platen of the frame supplies the load to the upper piston.

The sample cavity is honed to within 0.0002-inch fit of a standard 0.25-inch carbide cylinder (Carboloy 883), which serves as a piston. The sample is placed between the deforming piston and a stationary hardened steel plug, which has a small hole to allow egress of the water expelled during the experiment. This plug and the high-pressure cylinder rest on a shouldered, hardened, steel stock which is seated in the lowermost of the two 2-inch thick, triangular supporting steel platens. Pressure on the ram is measured by a 10,000-p.s.i. U.S. "Supergauge" having an accuracy of ± 0.5% of the full-scale reading.

Patterson (1970) described a 10-kb, 1,000°C apparatus for measuring the stress–strain relationship of consolidated rocks and other competent materials under high pressure and temperature. Argon was used as the pressure medium. The pressure vessel is mounted with its axis vertical because it is easier to compensate for the influence of convection on the temperature distribution in the furnace. The cooling jacket is a steel sleeve, grooved on the inside, which circulates coolant in contact with the steel vessel. The coolant may consist of water plus a corrosion inhibitor or of a light oil, which circulates through a heat exchanger. The specimens are jacketed in annealed copper tubing having a wall thickness of 0.010 inch, sealed to the piston by force-fitted rings. Most of the high-temperature work has been done on specimens 1 cm in diameter and 2 cm long.

High-pressure dual-piston compaction apparatuses. The various previously described experimental autoclave devices used in the compaction of hydrated clays and sediments, whether to study the physical changes that take place in these materials under pressure or the chemistry of the fluids expelled, have one common characteristic: the expulsion of the fluid is unidirectional, except in the case of consolidometers. The equipment described by Chilingar and Knight (1969) and Rieke et al. (1969) utilizes the consolidometer principle of two-directional fluid flow, while using a dual, linear piston pressure application (Figs.208–210).

As shown in Fig.208, the high-pressure apparatus consists of four principal parts: (1) top cap normally constructed from AISI-SAE 4340 or Armco's 17-4 PH steel; (2) hollow, thick-walled high-pressure cylinder machined out of AISI-SAE 4340, 320, 17-5 PH, or 300 CVM steel; (3) tension member which is machined out of AISI-SAE 4130, 4140, or 1021 steel; and (4) a 20-ton Blackhawk ram actuated by a hand-operated hydraulic jack (Enerpac pump P-39), which can deliver line pressures up to 10,000 p.s.i.

Inasmuch as this equipment proved to be very reliable at pressures up to 150,000 p.s.i., a detailed procedure for assembling and operating it is presented here in an outline form as follows:

(1) In assembling the device for a high-pressure experiment, the upper cap is screwed off, the pressure cylinder is removed and all parts are thoroughly cleaned with cotton. The walls of the cylinder are then lubricated by putting a small amount of noncorrosive hydraulic jack oil mixed with graphite or "Molykote"* Type G (a grease composed of MoS_2 and mineral oil) on the end of a cleaning rod and using it as a swab. In case any of the lubricating compound is rubbed off while introducing clay or cotton, it should be replaced with an additional amount of compound.

(2) The cylinder is then turned upside down and the cold-rolled steel 45° packing ring, drilled, hardened steel disk, and the lower piston are inserted into the borehole with an aid of a cleaning rod (a small diameter piston) and, if necessary, hammered with a rubber mallet.

(3) Then the Braun stainless-steel, micrometallic filtering disk having an average pore size of 5 μ is introduced on top of the hardened steel disk. These filtering disks warp with

* Trademark: The Alpha-Molykote Corp., Stamford, Connecticut.

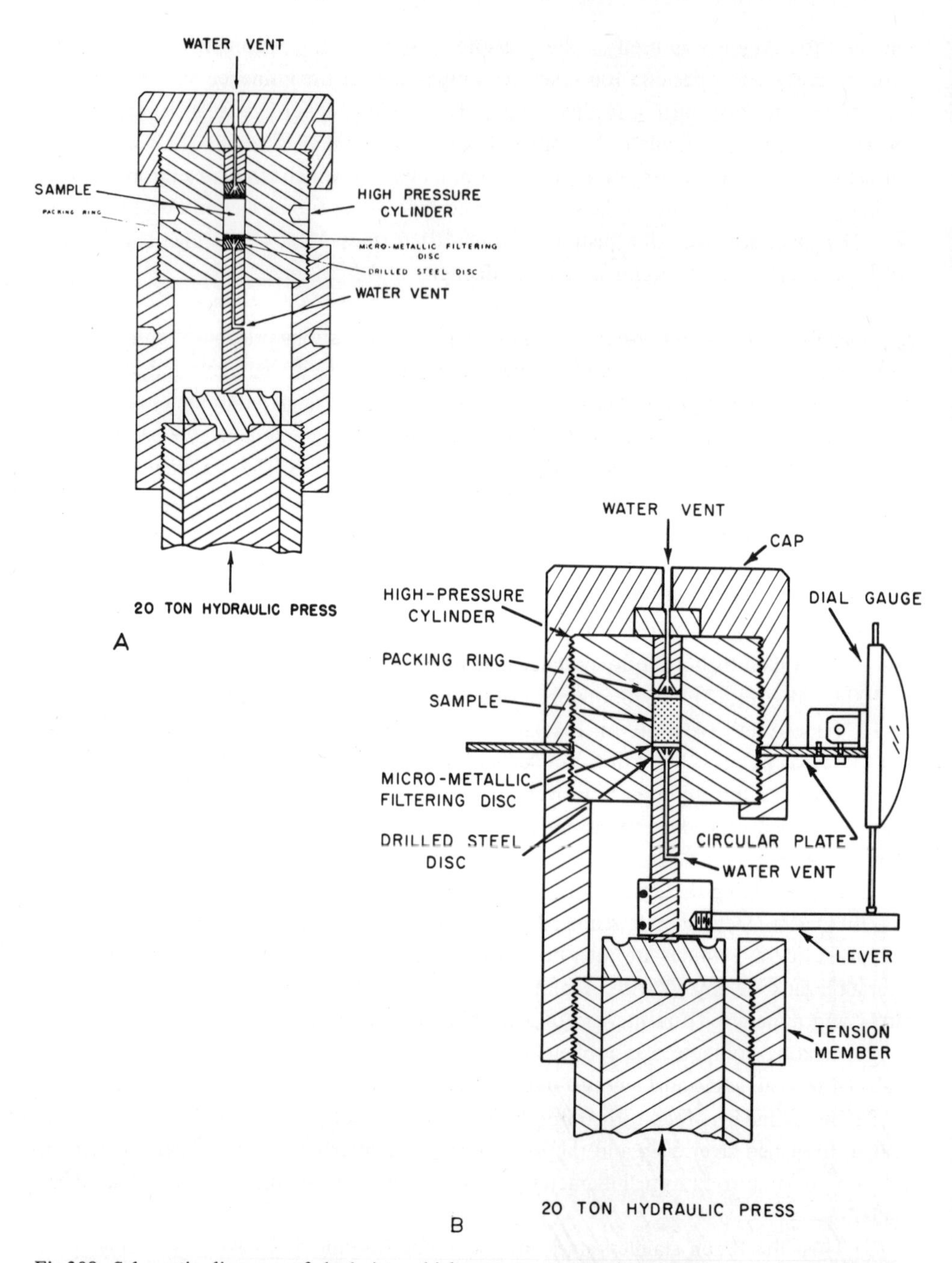

Fig.208. Schematic diagrams of dual-piston high-pressure compaction apparatus. A. Standard equipment. B. Equipment with attachment (Ames dial gage) to measure changes in height of clay components. (After Chilingar and Knight, 1960, fig.1, p.103, and Rieke et al., 1969, fig.1, p.819. Courtesy of American Association of Petroleum Geologists.)

Fig.209. High-pressure autoclave devices used at the Petroleum Engineering Department, University of Southern California. A = two dual-piston compaction units; B = 500,000-p.s.i. compaction unit (see Fig.207); C = blackhawk hand pump.

Fig.210. Extrusion unit used by the writers to remove sample from high-pressure chamber.

use; therefore, if they are reused, the concave side should face the clay sample. It is preferable, however, to replace them after each run. Disks commonly crack when subjected to pressure due to strain-hardening of the sintered metal. A small wad of cotton (about ½ inch in diameter) is packed down with the cleaning rod on top of the filtering disk, and a 2- or 4-cm^3 clay sample is introduced above it. Another (a) cotton wad, (b) filtering disk, (c) ring, (d) steel disk, and (e) upper piston are placed above the clay sample in the order indicated. The cotton should not contain lubricating compound which makes it impermeable to water. An attempt was made to use sand instead of filtering disks; however, the sand filters crushed at pressures around 3,600 p.s.i. In addition, clay had a tendency to squeeze out into the sand on application of pressure.

(4) The thick-walled hollow cylinder is then placed on the tension member so that the lower piston engages against the ram. The cylinder is turned until it is locked in place. Subsequently, the cap is replaced.

(5) Upon closing the hand-operated jack valve, the pressure is slowly applied to the sample. In case of some clays, such as montmorillonite hydrated in distilled water or attapulgite hydrated in sea water, by-passing (squeezing-out through the water-vent hole) presents a problem. This is owing in part to a poor fit of the micrometallic filtering disk, especially if these disks are being reused. A considerable portion of the free water, however, can be pressed out at low pressures (up to 500 p.s.i.) without by-passage. Thus, the subsequent increase of pressure to a desired value does not cause any squeezing-out of the clays.

(6) After a certain time interval (the tests at each pressure are repeated until attainment of equilibrium) the pressure is slowly decreased by opening the jack valve slightly in order to prevent bouncing of the gauge needle. It is imperative to remove the excess water present in the water-vent opening of the piston with a syringe, because some water may be sucked back into the clay sample upon the release of pressure.

(7) Upon removing and weighing the sample (in a closed weighing bottle), it is cut into small pieces and placed into an electric vacuum oven with a hydraulic thermostat which controls the temperature. The samples are then dried to a constant weight at a desired temperature (95°–105°C). Possibly, several different drying temperatures should be used.

(8) Both rings should be replaced after each run, because the worn rings cause galling and scouring inside the cylinder.

Cebell and Chilingar (1972) presented detailed procedure for measuring the height of clay compacts at different pressures. The Ames gauge provides sample height to an accuracy of 0.0001 inch (see Fig.208B).

Triaxial and hydrostatic compaction apparatuses

In 1953, Hall employed a hydrostatic pressure cell to study pore-volume changes in sandstone and limestone cores (Fig.211). Since then, a number of authors (Fatt, 1953, 1958b; McLatchie et al., 1958; Wyble, 1958; Glanville, 1959; Dobrynin, 1962; Gray et

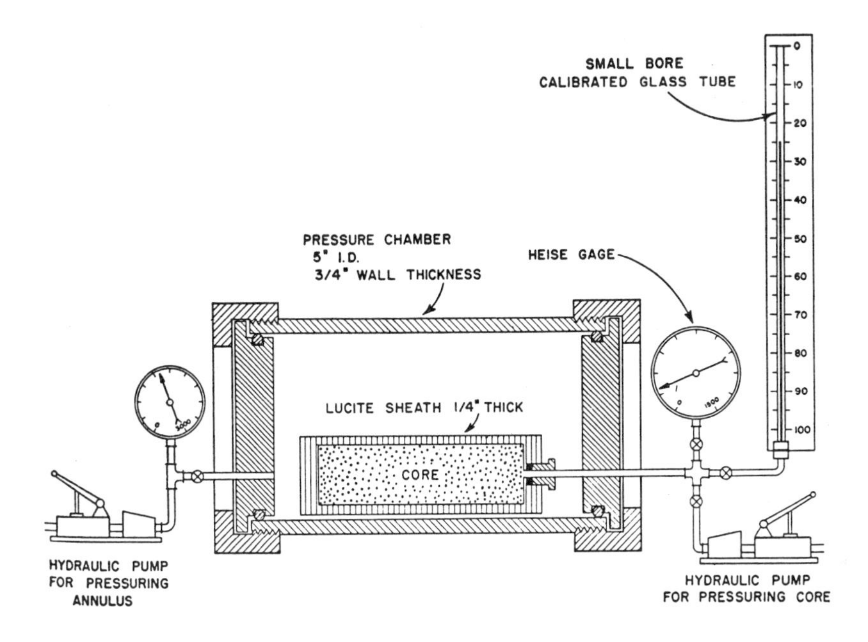

Fig.211. Diagram of equipment used for rock compressibility tests by Hall (1953, fig.1, p.309).

al., 1963) have considered the effects of overburden pressure changes, as simulated in hydrostatic high-pressure cells, on reservoir-rock porosity, absolute and relative permeabilities and resistivity. Serdengecti and Boozer (1961) also described a triaxial cell which was used to study the effects of strain and temperature on the behavior of rocks under triaxial compression.

Sawabini et al. (1971) described a triaxial compaction apparatus which was designed to accommodate both clay and sand samples composed in the laboratory and those taken from oil-well cores.

Triaxial loading in their apparatus is hydraulically applied to a sample mounted in a closely-fitted rubber sleeve. The sample is attached to a filter and a drain system. Internal fluid pressure can be reduced and fluids produced from the sample via the drain.

Schematic diagrams of their compaction apparatus are presented in Figs.212 and 213. Hydraulic fluid (Pydraul 60, a product of the Monsanto Company) is used to apply a uniform pressure on all sides of the cylindrical rubber-sleeved sample with the aid of a hydraulic hand pump. The rubber sleeve (ethylene-propylene tubing, 2-inch I.D. × 1/16-inch wall thickness, 60K5 type) is a product of the West American Rubber Company, Orange, California. The equipment consists of the following main parts: (1) pressure vessel, (2) electric oven, (3) control section, and (4) swing arm and stand. Fig.213 shows the pressure vessel which houses the sleeved core sample and the hydraulic fluid which surrounds the sample. The lower bushing, which is fastened to the pressure vessel, contains the pressure assembly unit for measuring the pore pressure of the sample being compacted.

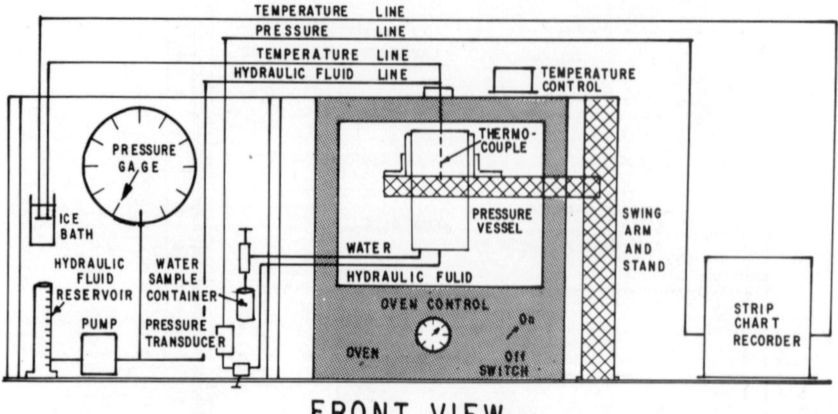

Fig.212. Schematic diagram of assembled triaxial compaction apparatus. (After Sawabini et al., 1971.)

The upper bushing was designed as the upper end plug for the vessel and contains conduits for both the temperature thermocouple element and the hydraulic fluid. A filter disk holder is placed below the sleeved sample and holds the micrometallic filter (20-μ, Pall Corp., Long Island, N.Y.). The electric oven is large enough to house the pressure vessel, its associated parts, and connections. The control section consists of (1) a number of devices to measure, control and adjust the liquid pressure applied to the sleeved sample, (2) the reference ice bath and temperature-variable transformer, and (3) the fluid collection system. The swing arm and stand is a mechanical device to lower, raise, move-in and roll-out the pressure vessel in and out of the electric oven. A circular shield is placed around the pressure vessel and a steel plate is placed on top of the electric oven for safety purposes.

The compaction apparatus described by Sawabini et al. (1971) was designed to accomodate an internal pressure inside the pressure vessel of up to 40,000 p.s.i. and a temperature of up to 212°F. The main pressure vessel and the end glands were made from AISI E-4340 alloy steel (3-way upset forging) which was normalized and tempered to best machining condition to a hardness of approx. Rockwell 30. Mar quenching (salt bath at 1,000°F) was used in heat treatment to reduce the thermal shocking. After machining to the desired dimensions, it was heat treated to Rockwell 38 to develop an ultimate tensile strength of 170 kp.s.i. The pressure vessel was then subjected to ultrasonic and magnetic inspection to detect any material flaws. After that, the vessel's bore was honed and polished.

E-4340 steel was selected because it met the required design strength and is relatively inexpensive and highly reliable. The high temperature and corrosion limitations were not critical in the pressure-vessel design. The calculated safety design factor was 1.66.

The end bushings were made of 15-5PH stainless steel. After machining, the steel was heat-treated to a condition of H-1025 to develop an ultimate tensile strength of

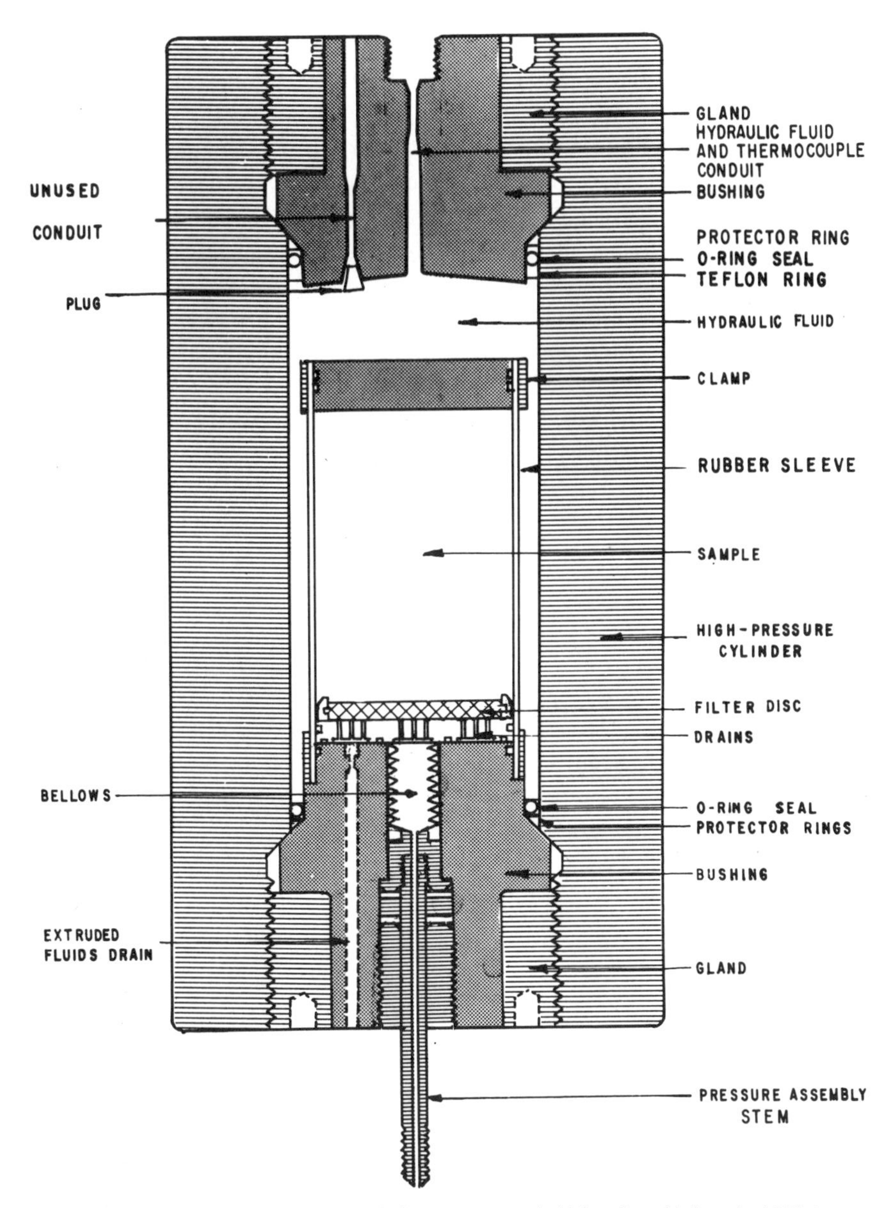

Fig.213. Schematic diagram of assembled pressure vessel. (After Sawabini et al., 1971.)

155 kp.s.i. The 15-5PH stainless steel was selected because of its resistance to corrosion, high strength and moderate ductibility properties: ultimate tensile strength = 170,000 p.s.i.; 0.2% tensile offset = 155,000 p.s.i.; reduction of area = 58%; and elongation = 16%. The pressure vessel has an elastic strength of 67,000 p.s.i. (internal pressure).

The 15-5PH stainless steel was heat-treated to 1,025°F for 4 hours and then allowed to cool in an atmospheric environment. This gave rise to the following properties: ultimate tensile strength = 155,000 p.s.i.; 0.2% tensile offset = 145,000 p.s.i.; reduction of area = 45%; and elongation = 12%.

In testing oil-well core samples, sections ranging from 4 to 12 inches in length are cut from the original 4 inches in diameter long cores and preserved for future testing in the laboratory by wrapping in aluminium foil and dipping them in a small tank containing liquid plastic material. An 1/8-inch thick coat of clear plastic is formed around the sample which prevents evaporation losses. Later, when the sample is ready to be tested, a smaller core (1 7/8 inch in diameter and 2–3 inch long) is cut from the preserved sample with a diamond-head core drill. Then, the core sample is saturated under vacuum with formation water for 24 hours.

Bulk and/or pore volume are measured prior to placing the core sample inside the rubber sleeve (Amyx et al., 1960). Then the sleeved sample and its associated parts along with the pressure assembly unit are inserted and fastened into the pressure vessel.

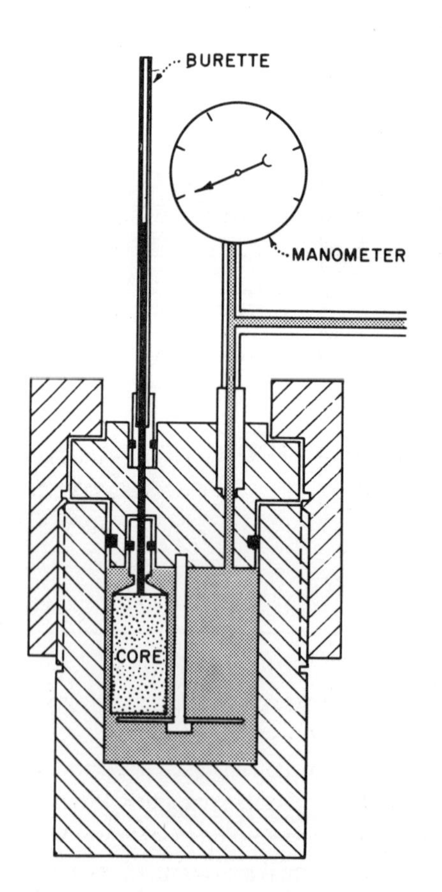

Fig.214. Hydrostatic compaction cell. Burette, 0.1 cm^3 per 6 cm. (In: Teeuw, 1970, fig.3.)

Design provisions were made to utilize a non-petroleum hydraulic fluid as the driving force to create a liquid pressure of up to 40,000 p.s.i., which acts on the sleeved sample from all directions. After pouring the amount of fluid necessary to cover the sleeved sample, the upper bushing and gland along with the thermocouple element are inserted and fastened to the pressure vessel.

The closed pressure vessel is then moved and placed inside the electric oven, in which the temperature can be varied from room temperature to 212°F. After connecting the appropriate hydraulic fluid and extruded-solution lines, the sample is compacted at various overburden pressures and temperature ranges. The volumes of hydraulic fluid added and the volumes of pore fluid expelled are recorded at various applied pressures.

Heuer et al. (1965) described an excellent triaxial apparatus for measuring the compressibility of porous materials. The confining fluid pressure is incrementally increased by steps, while the internal fluid pressure is maintained essentially constant. The measured volume of internal fluid, which is displaced from the sample after each incremental increase in confining fluid pressure, is a direct measure of the net change in pore space. This apparatus functions substantially automatically to maintain a constant pressure in the measuring system.

Teeuw (1970) described a well-designed compaction cell for measuring the pore compressibility of consolidated rocks (Fig.214). A thin jacket (elastometer or metal foil) is slipped around the sample and, after being pressed around the core, fits it tightly. There is room for four cores, each having a length of 50 mm and a diameter of 25 mm. Teeuw (1970) also described a triaxial compaction cell in which the vertical and lateral stresses can be varied independently.

SOME SALIENT FEATURES OF CALIBRATING UNIAXIAL COMPACTION APPARATUSES

As pointed out by Th. Rosenqvist (p.206) and R.M. Quigley (p.212) in the *Proceedings of the International Clay Conference in Tokyo, Japan* (1969, vol.2), the generation of high pressures in clays on using uniaxial-compaction apparatus involves uncertainties in the magnitude of the pressure produced. These uncertainties arise from the modification of the applied load by forces owing to internal friction and maintaining the gaskets in an expanded position against the borehole wall. Piston friction against the pressure cylinder wall has several possible sources: (1) misfit of piston and borehole; (2) expansion and deformation of the piston and wall under pressure; (3) type and shape of the internal parts; (4) degree of piston advance into the borehole; and (5) surface conditions of cylinder and piston. Accurate pressure calibrations of such a high-pressure system on an absolute basis is difficult and requires either the elimination of these frictional forces or an estimate of their magnitude. Corrections should also be made for clay compact rebound and compression of steel pistons in height-versus-pressure studies (Cebell and Chilingarian, 1972).

Inasmuch as many investigators neglect the side friction in reporting their results, the salient features of calibrating the compaction apparatuses are presented here.

Calibration of high-pressure linear piston devices

There are several methods by which the true sample pressure may be experimentally determined in a linear-piston device.

In any type of linear-piston apparatus, pressure is generated by a piston of circular cross section moving in a borehole and transmitting a load quasi-hydrostatically to the sample. In some cases, this pressure is transmitted to the sample via solid-enveloping media of low shear strength, i.e., silver chloride, talc, etc., as in the case of the multiple-piston devices. Inasmuch as these solids are not perfectly "hydrostatic" (namely, transmit pressure equally in all directions), it is difficult to determine the pressure at points within the solid material.

Pressure calibration of high-pressure devices on an accurate absolute basis is difficult. The primary cause of this difficulty is that part of the piston's thrust force (load) is used in maintaining gaskets in an expanded position as well as the internal friction of the pressure transmitting material (Decker, 1965, p.157). One can simply calculate a pressure by knowing the load and the dimensions of the piston's cross-sectional area; but inasmuch as the over-all force/area ratio is not the true pressure on the piston face, or in the center of the pressure chamber, other techniques of pressure determination must be adopted. Very high-pressure apparatuses (> 200,000 p.s.i.) are commonly calibrated by noting the ram loading at which certain isomorphic transitions (evidenced by electric or volumetric discontinuities) occur in certain metals such as barium, bismuth, cesium, tellurium and thallium (Table XLIX).

TABLE XLIX

Phase transitions in barium, bismuth, cesium, tellurium and thallium. Experimental data for solid–solid phase transition points in various pure metals at high pressures. Based both on the studies of volumetric and electric discontinuities in the metals (Kennedy and La Mori, 1962; Haygarth et al., 1967)

Metal	*Pressure (mean value in bars * at 25°C)*
Barium I–II	55,000 ± 500
Bismuth I–II	25,405 ± 80
Bismuth II–III	26,962 ± 195
Cesium I–II	22,275 ± 600
Cesium II–III	41,962 ± 1,000
Tellurium I–II	40,160 ± 3,600
Thallium II–III	36,690 ± 100

* To change bars to p.s.i. multiply by 14.5038.

There are several other ways in which the true sample pressure may be experimentally determined. A load cell or a solid which acts hydrostatically also can be utilized. In either situation, corrections must be made for the frictional effects in the pressure system. When these frictional effects are taken into consideration then the linear-piston device becomes an absolute pressure device. One can also use X-ray diffraction in measuring lattice changes within the solid in response to pressure (Decker, 1965). Maximum pressure attainable in some high-pressure apparatuses (e.g., Fig.208) depends on the elastic properties of the hollow, thick-walled cylinder and associated parts, whereas the maximum attainable pressure in others (e.g., Fig.207) depends mainly on the crushing strength of the carbide pistons.

In the calibration of the 150,000-p.s.i. dual-piston high-pressure apparatus, the force at the top end of the upper piston (Fig.208) can be measured by a Baldwin SR-4 load cell having an accuracy of 0.1% (Chilingar and Knight, 1960). Such load cells are electronic transducers which translate changes in force or weight into voltage changes. The change in voltage produces in the X-Y recorder a repeatable deflection which is calibrated directly in terms of the load applied to the cell. Series of force measurements can be taken and calculated to the stresses at the piston face. Then these measurements should be compared with similar stresses at the bottom of the lower piston as calculated from the product of the gauge pressure and the ram/piston area ratio. Chilingar and Knight (1960) used an oil-pressure gauge (U.S. "Supergauge") with an accuracy of ± 0.5% of full-scale reading. The oil-pressure gauges should always be calibrated against a certified pressure standard (e.g., National Bureau of Standards in U.S.A.).

High-pressure apparatuses can also be calibrated by using silver chloride. Silver chloride has been used as a pressure-transmitting solid in many high-pressure experiments (Tydings and Giardini, 1963). Hall (1958) stated that silver chloride produces a more homogeneous pressure than does pyrophyllite or talc. "Baker" reagent-grade material can be used in the preparation of pills by cold-pressing the silver chloride up to 28,000 p.s.i. ($\approx$ 2 kb) in a steel pill press (e.g., Fig.210).

All the internal parts of the media, which transmit the pressure to the silver chloride pill, have inherent strength of their own. Consequently, the pressure on the sample is non-hydrostatic and also not equal to the force per unit area on the face of the piston. The silver chloride inside the high-pressure cylinder when exposed to the force of the oil-hardened piston does not transmit this force uniformly. This situation is termed a quasi-hydrostatic condition. The pill pressure is not the same as the gauge pressure or the pressure calculated on the piston's face.

When the piston is advancing in the borehole during the compressional cycle, both friction and shear strength of the silver chloride cause the pressure in this medium to be less than the calculated pressure. At the end of the run, with the piston retreating from the borehole during the decompression cycle, both friction and shear strength give rise to a negative correction. It is often stated that friction in hydrostatic rams is "symmetrical", i.e., it is the same with both increasing and decreasing pressure (Tydings and Giardini,

1963, p.204). Kennedy and La Mori (1962) confirmed this finding. The corrected oil-gauge pressure is simply expressed as:

$$p_{gt} = \frac{p_c + p_d}{2} \tag{8-1}$$

where p_{gt} is the true gauge pressure corrected for the double-valued friction, p_c is the calculated pressure on the compression stroke (increasing force), and p_d is the corresponding pressure from the decompression stroke (decreasing force). It can be solved graphically by plotting a line half-way between the compressional and the decompressional curves (Fig.215A). The mean of the upstroke and downstroke hydraulic-fluid pressures (eq.8-1) and/or their difference is termed the double-valued friction.

Pressure on the silver-chloride sample must be applied slowly by advancing the piston, which is actuated by a hydraulic ram. The hydraulic fluid pressure actuating the ram is read on the pressure gauge. Pressure on the silver-chloride sample is increased by predetermined increments, until the maximum desired gauge pressure is reached. This completes the compression stroke, after which the pressure in the chamber is reduced by bleeding fluid from the ram stepwise through a release valve on the jack. Oil pressure is recorded at the predetermined intervals on both the compression and decompression strokes with a time interval of 10 min between pressure readings. The pressure is raised after attaining equilibrium at each point. Pressure on the silver chloride is released in increments, with reproducibility at each predetermined pressure point being verified by repeated runs; a graph is then drawn such as the one shown in Fig.215A. All the calibration runs must be repeated several times to determine any systematic variation in the piston deflection. Reproducibility is important to insure proper instrument calibration. The advance of the piston in the borehole can be measured by an Ames indicator dial gauge and lever-arm assembly having a precision of 0.0001 inch (see Fig.208B). Piston displacements are read from this gauge, which is sensitive to 0.00001 inch by estimation between the 0.0001-inch marks.

The raw data is given in the form of piston-displacement versus gauge-pressure curves. Fig.215B shows the results for a typical upstroke cycle through 5,000 p.s.i.g. for a 17-5 PH steel cylinder. The double-valued friction was calculated for every run and several runs were averaged together. The values of the true sample pressure were found by multiplying the true gauge pressure by the pressure factor, which is the square of the ratio of the diameter of the hydraulic ram piston to the effective diameter of the piston at a given pressure. The effective piston diameter was measured before and after the piston was used in the calibration experiment. Due to the elastic properties of the alloy, of which the piston is composed, there was no measurable difference between the initial and final readings (Rieke, 1970, p.131).

The hysteresis loop as illustrated in Figs.215A and 216A probably can be attributed to the following causes (I. Getting, personal communication, 1968): (1) At low pressures

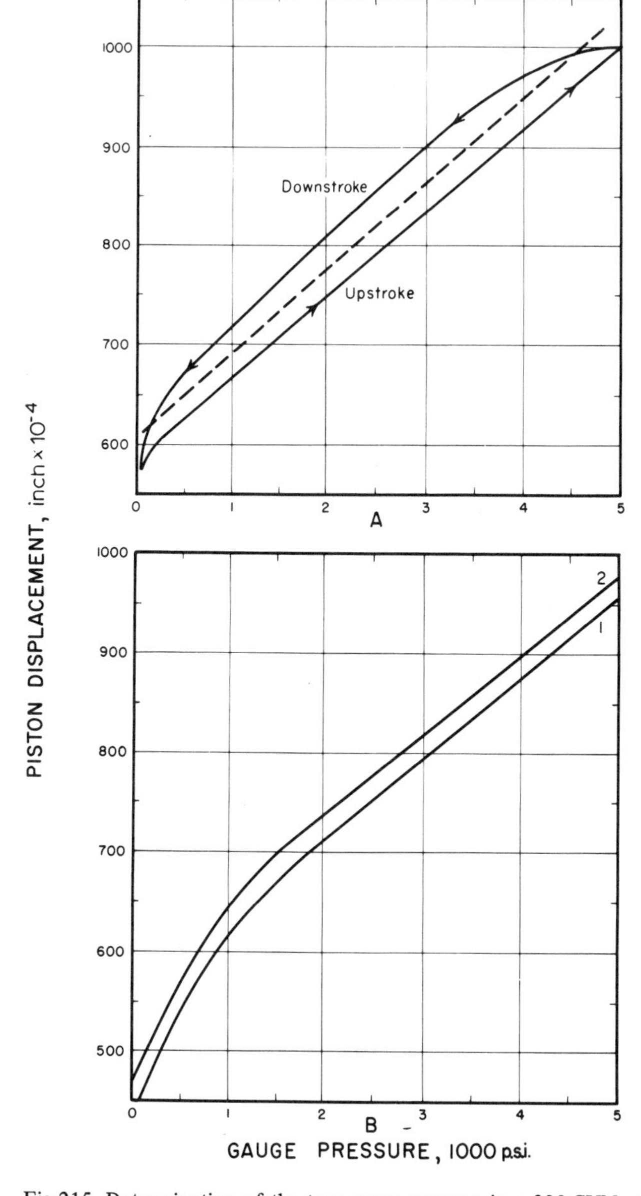

Fig.215. Determination of the true gauge pressure in a 300 CVM steel pressure cylinder and creep in silver chloride at 5,000 p.s.i. (After Rieke, 1970, p.128.)

A. Typical hysteresis curve for a 300 CVM steel (Rockwell hardness *c* scale value $Rc = 50$) high-pressure cylinder. The mean of the linear portion of the upstroke and downstroke curves is represented by the dashed line.

B. Curve *1* is the virgin upstroke curve of a silver chloride sample that has not been subjected to high pressure. Curve *2* is a composite of 3 runs after the silver chloride sample was subjected to 5,000 p.s.i.g. for 24 hours. Pressure cylinder was built from 17-5 PH steel.

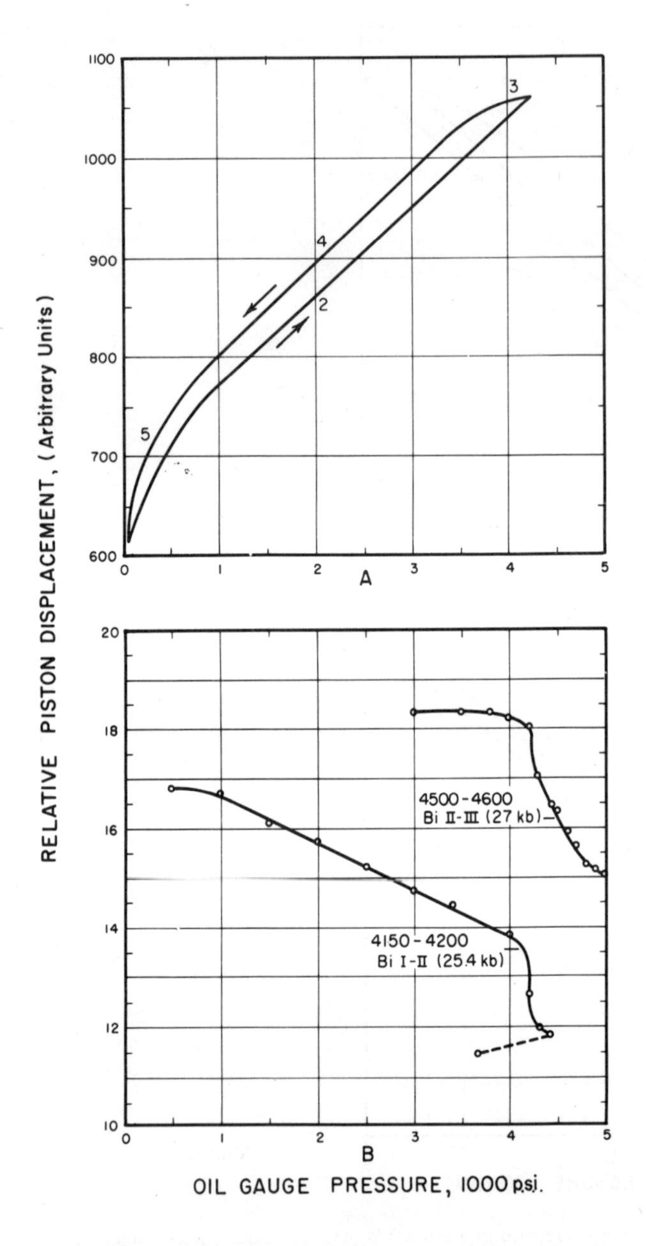

Fig.216. Piston displacement in various high-pressure cylinders. (After Rieke, 1970, p.133.)
A. Hysteresis loop of piston displacement in a 150,000-p.s.i., 17-5 PH steel pressure cylinder. The curve's shape may be explained by (*1*) piston friction, (*2*) compressibility and elasticity of the pressure system, (*3*) piston "hang-up", (*4*) rebound in the pressure system, and (*5*) piston friction.
B. Relative piston displacement data plotted versus oil-gauge pressure as partial upstroke cycles through the Bi I–II and Bi II–III solid–solid phase transition. Data was recorded by X-Y recorder (piston displacement versus time).

during the upstroke, the piston would not move because of friction between the internal parts and the cylinder's borehole surface. (2) This straight-line portion is due to the compressibility and compliance of the pressure system. (3) "Hang-up" of the piston and associated parts due to friction occurs at the beginning of the downstroke. The rounding of this upper portion of the hysteresis loop is an indication that the sample is not acting hydrostatically. (4) The straight-line portion on the downstroke is owing to rebound of the system. (5) Curvature in the lower portion of the decompression curve is probably due to the friction in the internal parts of the equipment and the silver chloride. Cohen et al. (1966) stated that friction uncertainties can be reduced by using larger diameter pistons, because the circumference-to-area ratio decreases with increasing piston diameter. The friction corrections become proportionally smaller with larger diameter pistons.

Calibration of very-high-pressure apparatuses

For the purpose of calibrating very-high-pressure apparatuses (e.g., up to 500,000 p.s.i.), reversible phase transformations which can be reproduced accurately are most desirable. The pressure capability of high-pressure and/or high-temperature apparatuses is determined by the detection of discontinuities in the electrical resistance or volume of metals at transition pressures. Ideally, solid–solid phase transitions should (1) be highly reproducible, (2) take place instantaneously, (3) involve a detectable heat of transformation, and (4) be easily obtainable in a pure-metal state. The pressure corresponding to phase transitions in barium, bismuth and thallium must previously be measured in a linear-piston apparatus where corrections could be estimated for mechanical and internal friction. Haygarth et al. (1967) described the technique used in the determination of phase transitions in metals with a single-stage piston cylinder apparatus.

When using a solid pressure-transmitting medium (e.g., bismuth) the calibration will yield different results for different samples and sample shapes. The true sample pressure is only as valid as the experimental determinations which have defined the axial force applied to the sample in terms of pressure. The transition pressure values are periodically revised, because of the use of metals of greater chemical purity, different pressure-vessel design, and improved design of the overall experiments. An investigator should check the current literature for the more recent accepted transitional values. The universal high-pressure scale (Bundy, 1962, p.19) has been used by many investigators. The calibration is valid only for runs of continuously increasing loads. If the loading is cycled, the pressure at points internal to the solid medium is not known.

Pressure on the sample in the 500,000-p.s.i. high-pressure compaction apparatus, shown in Fig.207, was calibrated by measuring the piston displacement and volume change which takes place during the transition of bismuth from one phase into another (Chilingar et al., 1963). Bismuth was inserted in the borehole of the composite, high-pressure plate and the distinctive abrupt shift in volume at two definite transition pressures (bismuth I–II and bismuth II–III) was measured. Inasmuch as the carbide pistons

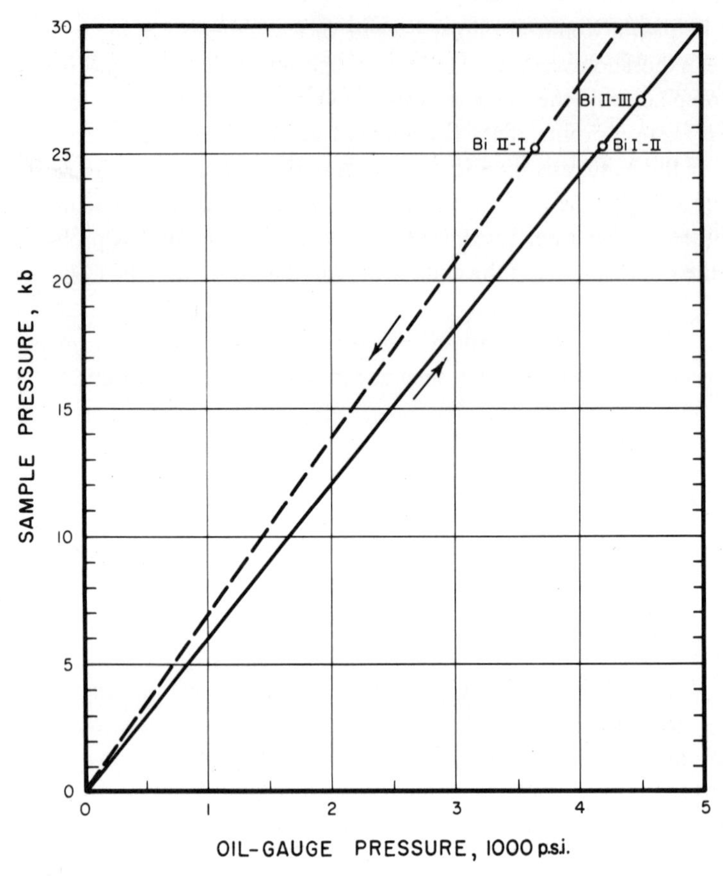

Fig.217. Calibration of the 500,000-p.s.i. high-pressure cylinder using Bi I–II and Bi II–III phase transitions. Solid line represents the upstroke (increasing ram force) and the dashed line represents the downstroke (decreasing ram force). The Bi II–I phase transition on the downstroke occurs at a slightly lower oil-gauge pressure than on the upstroke. The values of both upstroke and downstroke pressure were plotted and linearly extrapolated to zero. (After Rieke, 1970, p.138.)

used were cylindrical in shape, a relative piston displacement was utilized rather than the volume change during the transition of bismuth. The piston displacement is a relative measure of the volume change of a substance. The known transition pressures of bismuth have been determined previously in a linear piston apparatus in which the pressure was calculated accurately from the force and area. At the time of calibration, the main bismuth (I–II) transition was determined to have taken place at 368,010 ± 290 p.s.i. (Kennedy and La Mori, 1962, p.852).

Piston displacement was plotted as a function of time by using an X-Y recorder and the oil-gauge pressure was marked accordingly on the graph. The bismuth I–II transition was evidenced by a large change in the piston displacement and was readily determined; whereas the bismuth II–III transition was very vague and difficult to determine. Any piston friction at this point in the system may be relieved by simply rotating the small carbide piston until there is no further change in the pressure gauge readings. The registered pressure at the discontinuity is the solid–solid phase transition pressure of bismuth. The relative displacement of the piston can be plotted versus the oil-gauge pressure (Fig.217). The reverse transition on unloading does not occur until the thrust load is slightly lower than the value at which it took place on the loading cycle. This lag is not due to any appreciable lag in the behavior of the bismuth sample, but probably to the slight reverse friction and differential unloading effects in the pressure system. Because of this close agreement between the two results, it can be said that the friction is "symmetrical".

REFERENCES

Adams, F.D. and Coker, E.G., 1910. An experimental investigation into the flow of rocks. *Am. J. Sci., 4th Ser.*, 29(174): 465–487.

Adams, F.D. and Nicholson, J.T., 1901. An experimental investigation into the flow of marble. *Phil. Trans. Roy. Soc. Lond., Ser.A*, 195: 363–401.

Adams, L.H. and Williamson, E.D., 1923. The compressibility of minerals and rocks at high pressures. *J. Franklin Inst.*, 195: 475–529.

Adams, L.H., Williamson, E.D. and Johnston, J., 1919. The determination of the compressibility of solids at high pressures. *J. Am. Chem. Soc.*, 41: 12–42.

Amagat, E.H., 1893. Mémoires sur l'élasticité et la dilatabilité des fluids jusqu'aux très hautes pressions. *Ann. Chim. Phys.*, 29(6): 68–136.

Amyx, J.W., Bass Jr., D.M. and Whiting, R.L., 1960. *Petroleum Reservoir Engineering.* McGraw-Hill, New York, N.Y., pp.43–56.

Barus, C., 1892a. The compressibility of liquids. *U.S. Geol. Surv. Bull.*, 92: 96 pp.

Barus, C., 1892b. Mechanism of solid viscosity. *U.S. Geol. Surv. Bull.*, 94: 138 pp.

Barus, C., 1892c. The volume thermodynamics of liquids. *U.S. Geol. Surv. Bull.*, 96: 100 pp.

Bridgman, P.W., 1909. The measurement of high hydrostatic pressure, I. A simple primary gauge. *Proc. Am. Acad. Arts Sci.*, 44: 201–217.

Bridgman, P.W., 1918. Stress–strain relations in crystalline cylinders. *Am. J. Sci.*, 45: 269–280.

Bridgman, P.W., 1931. Compressibility and pressure coefficient of resistance, including single crystal magnesium. *Proc. Am. Acad. Arts Sci.*, 66: 255–271.

Bridgman, P.W., 1937. The resistance of nineteen metals to 30,000 kg/cm^2. *Proc. Am. Acad. Arts Sci.*, 72: 157.

Bridgman, P.W., 1958. *The Physics of High Pressure.* Bell Ltd., 2nd ed., 445 pp.

Bundy, F.P., 1962. General principles of high-pressure apparatus design. In: R.H. Wentrof Jr. (Editor), *Modern Very High Pressure Techniques.* Butterworths, London, England, pp.1–24.

Bundy, F.P., 1963. Direct conversion of graphite to diamond in static pressure apparatus. *J. Chem. Phys.*, 38(3): 631–643.

Buneeva, A.N., Kryukov, P.A. and Rengarten, E.V., 1947. Experiment in squeezing out of solutions from sedimentary rocks. *Dokl. Akad. Nauk S.S.S.R.*, 57(7): 707–709.

Cailletet, M.L., 1872. Compressibilité des liquides sous de hautes pressions. *Compt. Rend.*, 75: 77–78.

Canton, J., 1762. Experiments to prove that water is not incompressible. *Phil. Trans. Roy. Soc.*, pp.640–643.

Carter, N.L., Christie, J.M. and Griggs, D.T., 1964. Experimental deformation and recrystallization of quartz. *J. Geol.*, 72(6): 687–733.

Cebell, W.A. and Chilingarian, G.V., 1972. Some data on compressibility and density anomalies in halloysite, hectorite and illite clays. *Bull. Am. Assoc. Pet. Geologists,* 56(4): 796–802.

Chilingar, G.V. and Knight, L., 1960. Relationship between pressure and moisture content of kaolinite, illite and montmorillonite clays. *Bull. Am. Assoc. Pet. Geologists,* 44(1): 101–106.

Chilingar, G.V., Rieke III, H.H. and Robertson Jr., J.O., 1963a. Relationship between high overburden pressures and moisture content of halloysite and dickite clays. *Geol. Soc. Am. Bull.,* 14(8): 1041–1048.

Chilingar, G.V., Rieke III, H.H. and Robertson Jr., J.O., 1963b. Degree of hydration of clays. *Sedimentology,* 2(4): 341–342.

Cohen, L.H., Klement Jr., W. and Kennedy, G.C., 1966. Investigation of phase transformations at elevated temperatures and pressures by differential thermal analysis in piston-cylinder apparatus. *J. Phys. Chem. Solids,* 27: 179–186.

Comings, E.W., 1956. *High Pressure Technology.* McGraw-Hill, New York, N.Y., 572 pp.

Deal Jr., W.E., 1962. Dynamic high-pressure techniques. In: R.H. Wentrof Jr. (Editor), *Modern Very High Pressure Techniques.* Butterworths, London, pp.200–227.

Decker, D.L., 1965. Equation of state of NaCl and its use as a pressure gauge in high-pressure research. *J. Appl. Phys.,* 36(1): 157–161.

Dobrynin, V.M., 1962. Effect of overburden pressure on some properties of sandstones. *Trans. Am. Inst. Min. Metall. Engrs.,* 225: 360–366.

Fatt, I., 1953. The effect of overburden pressure on relative permeability. *Trans. Am. Inst. Min. Metall. Engrs.,* 198: 325–326.

Fatt, I., 1958a. Pore structure in sandstones by compressible sphere-pack model. *Bull. Am. Assoc. Pet. Geologists,* 42(8): 1914–1923.

Fatt, I., 1958b. Compressibility of sandstones at low to moderate pressures. *Bull. Am. Assoc. Pet. Geologists,* 42(8): 1924–1957.

Flint, E.P., 1968. The Hannay diamonds. *Chemistry and Industry,* pp.1618–1627.

Gann, D.P., 1965. *Changes in Ionic Concentration of Effluent from Compaction of Clay.* Thesis, Univ. Houston, Houston, Texas, 96 pp.

Giardini, A.A., Tydings, J.E. and Levin, S.B., 1960. A very high-pressure high-temperature research apparatus and the synthesis of diamond. *Am. Mineralogist,* 45(1–2): 217–221.

Gibson, C.E., 1928. The influence of pressure on the high-low inversion of quartz. *J. Phys. Chem.,* 32: 1197–1205.

Glanville, G.R., 1959. Laboratory study indicates significant effect of pressure on resistivity of reservoir rock. *J. Pet. Tech.,* 216(11): 20–26.

Gray, D.H., Fatt, I. and Bergomini, G., 1963. The effect of stress on permeability of sandstone core. *Soc. Pet. Eng. J.,* 2: 95–100.

Griggs, D.T., 1936. Deformation of rocks under high confining pressures. *J. Geol.,* 44: 541–577.

Hall, H.N., 1953. Compressibility of reservoir rocks. *Trans. Am. Inst. Min. Metall. Engrs.,* 198: 309–311.

Hall, H.T., 1958. Some high-pressure, high-temperature apparatus design considerations: Equipment for use at 100,000 atm and 3000°C. *Rev. Sci. Instr.,* 29(4): 267–275.

Hall, H.T., 1961. The synthesis of diamond. *J. Chem. Educ.,* 38(10): 484–489.

Hannay, J.B., 1880. On the artificial formation of the diamond. *Proc. R. Soc. Lond.,* 30: 188–189.

Hartmann, M., 1965. An apparatus for the recovery of interstitial waters from recent sediments. *Deep-Sea Res.,* 12: 225–226.

Haygarth, J.C., Getting, I.C. and Kennedy, G.C., 1967. Determination of the pressure of the barium I–II, transition with single-stage piston-cylinder apparatus. *J. Appl. Phys.,* 38(12): 4557–4564.

Heard, H.C., 1963. Effect of large changes in strain rate in the experimental deformation of Yule Marble. *J. Geol.,* 71(2): 162–195.

Heard, H.C., Turner, F.J. and Weiss, L.E., 1965. Studies of heterogeneous strain in experimentally deformed calcite, marble and phyllite. Univ. Calif. Press, Berkeley, Calif., *Publ. in Geol. Sci.,* 46(3): 81–152.

Heuer Jr., G.J., Knutson, C.F., Cavanaugh, R.J. and Bright, C.W., 1965. Method and apparatus for measuring compressibility of porous material. *U.S. Patent* 3, 199, 341.

Johnson, J. and Adams, L.H., 1912. Density of solid substances. *J. Am. Chem. Soc.*, 34: 565–584.

Kazintscv, E.A., 1968. Pore solutions of Maykop Formation of Eastern Pre-Caucasus and methods of squeezing of pore waters at high temperatures. In: G.V. Bogomolov et al. (Editors), *Pore Solutions and Methods of Their Study (A symposium)*. Izd. Nauka i Tekhnika, Minsk, pp.178–190.

Kennedy, G.C., 1950. Pressure–volume–temperature relations in water at elevated temperatures and pressures. *Am. J. Sci.*, 248(8): 540–564.

Kennedy, G.C. and Holser, W.T., 1966. Pressure–volume–temperature and phase relations of water and carbon dioxide. *Geol. Soc. Am., Mem.*, 97 (*Handbook of Physical Constants*), pp.371–383.

Kennedy, G.C. and La Mori, P.N., 1962. The pressures of some solid–solid transitions. *J. Geophys. Res.*, 67(2): 851–856.

Kick, F., 1962. Die Principien der mechanischer Technologie und die Festigheitslehre. *Z. Ver. Dtsch. Ing.*, 36(32): 919–923.

Kryukov, P.A., 1961. Toward procedures of squeezing out of solutions from sedimentary rocks. *Gidrokhim. Inst. Novocherkask., Gidrokhim. Mater.*, 33: 191–197.

Kryukov, P.A., 1964. *Soil, Mud and Rock Solutions* (Diss. Abs.). Akad. Nauk S.S.S.R., Inst. Geochemistry and Anal. Chem. of V.I. Vernadskiy, Moscow, 56 pp.

Kryukov, P.A. and Zhuchkova, A.A., 1963. Physical-chemical phenomena associated with driving out of solutions from rocks. In: *Present Day Concept of Bound Water in Rocks.* Izd. Akad. Nauk S.S.S.R., Laboratory of Hydrogeological Problems of F.P. Savarenskiy, Moscow, pp.95–105.

Kryukov, P.A., Zhuchkova, A.A. and Rengarten, E.V., 1962. Change in the composition of solutions pressed from clays and ion exchange resins. *Dokl. Akad. Nauk S.S.S.R.*, 144(6): 1363–1365.

Lambe, T.W., 1951. *Soil Testing.* Wiley, New York, N.Y., 165 pp.

Liander, H. and Lundblad, E., 1960. Some observations on the synthesis of diamonds. *Arkiv. Kemi.*, 16(9): 139–149.

Lisell, E., 1903. *Om Tryckets Imflytande på det Elektriska Ledningmototånded hos Metallar samt en ny Metod att Mäta Hogä Tryck.* Upsala Univ. Publ., no.1.

Macey, H.H., 1940. Clay–water relationships. *Proc. Phys. Soc. (Lond.)*, 52(5): 625–656.

Manheim, K.T., 1966. A hydraulic squeezer for obtaining interstitial water from consolidated and unconsolidated sediments. *U.S. Geol. Surv., Profess. Pap.*, 550-C: 256–261.

McKelvey, J.G. and Milne, I.H., 1962. Flow of salt solutions through compacted clay. *Clays Clay Miner. Proc. Natl. Conf. Clays Clay Miner., 9(1960)*, pp.248–259.

McLatchie, A.S., Hemstock, R.A. and Young, J.W., 1958. The effective compressibility of reservoir rock and its effects on permeability. *Trans. Am. Inst. Min. Metall. Engrs.*, 213: 386–388.

Newton, M.S. and Kennedy, G.C., 1965. An experimental study of the P–V–T–S relations of seawater. *J. Mar. Res. (Sears Found. Mar. Res.)*, 23(2): 88–103.

Patterson, M.S., 1970. A high-pressure, high-temperature apparatus for rock deformation. *Int. J. Rock Mech. Min. Sci.*, 7(5): 517–526.

Pomeroy, R.D., Lacey, W.N., Scudder, N.F. and Stapp, F.P., 1933. Rate of solution of methane in quiescent liquid hydrocarbons. *Ind. Eng. Chem.*, 25(9): 1014–1019.

Presley, B.J., Brooks, R.R. and Kappel, H.M., 1967. A simple squeezer for removal of interstitial water from ocean sediments. *J. Mar. Res. (Sears Found. Mar. Res.)*, 25(3): 355–357.

Rieke III, H.H., 1970. *Compaction of Argillaceous Sediments (20–500,000 p.s.i.).* Diss., Univ. Southern Calif., 682 pp.

Rieke III, H.H., Ghose, S.K., Fahhad, S.A. and Chilingar, G.V., 1969. Some data on compressibility of various clays. *Int. Clay Conf., Tokyo, Japan,* 1: 817–828.

Robertson, E.C., 1967. Laboratory consolidation of carbonate sediment. In: A.F. Richards (Editor), *Marine Geotechnique, Int. Res. Conf. Mar. Geotech.* Univ. Illinois Press, Urbana, Ill., pp.118–127.

Sage, B.H., and Lacey, W.N., 1948. Apparatus for determination of volumetric behavior of fluids. *Trans. Am. Inst. Min. Metall. Engrs.*, 174: 102–120.

Sage, B.H., Schaafsma, J.G. and Lacey, W.N., 1934. Phase equilibria in hydrocarbon systems, V. Pressure–volume–temperature relations and thermal properties of propane. *Ind. Eng. Chem.*, 26(11): 1218–1224.

Sawabini, C.T., Chilingar, G.V. and Allen, D.R., 1971. Design and operation of a triaxial, high-pressure compaction apparatus. *J. Sediment. Petrol.*, 41(3): 871–881.

Serdengecti, S. and Boozer, G.D., 1961. The effects of strain and temperature on the behavior of rocks subjected to triaxial compression. *Proc. 4th Symp. Rock Mech.*, pp.83-97.

Shepard, F.P. and Moore, D.G., 1955. Central Texas Coast sedimentation. *Bull. Am. Assoc. Pet. Geologists,* 39: 1463–1593.

Siever, R., 1962. A squeezer for extracting interstitial water from modern sediments. *J. Sediment. Petrol.,* 32(2): 329–331.

Standing, M.B., 1952. *Volumetric and Phase Behavior of Oil Field Hydrocarbon Systems.* Reinhold, New York, N.Y., 122 pp.

Tait, P.G., 1889. Report on some of the physical properties of fresh water and sea water. *Rep. Voyage of H.M.S. Challenger, Phys. Chem.,* 2: 1–76.

Taylor, D.W., 1948. *Fundamentals of Soil Mechanics.* Wiley, New York, N.Y., 700 pp.

Teeuw, D., 1970. Prediction of formation compaction from laboratory compressibility data. *Soc. Pet. Eng. Am. Inst. Min. Metall. Engrs., 45th Ann. Fall Meet., Houston, Tex.,* Paper no.2973: 8 pp.

Terzaghi, K., 1925. *Erdbaumeckanik auf bodenphysikalischer Grundlage.* Deuticke, Leipzig, 399 pp.

Tsiklis, D.S., 1965. *Technique of Physicochemical Investigations at High and Superhigh Pressures.* Izd. Khimiya, Moscow, 3rd ed., 415 pp.

Tsiklis, D.S., 1968. *Handbook of Techniques in High Pressure Research and Engineering.* Plenum, New York, N.Y., 504 pp.

Tydings, J.E. and Giardini, A.A., 1963. A study of stress homogeneity in cylindrical cavities at high pressures. In: A.A. Giardini and E.C. Lloyd (Editors), *High-Pressure Measurement.* Butterworths, Wash., D.C., pp.200–220.

Van der Knaap, W. and Van der Vlis, A.C., 1967. On the cause of subsidence in oil-producing areas. *Proc. 7th World Pet. Congr., Mexico City, Mexico.* Elsevier, Amsterdam, 3: 85–95.

Von Engelhardt, W., 1967. Interstitial solutions and diagenesis in sediments. In: G. Larsen and G.V. Chilingar (Editors), *Diagenesis in Sediments.* Elsevier, Amsterdam, pp.503–521.

Von Engelhardt, W. and Gaida, K.H., 1963. Concentration changes of pore solutions during the compaction of clay sediments. *J. Sediment. Pet.,* 33(4): 919–930.

Von Karman, Th., 1911. Festigheitversuche under alkeitigem Druck. *Z. Ver. Dtsch. Ing.,* 55: 1749–1757.

Wentrof Jr., R.H., 1962. *Modern Very High Pressure Techniques.* Butterworths, London, 233 pp.

Westman, A.E.R., 1932. The effect of mechanical pressure on the inhibitional and drying properties of some ceramic clays, I. *J. Am. Ceram. Soc.,* 15(10): 552–563.

Wilson, W.B., 1960. Device for ultra-high pressure high-temperature research. *Rev. Sci. Instr.,* 31(3): 331–333.

Wyble, D.O., 1958. Effect of applied pressure on conductivity, porosity, and permeability of sandstones. *Trans. Am. Inst. Min. Metall. Engrs.,* 431–432.

Yoder Jr., H.S., 1950. High-low-quartz inversion up to 10,000 bars. *Trans. Am. Geophys. Union,* 31(6): 827–835.

Appendix A

CONVERSION FACTORS

Multiply ⟶	by ⟶	*to obtain*
To obtain ⟵	by ⟵	*divide*
acres	43,560	square (sq.) feet (ft.)
ac	4,046.8564	square meters (m^2)
ac-ft.	43,560	cubic (cu.) feet (ft.)
ac-ft.	$3.259 \cdot 10^5$	gallons (gal., U.S., liq.)
atmospheres (atm)	33.947	ft. H_2O, 62°F
atm	33.899	ft. H_2O, 39.2°F
atm	29.9213	inch Hg, 32°F
atm	760	mm Hg, 32°F
atm	14.6960	pounds per square inch (p.s.i.)
bars (b)	$9.8692 \cdot 10^{-1}$	atm
b	$7.50062 \cdot 10$	cm of Hg (0°C)
b	$1 \cdot 10^6$	dynes/cm^2
b	1.01972	kg/cm^2
b	14.5038	p.s.i.
barrels (bbl), oil	5.614583	cu. ft.
bbl, oil	0.159	m^3
bbl, oil	42	gal. (U.S., liq.)
British thermal unit (B.t.u.)	777.649	foot-pounds (ft.-lb.)
B.t.u.	$2.9287 \cdot 10^{-4}$	kilowatt-hours (kWh)
B.t.u./min	17.5725	watts (W)
cm	0.3937	inch
cm Hg, 0°C	$1.315789 \cdot 10^{-2}$	atm
cm Hg, 0°C	$1.93368 \cdot 10^{-1}$	p.s.i.
cm/sec	1.968504	ft./min
cm/sec	$1.033 \cdot 10^6$	md
centipoise (cp)	0.01	poise (g cm^{-1} sec^{-1})
cp	$6.72 \cdot 10^{-4}$	lb. $ft.^{-1}$ sec^{-1}
cu. cm (cm^3)	0.0610237	cu. inch
cu. cm (cm^3)	0.033814	fluid ounces (fl. oz.)
cu. ft.	1,728	cu. inch
cu. ft.	0.03704	cu. yards
cu. ft.	7.4805	gal. (U.S.)
cu. ft. H_2O, 39.2°F	62.4262	lb.
cu. ft. H_2O, 60°F	62.366	lb.
cu. ft. H_2O	7.4805	gal. (U.S.)
cu. ft. H_2O	6.232	gal. (Imperial)

Multiply ⟶	by ⟶	*to obtain*
To obtain ⟵	by ⟵	*divide*
cu. ft./min	7.4805	gal./min
cu. ft./min, 39.2°F	62.4262	lb. of H_2O/min
cu. inch	0.004329	gal. (U.S.)
cu. inch	16.387	cm^3
cu. m (m^3)	264.2	gal. (U.S.)
ft.	30.48	cm
ft.	0.30481	m
ft.	0.3594	vara (Texas)
ft.	0.3333	yards (yd)
ft. H_2O, 60°F	0.4331	p.s.i.
ft./min	0.508	cm/sec
gal. (Imperial)	4.546	liters (l)
gal. (Imperial), H_2O, 62°F	10.0	lb.
gal. (U.S.)	0.8327	gal. (Imperial)
gal. (U.S., liq.)	$1.3368 \cdot 10^{-1}$	cu. ft.
gal. (U.S.)	3,785.4	cm^3
gal. (U.S.)	231	cu. inch
gal. (U.S.) H_2O, 39.2°F	8.345	lb. (H_2O)
gal. (U.S.) H_2O, 60°F	8.337	lb. (H_2O)
grams (g)	980.7	dynes
g	15.432	grains (gr)
g	$3.527 \cdot 10^{-2}$	oz. (avdp)
g	$2.2046 \cdot 10^{-3}$	lb. (avdp)
g/cm^3	$3.6127 \cdot 10^{-2}$	lb./cu. inch
hectares	2.471	ac
hectoliters (hectol)	100	l
hectol	3.531566	cu. ft.
hectol	26.41794	gal. (U.S.)
inch	2.54	cm
inch Hg, 32°F	$3.3421 \cdot 10^{-2}$	atm
inch Hg, 32°F	1.133	ft. H_2O, 39.2°F
inch Hg, 32°F	7.85847	oz./sq. inch
inch Hg, 32°F	70.73	pounds/sq. ft. (p.s.f.)
inch Hg, 32°F	0.4912	p.s.i.
inch H_2O, 39.2°F	$2.458 \cdot 10^{-3}$	atm
inch H_2O, 39.2°F	0.07355	inch Hg, 32°F
inch H_2O, 39.2°F	0.578	oz/sq. inch
inch H_2O, 39.2°F	5.202	p.s.f.
inch H_2O, 39.2°F	$3.613 \cdot 10^{-2}$	p.s.i.
Joule (absolute) (J)	$9.48451 \cdot 10^{-4}$	B.t.u.
J (absolute)	$7.37562 \cdot 10^{-1}$	ft.-lb.
kilograms (kg)	2.20462	lb. (avdp)
kg/cm^2	$9.8066 \cdot 10^{-1}$	bar
kg/cm^2	32.843	ft. H_2O, 60°F
kg/cm^2	14.2233	p.s.i.
kg/m^3	$6.243 \cdot 10^{-2}$	pounds/cu. ft.
l	$3.531 \cdot 10^{-2}$	cu. ft.

Multiply ⟶	by ⟶	*to obtain*
To obtain ⟵	by ⟵	*divide*
m	39.37	inch
miles	63,360	inch
miles	1.609	kilometers (km)
miles	5,280	ft.
millidarcy (md))	$9.6 \cdot 10^{-7}$	cm/sec
milliliter (ml)	1.000028	cm^3
millimeters (mm)	$3.937 \cdot 10^{-2}$	inch
Newton	$1.000 \cdot 10^5$	dynes
Newton	$2.248089 \cdot 10^{-1}$	lb.
pounds (lb.)	7,000	grains (gr)
lb.	453.592	g
lb. H_2O, 39.2°F	26.68	cu. inch
lb. H_2O, 39.2°F	0.1198	gal. (U.S.)
lb. H_2O, 39.2°F	$9.98 \cdot 10^{-2}$	gal. (Imperial)
lb./cu. ft., 39.2°F	0.1337	lb./gal.
lb./cu. ft.	$1.602 \cdot 10^{-2}$	g/cm^3
p.s.i.	$6.8046 \cdot 10^{-2}$	atm
p.s.i.	$6.8947 \cdot 10^{-2}$	bar
p.s.i.	27.6807	inch H_2O, 39°F
p.s.i.	2.036	inch Hg, 32°F
p.s.i.	$7.031 \cdot 10^{-2}$	kg/cm^2
specific gravity (sp. gr.) (60°F)	8.33727	lb./gal.
sp. gr. (60°F)	350.5	lb./bbl (60°F)
sq.cm (cm^2)	0.155	sq. inch
sq.cm/sec (cm^2/sec)	3.155815	$km^2/10^6$ year
sq. inch	$6.944 \cdot 10^{-3}$	sq. ft.
sq. m (m^2)	10.764	sq. ft.
temperature [1] (°F −32)	0.5555	temp. (°C)
temperature (°C/km)	$5.48641 \cdot 10^{-4}$	temp. (°F/ft.)

[1] (°C × 1.8) + 32 = °F; (°F + 40) ÷ 1.8 − 40 = °C; °K (Kelvin) = °C + 273.16; °R (Rankine) = °F + + 459.688.

REFERENCES INDEX

Figures in italics refer to reference lists

SUBJECT INDEX